W9-AVG-870

·GEOGRAPHY·UNBOUND·

· G · E · O · G · R · A · P · H · Y ·

French Geographic Science from

ANNE MARIE CLAIRE

THE UNIVERSITY OF CHICAGO PRESS

·U·N·B·O·U·N·D·

Cassini to Humboldt

GODLEWSKA

CHICAGO AND LONDON

G
97
.G63
1999

ANNE MARIE CLAIRE GODLEWSKA is associate
professor of geography at Queen's University,
Canada. She is author of *The Napoleonic Survey of
Egypt* and coeditor of *Geography and Empire.*

The University of Chicago Press, Chicago 60637
The University of Chicago Press, Ltd., London
© 1999 by The University of Chicago
All rights reserved. Published 1999
08 07 06 05 04 03 02 01 00 99   1 2 3 4 5
ISBN (cloth): 0–226–30046–3
ISBN (paper): 0–226–30047–1

Library of Congress Cataloging-in-Publication Data

Godlewska, Anne.
    Geography unbound : French geographic science from Cassini to
Humboldt / Anne Marie Claire Godlewska.
        p.   cm.
    Includes bibliographical references (p.   ) and index.
    ISBN 0-226-30046-3 (cloth : alk. paper).—ISBN 0-226-30047-1
(pbk. : alk. paper)
    1. Geography—France—History—18th century.   2. Geography—
France—History—19th century.   3. Geography—Philosophy—
History—18th century.   4. Geography—Philosophy—History—19th
century.   I. Title.
G97.G63   1999
914.4—dc21   99-34189
                                                            CIP

♾ The paper used in this publication meets the minimum requirements of the American
National Standard for Information Sciences—Permanence of Paper for Printed Library
Materials, ANSI Z39.48–1992.

For my parents

Contents

LIST OF FIGURES   *ix*

ACKNOWLEDGMENTS   *xi*

INTRODUCTION   *1*

Part One / Geography's Fall

ONE   *The Nature of Eighteenth-Century Geography: Cartographic and Textual Description*   *21*

TWO   *Geography's Loss of Direction and Status*   *57*

Part Two / Reaction and Continuity

THREE   *Universal Description*   *89*

FOUR   *The Powerful Mapping Metaphor*   *129*

FIVE   *Handmaiden to Power*   *149*

Part Three / Innovation on the Margins

SIX   *Explaining the Social Realm*   *193*

SEVEN   *Innovation in Natural Geography*   *235*

EIGHT   *Tough-Minded Historical Geography*   *267*

CONCLUSION   *305*

NOTES   *315*

REFERENCES   *367*

INDEX   *433*

Figures

FIG 1   "The vanishing grid." Jean-Baptiste-Bourguignon d'Anville, *Partie occidentale du Canada et septentrionale de la Louisianne, avec une partie de la Pensilvanie* (Venise: Par P. Santini, 1775). Author's collection. Photographed by Brandon Beierle.

FIG 2   "A Brittany sheet from the Cassini map of France." César-François Cassini. *Carte générale de la France*, vol. 1, *Partie du nord, Diocèse de Quimper* (Paris, 1750–1815). Reproduced by permission of Harvard University Libraries, Harvard University.

FIG 3   "A comparison of the seas of the world." Alexander von Humboldt, *Atlas zu Alexander von Humboldt's Kosmos: in zweiundvierzig Tafeln* (Stuttgart: Verlag von Krais & Hoffmann, 1851), plate 18. Reproduced by permission of Houghton Library, Harvard University.

FIG 4   "The monumental importance of Egypt." Frontispiece of *Description de l'Égypte, ou, Recueil des observations et des recherches qui ont été faites en Égypte* pendant *l'expédition de l'armée française, publié par les ordres de Sa Majesté l'empereur Napoléon le Grand* (Paris: Imprimerie impériale, 1809–1828). Reproduced by permission of Houghton Library, Harvard University.

FIG 5   "Military geographers working in the field." Frontispiece from Colonel Henri-Marie-Auguste Berthaut, *Les Ingénieurs-géographes militaires, 1624–1831, étude historique*, vol. 1 (Paris: Imprimerie du Service géographique de l'armée, 1902). Reproduced by permission of Harvard University Libraries, Harvard University.

FIG 6   "Paris at the time of Chabrol de Volvic's administration." Frontispiece from David Carey, *Life in Paris; comprising the rambles, sprees, and amours of Dick Wildfire . . .* (London: Printed for J. Fairburn, 1922). Reproduced by permission of Houghton Library, Harvard University.

FIG 7   "The geography of plants." Géographie des plantes équinoxiales. Tableau physique des Andes et Pays voisins, from *Voyage de Humboldt et Bonpland. Voyage aux régions équinoxiales du nouveau continent.* Cinquième partie, *Essai sur la géographie des plantes* (Paris: Theatrum orbis terrarum, 1805). Facsimilé intégral de l'édition Paris 1805–1834 (Amsterdam and New York: De Capo Press, 1973). Reproduced by permission of Blacker-Wood Rare Books, McGill University.

FIG 8   Alexander von Humboldt, "Cartes des diverse routes par lesquelles les richesses métalliques refluent d'un continent à l'autre." Reproduced by permission of Harvard University Libraries, Harvard University.

FIG 9   A set of four graphs: Alexander von Humboldt, "Produit des mines de l'Amérique depuis sa découverte; Quantité de l'or et de l'argent extraits des mines du Mexique; Proportion dans laquelle les diverses parties de l'Amérique produisent de l'or et de l'argent; Proportion dans laquelle les diverses parties du Monde produisent et de l'argent." Reproduced by permission of Harvard University Libraries, Harvard University.

FIG 10   Alexander von Humboldt, "Tableau comparatif de l'étendue territoriale des Intendances de la Nouvelle-Espagne" (I), together with "Etendue territoriale et Population des Métropoles et des Colonies en 1804" (II). Reproduced by permission of Harvard University Libraries, Harvard University.

FIG 11   Alexander von Humboldt, "Esquisse géognostique des formations entre la Vallée de Mexico, Moran et Totonilco." Reproduced by permission of Harvard University Libraries, Harvard University.

FIG 12   Alexander von Humboldt, "Tableau physique de la Nouvelle-Espagne. Profile du Chemin d'Acapulco à Mexico, et de Mexico à Veracruz." Reproduced by permission of Harvard University Libraries, Harvard University.

FIG 13   "Syracuse as reconstructed by Letronne." Jean-Antoine Letronne. *Essai critique sur la topographie de Syracuse au commencement du cinquième siècle avant l'ère vulgaire* (Paris: Pélicier, 1812). Reproduced by permission of Harvard University Libraries, Harvard University.

Acknowledgments

It is a pleasure after years of work on a manuscript to be able to thank those who supported, guided, encouraged, and cajoled. I owe a great deal to the Social Sciences and Humanities Research Council of Canada. Without their support and fair-minded patience, it would have been extremely difficult, if not impossible, to write this book. I also owe the Advisory Research Committee at Queen's a significant debt of gratitude for always responding favorably to my heart-rending pleas for financial assistance. Support of another kind was forthcoming from my colleagues in the Geography Department at Queen's, both faculty and staff, who sometimes spoke on my behalf, sometimes made it easier for me to escape to the library in the summer, and always provided an atmosphere of pleasant collegiality in which to think and write. Finally, very early on in the planning stages of this work, the Hermon Dunlap Smith Center for the History of Cartography at the Newberry Library provided me with access to their collection and a summer of intellectual bliss.

This book is very much mine and so are all its shortcomings and errors. I did receive some significant guidance and criticism from colleagues and friends. My friend Paul Claval listened with interest and support to my ideas, suggested further references, and majestically contained his impatience at my publishing speed. Colleagues who heard me present or read bits and pieces of the book responded with ideas and criticism that have proved invaluable, even if I have not always taken their advice. These include Daniel Nordman, Bernard Lepetit, Michael Heffernan, David Livingstone, Denis Cosgrove, Robert Fox, Ingo Schwarz, Nicolaas Rupke, Derek Gregory, Neil Smith, Cole Harris, Christian Jacob, Matthew Edney, and Patrice Bret. I would especially like to thank Charles Withers, Paul Claval, Peter Goheen, Paul Datta, and an anonymous reader for a very close and suitably critical reading of the manuscript. Librarians have been some of my most appreciated guides. The collection and staff at the Bibliothèque nationale, including the book runners and the security guards, provided me with a home away from home which, although it may have been cold in the winter and hot in the summer, was always warm. Equally helpful were the archivists at the Service historique de l'armée de terre. Shirley Harmer at Queen's University provided sure assistance at home base, as did the general reference librarians at Queen's. Penny Kaiserlian, Associate Director of the University of Chicago Press, provided me with sage advice early on in the writing of this book. Her staff, including Russell Harper, Mary Laur, and David Aftandilian, have done all they can to present my thoughts in the best possible way. Lys Ann Shore proved a very careful and imaginative indexer. My students Michael Imort and Robert Davidson helped me to tidy up the bibliography. I have presented my evolving ideas about the thesis of this book to colleagues, graduate students, and undergraduate students at conferences and at Queen's University, Paris-IV, the University of Syracuse, York University, the University of Oklahoma, the University of Toronto, McGill University, and the University of British Columbia over the last ten years. Their comments and observations have been invaluable to me.

My friends and family have encouraged me all the way through the research and writing of this book. My dear friend Walter Schatzberg saw the greatest number of iterations and managed to appear excited and interested to the end. Wendy Craig read parts of the manuscript, attacked my writing style precisely where it needed to be savaged, and kept my thinking in proportion by regularly launching me into fits of laughter. Peter Goheen was always willing to talk about ideas and rescued me from my moments of self-doubt with gentle words firmly delivered. Others whose friendship lent me support include Evelyn Peters, Sheila MacDonald, Michael Imort, and Nancy Wood. My family encouraged and cajoled and provided the kind of support that has made me among the happiest and most privileged of people. My gratitude to them is beyond expression.

Introduction

> The end of Classical thought and of the episteme that made general grammar, natural history, and the science of wealth possible—will coincide with the decline of representation, . . . And representation itself was to be paralleled, limited, circumscribed, mocked perhaps, . . . posited as the metaphysical converse of consciousness. Something like a will or a force was to arise in the modern experience—constituting it perhaps, but in any case indicating that the Classical age was now over, and with it the reign of representative discourse, the dynasty of a representation signifying itself and giving voice in the sequence of its words to the order that lay dormant within things. —Foucault, *The Order of Things* (1971), 209

This book is about the nature of geography two hundred years ago. It is not a disciplinary history.[1] It has not been written as a reassuring reconstruction of the activities of great men or women long past.[2] It is also not a classical history of ideas, searching for elements of contemporary thought in past societies.[3] Nor is it a normative history of "our discipline" seeking to establish the correct path to the future through the insights of the past.[4] Nor is it even a reconstruction of the "tradition" of geography seen as stretching from today back into the mist of the Anglo-Saxon (or non-Anglo-Saxon) past.[5] It is an argument about what the discursive formation of geography was and was not in France about two hundred years ago. It is Foucaultian in its focus in that it seeks to explore one of those "curious entities which one believes one can recognize at first glance, but whose limits one would have some difficulty in defining," in this case the discursive formation called geography in the late eighteenth and early nineteenth centuries.[6] It is also Foucaultian in that it wants to get at changes that alter discursive patterns: change that alters the objects of the study of particular discursive formations, the operations they may engage in, the apparent relevance and importance of their concepts, and the theoretical options open to thinkers; change that alters the boundaries of a given discursive field and the means by which it reinvents itself; and change that overturns hierarchies in the intellectual division of labor and consequently the orientation of research. Change is a difficult concept: it seems to immediately catapult one into evolutionary modes of explanation, into a focus on the theme of becoming rather than on the "analysis of transformations in their specificity."[7] I have tried to avoid that trap in this book. I am not sure that I have always entirely succeeded. The argument in this book is also Foucaultian in that it posits a sea change at the end of the eighteenth century and in that it sees modern geographic thought as emerging—certainly not fully formed—in that period. It differs from Foucault's abstract approach in *Les Mots et les choses* and in *Politics and the Study of Discourse* in that it is fundamentally interested in how the individual experienced, helped to create, and negotiated discursive formations.[8]

Up to the end of the eighteenth century the discursive formation of geography revolved around exactly what its name suggests: earth description. What the describable earth encompassed was substantially determined by contemporary understanding of the nature of the earth and its motivating forces. In Classical Greek thought, geography and astronomy were profoundly linked. In Medieval and Renaissance Europe, the cosmos studied by geography necessarily included the divine will which bestowed fundamental motivation and meaning upon the earth and all of the creatures living there. Thus, for a considerable period stretching, for some geographers, well into the nineteenth century, geography and theology were closely tied.[9] In the course of the eighteenth century with the discovery of geology, the geologic structures of the earth, and geologic time, geography began to dip beneath the surface of the earth. Thus, throughout the history of geography, earth description tended to be broadly inclusive, expanding in focus in accordance with contemporary interests and understandings about what was important about the earth. So that at different times, it encompassed not only the earth's natural and human topography, but the air above it, the oceans alongside it, and even the layers beneath it—not to mention all of the creatures flying, swimming, and burrowing there.

It is important to note, however, that while the breadth of these geographies was enormous, its approach was limited. Geography was description. That description could take a number of forms, but earth measurement, sketching, and literary depiction were its primary and most ancient tools. In all periods up to the end of the eighteenth century, the ultimate purpose of geography was, using those tools, to create a unified, true, comprehensive, and complete picture of the earth or of parts of the earth. This, it was believed, would, in and of itself, expand knowledge.

This is not to argue that there was a lack of ideology or bias or even that there was no worldview behind the writings of geographers. Pure description, in that sense, does not exist: we are incapable of functioning without presuppositions. But there is a large difference between untested suppositions used for daily functioning (which often happily allow the coexistence of eclectic and contradictory "truths" and erratic and vague demarcations between relevant and even irrelevant phenomena or processes) and the deliberate theory formation and testing that characterizes modern scientific thought (which, whether it succeeds or not, is generally trying to avoid self-contradiction). Theory, in this sense, establishes ideal arguments about the world or some aspect of it—intellectual models—which are then explored for the meaning and insight they can give. Theories can closely approximate the world or they can seem far removed from it and yet be useful tools for the exploration of understanding. Part of the growing sophistication of scientific thought through the nineteenth and twentieth centuries revolves around the fact that theories have been increasingly understood as constructions whose relationship to reality is problematic. It is argued in this book that the geographers of the eighteenth century were describing the world without critically examining the theory upon which their descriptions were based. They were more interested in locating phenomena than in understanding them. They were engaged in the project of location and in the development of a language of location but not in the project of validating theories of location. Nor did they recognize the theories emerging from other discursive formations as in any way different from their own descriptions. Instead they were preoccupied with erecting a spatial classification scheme whose relationship to understanding was unstated and essentially beyond criticism. It is difficult to conceive today of just how nonexplanatory geography then was. It was not geography's function to explain the forms it found, the flow of rivers, the shape of coasts, the distribution of plants and animals, the variance in human ways of life . . . but to describe and locate these. While, for example, geographers periodically commented on the effect of a particular climate on a given people, this was never expressed as a problem to be explained or investigated by geographers but as a state of being. That is, the relationship was simply described.

Up to the end of the eighteenth century, then, from astronomy to theology to geology, geography expanded to accommodate enlarging conceptions of the cosmos without fundamentally changing. Geography described, and description, whether mathematical, graphic or literary, was understood to lead to understanding and enlightenment. When, in the context of the Revolutionary and Napoleonic wars, the growth of state statistics, and the rise of modern economic thinking, geographers began to devote more attention to the description of human activity on the surface of the earth, there was every reason to think that geography would continue to be what it had always been, a respected descriptive, universalizing, and inclusive science. Indeed, with the growth in interest in both physical and biological nature and increased awareness of the social realm, geography's descriptive purview seemed about to expand dramatically.

Instead, geography entered a period of uncertainty and extended dormancy which lasted until at least the 1870s (and some would argue that it has still not entirely emerged from this hibernation). Geography's loss of direction, documented in chapter 2, was manifested in a loss of purpose, uncertainty about what to teach both junior and advanced students, discomfort with the role and nature of theory, loss of status, a jettisoning of both the most technical and the most theoretical aspects of the field, a proliferation of exclusivist definitions of the discursive formation, and general uncertainty about how to assess research quality.

The nature of the malaise evident in the discourse, the inherently conservative reaction of many geographers to the challenges emerging from other discursive formations, the survival of the weakened field through a momentum more structural and circumstantial than substantive, and the new and stimulating directions pointed out but largely left unexploited by scholars working at the margins of the field are the subject of this book. Because discursive formations are fluid and not easily dissociable, and because important change was taking place in a number of arenas of academic and social thought in the period covered by this study, the relevant contexts of geography are many. Thus, developments in astronomy, natural history, geology, archaeology, government administration (the nursery of the social sciences), and, of course, geodesy and mapping are touched upon. Though perhaps not in sufficient detail to satisfy historians of those fields. The context that this book explores is the context of geography. Thus the ideas of Malte-Brun, Balbi, Cassini, Jomard, etc., are contextualized not to construct a broad background but as "part of an attempt to 'explain' the position taken by an individual by means of a full account of the possible positions available."[10] This history of this geography will be in part explanatory and in part descriptive. The complexity of the conjunction of events, personalities, and ideas demands explanation—no matter how tentative—to allow us to progress beyond the detail of individuals, books, institutions, and opinions to a general understanding of the role of geography in post-Revolutionary France. And there is much to explain.

It seems paradoxical that just as the scientific world began to focus on systematic exploration of the nature and age of the earth, and the organization of life on earth, geography—a field which had long been closely associated with these concerns—should lose direction. The rise to prominence of natural history and its evolution into biology and geology in the late eighteenth and nineteenth centuries entailed a shift from a focus on the mostly static depiction of form and location to an increasing appreciation of the mostly dynamic behaviors, interactions, and movements of living and nonliving phenomena. It also entailed a shift from the domination by the distant past (as laid down by God or long-past [and hence essentially not investigable] physical forces) to the immediate, interactive present moved by forces very evidently still active. Both of these shifts reduced the value of the static geographic map. Instead, it became necessary to move from the possibility of building one kind of model at different scales to that of building many different kinds of models each with a logic dictated by the dynamic, immediate, interactive phenomena under study (thematic maps and scientific models). As Bruno Latour has pointed out, each of these maps or models, from Mendeleev's periodic table to Reynolds's turbulence coefficient, created a new and distinct space and time.[11] This space and time was the product of an intellectual construct or classification, the logic of which was related to the reshuffling of connections between elements (such as chemical elements or plant genera) which was made possible by the separation of these from their geographic context and their representation by mobile, stable, and combinable inscriptions. The space and time created by this reshuffling was and is accessible only to those trained to read the inscriptions and able to understand the logic behind the reshuffling. The space and time of chemistry, geology, hydraulic engineering, and sociology is not necessarily closely related to that of geography. Consequently, geographers were and are no more able to describe and map these space/time relations than any other noninitiate.

The aim of geography was the description of the earth. The aim of the new sciences was the explanation of selected aspects of the earth's (or its inhabitants') functioning. Geography's holistic but nontheoretical, nonexplanatory, and unanalytical vision of the earth, together with its static and geographic representational system, could not incorporate these sciences into its descriptive system. Latour seems to suggest that in the late twentieth century perhaps only mathematics is capable of describing the world in inscriptions which are sufficiently mobile, stable, and combinable to allow exchange across the modern sciences and social sciences. This may be the case, but there is little doubt today that geography cannot play the holistic, yet theoretical and scientific, unifying role that Humboldt sought in his *Cosmos*.[12]

It also seems paradoxical that geography lost a sense of direction just as Europe moved into the most dynamic and aggressive phase of its age of exploration. Barbara Stafford, in my view confusing geography and exploration, has described the geography of this period as a speculative or theoretical science because geographers theorized about the form and content of the earth's continents when these were substantially unknown.[13] It is possible to see geography as theory-driven in cases where exploration of location and form were driven by geographic hypotheses formulated by geographers. An example of this is Philippe Buache's highly conjectural system of mountain chains and river basins. In 1752 Buache proposed a predictive description of the world's mountains arguing that they formed a regular structural framework which could be used to locate all of the mountains and rivers of the globe.[14] This did suggest theoretical possibilities for geography, as the somewhat barbed response to his theory at the Academy of Sciences indicates:

> This approach to our globe opens a whole new career for geography. It is perhaps more interesting to know the direction of mountain chains . . . than to recognize the ancient limits of a country or empire no longer in existence.[15]

Similarly, it was theories about the location and nature of El Dorado that led to the exploration of the interior of Peru, Venezuela, Colombia, and the Orinoco and Amazon basins; the quest for Prester John that directed European attention to the Middle East, the Far East, and ultimately to Ethiopia; and the myths of Cíbola, Quivira, etc., that led to the exploration of parts of Central and North America. It was, however, also toward the end of the eighteenth century that exploration driven by theories concerning location began to seem irrelevant end even irresponsible. The voyages of La Pérouse (1785–1788), Bouguer and La Condamine (1735–1743), Maupertuis (1735–1736), Cook (1768–1771, 1772–1775, 1776–1780), and Bougainville (1766–1769) marked the advent of the age of state-funded, large-scale, team-structured exploration incorporating the talents of many specialists. The opinions of cabinet-bound geographers working from interpretations of previous accounts, old maps, and sometimes the views of the ancients on the nature of a world they had never seen, became increasingly obsolete and scarcely interesting. Explorers turned more and more to the results of other expeditions rather than to the speculations of geographers which were not based on any explanation of the shapes, forms, and locations described but on compilations of earlier descriptions. The speculative descriptions of geographers were of some interest only as long as the shape and articulation of the world was unknowable.

Geographers, variously aware of the malaise and confusion they were witnessing, struggled to maintain the profile and quality of their work. Some sought validation for their work through the renovation and extension of well-tried traditional approaches: principally literary description, mapping, and the harnessing of geography to the will and needs of the state. To a very large extent these failed to win geographers recognition in scientific circles. There were other geographers and scholars with strong geographic interests who explored the possibility of rendering geography more explanatory without abandoning its integrative and universalizing role or its strong spatial focus. Already early in the nineteenth century there are signs of these efforts in subfields as varied as what today might be termed incipient anthropological, physical, historical, and social scientific geography. For a variety of reasons, these efforts did not "save" geography, which continued to struggle with its identity, purpose, and place in the "connaissances humaines" well beyond the period covered by this study.[16]

This book builds on the already considerable literature positing that a profound shift took place in the nature of scientific thought at the end of the eighteenth century.[17] In the 1960s, arguing that historians of science and of thought were too focused on the biographies of great thinkers, on the progressive linearity of thought, on the success and failure of concepts, and on seeking unitary causes for hopelessly complex events and developments, Michel Foucault identified a new focus of analysis: the episteme.[18] Foucault saw the episteme—the product of research into the archaeology of thought—as the fundamental regularity to knowledge, determining the possible and the impossible dimensions of scientific thought. Foucault was particularly interested in the transition points between three epistemes: the pre-Classical (ending approximately in the mid–seventeenth century), the Classical (ending at the end of the eighteenth century), and the Modern (still in effect today).

The centerpiece of Foucault's analysis was language, but his study ranged widely from literature to art to biology (or natural history), to economics, psychology, sociology, history, ethnology, psychoanalysis, and medicine, encompassing, in fact, the full panoply of Western intellectuality focused on the study of "man." Within the pre-Classical episteme, it was the adjacency, similarity in form, and sympathy of phenomena which were understood to provide all clues to the meaning and message left by God for humans to discover in all of creation. Thus language, under this episteme, was a symbolic—although warped by Babel—representation of the world, and the words of which it was composed were to be itemized (not analyzed) in terms of their adjacency, form, and sympathy. The Classical episteme, marked as beginning with Don Quixote, rejected resemblance as "the fundamental experience and primary form of knowledge,"[19] regarding knowledge instead as existing only through and as a result of the comparative analysis of identity

and difference, structured and given meaning through measurement and
the imposition of order. In this episteme, language is not an hermeneutic
sign or form of truth, but a transparent semiology. In the transition to the
third, "Modern," episteme, the surficial unity of the known world breaks up
and unity and insight is to be found only through relationships established
through an analysis of invisible, interior, and functional relationships. In this
episteme, "mathesis" and the taxonomic order of appearances—surficially
determined identity and difference—are not only relatively unimportant
but deceiving. Foucault's point is that much of what is interesting about the
history of thought is inexplicable outside of the context of the dominant
episteme. However, he does recognize the value of multiple levels and
methods of analysis.[20] The action covered in this book takes place in precisely
the period identified by Foucault as transitional between the Classical and
Modern episteme.

I certainly did not begin my research into early-nineteenth-century
French geography by looking for an archaeological explanation for geography's position in early-nineteenth-century France. Nevertheless, as I
explored the nature of French cartography and geography of that period,
I found significant and thought-provoking resonance between Foucault's
grand perspective and my own much smaller-scale and smaller-scope study.
I agree profoundly with his impatience with simple-minded causal explanations for hopelessly complex constellations of events, conceptions,
and interactions. I also feel that the complexity of events, concepts, and
relationships demands an explanation which reflects the complexity and
multifarious nature of knowledge and human existence. Indeed, in my
view, there are *at least* three foci of analysis appropriate and essential to the
history of thought: the conceptual, the sociological, and the epistemological.
Although analysis of the past is often written with the assumption that
there is only one focus of analysis—and specialists differ with respect to
their preferred focus—the levels are interdependent, and truly explanatory
history demands an interactive exploration of aspects of all three levels.

Writing in the 1960s and '70s, Foucault felt that most history was focused
on concepts. He was consequently seeking to diversify and deepen historical
analysis. There are, however, many aspects of conceptual history which are
entirely unexplored. This is especially true of the discursive formations of
geography and, for reasons elaborated on below, of the geography practiced
in the early nineteenth century. Conceptual history focused on geography
might explore the nature of geography practiced by geographers or those
writing "geographies." What were the kinds of questions they asked? Does
there appear to have been a wide consensus among geographers about those

ideas? What failed to draw their interest? What contemporary scholars did they read and what did they take from them? How did other scholars respond to their works and ideas? Did geographers generate any particularly powerful and influential concepts? Part of a good, solid conceptual history is the recovery of forgotten ideas and personalities. These abound in the history of nineteenth-century geography. We have, for example, entirely neglected one of the important players in the development of a non-race-based ethnogeography, Adrien Balbi. He has attracted more attention from anthropologists, linguists, and even historians of French identity than from historians of the field in which he understood himself to be practicing. This is equally true for Conrad Malte-Brun, Jean-Antoine Letronne, Edme-François Jomard, Louis Vivien de Saint Martin, André de Férussac, Charles Athanase Walckenaer, and a large number geographers less known to other fields. What was the aim of their research? Were they using an established methodology or research tailored to the problem at hand? What constituted original and derived research for them? What was the structure and course of their reasoning? What were their key ideas? Which were the decisive influences shaping not only their ideas but the method and presentational form of their work? Why, for example, was geography so seduced by classics in the first half of the nineteenth century? What if anything does this have to do with the humanism of the period? What was, and what has been, the impact of their ideas and approaches to problems? These often apparently banal questions are the foundation of conceptual history.[21] Answering them requires extensive research. Sadly, they have not been asked for even the best-known geographers of this period.

As has been so often pointed out, ideas and concepts do not have lives of their own. No matter how disembodied historians have made them seem, they do not float across the ages over the heads of thinkers and practitioners but are the products of not only individual thinkers but of social organizations, networks of scholars, and of acrimonious and often highly personal debate. They do not have a life apart from the technologies, cultures, economies, and polities which spawn, foster, or oppose them. The second focus of historical analysis, the sociology of thought is, then, essential to an explanatory history of thought. The sociology of thought is integral to Foucault's analysis but expressed in very abstract terms precisely because his attention is riveted on discursive formations largely to the exclusion of the individual. I am not arguing that all social dimensions of any given question must be identified, any more than all ideas spouted by an author must be taken equally seriously. The social context to be explored depends on the central question posed. The sociology of thought might explore the formal and informal structure within which individuals function and present their ideas. In geography, for example, how and where did geographers meet? Where did they publish? For whom did they write? How many journals might they have published in? What were the relations between these, and what criteria did particular journals use for the acceptance or refusal of submitted articles? To what extent did the societies and associations to which geographers belonged direct and control their research? And by what mechanisms? Were there subjects or themes that they rejected out of hand? What role did they play in the transmission or blockage of ideas? What were the relations between these societies and journals and secular power? As anthropologists and linguists have shown us, even informal and relatively unrecognized structures can sometimes have a determining influence.

The way in which I have asked these questions implies a static view of the sociology of thought. However, it is important to remember that institutions change over time and in the face of circumstances as subtly as do the thinkers which both form them and provide their context. It is especially important to bond these structural and institutional questions to the geographic ideas and research endeavors which gave them life and meaning. A social history of an institution which substantially ignores the ideas and debates which motivated it is like the aesthetic study of architecture apart from its immediate function, structural integrity, and larger social role.[22] Ultimately all it offers is arbitrary judgment far more reflective of the society offering the judgment than any past aesthetic, architecture, or society. It is for this reason that I have found myself so dissatisfied with the history of geographic institutions written by historians: it is virtually impossible to strike a balance between the context, the social dimensions of research, and the realm of ideas in an institution such as the Société de géographie de Paris if one does not explore the changing nature of the geography being practiced or excluded by that society.[23]

Of course, the sociology of geography alone is inadequate to describe the discursive formation of geography at the end of the eighteenth and the beginning of the nineteenth century. This was a period of ragged edges in discursive formations. Thus to understand the society of geographers, one must understand the society of geologists, statisticians, historians, proto-linguists, proto-archaeologists . . . , not to mention the larger society. It is a massive project to truly reconstruct even one discursive formation in this period. I have sought to provide contextualization adequate to the subject at hand in this book. But I am well aware that further contextualization would greatly enrich the subject.

Without the marriage of the conceptual and the sociological, one cannot begin to enter into the third and perhaps most explanatory level of analysis: the epistemological. Epistemological questions are focused less on the birth and existence of concepts or on the sociology of thought than on that more volatile dimension opened by Foucault and explored by a large number of successors: the episteme, the zeitgeist, the paradigm, the exemplar, and the logic of the discourse (among other epithets). These kinds of issues are stimulating, suggestive, highly provocative, virtually impossible to prove, extremely difficult to argue, and utterly meaningless when ungrounded to individuals, institutions, and concepts. The key epistemological questions are, What kind of thought was possible or impossible under the reign of a particular episteme? What sorts of subjects and preoccupations were deemed worthy of attention by, for example, the geographical engineers? And why? Why might a theory such as Buache's system of mountains and river basins be so convincing to field scientists who had had ample evidence to note its inconsistency with the actual shape and patterns of mountain chains and river basins?[24] Why would a geographer like Jomard so persistently use mapping to explore questions with little or no geographic-spatial dimension? Is the answer to both of these questions the cartographic model or, more broadly, the descriptive model of thought which seems to have dominated the geography of the time? Or is the explanation the "pan-mathematics of Descartes, the self-deception of believing that the multiplicity of the world could be overcome by mathematical ordering," as Cassirer would suggest?[25] But why the tenacious hold to these approaches? What we are seeing there, I would argue, is some of the undermining strength of the bonds of discursive formations acting on an extremely subtle—indeed epistemological—level. On a level, in fact, that contemporaries as involved participants could not have understood. The answer to these questions, then, lies in a zone where concepts, the sociology of thought, and epistemology interact. Similarly, we can ask, What gave reigning ideologies (in the oldest sense of the term) their power and hold? What and why is the form of their resonance? How did the ideologies facilitate and hinder exchange? How did they filter ideas and types of scholarship or scholars? Did the eclectic character of the Société de géographie in approximately the first sixty years of its existence reflect an absence of competing ideologies—or a lack of recognition of these? Behind these questions lie others still more profound and less answerable: what is intellectual creativity, where does it come from, what does it need to find realization, and what is its relationship to discursive formations?

The focus of this book is on a discursive formation which was the representational discourse par excellence, which moved easily from "mathesis" to order, which in the course of the eighteenth century and into the early nineteenth century had developed one of the best universally valid constructed languages (cartography), and which unified the sciences having anything to do with the earth or the cosmos through the mapping of identity and difference in space. In fact, the early-nineteenth-century discursive formation of geography (in France)—at least into the 1830s—was one formed by pre-Classical and Classical modes of thought. There was, at least initially, virtually no place for such a field within the Modern episteme. The extraordinary importance accorded adjacency—which is almost definitional of geography—had already waned with the Classical episteme. The unity geography offered and proclaimed through representation or description seemed empty and futile in the face of the fragmentation and explosion of the epistemological field. Geography's language became so accessible, unanalytic, and everyday as to be general or part of the basic equipment of all scientists. Yet there were still geographers who sought to practice geography as had their predecessors and who puzzled over the diminishing respect the field gathered to itself. Here we will look at the fate of a discursive formation which once spanned much of the terrain later claimed by the empirical earth sciences and incipient social sciences—one that was crippled and left behind by the changes. The aim is not to ponder the scientific failures along the side of the road to some sort of truth but to understand the nature of the change not just through what was retained or added but through what was cast off or left behind. Geography, it will be argued here, provides an excellent vantage point on both the nature of the new episteme and the old because it so completely embodied the old. There are both discontinuities and long-duration continuities in the history of thought. In early-nineteenth-century geography, we can see some of their interplay and their consequences.

I also describe the nineteenth-century epistemic change with a slightly different vocabulary than Foucault. His characterization of modern science as focused on "internal" and "invisible structures" in terms of their "functions" is useful for understanding the new episteme as it is manifested in biology, literature, and economics. But the vocabulary for explaining an absence of something to be found in other sciences must be different. It is not so much that geography failed to find interior spaces, or develop functional explanations, but that it could not give up surficial *description* for *explanation* of interior functions without fundamentally altering its nature, its composition, and, above all, its unifying scope. This difference in vocabulary has significant implications in drawing the line between the Classical and Modern epistemes. For example, while Foucault sees Linnaeus's "system" and Adanson's "method" as "two ways of defining identities by means of a general grid of differences" (and, thus, both firmly in the Classical episteme), I see them as fundamentally different and as belonging to two different epistemes. For while Linnaeus was concerned with mostly visible and external features of plants, his classification was explanatory and theory-laden description, focused on the reproductive "function."[26] Adanson's method was a plant classification that made an ideal of pure description, impossibly forbidding the prior imposition of theory or argument.[27] Fundamentally, Adanson wanted to map and mathematize the plant world before coming to any conclusions about it. His view of the world was much closer to that of early-nineteenth-century geographers than to eighteenth-century natural historians like Linnaeus or Buffon. Thus, I highlight the stuttered and gradual shift between the Classical and Modern episteme which nevertheless left geography substantially behind its sister sciences by the 1830s.

Despite Foucault's focus on the representational basis of science in the eighteenth century, he is relatively insensitive to the different modes of representation employed in science. Thus, apart from his analysis of the Meninas, he rarely mentions and does not take into account the wide variety of pictorial media such as maps, graphs, trees, diagrams, and their changing function in science within and across the epistemes. As a result, he misses an important consequence of the changes taking place in the shift from Classical to Modern thought: while representation was being abandoned as the guiding ideal of scientific thought, many of the sciences moved toward a far greater use of scientific illustration. This is neither paradoxical nor contradictory to Foucault's argument that the sciences were becoming more focused on invisible interior structures and their functional roles.[28] Indeed, the graphics and cartographics employed by discursive formations such as geology, and later climatology, botany, zoology, epidemiology, etc., were—as Martin Rudwick has pointed out for geology, as Jane Camerini has pointed out for plant biology, and as Cambrosio, Jacobi, and Keating have pointed out for immunology—theory-loaded.[29] Together with sections, sketches, and landscapes, maps were used extensively in the development, elaboration, and testing of theory. Indeed as LeGrand has shown for a later period, such graphics and cartographics could even replace experimentation in the testing of theory.[30] Thematic cartography, then, came to be because the use of maps in representing what could not be seen but could be hypothesized required the development of a new and extensive symbolic vocabulary.[31]

Focusing on the use of representation in science also adds dimension to the dilemma within which geography found itself. As we will see in chapters 1 and 2, ironically, just as mapping was developing a consistent, systematic, and comprehensive mode of expression[32]—or just as it was perfecting description—it could no longer win geographers intellectual status among astronomers, physicists, and analysts of literary and ancient culture. Yet, scholars in a wide variety of fields began to pay attention to maps, using their topographical content as incidental background for a larger and more significant intellectual argument, about which geographers (or those who chose to remain descriptive geographers) had little to say.

The argument contained in this book is presented in three parts. In section 1 the focus is on geography's loss of direction and status in France toward the end of the eighteenth century. Chapter 1 details the depth of geography's commitment to description and discusses the degree to which there was a sense of identity in eighteenth-century geography. Those institutions which gave geography a home and a means of dissemination are described, including the Encyclopédie, the Academies of Science and Inscriptions, the Jesuit colleges, and the military colleges. This chapter also reconstructs the sense of identity, community, and tradition among geographers such as Jean-Baptiste Bourguignon d'Anville, Didier Robert de Vaugondy, Guillaume Delisle, Joseph Nicolas Delisle, Philippe Buache, César François Cassini, Nicolas Desmarest, and Antoine Augustin Bruzen La Martinière. Finally, it explores the kind of work geographers published in the eighteenth century, their principal intellectual concerns, and the relationship of those to the major intellectual debates of the period, such as that over the size and shape of the earth. Chapter 2 explores the signs and evidence of a loss of direction and status in the two most important branches of eighteenth-century geography: descriptive cabinet geography and large-scale field cartography. It looks at a moment in the careers of the cabinet geographers Jean Nicolas Buache de la Neuville and Edme Mentelle in which they were informed by colleagues in one of the highest institutions of higher education that their geography had nothing to offer modern science. It then reviews and analyzes the even more spectacular casting out of large-scale cartographer Jacques Dominique Comte de Cassini from the halls of academe. This story is all the more compelling as Cassini did not go quietly.

Section 2, entitled "Reaction and Continuity," is devoted to the intellectual reaction to this loss of status. Here, I explore the traditionalist solutions: those which were seen to be sound responses because they emulated traditional approaches. Chapter 3 looks at the tradition of universal geographies, or descriptions of the surface of the earth. By the nineteenth century this genre was dealing with an explosion of knowledge about the earth which rendered its descriptive task impossible. Here I try to understand the rationale for the genre and to capture its essence by going back to Strabo's geography and bringing the reader back through universal geographies as they were written in the seventeenth and eighteenth centuries. The core of the chapter focuses on Malte-Brun's universal geography as this was said both at the time and by historians of geography to have reestablished the genre. Included in this discussion are the remarkable works of Alexander von Humboldt, *Cosmos* (1849–60), and Le Père Jean François, *La Science de la géographie* (1652). Chapter 4 describes Edme-François Jomard's attempt to apply a geographic method, mapping, to the decipherment of hieroglyphs. The method was entirely inappropriate but made sense in terms of the traditions prevailing within the discursive formation and had significant impact on the field of Egyptology. This is a suggestive chapter because it explores both the power and limitations of the mapping metaphor, which arguably is still central to modern geography. Chapter 5 discusses the old alliance between geography and secular power, particularly as it was manifested in the Napoleonic period, when geography's association with the state was as close as it was to become in colonial Europe in the late nineteenth and early twentieth centuries, or in Nazi Germany. It examines the degree to which this close relationship modified the nature of geographic research through both warfare and major scientific expeditions and the impact of state influence on the thinking and decisions of individual scholars, in particular the two "natural" geographers André de Férussac and Bory de Saint-Vincent.

Section 3, entitled "Innovation at the Margins," considers the innovative steps taken by intellectuals, invariably on the margins of the discursive formation, to explore a theoretical and explanatory geography. All of these scholars had some sort of identification or link with geography and geographers in the course of their careers but, for the most part, functioned outside the discursive formation. Chapter 6 looks at the rare forays into something resembling a social scientific geography. The Napoleonic era and its immediate royalist aftermath, although providing opportunities to study society, was unfriendly to the kind of social criticism that social science implies. When Napoleon crushed the second class of the Institut de France, he also ended the possibility of a social scientific geography at least until sometime in the 1830s and arguably until the late nineteenth century. Nevertheless, the early work of Comte Constantin-François Chasseboeuf de Volney and the statistical study of Paris carried out twenty years later by Comte Gilbert-Joseph-Gaspard de Chabrol de Volvic and his geographic associates suggest that some of the elements of a social scientific geography were already present in the early nineteenth century. The statistical and ethnographic geography of Adrien Balbi, set squarely within the geographic discourse, serves as a backdrop and point of comparison with this more innovative work. Chapter 7 examines the activities of another scholar working on the margins of early-nineteenth-century geography, Alexander von Humboldt. Although he is now considered synonymous with nineteenth-century geography in most histories of geography, largely for the synthetic effort of his *Cosmos,* in his work on physical science he functioned on the edges of the discursive formation. In that realm he explored ideas emerging in mineralogy, geology, and biology which, although little understood by French geographers in his own day, are now considered to be at the core of the modern field of geography. These include the idea that landscapes and natural regions are worthy objects and scales of analysis; the insight that interrelationships between realms (for example plants and rocks, or climate and soil) reveal as much about the functioning of the natural world as the study of the classification of rocks or plants or climates; and the realization that comparison of like phenomena across space can reveal much about the past and present nature of the earth. Also explored in both chapters 6 and 7 is experimentation with a new mode of graphic expression more capable of elucidating the theories and explanations these innovators were exploring. Chapter 8 describes the integrity and quality of Jean-Antoine Letronne's highly innovative work in historical geography. It does so by placing Letronne's work within the context of erudite historical geography as it was practiced by Jean-Denis Barbié du Bocage, Pascal-François-Joseph Gosselin, and Baron Charles Athanase Walckenaer. Letronne was thoroughly engaged in the intellectual debates of his own time, which revolved around a renewed fascination with the ancient origins of Western science, religion, and philosophical thought. Driven by intellectual curiosity, Letronne abandoned the map, the narrow sense of geography as topological information, and a descriptive writing mode. As many of his non-geographer contemporaries were beginning to ask questions about the nature of society and government, Letronne turned that curiosity on the ancient world. Yet, while Letronne moved well beyond the bounds of geography's traditional interests, he never lost his interest in geography and saw, and described, many of his works as contributing to the geographic discourse. Still, Letronne was from at least the 1820s marginal to geography, cited superficially and rarely by most geographers. By the end of his career, he was scarcely regarded as a geographer at all, as is clear from the total silence regarding him from historians of geography.

An explanation of my tendency to biography is necessary. This book is focused on a period that has been largely ignored. It has been necessary to rebuild nineteenth-century geography from the ground up. Going back to the key thinkers and rebuilding the discursive formation through patterns of citation, mention, idea commonality, and expressions of territoriality is necessary because so very little has been written about the geography of this period. This relatively clean slate and my method of reconstructing geography will allow us to avoid—as much as is possible—the imposition of twentieth-century conceptions of disciplinarity on earlier structures of intellectual organization. After all is said and done about concepts, epistemes, and the sociology of knowledge, I still believe that it is essential to begin with and return to the individual "scientist" and his or her encounter with the world, with ideas, with institutions, and with colleagues.[33] I do not, however, see the scientist as the lone inquiring mind, but as a "node in a network."[34] This book repeatedly returns to individuals and the constraints their circumstances and personalities placed on them. If in the past, most readers of history were interested in scientists' genius and the remarkable coherence of their thought, today we tend to be more interested in the constraints they encountered and the impact of those on the quality and nature of their work. We also *expect* to find incoherences, confusion, and miscommunication. Constraints, which can be institutional, physical, political, economic, or psychological . . . are, I believe, clearest where they were experienced, at the level of the individual's struggle to understand, contribute, and create.

One final point: the focus of this study is geography in France from roughly 1760 to the mid–nineteenth century. Arguably, in this period French geographers were some of the best-known and best-respected geographers in Europe. Further, France was a center of innovation in precisely those sciences that were moving away from relatively uncritical description to theory and explanation. There was increasing communication across national borders in many fields in this period. Nevertheless, most geographers worked within a national frame which admitted foreign scholars but was structured by a national tradition and often by national needs. Consequently, France serves as a sort of microcosm of developments in both European geography and the sciences that formed its intellectual context.

·PART·ONE·

Geography's Fall

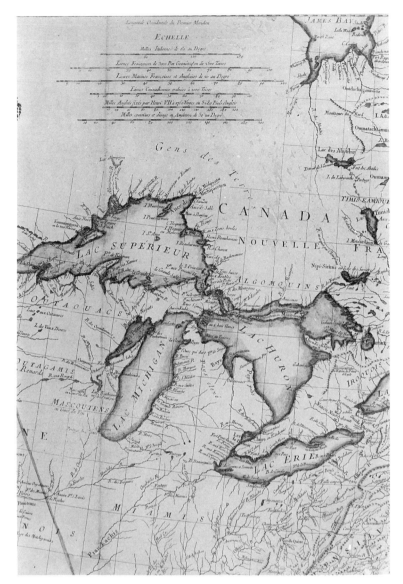

FIG. 1.   "The vanishing grid." One of the more ingenious and suggestive ways of expressing uncertainty developed by eighteenth-century cartographers: d'Anville developed the trick of suddenly ending the graticule at the border of sure topological knowledge to make clear that while relief was little known in general, it was here that one stepped beyond the acceptable bounds of Western science (represented by the locational grid).

The Nature of Eighteenth-Century Geography: Cartographic and Textual Description

Under whatever form one considers it, the language of geographers bears witness to a stage of geographic consciousness [absorbed with] determining the location of objects, and qualifying them with growing rigor thanks to the enrichment, the precision and the progressive harmonization of the means of expression. Sufficiently absorbed by these problems, the geography of that epoch described without explaining. —De Dainville, *Le Langage des géographes*, (1964), 336

GEOGRAPHY DESCRIBES THE EARTH

Most scholars looking at eighteenth-century geography have described it as essentially cartography or nothing if not mapping. Unquestionably, to the retrospective eye, the excitement of eighteenth-century geography lies in the great developments in mapping science of that period: in the solution of the longitude, in the increasingly accurate measurement of the earth, in the unfolding on maps of the great continental interiors of the New World and of some parts of the Old . . . . Nevertheless, the defining characteristic of the geography of this period was not the quest for mathematical accuracy, nor the demonstration of exploratory courage or curiosity, nor the search for pictorial beauty—no matter how much these were in fact striven for and even attained. The overriding aim of sixteenth-, seventeenth-, and eighteenth-century French geography was the reflection by word and symbol of the unity and coherence of the world. In the eyes of eighteenth-century geographers it was geography's major achievement to have replaced a chaotic vision of the world as "a pile of debris and a world in ruin" with a world in which "order and uniformity are perceived, [and] where the general relationships become apparent under our footsteps."[1] Eighteenth-century geography, then, was both mapping and much more than mapping. It was an attempt to understand the cosmos—to seize its meaning—*through description*, and that attempt found both graphic and textual expression.

It is difficult for a modern thinker to accept the concept of understanding without explanation, but this was the geographic ideal.[2] Geography's mission was to reveal the earth, broadly defined, by describing it. It was the philosopher's task to discourse on the nature of the earth and explore the causes of that particular nature. Although often taught under the rubric of mathematics from at least the sixteenth century, neither was geography synonymous with mathematics. Measurement and calculation, whether carried out in the office or in the field, were ancillary tools; they did not define geography. Thus, the engineer might take measurements in the field, and the astronomer might calculate and predict the movements of the stars, but it was the geographer's task to gather together those measurements, and to translate them into the synthetic pseudo-landscape of a map or text.

The geographer might conduct mapping in the field, and many eighteenth-century geographers had both field and office experience of mapping. But the geographer's essential task was to collect information of many types and from multiple sources, to critically analyze that information and, through insight into the connections between phenomena and the patterns they formed, to creatively perceive the essence of reality. The path to insight was not some sort of numerological mysticism, but precise definition, rational classification (based on surficial location and appearance), scrupulously honest judgement, and experience in the practice of the trade. Consequently, not every mapmaker or author of a geographic text was a geographer. Those who reproduced the maps or descriptive texts of others without struggling to creatively perceive reality were not scholars but editors, mere copyists, or worse: thieves, charlatans, and cheats. This creative perception of reality was not the easy task that it may now seem to us: until the very end of the eighteenth century, Europeans and European geographers lacked the tools, the concepts, and especially the vocabulary to describe the earth completely, consistently, and accurately.

On no account was it the geographer's task to explain the earth either historically or ontologically. It was not geography's role to explore the unknown but to banish it. As such geography was not a theoretical subject and brooked no "unknown principles." Robert de Vaugondy, an eighteenth-century map and globe maker, declared geographer, and author of numerous important publications in geography, including the most influential of eighteenth-century histories of geography, was unequivocal about this:

> I would say, with La Martinière, that "topography," which is the description of a particular place, and "chorography," which is that of a region, gave birth to "geography," which must not be considered a science with still unknown

principles but according to the correct meaning of its name, [it must be understood as] . . . "the description of the earth."[3]

For Robert de Vaugondy this meant that physical geography, by virtue of its more systematic approach and its tendency to follow curiosity into the depths of a question, was not really geography.

> It is of positive geography that I speak here. That is to say, of that science which relates the natural division of the earth with its political division. Physical geography, which is a subject of study for some people, is too systematic to find any place here [in this book on the history of geography]. It pricks the curiosity more than procuring any real advantage. The structure of the globe is worthy of the attention of the physicist. The knowledge that he will have of it may even extend his understanding of principal phenomena. But the geographer must only focus on the surface of the earth; how many other functions should he attempt to fulfill![4]

Was Robert de Vaugondy worried about admitting the unseen below the surface of the earth to geographic inquiry precisely because it would necessitate theory-based study of the unknown and significant deviation from the descriptive mode of the map and the traditional geographic text?

This is very probably the case, but Robert de Vaugondy was in no way declaring anything that would surprise contemporary geographers. From the sixteenth century, and inspired on the one hand by Strabo and on the other by Ptolemy, geography in France had been taught under only two rubrics: rhetoric and mathematics. In both realms the geography taught and practiced was deliberately and exclusively descriptive. As early as the sixteenth century, the pedagogical regulations of the Jesuits,[5] perhaps the most significant keepers and shapers of the geographic endeavor in France, forbade the confounding of descriptive geography and general theories (or "Philosophorum more"). Unsurprisingly then, Le Père Philippe Briet, author of one of the most influential geographies of the seventeenth century, found no room in his geography for explanations and philosophical discourse:

> Geography is not a matter of intellectual activity ["affaire d'intelligence"], one only needs eyes to understand it. For myself, I would speak of geography in a manner that allows me to show everything that I say: there would be nothing on the map that I do not describe in my text and nothing in my text unrepresented on my map.[6]

Understanding in geography was to come, not from philosophical discussion and theoretical explanation, but from order of representation of surficial phenomena and a certain flexibility in that depiction (whether textual or cartographic). Le Père Philippe Labbé, in his 1662 *La Géographie royalle*, suggested how such description might be a powerful learning tool:

> I would warn the reader that it is not necessary that all of these points and principal locations . . . be considered in the order in which we have given them . . . as it is often more suitable to mix them up and to put some in front and others behind, according to the disposition of he who undertakes to describe and tell them.[7]

Shuffling order was a world away from playing with theories and hypotheses—the danger of which Galileo had so clearly demonstrated to all of the servants of the Church.[8] And how could anything below the surface of the earth be explored without theory and hypothesis? No, in Le Père Jean François's view it was geography's mission to

> limit itself to the surface of the earth, which serves as home to men, in order to show its diversity in both nature and art.[9]

Still, geography was expected to provide insight into and understanding of God's creation. The words of Le Père Pierre Coton (1564–1626) give a strong sense of the hope for the illumination that an informed sense of the meaning underlying clear description (meaning-filled letters) might provide:

> Do not allow, Architect of the world, that my eyes should become like those of small children who entertain themselves by looking at illuminated letters and who play with these instead of considering what they signify. Illuminate my eyes.[10]

Indeed, for much of the Classical period and until the gradual rupture between scientific and religious thought (perhaps sealed in France and for French geography with the dissolution of the Jesuit order and the end of the Jesuit dominance of secondary education), geography is best understood as a creative, an erudite, and "a magnificent poem to the love of God."[11]

Toward the end of the eighteenth century, perhaps as a result of the observations made and questions raised by mathematicians and the successors to the Aristotelian philosophers, the physicists, and increasingly explored by natural historians, a few geographers began to wonder whether a more subtle conception of the relations between geography, physics, and natural history might be entertained. Certainly there is evidence of a more nuanced sense of the nature and function of geography in the writings of the physical geographer Nicolas Desmarest. In contrast to Robert de Vaugondy, his description of the purview of physical geography in the *Encyclopédie* has a surprisingly (and deceptively) modern ring to it. For Desmarest, physical geography was the fruit of the improvement and coming together of physics and geography. The resultant physical geography, he argued, might include:

> the organization of the globe: in which the aim is insight into the principal operations of nature, in which one discusses their influence on the particular and subaltern phenomena, and in which, through a continuous chain of facts and reasoning, an explanatory plan is formed.[12]

Or it could "be more wisely limited to the establishment of analogies and principles."[13] Yet a careful reading of Desmarest reveals that the distance between Desmarest and Robert de Vaugondy was less great than a modern reading of his words would make it seem. For even in the nascent physical geography, so summarily cast out by Robert de Vaugondy, the essence of geography was the observation of facts, their informed, critical, and yet creative combination, and finally the generalization of the results. For Desmarest, then, physical geography had the same sort of scaled structure and function which Robert de Vaugondy ascribed to the trinity of topography, chorography, and geography with geography both generalizing and encompassing the others. Nor did Desmarest expect physical geography to abandon its topological focus as it was the geographer's role to "distribute by region and territory that which the naturalist described and arranged by class and by order of collection."[14] It was the physical geographer's task to search for regularity and repetition of pattern—in order to avoid a focus on isolated phenomena and to concentrate instead on multiple or analogous occurrences of the said phenomena in different parts of the world. For Desmarest and in contrast to Robert de Vaugondy, analogy or pattern could be derived not only from exterior form and the physical context of phenomena but also from interior masses or configurations. But for Desmarest as well as for Robert de Vaugondy, it was the geographer's task to fit all of the observed phenomena into an edifice—not an edifice built on selective theory but an open structure, like Adanson's classificatory system, always ready to accept new information. Speaking of the general division of the earth by physical feature, he commented:

> . . . these general divisions would be more appropriate . . . and more varied in relation to our limited and imperfect knowledge of certain complicated subjects . . . than those truncated views which the imagination has given the form and appearance of a theory. These tables would be open, like the archives of discoveries and the depository of our acquired knowledge, to all those who would have the zeal and talent to enrich them anew.[15]

The final product of physical geography was not necessarily a map but would be map-like in its fundamentally descriptive nature, its concern with topological order, and its hierarchy of scaled information.

THE STRUCTURES OF EIGHTEENTH-CENTURY GEOGRAPHY

There is every evidence that geography was considered one of the most important endeavors of the eighteenth century. It is clear from the space that it was accorded in that résumé of all scientific knowledge, the *Encyclopédie,* that it had the respect and attention of the scholarly world. Perhaps even more significantly, several of its most accomplished practitioners found representation in either the "Académie Royale des Sciences" or the "Académie Royale des Inscriptions et Belles-Lettres." The community and structures of geography were also very much in evidence at the secondary level. Geography was central to secondary education until the dissolution of the Jesuit order and continued to be taught in other religious schools such as those of the Oratorians, whose geography instruction was arguably modeled on that of the Jesuits. At a time when universities were few and taught a scarcely more advanced curriculum than secondary schools, this amounted to the highest level of instruction in the land. It was a level of instruction deemed appropriate, until the establishment of military colleges, to the formation of officers, engineers, scholars, missionary explorers, and noblemen. But religious colleges were not the only educational institutions in which geography was represented. From approximately the middle of the century geography was also taught in the military academies, and there are indications that commercial geographers, or those not salaried by the state, and even state-sponsored geographers, engaged in apprenticeship training and tutoring.[16] On a more professional level, early in the century a special geographical corps was established, known variously as geographical engineers, topographical engineers, or engineers of the camps and armies. Their function was mapping, providing strategic intelligence to senior officers and, through what was to become the Dépôt de la guerre, the collection and coordination of terrain information of general value to military operations. Less tangibly, but just as importantly, there is evidence in the writings of many geographers of a strong sense of community, shared standards, and professional turf. Finally, geography was sufficiently self-conscious to warrant the writing of histories of geography.

It is clear that geography was deemed important by the editors of the *Encyclopédie*. They devoted over eighteen densely packed, double-columned pages to geography as an endeavor (approximately eighteen times the space accorded geodesy, four times that for botany and twice that for astronomy). At the same time, geology, while recognized as one of the "connaissances humaines," received no textual entry of its own. However, even this generous coverage of geography was deemed inadequate by Diderot, who used this insufficiency, along with others, to argue for a supplementary or new edition. In addition, on the elaborate tree of understanding which graphically summarized the *Encyclopédie*, the "Essai d'une distribution généalogique des sciences et des arts principaux," geography occupied a full quarter of the space accorded to the subfields of mixed mathematics, which also included astronomical geometry, optics, the analysis of hazards, mechanics, and dynamics.[17] It received further graphic development in Robinet's *Supplément à l'Encyclopédie*, where it was divided into topography, chorography, and three types of universal geography: absolute (description of the earth and its inhabitants), relative (differences in conditions in different parts of the globe), and comparative (information necessary for movement across the globe).[18] Finally, geography was accorded a total of thirteen volumes in the 192-volume *Encyclopédie méthodique*, including a two-volume atlas.[19] Despite the implied or stated criticism of the treatment of geography in the previous collection by each subsequent *Encyclopédie* or *Supplément*, there is substantial agreement among them about the fundamentally descriptive nature of geography. Their ongoing concern to correct and elaborate on previous characterizations also suggests that geography was considered important enough to warrant careful definition and description in text, exhaustive figurative tables, and discussions both textual and figurative of the place of geography in the "connaissances humaines." There is also a more profound philosophical link between the various *Encyclopédie*s and geography. Geography was to the earth what the *Encyclopédie* was to human knowledge. Just as the *Encyclopédie* sought to define and describe all human knowledge in an orderly textual manner and then to express this in graphic form as a tree of understanding, geography sought in text and map to define and describe the world.

Another sign of recognized intellectual status, the "Académie Royale des Sciences," from its very foundation in 1666, had among its members individuals involved in what the late eighteenth century would deem geographic research, that is, in some form of earth description. These included scholars as varied as Sebastien Le Prestre Seigneur de Vauban (1633–1707), the great fort and siege engineer; Jean Dominique Cassini (I) (1625–1712), the Italian astronomer lured to France by Jean Baptiste Colbert, who worked in the intermediate zone between astronomy, geodesy, and geography; Philippe de La Hire (1640–1718), the geodesist who worked with Picard on the Carte de France; l'abbé Jean Picard (1769–1828), who devoted his life to determining the size and shape of the earth; and Jacques Ozanam (1640–1717), the mathematician who composed tables of sines, tangents, and secants and who wrote treatises on mapmaking and on the application of mathematics to siege warfare. Although these individuals were considered intellectual predecessors by the geographers of the late eighteenth century, for the most part they did not call themselves geographers and were not classed as such in the Academy. In fact, in the period of Vauban, La Hire, Cassini, and Ozanam, the Academy had no sections at all until 1699, after which it had only six sections, devoted to geometry, mechanics, astronomy, chemistry, botany, and anatomy. This does not mean that the Academy restricted its purview to only those fields. Under Jean-Baptiste Colbert, Minister of Finance to Louis XIV, the Academy initiated three major geographic projects: the establishment of the Paris meridian, the production of a large-scale general map of the realm, and a map of the coasts and major littoral waterways of France (which culminated in the remarkably accurate and detailed *Neptune François* in 1694). In addition, as Roger Hahn has shown, a Société des Arts (ca. 1720s), whose early regulations claimed as part of its own brief "geography, navigation, mechanics, and civil and military architecture, but without neglecting any of the other arts, be they useful or simply pleasurable," was considered to be invading the Academy's turf and was duly eliminated.[20] In the course of the eighteenth century, a significant number of members were elected to the Academy who either described themselves as geographers or who came to be known as such, including Guillaume Delisle, Jacques Cassini (II), Joseph Nicolas Delisle, Philippe Buache (the first of the "associé géographes," so named in 1730), Jean-Baptiste Bourguignon d'Anville, (also a member of the Académie des inscriptions et belles-lettres), César François Cassini de Thury (III), and Nicolas Desmarest. Other members were demonstrably involved in earth description and mapping but saw their activity as too systematic, theoretical, or explanatory to fall under the rubric of geography.[21] This was the case, for example, with Luigi Fernando Marsigli's work on depth measurement and the mapping in the Gulf of Lyons, which work he classed as natural history.[22] Thus, despite the fact that a section specifically devoted to geography (and navigation) was not introduced until 1803, when Napoleon reorganized the recently founded successor institution to the Royal Academy, the Institut de France, a great deal of geography was practiced in the eighteenth century under the auspices of the most prestigious institution of higher research, the Académie Royale des Sciences.[23]

The Académie des inscriptions et belles-lettres, formerly called the Académie des inscriptions et médailles, was founded in 1663. It was a less prestigious institution than the Académie des sciences largely because roughly one-fourth of its members were "honorary," or named by senior government officials rather than elected by members of the Academy. It would be erroneous to imagine that it was less prestigious because French governments considered it a less useful organization: even in the Napoleonic period, it was the Académie des inscriptions that was most frequently assigned tasks by the government.[24] Nor did that Academy receive significantly less revenue from the government.[25] Moreover, the Academy included scholars of first rank whose primary interests were history, especially ancient and French history, archeology, religion, languages, inscriptions, medals, and geography of the more erudite sort. The Academicians presented and published papers which were discussed not just in the Académie des inscriptions but also in an annual meeting with the Académie des sciences.[26] Over the years of its existence, the Academy was home to a number of prominent geographers, or at least scholars who wrote works that geographers recognized as their own and cited, including, Nicholas Frérét,[27] Jean-Pierre de Bougainville, l'abbé Jean-Jacques Barthélemy, Jean-Baptiste Bourguignon d'Anville, Baron Guillaume-Emmanuel-Joseph Guilhem de Clermont-Lodève (Sainte-Croix), Marie-Gabriel-Florent-Auguste comte de Choiseul-Gouffier, and Pascal-François-Joseph Gosselin.[28] As such, the Académie des inscriptions contributed to geographic research, in part through approbation of its quality.

While the Académie Royale des Sciences and the Académie des inscriptions et belles-lettres represented achievement and excellence in scientific and humanities research for the second half of the sixteenth and all of the seventeenth and eighteenth centuries, up to 1762 the Society of Jesus had a virtual monopoly over both primary and higher education in France. The Université de Paris and its forty colleges were largely deserted by 1600. It had become embroiled in religious controversy and had little influence on either higher education or research in this long period. Thanks to the impressive and engaging work of François de Dainville, we know something about the purpose, substance, and rubrics of geographic education in the French Jesuit colleges, which shaped French geography. Jesuit geographic education was designed to serve many purposes. On a philosophical level, the aim of earth description was theological or, put differently, the study of the earth as a part of God's creation. On a more pragmatic level, geographic education was vital to the mission of the Jesuits: to their global strategic planning, as an instrument of negotiation and influence with other cultures, and in the forging of an alliance of power and knowledge with state authority. For all of these reasons, knowledge of the world, or a complete and accurate description of the world, was of considerable value to the Jesuits. It was the belief of the sixteenth-century Humanists, who set the course of Jesuit education, that there were two principal pathways to forming a complete picture of the world: description based on direct and careful observation (mathematics), and description derived from careful study and exegesis of ancient scholarship (rhetoric). (The mathematics taught was also strongly shaped by a focus on the practices of the ancients.) Consequently, geography as earth description was taught, with some overlap in material, under the rubrics of mathematics and rhetoric in the many Jesuit Colleges of France.[29]

It is the prejudice of moderns to dismiss rhetoric[30] and focus on the importance of mathematics in early geography. Arguably, it was rhetoric and the activities of the Jesuits in that realm that best explains what most puzzles moderns about the flavor and point of eighteenth-century geography both cartographic and textual. Rhetoric can be defined as the use of language as an art based on an organized body of knowledge. In ancient Greece it was considered to be the art of persuasion and was linked to political democracy and the practice of law. Not just a technical tool, it was judged by Aristotle to have moral force as an oratorical skill whose aim was to instruct, to move, and to delight, and which was expected to reveal truth and justice. Behind the teaching of rhetoric, both in ancient times and by the Humanists, was a concern with clarity of definition, logical argument, and an understanding of the psychology and capacities of the audience—all matters of concern to any communication but particularly critical to scientific communication. The art of classical rhetoric had a fivefold structure whose relevance to eighteenth-century mapmaking and literary geography is evident: collecting the material, arranging it, putting it into words, memorizing it, and delivering it. The Humanist rhetoric taught by the Jesuits was focused on imitating the literature initially of antiquity but ultimately also that of the more scientific and experiential present so as to absorb recent cultural and intellectual triumphs. Imitation emphasized style, memorization, and delivery. It was, thus, that geography came to be taught within rhetoric as ancient descriptions of the world, historical and modern maps, globes, and accounts of travel and exploration were expressed in words, memorized, and reproduced. This geography was not without a critical dimension: for example, understanding sacred geography required the critical exegesis of the Bible.

The comparison of many different texts and maps both modern and ancient would also inevitably lead to the correction of some of these sources and a quest for the most correct information. The focus of both ancient and modern geographic texts on peoples as well as boundaries also stimulated curiosity about the world and kept students interested through what must have been fairly boring sessions of dictation, memorization, and repetition.

Mathematics as taught in the Jesuit schools encompassed astronomy, geometry, measurement, and the study of the sphere, or cosmography, which would have included description of the circles, zones, and parallels of the earth and its division into 360 degrees; units of linear measure; latitude and longitude (and how these may be determined); and how to use an artificial globe. Cosmography would also have included the location, divisions, and descriptions of continents, seas, winds, and states, or more properly "geography." These were considered the foundation to what a future holy father, military officer, magistrate, or state official might need to be able to learn the principles of navigation, the manipulation of astronomical instrumentation, the collection and digestion of geographic information, and the making of maps. Also taught under the rubric of mathematics, and of value in understanding the principles of the sphere and of land measurement, was practical geometry. This was concerned with the measurement of vertical and horizontal lines, the calculation of inaccessible heights and distances, the calculation of angles and scale, the measurement of surfaces, and the sectioning and portioning of spheres.[31]

Geography was taught under the rubric of mathematics and rhetoric until the closing of the Jesuit colleges. This contributes to an explanation for the remarkable coexistence in the eighteenth century of a mathematical tradition of geography increasingly married to direct observation and rigorous measurement and capable of producing the Cassinis, the Delisles, and the Robert de Vaugondys, and a more literary tradition, fundamentally rooted in the study of the ancients, the reproduction of historical geographies, and the study of the measurement systems of the ancients, and capable of producing an Antoine Augustin Bruzen de la Martinière, a Jean-Baptiste Bourguignon d'Anville, or a Pascal-François-Joseph Gosselin. Remarkably different in approach and increasingly different in standards and philosophy, the two types of geography as taught in the schools and as practiced still shared much, including many of the same sources and some of the same preoccupations. Thus, a geographer fixated on the exact interpretation of the metrical systems used by the ancients was no less concerned with producing a universal cartographic coverage with accuracy of latitude and longitude than the mathematical geographer taking measurements in Peru.[32]

The Jesuits provided only rudimentary training for future military officers. In 1587 de la Noue expressed the conviction that military officers "should be taught mathematics, geography, fortifications, and several vernacular languages" at a higher level.[33] But it was some time before the King of France had an organized army and specialized corps which would benefit from advanced training. Under Richelieu, Le Tellier, and Louvois, plans were laid for military colleges in which officers were to be taught "to manage a horse, to handle arms, and to be instructed thoroughly in morals, mathematics, fortification, logic and physics, and more specifically in the French language, geography, and history."[34] These did not come to fruition. At the time of Vauban and Louvois, specialized engineering and mapping corps were formed, including the corps du génie and the ingénieurs des camps et armées.[35] Still, these officers received no training beyond field instruction, what they could glean from practical treatises on fortifications, engineering, and mapmaking, and what they had learned in the Jesuit, Jansenist, or Oratorian colleges. It was the eighteenth century before military and specialized engineering schools were permanently established and began advanced technical training in geography and mapmaking. Frederick B. Artz suggests that it was in part the demand that the newly established corps of civil engineers (ponts et chaussées, 1716) produce maps of all the great highways of the kingdom that increased the need for more and more highly trained civil engineers and led in 1775 to the foundation of the École des ponts et chaussées.[36] Schools for the training of artillery officers began to appear in the 1720s. They offered a highly technical education with an emphasis on geometry and drawing. In 1751 the École royale militaire was founded and offered a comprehensive course of studies including horsemanship, fencing, dancing, French, Latin, Italian, German, English, history, geography, drawing, and elementary and advanced mathematics and physics—including mechanics, hydraulics, and the principles of fortification. The best of the military colleges was the *École du corps royal du génie* at Mézières, which from 1751 had rigorous entrance and annual exams and from 1777 a fixed curriculum. Students graduating from the first year were expected to understand "the science of shadow and perspective" and the drawing of plans and maps, but mapmaking was also taught in the second year. There is every sign then that geography was a central part of the basic military training of most officers and certainly of the special engineering and mapping corps by the mid-eighteenth century.

Throughout this period there was also growing focus on the gathering of geographic information for military purposes, in particular in the form of maps. This led to the establishment in 1688 of the Dépôt de la guerre and its gradual entrenchment as center of geographic information,[37] a coordination center for the ingénieurs géographes (successors to the ingénieurs des camps et armées), and increasingly a center of instruction for these mapping engineers. Mathematics and language masters were established at the Dépôt as early as 1769. The eighteenth-century institutionalization of the Dépôt as an instructional center culminated in Calon's attempt in 1793 to attract specialists in astronomy, ancient geography, physical geography, historiography, military topography, engraving, and drawing to the Dépôt for the purposes of instruction, consultation, and work on particular projects.[38] It would seem, then, that in the course of the eighteenth century geography's value in military education was increasingly recognized.

It is clear from the presence of geographers and geographic research in the Academy of Sciences, in the Jesuit schools and scholarship, in the pages of the *Encyclopédie*, and in the military that eighteenth-century geography was supported by some powerful and deep structures of long standing. Still it is important to ask, To what extent was there a self-conscious community of geographic scholars and teachers? In a book not primarily focused on the nature of eighteenth-century geography it would seem excessive to survey all of the written and cartographic works of all known geographers for evidence of communication, exchange, and conflict with other geographers. A more vague sense of community, shared standards, and professional turf can, however, be gathered using the history of geography written by Robert de Vaugondy and the writings of some of the key French geographers named in that work.

A GEOGRAPHIC COMMUNITY?

In 1755 Didier Robert de Vaugondy published his *Essai sur l'histoire de la géographie*. This was certainly not the first history of geography written in the eighteenth century, but it was the most influential.[39] In it he recounted the history of Western exploration of the world, the nature of ancient and Medieval understanding of the earth, recent developments in astronomy and the theory of map projections that had fundamentally advanced the field, and an assessment of the state of the art in Germany, England, Spain, Italy, Sweden, Russia, and finally France. There is little doubt from the individuals listed as geographers in all of the countries surveyed by Robert de Vaugondy that he saw two major sorts of work as fundamentally geographic: descriptive cosmographies or geographies, and maps and globes. The individuals described as contemporary French geographers included mathematical cartographers, maritime geographers, terrain mappers and geodesists, cabinet map compilers, and what later came to be known as descriptive geographers.

The following French geographers were listed by Robert de Vaugondy as prominent, active, and highly respected geographic practitioners: Guillaume Delisle (1675–1726), Philippe Buache (1700–1773), Jacques Nicolas Bellin (1703–1772), Didier Robert de Vaugondy himself (1723–1786), the Cassini dynasty—from Jacques-Dominique Cassini (1625–1712) to Jacques Cassini (1677–1756) to César François Cassini (Cassini de Thury) (1714–1784)[40]—Bruzen de la Martinière (1662–1746), a descriptive geographer and author of the 1739 *Le Grand Dictionnaire géographique, historique et critique,* and Jean-Baptiste Bourguignon d'Anville (1697–1782). This does not exhaust Robert de Vaugondy's list of French geographers, which also includes, in addition to the Cassini family, a number of terrain geographers/geodesists (including Jean Picard, Giovanni Battista Riccioli, and Gabriel Philippe de La Hire) and, in addition to Bellin, a number of nautical geographers (including Joseph Sauveur and Jean-Baptiste-Nicolas-Denis d'Après de Mannevillette). Although not a matter discussed by Robert de Vaugondy, a number of these geographers also from time to time made their living by teaching geography, including Robert de Vaugondy himself,[41] Guillaume Delisle,[42] Philippe Buache,[43] and the Cassinis (if we consider the apprenticeship training of engineers as instructional).[44] These selected geographers,[45] then, offer a representative sample in terms of their skills, interests, activities, and cartographic and textual production.

There is no question that these geographers did form something of an informal community. They cited each other. They judged each others' work and sometimes acted as arbiters in matters of commercial or more scholarly dispute between their colleagues. It is often clear that they knew each other personally, even if that familiarity did not always breed respect and friendship. Delisle was the most frequently cited geographer and to some extent appears to have been a touchstone of quality and integrity. Bruzen de la Martinière, writing in the 1730s, claimed that the news of Delisle's death "significantly afflicted me."[46] Robert de Vaugondy compared his own efforts at the estimation of the total area of Paris with that of Delisle and was comforted to be able to explain the differences away.[47] Even as late as 1763, approximately sixty years after Delisle's depiction of the area, Bellin felt the need to fully explain the contrast between his own depiction of Rio Negro on the coast of Guiana and Delisle's. There is little doubt that Bellin knew Delisle personally as he avowed that at one point he had explored Delisle's "cabinet" for material which had later served in the composition of his *Description géographique de la Guyane*.[48] Delisle, in turn, referred to a number of the geographers mentioned by Vaugondy. As a student of Jacques-Dominique Cassini's, Delisle had considerable contact with the Cassini family, and it was to the elder Cassini that he addressed his published explanations for the decisions that lay behind some of his more controversial maps. As might consequently be expected, Delisle referred frequently in his writings to Cassini and his colleagues at the Académie Royale des Sciences—La Hire, Riccioli, Jean Picard, Pierre Louis Moreau de Maupertuis, and Charles Maria de la Condamine—as did every other eighteenth-century geographer.[49]

D'Anville was almost as frequently cited in the writings of the geographers dating from the second half of the eighteenth century—although he was often cited less appreciatively. While Bellin acknowledged use of d'Anville's published maps,[50] Robert de Vaugondy engaged in a severe critique of the geographer which extended into an acrimonious personal correspondence. He himself was the victim of a similar critique from Philippe Buache and Jean Nicholas Buache de la Neuville.[51] These geographers, then, knew each other's work and formed a community if only through their common citation of the endeavor's key figures.

The patterns of citation do not alone capture the strength of the intellectual bond between these men. They saw themselves as gatekeepers: as both representative of the geographic endeavor and responsible for defending its reputation and quality. In the case of the Cassinis, Delisle, and to some extent Robert de Vaugondy (a distant relation of Nicolas Sanson), this must have been accentuated by the familial, almost dynastic, nature of their association with geography. Thus, Delisle devoted 245 pages of his *Introduction à la géographie* to an explanation of the earth as an astronomical sphere:

> . . . in order to not fall into the error described by a contemporary scholar-mathematician who complains that most geographers have described the principles of this science so superficially that they provide only a very slight and imperfect idea of it.[52]

He similarly defended geographers from the complaint by a learned priest that there seemed to be no agreement in the longitude that geographers assigned to the city of Paris.[53] But while he defended geographers from criticism from non-geographers, Delisle was upset by the lack of scientific rigor and professionalism of some well-known mapmakers (whom Bruzen de la Martinère characterized as no more than "picture vendors"[54]) and openly expressed his wish to Cassini that "those not initiated into the mysteries of Geography, not involve themselves in the making of maps."[55] But the concern to keep geography unsullied by the hands of amateurs was not restricted to the dynasties. Robert de Vaugondy sought to defend Delisle from the criticism of a certain Vincent de Touret, who claimed that Delisle ought to have employed a single projection for all of his maps. The words he used to characterize the critic and the criticized clearly convey his sense of the respect due a geographer.

> The disproportion that is to be found between the aggressor and the aggressed scholar mean that I scarcely have to say anything more.[56]

Elsewhere, Vaugondy slammed map copyists as one of the field's great plagues and made it clear that poor-quality maps were beneath discussion.[57] Finally, Bruzen de la Martinière found the absence of any entry under "geography" in an earlier dictionary of geography utterly unacceptable.[58]

Geographers were not always so fierce in their identification with the field. It often took a more healthy and reflective form as, for example, when Delisle demonstrated enough faith in the methods, training, and critical faculties instilled by geography to challenge the Academy of Science's views on the location of particular physical features.[59] Similarly, Vaugondy's declaration that geographers had insight and training that would allow them to use sources unreliable for laypeople speaks of both assurance and a sense of turf. Perhaps it was Bellin who placed the highest stakes on his faith in geography when he informed navigators that should they find disagreement between their own observations and those of a geographer, they must go back and check and recheck their observations.[60]

There is little doubt, then, that in eighteenth-century France, geography, although perhaps not endowed with what today we associate with a full disciplinary structure, was a recognized field with representation and activity at multiple levels of instruction and scholarly research and production. A number of institutions supported or advanced geography, including the honorary or financial entitlements of "Géographe du roi" and "Premier géographe du roi," "Géographe de la ville de Paris,"[61] and the position of "Censeur royal" for geographic works, held by Jacques Nicholas Bellin from 1745 to 1772, by Robert de Vaugondy from 1773 to 1787, and by Dupain Triel from 1785 to 1787.[62] But it would be misleading to suggest that geographers formed a tightly knit community, that they were an organized profession, or that many found the means of earning a regular salary analogous to the professorships of today. All but the most successful practicing geographers, or those working within the orders, had to scramble for a living and don many hats from scholar to commercial mapmaker, to engraver, to editor, to private tutor. In addition, while there were significant centers of geographic activity beyond the Jesuit colleges, there was little coordination between them and much of the contact and exchange that did take place was informal.

THE NATURE OF THE GEOGRAPHIC TEXTS PRODUCED BY GEOGRAPHERS

Le Père Philippe Briet's seventeenth-century declaration of the insepara-bility of geographic text and map captures the essence of geography as it was practiced until the late eighteenth century (see text at note 6 above). It is true that after the seventeenth century there was an increasing physical separation of map and text as maps were no longer produced surrounded by or backed with text, largely because the technology of map and text reproduction were pulling them apart. But technology's and geography's missions were not consonant in this regard as geography relinquished neither map nor text from its purview. In the course of the eighteenth century the strides made by mathematical or observation-based geography began to put considerable distance between mathematical geography and erudite or text-based geography, creating a rift which was soon to break the old geography asunder. Still, in the eighteenth century, geography straddled map and text and measurement and erudition. It is perhaps for this reason that both the texts and maps produced by the geographers of this period seem so peculiar to us. In form and function they share a single descriptive purpose which makes the texts read like maps and the maps difficult to evaluate without their textual context. The texts produced by geographers were of three principal sorts: textual descriptions of the earth perhaps—but frequently not—illustrated by maps, textual explanations of the decisions behind the construction of a given map, and guidebooks to cartographic data collection or representation.[63]

All of these textual genres are far less accessible to us today than are, for example, the exploration accounts produced contemporaneously to them. Travel accounts were written for the general public and were designed to awe and entertain, while conducting the reader on a tour of the unknown and the unfamiliar. That genre is still flourishing today in the modern travel account (which now seeks to make the familiar exotic) but also in magazines like *National Geographic*, in film, in literature and even in anthropology. Consequently, particularly for an historically inclined reader, the eighteenth-century travel account is still accessible and may even entertain or titillate. Geography texts, on the other hand, were part of a more specialized form of scientific expression. They were designed to fulfill a scientific purpose for which we no longer strive and which we do not entirely understand. Thus, for example, the textual descriptions of the earth were at once an attempt at systematic description and an attempt at classification. The author would generally begin the description at the level of the continent and then describe smaller and smaller features of the world, either physical or human. Reminiscent of the map by its scaled structure, this form of description was increasingly seen by geographers as analogous to natural history classifications and they therefore sought scientific status through completeness of description. These texts read like maps but have none of the hierarchical flexibility and dream-inducing pictorial beauty of maps. In form and function, they are nevertheless closely linked to the map. The geographic dictionary offered an alternate form of the same basic genre, ordered alphabetically rather than geographically. In this form, the focus was on definition, but classification within and across alphabetical categories was also a major concern in dictionaries like Bruzen de la Martinière's.

After geography texts, the second textual form, the explanation of the decisions behind the construction of a map, is entirely inseparable from the cartographic genre. In these texts, designed to be read while consulting the map under discussion, the geographer would describe, explain, and justify his or her sources; declare the aim of his or her representation and the approach the reader should take to it; perhaps justify the limits, borders, and divisions that he or she chose to depict; discuss the standards of measurement used; perhaps provide something of a history of what was depicted or a history of its depiction; and, above all, explain how he or she had decided upon the location of key features. These works were designed to give the map a textual discursiveness otherwise impossible in cartography and, like footnotes, to link the map to its larger cartographic and textual context.

The third textual form, the guide to the collection and representation of data, was a genre written by geographers or geometers and exclusively for geographers or geometers. These were manuals of observation, measurement, and graphic representation. Their purpose was to encourage and facilitate the reliable, correct, and consistent production of maps, especially of survey-based, large-scale maps. Throughout the eighteenth century, then, maps and text worked hand in hand to fulfill the common goal of scientific—meaning accurate, detailed, and reconstructible—description.

THE EIGHTEENTH-CENTURY MAP:
MEASUREMENT AND ERUDITION

An eighteenth-century geographer, faced with the task of making a map, rarely found him- or herself either in possession of complete survey information or able to conduct his or her own field measurements. After all, there was an entire world to depict and very little of it had been measured. The great expeditions of Philip Carteret, Samuel Wallis, Baron Philippe Picot de la Peyrouse, Captain James Cook, and Nicolas Baudin were carried out only after mid-century. Great overland expeditions were few in the eighteenth century. The Lapland and Peru expeditions to measure the arc of the meridian in northern and equatorial latitudes took place only in the 1740s and '50s. Much of the world, then, was known exclusively through the accounts and sketches of travelers and missionaries who, while perhaps interested in mapping the terrain, generally had much else to worry about and to achieve. Indeed, the detailed, careful, and rigorous survey carried out with the best available instrumentation was a child of the seventeenth century and was largely restricted to European territories. A geographer with the best possible sources at his or her disposal and fortunate enough to have some field measurements, until the middle to late nineteenth century and for most parts of the world, would have had to mix these with information of lesser quality to arrive at anything approaching a complete and comprehensive picture of the part of the world he or she wished to depict. This is why mathematically inclined astronomer/geographers such as Cassini I and Delisle were as interested as the more erudite geographers in assessing the value of the measures used by the Romans.[64]

Until the nineteenth century many parts of the world, including most of Africa and Asia, and large parts of North and South America, either had never been visited or had not been seen by many Europeans since the fall of Rome. Geographers, who had been taught in the Jesuit and other religious schools (or by tutors trained in them) both to respect the knowledge of the ancients and how to analyze their texts, were more inclined than we would be today to believe in and use ancient descriptions of the Middle East and parts of Africa and the Far East. Thus, until the introduction of the national surveys, the geographer's task invariably involved three steps. The first was the gathering together of all possible textual and graphic sources and assessing their topological and orthographic accuracy, gauging their completeness, and understanding their terms of reference such as the standards of measure used in calculating distances and any systematic sources of distortion. At a time when public libraries did not exist and when access to government information and archives was considerably more restricted, this could involve a great deal of correspondence, wide and careful reading, travel, and local legwork. Most geographers gradually accumulated impressive libraries of their own. D'Anville's collection of almost ten thousand maps later became the foundation of the French national map collection in the Bibliothèque nationale, Paris. The second step was to compose a base of information either cartographic or textual by comparing the fully interpreted sources. In the case of Robert de Vaugondy, this consisted of choosing the most accurate possible base map, correcting it according to the most recently available astronomical coordinates, and adding information to that frame from other carefully corrected sources. More questionable details would be added last.[65] This process of critical comparison was frequently difficult and complicated. For example, a geographer wishing to make a map of Egypt toward the end of the eighteenth century would have had to cobble one together from a few ancient Arabic and Byzantine voyage accounts and geographies, the published travels of Herodotus, the observations of Pliny, Ptolemy, Strabo, and Eratosthenes, among others, biblical accounts of the travels of Moses, the travel log of a missionary, a rough sketch map made by an admiral some years earlier, a series of previous cartographic compilations of varying quality, and the sketches and text made by a traveler with no geographic training to speak of. Which bit of information to take from which source and how to balance it against a conflicting opinion was a matter for refined informed judgement. The geographer's final step was to give the information the particular form he or she sought: the coverage, depth of detail, divisions to be recorded, etc.

Geographers involved in national surveys, such as the Cassinis, were engaged in surprisingly similar activities. In contrast to the cabinet geographers, they certainly had to "lever un plan," or work in the field measuring angles and lines, inclination to the meridian, and distances from the meridian and the perpendicular; check the local toponymy with local officials; and produce rough sketches of the topography. However, there was also a very important cabinet dimension of their work. It was back in the Paris office that Cassini III and his assistants, for example, were engaged in "faire un plan," that is in recalculating and redrawing all the angles and lines already measured in the field and matching the topography to the locations so that the figure drawn on the paper would be as close to the terrain as possible.[66] Thus, their mapping necessitated critical office work and calculation to check the accuracy of the survey measurements and to iron out discrepancies between surveyors and between surveyors and some of their sources of information.[67] In some cases such discrepancies ultimately required a return to the field. Even the plane-table work of the ingénieurs géographes required office compilation including checking, rectification, and some measure of fudging. Arguably, then, the critical office work required by survey mapping and by erudite mapping was essentially the same type of operation, and while the field-survey-based map demanded an altogether different type of fieldwork and work with instrumentation, significant personnel and equipment coordination, fund-raising, and political maneuvering, the erudite geographer had an equally mammoth job obtaining, coordinating, and manipulating an extraordinary variety of often incomplete, inaccessible, or garbled sources.

THE MAJOR CONCERNS OF
EIGHTEENTH-CENTURY GEOGRAPHY

Most historians of the eighteenth and nineteenth centuries imagine the geography of the period to have been principally concerned with either mapping or global exploration. As already described, most geographers collected the results of those explorations, assessed them, and rendered them into holistic pictures or descriptive texts. However, with the exception of the topographical engineers, they were not themselves, for the most part, engaged in exploration, and, while they often influenced it through the gaps left on their maps, they had relatively little control over the process. Indeed, it would seem that apart from the daily business of keeping the family alive and fed, the concerns of most geographers revolved around problems of representation; in particular: the development of a language of representation sufficiently simple to be widely understood and rich enough to fully express a growing knowledge about the world; the development of and capitalization upon location-determining technology and the improvement of cartographic processes so as to increase representational accuracy; and, finally, testing the limits of representation.

The Language of Representation

The earliest stage in the development of any science—human, natural, or physical—is the development of a language sufficiently rigorous to allow all practitioners to understand and use the terminology, complete enough to suggest areas of interest within the relatively little known, and sufficiently supple to allow the easy development of subtleties in meaning and expansion in vocabulary as new phenomena are encountered and understood. Again,

François de Dainville, in his *Le Langage des géographes*, is the only historian of geography who has explored this most important stage in geographic thought, the development of the geographic language. Yet to ignore this development for a focus on geographic discoveries (which belonged to many sciences) is to look through the map at its content ignoring the form, the order, and the poetry of the map, which is the first language of geographers. From the sixteenth century to the nineteenth century, geography was substantially preoccupied with the development of a scientific language, again both textual and graphic. We can identify strong general trends in that development which culminated in a language very similar to, although not yet as rich, as our own by the end of the eighteenth century: the multiplication of terms as new phenomena were encountered, the refinement and precision of meanings as the nature of phenomena and the relations between them became apparent, the generalization of local or foreign terms, a movement away from the figurative[68] and toward the geometric, a simplification yet expansion and increasing uniformity of graphic expression, and, perhaps most importantly, the elaboration of a hierarchical and structured perspective.

By the beginning of the eighteenth century, a substantial proportion of the geographic vocabulary was already well established. Geographers had words and cartographic symbols for frontiers, a large variety of administrative divisions both secular and ecclesiastical, towns and settlements, the sea and its dangers, the coasts and their aspects, the commonplaces of agriculture, small-scale industrial production and trade, the rough location of natural obstacles or passages such as mountains, marshes, and rivers, and the location and even rough characterization of forests. However, the eighteenth century witnessed a rapid growth in geographic vocabulary in areas which had only just begun to draw the attention of the European mind. As Marjorie Nicolson and Numa Broc have shown us for mountains and Jean Roger has shown us for nature in general, it was in the eighteenth century that Europeans began to look at the natural world around them with a new curiosity.[69] As the proliferation of geographic encyclopedias, dictionaries, and handbooks toward the end of the seventeenth century and at the beginning of the eighteenth century suggests, the development of this new vocabulary was conscious, deliberate, and increasingly directed. New hydrological terms, born of a closer look at coasts and the shelter they could provide, gave birth to the neologisms of "crique,"* "recran,"†

---

*A coastal creek with the possibility of a good small natural harbor.
†A smaller coastal creek.

and "récif."* Increasing attention to rivers as a means of transport and a factor in war created the distinction between the "affluent"† and the "confluent,"‡ a much more careful graphic depiction of river courses, and, as the activity of the engineers increased, a river management vocabulary.§ The greatest expansion took place not in neologisms but in the assignment of single and precise meanings to a multiplicity of general, regional, or foreign terms. Such was the case with "plage,"** which became defined by three characteristics: low tide, extensive coast with neither port nor promontory, and a long, shallow approach. "Rade"†† became distinguished from "port," "canal"‡‡ from "lit,"§§ "source"*** from "fontaine,"††† "embouchure"‡‡‡ from "bouches,"§§§ "digue"**** from "levée"†††† from "chaussée,"‡‡‡‡ and "rives"§§§§ from "rivages."***** "Désert" acquired a definition based on its physical characteristics (lack of rainfall) rather than its earlier definition based entirely on the human consequences: an absence of inhabitants.[70] Some terms, such as "torrent," acquired an almost mathematical definition by the end of the eighteenth century.†††††

An even larger number of neologisms appeared in physiography, and relief depiction. Geographers became concerned to go beyond the mere indication of mountain location to suggest something of their character. By the end of the seventeenth century, some geographers picked out characteristic peaks and remarkable summits. Nevertheless, many large-scale

---

*A chain of underwater rocks.
†The river water tributary to another river.
‡The meeting place of two rivers.
§Including words such as "épi," or a construction on the side of a river designed to help control its flow.
**Strand.
††"Road for ships," as an area of water which is calm and sure for navigation.
‡‡As an artificial construction.
§§As a natural river bed.
***As the source water itself together with the location from which it emanates from the earth.
†††As the basin into which source water flows and is held.
‡‡‡The mouth of a river.
§§§The individual mouths in a river which has a delta formation and many mouths.
****A dike against sea water.
††††Masonry walls against floods.
‡‡‡‡The consolidation of river banks which in marshy territory may allow passage.
§§§§The banks of a river.
*****The coastal area of a sea or a lake washed by waves.
†††††A river whose water has a slope of 3 mm per meter.

seventeenth-century maps were mute as far as relief was concerned. The eighteenth century replaced this silence with more attention to mountain toponymy[71] and worry as to how best to describe mountains whose structure and nature were not understood. Some maps, such as Bourçet and d'Arçon's "Carte des frontières Est de la France depuis Grenoble jusqu'à Marseilles" (1778), sought to reveal the chain-like structure of mountain ranges. Perhaps the most successful eighteenth-century depictions were hachures which simulated the course of rain on the mountainside (with heavier lines at the crest and lighter lines at the foot) combined with shading as if the map were lit from an oblique source. This method, further developed by Bacler d'Albe's cross-hatching in his thirty-sheet map of Italy[72] and by many of the best maps produced by the ingénieurs géographes, created a remarkably tangible impression reminiscent of a relief model. There was also, however, a profound shift in the aim of cartographic depiction: geographers did not just want to depict what was known but to give a clear sense of geographic ignorance and above all to separate hypothetical from true geography. Philippe Buache covered his "Cartes des nouvelles découvertes"[73] with broken lines and warning legends to indicate the various competing hypotheses about the nature of the coast and terrain of what was to become Pacific Canada and Russia. Geographers became increasingly adept at conveying a "scientific" lack of knowledge. Blondeau's engraving of the results of the voyages of Captain Vancouver depicted a detailed coastline with an elaborate relief immediately adjacent to the coast, behind which, for at least a third of the space covered by the map, stretched an entirely blank continental interior. D'Anville, distinctly uncertain of the topography of many of the regions he was mapping, developed the trick of suddenly ending the graticule at the border of sure topological knowledge to make clear that while relief was little known in general, it was here that one stepped beyond the acceptable bounds of Western science (represented by the locational grid).[74]

A focus on mountains and relief also created a whole new non-graphic geographic vocabulary. New words, including "crête,"* "saillies,"† "ravin,"‡ "braye,"§ "plateau," "défilé,"** "partage des eaux,"†† "moraine," and "trem-

---

*As the highest portion of a mountain.
†As an abrupt rise in the mountainside.
‡As the cut made in a mountain- or hillside by fast-flowing water.
§As the swath left by letting cut wood run down the side of a mountain.
**As a thin single-file passage through the mountains.
††As the line of division between two distinct water basins—confused from the time of Buache until well into the nineteenth century with the crestline.

blement de terre,"* were all born of eighteenth-century experience and attention. As with river and coastal hydrography, many older terms acquired more refined definitions. This was the case with "cime," which became distinguished from "sommet";† "glacière," which made room for "glacier";‡ "bassin," which expanded to cover the whole drainage system of a major river; "vallon,"§ which precipitated from "vallée";** "cavernes,"†† which were distinguished from "grotto,"‡‡ or "caves," and so on.[75]

It is difficult to make hard and fast statements about changes in the symbology of eighteenth-century cartography as there was considerable variation in quality, form of expression, and vocabulary across the cartographic scales and genres. However, there was a decided trend toward geometric rather than pictorial depiction in, for example, town symbols. In general, the map was evolving through the eighteenth century from a mixed-perspective document with mountains, bridges, forests, and towns depicted as seen from the side and rivers, coasts, and borders depicted as seen from above to a unified composition with all features expressed in a consistent symbology derived from the appearance of the object from above. This unified perspective was not truly achieved until the replacement of hachures in the depiction of relief with contours in the nineteenth century. The trend, however, is clear and was a stated aim of the 1802 Paris "Commission chargé . . . à la perfection de la topographie." On a more philosophical level, the striving for and achievement of this consistent cartographic perspective can be seen as a declaration of the unity, objectivity, and comprehensiveness of the Western scientific vision.

The most important and characteristic development in the eighteenth-century language of geographers was the imposition of a systematic hierarchical order on geographic phenomena. Thus, it was Ozanam (1716) and Buchotte (1721) who advocated mapping at five distinct and fixed scales for maps of a country, a region, a county, a place, and an emplacement. Later in the century, the Dépôt de la guerre expanded the number of scales to be employed for different sorts of mapping to ten and broke these into four principal classes. Lettering too, which had been somewhat haphazardly

---

*Earthquake.
†*Cime* being the height of first order in a range.
‡*Glacier* becoming our glacier and *glacière* becoming an icehouse.
§Enclosed lowlands surrounded by mountains.
**A semi-enclosed lowland between two mountains but with a river both entering and issuing.
††Huge natural caves.
‡‡Generally artificial and ornamental.

applied in the seventeenth century, became increasingly ordered according to a hierarchy of importance. Similarly, towns and roads developed an hierarchical symbology according to their size and importance which was increasingly determined by some sort of numerical value: area covered, population size, speed of travel. The greater frequency of legends, their occupation in many cases of an entire map sheet in an atlas, and their extraordinarily detailed elaboration of the language of description all point to a scientific language approaching maturity. By the end of the eighteenth century, geographers were no longer merely translating landscape into text and cartography but translating cartography across language and culture as in the multilingual (French and Arabic) maps of Egypt produced by Napoleon's scholars or the multilingual (Polish, Arabic, and French) map of Poland produced by Rizzi-Zannoni.[76] The intellectual-hierarchical order of the map was to be a world order.

The Accuracy of Representation

Topological accuracy was the primary preoccupation of mathematical geography, or of the mapmaking wing of geography, prior to the end of the eighteenth century. Major strides were made in this area in the course of the century. The scientific method, mathematics, and the national economy had begun to intrude into everyday life, bringing a radical increase in both the accuracy and the scale of coverage of property mapping, from the mapping of small properties to huge estates and forest lands, diocesian maps, and city plans. Positional information grew increasingly reliable and topographic detail increasingly complete. Geographers worked concentratedly at accelerating these developments, and their contributions were significant in three distinct realms: in the accuracy of measurement, in the refinement of critical compilation methods, and in significant improvements in the technology of map reproduction.

Field mapping accuracy was a direct result of improvements in the technology of long-distance observation. This technology was developed in France with considerable state encouragement and in conjunction with both geographers and astronomers. The founding of the Academy of Sciences and the Observatoire de Paris, the attraction to France of both Jacques-Dominique Cassini and Christian Huyghens, and the attempted recruitment and payment of foreign instrument makers were all designed to bring about major improvements in both terrestrial and hydrographic mapping. The tangible result of these initiatives was significant improvement in the accuracy of instrumentation used in mapping in the latter third of the seventeenth century, including the development of the screw micrometer, the placement of telescopic sights on instruments such as the quarter circle, the development of a finely engraved vernier, the pendulum clock, the spring clock, and the barometer. Theoretical breakthroughs, particularly those in optics, also significantly improved the design and accuracy of instrumentation. Research of particular importance was that conducted into vibration and undulation, the causes of the propagation of light, the effects of concave and convex lenses, the laws of reflection and refraction, and the theoretical explanation for optical illusions. In most instances geographers neither carried out the theoretical research nor constructed the instruments but applied both theories and instruments in the field.

Geographers substantially improved the accuracy of field mapping by incorporating the principles of geometry into mapping, through the training of personnel in instrument use and care, and by developing strict rules of procedure. The mapping carried out on the topographic survey of France by Cassini III best captures this effort to attain accuracy. The essence of this map was a geometric frame covering all of France. By a combination of astronomical readings, ground measurements, and the observation of extensive triangulation nets, Cassini III and his predecessors established the skeleton for the map of France. Field mapping was then carried out to fill in an adequate amount of topographic detail. As no mapping venture on such an enormous scale had ever been carried out in France, Cassini was obliged to employ engineers with minimal training and to provide instruction in the field. The engineers were taught to use and readjust the angle measuring instruments, to judge the measurements that would require verification, to construct and check triangles of verification, to chose a location from which to make observations, to interview local officials, to take sightings on significant objects, to choose such objects, to record measurements unambiguously, to draw a map from measurements and sketches taken in the field, and to register place names.

Field mappers were not the only geographers who strove for accuracy. Over the course of the eighteenth century, geographers primarily occupied with compiling maps from secondary sources also worked to achieve a new accuracy. Earlier in the century, Guillaume Delisle transformed compilation mapping by constructing his maps around as many astronomically determined locations as possible. Approximately fifty years later, it was d'Anville who once again took compilation cartography to new levels of accuracy. As has already been described above, d'Anville was as concerned as Delisle to use the best available sources. His contribution lay in an extraordinary fastidiousness in the critical comparison of data and careful analysis of the measurement systems of the sources—whether ancient or modern—that he employed. The innovation that made his reputation and that has since become a standard feature of humanities research—still all too rarely used for mapping—was the explanatory footnote; d'Anville's many and voluminous compilation memoirs functioned as extended cartographic footnotes. As a scholarly genre, the days of the compiled map were numbered by the end of the eighteenth century. Although the care and responsibility taken by d'Anville in the construction of his maps increased the accuracy of compilation cartography, the reputation for arid scholarship which it engendered was to haunt geography for at least the first two decades of the nineteenth century.

Significant strides were also made during the eighteenth century in the accuracy of cartographic reproduction. Maps, once compiled, had to be engraved and printed. Geographers such as G. Delahaye and Cassini III worked hard to improve current engraving techniques. Over the course of the century, cartographic engravers increasingly worked in teams and specialized in the engraving of mountains, letters, cartouches, and lines. The stimulus provided by Cassini's *Carte de la France* resulted in the apprenticeship training of a large number of additional cartographic engravers. Precision, neatness of detail, and subtle differentiation of shading and light became hallmarks of the best French cartographic engravers. Neatness of line and contrast was the principal aim of Bellin's innovation of two-color printing for hydrographic charts. Through subtle color differentiation of land and sea, Bellin achieved unprecedented clarity, accuracy of line, and detail and legibility. Eighteenth-century geographers were not all equally successful in pushing the frontiers of accuracy. Nevertheless, accuracy was – their common preoccupation, as the esteemed accuracy of maps increasingly determined the map's value.

The Limits of Representation

It was in the course of the eighteenth century, and as a result of perhaps the greatest scientific debate of that century over the size and shape of the earth, that geographers tested, found, and reluctantly admitted the limits of representation and description. Many historians of geography and cartography have described geographers as taking part, as a result of their measurements, in the determination of the size and shape of the earth. They did take part in the debate, although the role and success of the geographers is frequently exaggerated in such accounts. What was more historically significant about the participation of geographers was what they tried, but could not accomplish, toward the resolution of the debate. Geographers sought to use their traditional tools of representation and description based either on erudition or measurement to determine the shape of the earth. Although historians of geography and geodesy now recognize that determining the shape of the earth through erudite reading of the ancients was doomed to failure, many still do not acknowledge that so too were more mathematical means of measuring and mapping the earth. The kind of accuracy needed to conclusively prove the shape and size of the earth by means of linear and geometric measurements did not exist in the time of Picard, Cassini I, Jean Richer, Maupertuis, and La Condamine. Without Newton's inverse-square law of gravitation and its prediction of the shape of the globe from the precession constant, scholars could have had little reason for choosing between one or another competing measure. Further, error in either direction would have been equally likely to have found acceptance. Linking the earth measurements engaged in by geographer/astronomers of the early eighteenth century with the experimental method, many historians of geography have failed to note the most significant difference between Sir Isaac Newton's theoretical and explanatory approach and that of the geographers. Newton was only indirectly concerned with the shape of the earth as symptomatic of its functioning within the larger solar system and as evidence of the universality of his gravitational law. It was the earth's role within the larger solar system which he sought to explain. The geographers, however, employed a fundamentally descriptive methodology: they sought to describe the earth by directly measuring its surface. In contrast to Newton, at the heart of their work lay no theoretical argument and no explanatory vision of the way the world, or the solar system, worked. The earth simply was, and as part of the general drive to greater and greater accuracy, it was there to be measured with all possible technology and care. While many modern historians of geography may have been reluctant to acknowledge this distinction, toward the end of the eighteenth century at least one geographer had already grasped the limited role geography and its traditional techniques could and did play in the debate.

The Debate

Debate over the size and the shape of the earth is at least as old as ancient Greece. Certainly attempts to measure the earth date back to Eratosthenes and the third century b.c. in the Mediterranean and to I-Hsing in the seventh century a.d. in China. In Europe, the size and shape of the earth had been of significance to mariners from at least the twelfth century.

Long-distance oceanic voyages represented huge investments in ships, personnel, and goods only too frequently lost at sea. As the accuracy of navigational instrumentation and cartography gradually increased, accurate knowledge of the value of each degree of latitude and longitude became potentially more important. The debate over the precise shape and size of the earth crystallized toward the end of the seventeenth century around the question of whether the earth was an oblate (pumpkin-like, as in flattened at the poles) or a prolate (egg-like) spheroid. Were the earth a perfect sphere, the value of degrees of latitude and longitude would be equal everywhere. Were the earth prolate in shape, the value of degrees of latitude, or the distance between parallels, would increase toward the poles. At the same time, the value of the degrees of longitude, or the distance between meridians, would decrease toward the poles. If the earth were an oblate spheroid and thus bulging at the equator, the value of degrees of latitude and longitude would increase at the equator while the value of a degree of latitude would decrease at the poles. As is so often the case, the interest in the question was not so much stimulated by the practical possibilities of its resolution as by more theoretical implications. It was the insight which the shape and the size of the earth could provide into the workings of the solar system which initiated the debate.

Newton's Perspective

The principal aim of Sir Isaac Newton's *Principia* was the uncovering of the principles (or the nature of the forces) guiding and sustaining the movements of the entire solar system. The *Principia* set aside the question of what had set planets and comets into motion to explore what might keep them moving as they were. According to Newton, it was the complex interactions of the gravitational forces of the bodies themselves which gave the solar system its particular shape and movement. If, however, gravitational attraction were capable of propelling the earth around the sun and the moon around the earth, then the gravitational force of the sun ought to distort the shape of the earth while the gravitational force of the moon and to some extent the sun should create visible and calculably regular fluctuations in the height of larger bodies of water. The phenomena and precise movements of the tides fit well into Newton's theory. The few exact linear or geometric measurements of the earth that had been undertaken by the 1680s, in particular those of Picard, Philippe de La Hire, and Jacques-Dominique Cassini, gave mixed results. Picard's measurements seemed to confirm Newton's hypothesis. Those of Cassini seemed to contradict it. In 1672 the French astronomer Richer was in Cayenne undertaking measurements to determine the distance between the earth and the planet Mars. Applying Kepler's third law, this measurement would serve to establish the scale of the solar system. Richer reported that the pendulum that he had adjusted in Paris did not keep correct time in Cayenne. Indeed, he had to significantly shorten the length of the pendulum to make it keep time. This implied that the force of gravity was lesser in Cayenne than in Paris. This in turn suggested that Cayenne was further from the center of the earth than was Paris, which pointed to an earth flattened at the poles and bulging at the equator. This confirmed Newton's hypothetical shape by means other than linear or geometric measurement.[77]

The Requisite Geometric Accuracy

A series of measurements were undertaken both in France and abroad from approximately 1700 to 1744. Many of these were primarily designed to serve as bases for the construction of large-scale topographic maps. In spite of the fact that such measurements were not yet sufficiently accurate or reliable to either confirm or refute Newton's hypothesis, they were used as ammunition in the debate. Between 1683 and 1700, Cassini I and then in 1701 Cassini I and Cassini II working in tandem measured an arc from Paris to Collioure (a small town on the Mediterranean near the Spanish border and not far from Perpignan). The value of a degree resulting from these measurements suggested a prolate spheroid. In 1718 Cassini II extended the arc from Paris to Dunkirk and still found proof of elongation. In 1733 Cassini II measured an arc from Paris to St. Malo (on the northwest coast of France). The measurement again confirmed elongation. In 1734 he extended the line to Strasbourg (on France's eastern border). He again found confirmation of an elongated earth. By their own calculations, to arrive at a correct solution, the Cassini measurements would have had to have been accurate to within 23.4 meters per degree or a total of approximately 128 meters over the total distance measured (in the case of the Strasbourg parallel). The checks they effected between astronomical readings and triangulation informed them that they might have committed an error of—at most—a few meters. In fact, they were off in the best case (that of the Paris to Strasbourg extension) by over 1.3 kilometers.[78]

By the 1730s many French scientists had come to believe that the Newtonian theory was correct and that something fundamental had gone wrong with the Cassini measurements. It was to finally resolve the issue by means of linear and geometric measurements that two expeditions were commissioned by the Academy of Sciences, one to Lapland, near the pole, and the other to Peru, near the equator. The aim was to measure the value of a degree of latitude in both of these places where the difference between the degrees should be the greatest. Both sets of measures confirmed Newton's hypothesis. However, what is important about the Lapland measurements, for the argument here, is that that expedition, arguably the one faced with the least difficult conditions, produced a value for the degree that was in error by 200 *toises*, or 390 meters.[79] Had that error been of an equal magnitude in the opposite direction, it would have erroneously refuted Newton's oblate spheroid. Arguably, then, the standards of accuracy attainable in linear and geometric measurement in the first half of the eighteenth century—even by the best geometers using the best equipment—were simply inadequate to confirm or refute Newtonian predictions.

What Role the Geographers?

The training of geographers and the equipment available to them for mapmaking were for the most part significantly below the level available for the Lapland and Peru expeditions. There is every reason, then, to expect that geographers would remain substantially aloof from the debate. The third generation of Cassinis, César-François Cassini (III), the most truly committed to geography of the Cassinis, had the wit to do so. Other geographers were not so wise. Jean-Baptiste Bourguignon d'Anville was among the most foolhardy. Thirty-eight years after a *Journal des savants* reviewer had declared the critical comparison of old maps and voyage accounts utterly inadequate to the determination of the location, in terms of latitude and longitude, of major topographic features,[80] d'Anville advocated the use of such sources not only for the rough determination of latitude and longitude on small-scale maps but for establishing the size and shape of the earth. With adequate care, d'Anville was sure that by taking measurements of maps, making analyses of scale and measurement variation, looking carefully at the effects of paper shrinkage, and comparing different maps and statements on the maps, he could measure degrees of latitude and longitude to sufficient accuracy. D'Anville believed that the Cassinis had underestimated the elongation of the prolate earth. Still, he felt that the kind of distortion created by the assumption of a prolate or oblate spheroid would have relatively little impact on the details on a map. He was sure, however, that it would have significant impact on the lines of latitude and longitude. This, to his mind, placed the question squarely within the geographer's purview. Other geographers were to claim even more than d'Anville. Later, in 1762, after the Newtonian hypothesis had clearly won out, Rigobert Bonne produced a small-scale map of the Mediterranean with a projection which purported to show the flattening of the earth.[81] Fundamentally, what drew d'Anville into the debate was less a sense of geography's turf than a profound discomfort, a discomfort likely shared by many trained in the Jesuit schools,[82] with theory, counterintuitive mathematics, and the use of instrumentation such as the pendulum clock, which produced results unconfirmable by the senses. The measurement efforts of the Lapland and Peru expeditions would lead to elucidation of the question, he was sure. It was the Newtonian methods that were questionable. He wrote about the benefits of both his method and the field measurement in the following terms:

> It is to be assumed in general that when we have measured the earth well, *just as it is today,* and by that *present measurement* and by *immediate observations,* we will *know positively* the figure and real dimensions of the earth. Then there will be less risk, from all points of view, in reasoning about the complicated effects of Gravity and the Centrifugal Force.[83]

D'Anville's discomfort with Newtonian physics was much more than a manifestation of scientific nationalism. What Newton's contribution to the debate over the shape of the earth was effecting was the alienation of geography, as earth description, from its very foundation, the measurable earth. This must have been particularly unacceptable to a geographer like d'Anville whose principal activity was the critical production of the best possible maps.

D'Anville's answer to Newton was written in 1735. However, he continued to produce maps based on his own determination of the shape and size of the earth into the 1740s. By the 1750s, and certainly by 1775, at least one geographer, Robert de Vaugondy, had concluded that the limits of geographic representation had been found. It was in his history of geography that Robert de Vaugondy first expressed his sense of those limits. Bravely, given the personality and status of d'Anville, he castigated the famous geographer for presuming to propose his own hypothesis for the size and shape of the earth. He accepted that astronomical readings were far from infallible and even frequently had to be abandoned in favor of terrain measurements in the course of map compilation just to make the pieces of the map fit together. Still, it was on the combined theoretic and empirical efforts of "la physique et la géométrie" that geography was to rely for the shape and the size of the earth:

> I do not think that anyone will ever accuse a geographer of imprecision when, in order to conform to ordinary graduation, Lombardy is accorded

a little surplus; just as it will not be difficult for him, in order to support a system, to reduce every one of the [other] spaces by a sufficient quantity to accommodate the reduction.[84]

It was the system that was paramount. Geographers, even the greatest of geographers, should expect to fill in the details on the sphere provided by physicists and geometers and to conform to that sphere by massaging their data.

Robert de Vaugondy was even clearer about the limits of geographic representation in his 1775 "Mémoire sur une question de géographie pratique." Responding to the debate between Rigobert Bonne and Giovanni Antonio Rizzi-Zannoni on the feasibility of depicting the flattening of the earth on maps, Robert de Vaugondy presented a memoir before the Royal Society which argued that to claim its feasibility was dishonest self-flattery. Graphic portrayal and the empirical techniques behind mapmaking were simply not of an accuracy to allow such depiction. To make a map purporting to show such distortion was to make the claim that they were.

> Let us not seek to blind the eyes of the public with an imposing display of calculations which, always supposing that they have been carried out, add nothing to graphic precision. It is overly assiduous to demand in practice the same precision presented to us in theory, the employment of which our sense organs reject.[85]

Geography could negotiate within a realm delimited by knowledge verifiable with the sense organs and by the feasibility of graphic expression. What lay beyond that, be it infinitesimally accurate calculations or theories about the system of the world, was not of geography's concern.

CONCLUSION

Geographers in the period up to the end of the eighteenth century were, then, almost entirely occupied with describing and representing; improving the accuracy of representation and ultimately testing the limits of representation. Still, it would be an overstatement to claim that geographers never engaged in explanation. There were a few exceptional geographers, such as Bernhard Varenius (1620–1680) and even earlier Le Père Jean François (1582–1668), who saw the need to borrow from physics and go beyond particularist description to a more general categorization of phenomena or even the beginnings of explanation. To a large extent, however, while the rare geographic author might be prepared to speculate or report on prevalent theories as to the cause of the tides, earthquakes, and storms, they did not investigate these in the field or, most of them, even critically.[86] Some maps were produced which might be described as semi-thematic or quasi-explanatory.[87] But again, their authors were involved less in elaborating an argument than in selective or refined description. Prior to the end of the eighteenth century, geography was committed to a descriptive approach which allowed the unified and integrated presentation of the surface of the earth and everything on it. That approach had achieved a great deal. Arguably, geographers had essentially resolved the most important of the problems deemed geographic by the end of the century. They had developed a sophisticated and reasonably consistent textual and graphic language, one which, with minor variations, has become a universal mode of cartographic expression. They contributed to a significant improvement in the accuracy of both maps and mapping through the testing and application of new technologies and methodologies. French geographers had undertaken and virtually completed the large-scale topographic mapping of France. Finally, together with physical scientists, officers, and servants of the state, a number of geographical engineers participated in a series of expeditions which sketched out and began to fill in the major features of the world. Toward the end of the century, as the physical processes behind the ongoing formation of the earth increasingly drew the attention of both scholars and the public; as the role of geography in shaping the locational patterns of plant and animal distributions began to become an issue; and as understanding of both the nature and importance of differences between peoples (and to some extent societies) began to dawn, descriptive geography, its achievements and the problems it traditionally addressed, began to seem elementary.

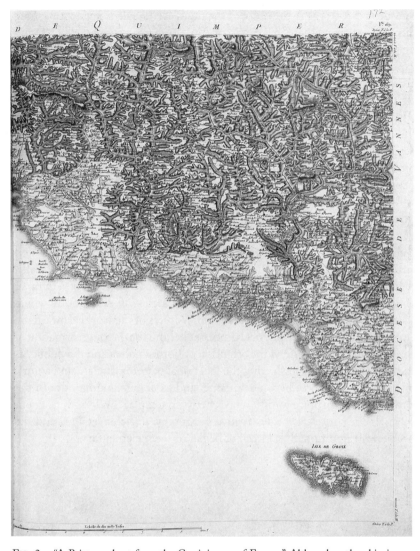

FIG. 2. "A Brittany sheet from the Cassini map of France." Although authorship is attributed to Jean-Dominique Cassini's father, it was Jean-Dominique who carried out the negotiations with the local authorities and saw to the completion of this sheet of the survey.

Geography's Loss of Direction and Status

One of the few tenable generalizations about "the nineteenth-century European mind" is that it was unsettled. Established beliefs, old ways of feeling and writing, received definitions, and comfortable limitations were disturbed by the generations of scientists, scholars, and travellers born around and after 1800.
—Richard Bevis, *The Road to Egdon Heath* (1999), part 4, "Sense and Sensibility," p. 159

Toward the end of the eighteenth century, the community of geographers entered a troubled period. This "trouble" consisted of a loss of direction and a perceived or real loss of status among the "sciences humaines." There is ample evidence of this in the writing of cabinet geographers in France and abroad.[1] In a sense, it is unsurprising: the achievements of terrain geographers, and in particular those of large-scale mappers such as the Cassinis, were increasingly rendering an important part of the work of cabinet geographers superfluous, except under the particular circumstances in which access to terrain was impossible. The achievements of field mappers and explorers increasingly deprived compilation cartographers of that part of their activity which in the eighteenth and early nineteenth centuries most readily defined them as geographers: mapmaking. At the same time, as the systematic sciences increasingly took shape and narrowed their focus, textual geographic description began to seem at best aimless and at worst irresponsibly ill-informed. One would expect, then, to see cabinet geography, of both a textual and a cartographic nature, gradually fade and finally entirely vanish in the face of an increasingly respected and powerful terrain geography. This is not what happened. With few exceptions cabinet cartography stagnated and became an almost entirely commercial venture. Textual descriptive geography either found inspiration in old formulas or assumed new forms. Either way, it quickly occupied the disciplinary heartland. Terrain geography (or large-scale mapping), which had been so successful in the eighteenth century, was soon banished from the institutions devoted to the advancement of science and was even gradually eased out of geography.[2] These developments cannot have been anticipated by contemporaries and are little understood by historians. In this chapter

I will explore the confusion and intellectual disorientation of both cabinet and terrain geographers.

A SENSE OF LOSS OF DIRECTION AMONG THE CABINET GEOGRAPHERS: BUACHE DE LA NEUVILLE, MENTELLE, AND THE ÉCOLE NORMALE

The sense of lost direction and status in geography echoes in the writings of the cabinet geographers throughout the first half of the nineteenth century but is perhaps clearest in the geography lessons and debates in the École normale and in their reception in the *Decade philosophique*.[3] The École normale was an institution of short duration. Born in January 1796, to resolve the educational crisis then facing France, it lasted only four months. There had been approximately two hundred Jesuit colleges in existence in France at the time of the expulsion of the Jesuit order in 1762. The sudden loss of this sizeable teaching corps caused significant disruption in secondary education, especially in the teaching of traditional subjects such as geography. Twenty-seven years later, the massive social change brought by the Revolution again hit secondary education hard. Teaching personnel were drained away from the schools by war, emigration, and the Terror. By 1794 the situation was dire, and the Comité de salut publique, through its Comité d'instruction publique, sought an immediate solution to the problem. The solution was both extremely pragmatic and highly idealistic. The plan was to establish a school for the training of teachers in Paris which would recruit students from the provinces. Once they had completed their course of studies, the students would return to the main towns of their provinces and teach their own students what they themselves had recently learnt. These students, in turn, would establish secondary schools in the various towns of each province. Thus, with a relatively small outlay of money, expertise, and effort, secondary education could be rapidly renovated. While the plan was pragmatic, the aims of those who designed the program of the school were idealistic and embodied the values and hopes of those inheritors of the Enlightenment belief in the perfectibility of "man," the possibility of social reform, and the Republic of the Arts and Sciences—the Ideologues. For men such as Dominique-Joseph Garat and Joseph Lakanal and for the school's supporters at the *Décade philosophique*, the École normale was to serve as the foundation for the creation of new scientifically informed citizenry, the basis of a new Revolutionary society. The strongest scientists in the fields represented in the newly established Institut de France were asked to prepare a course for the teaching of teachers. The plan for the normal school was derived from the successful recent experience with the course held in February–March 1794 in Paris to teach the general population how to make saltpeter (essential in the manufacture of gunpowder) from the potassium nitrate deposits in their own basements. The saltpeter experiment had been a resounding success, in that it provided an immediate solution to the gunpowder shortages suffered by the Revolutionary armies.[4]

An approach well suited to the transmission of straightforward techniques did not necessarily transfer well to the far larger and more complex, and inevitably disputed, aims of general education. Was the aim to teach teachers how to teach, or to teach basic or advanced material in a variety of fields to future teachers? If the former, then France's best scientists and scholars were almost certainly not what was needed. Neither were men who had practiced science and scholarship all their lives, rather than pedagogy, likely to be of much use in the teaching of basic material. If, on the other hand, the aim was to teach the highest state of the art of each field, then the students recruited would have to be well prepared for the lessons they would receive. The teaching corps were chosen for their brilliance, their past success, and their ideological sympathies. All (with the exception of Edme Mentelle) responded to the mixed aims by teaching their subjects from where they stood, at the pinnacle of their sciences.[5] The 1,400 students received in an auditorium, with seating for 700, ranged from illiterate peasants in their twenties and dilettantes in their fifties to some of the most prominent figures of the period, including the mathematician Jacques-Antoine-Joseph Cousin, the naturalist Comte Bernard Germain Etienne de la Ville de Lacépède, the navigator and geographer Admiral Bougainville, the astronomer Joseph-Jérôme Le François de Lalande, the natural historian Mathurin-Jacques Brisson, the editor Charles-Joseph Pancoucke, and some of the most brilliant minds of the next generation, including Jean-Baptiste-Joseph Fourier. The allowance paid the students while they were in Paris was insufficient for food and lodging. Consequently, very soon, most of the students from the provinces devoted their time to keeping body and soul together, leaving many of the classes virtually unattended. As a normal school, the École normale was a failure. However, some of the courses, in particular those of Gaspard Monge, Abbé René Haüy, and Constantin François Chasseboeuf de Volney, were a resounding intellectual success, in that new material of major scientific and intellectual importance was presented. The school also brought together scientists and scholars who otherwise might not have worked side by side and gave students, such as Fourier, access to some of the most brilliant minds of the period. If it was a failure as a pragmatic exercise, in many sciences it represented an intellectual milestone and still gives an idea of the stature and achievements of Revolutionary science and thought.

Jean-Nicolas Buache, known as Buache de la Neuville (1741–1825) and Edme Mentelle (1730–1815) were charged with the teaching of geography at the École normale. It would be fair, then, to consider them the best representatives of their field in 1794–95, on par with Abbé Roche Ambroise Cucurron Sicard in general grammar, Louis Jean Marie Daubenton in natural history, and Claude Berthollet in chemistry. Buache de la Neuville was, like Jacques-Dominique Cassini (IV) (see below), a product of ancien régime science. Arguably, his fame was derived less from his own achievements than from those of his more deservedly famous uncle, Philippe Buache, in cartography and the theory of mountain ranges and basins (see the Introduction and chapter 1). Nevertheless, Buache had an active academic life which involved a permanent position as geographic hydrographer (and ultimately conservator) of the Dépôt de la marine from 1789 to 1825. He entered the Academy of Sciences as an adjunct member in 1782 to fill the void left by the death of d'Anville. He became a full member of the Academy in 1785. His writings were relatively few and added little further luster to the name Buache. However, his opinions on cartographic and strategic matters were valued in both the old and the new regimes.[6] As conservator of the Dépôt de la marine, he had, as Mireille Pastoureau recounts, enough clout to override the best attempts of d'Anville's student Jean Denis Barbié du Bocage and the Ministry of the Interior to establish a permanent and untouchable collection of maps at the Bibliothèque nationale.[7] A member of the Academy of Sciences, he was named to the third class of the Institut de France on 29 brumaire an IV, at its foundation. He was also named to the Bureau des Longitudes. He was, therefore, considered by both the ancien régime and Revolutionary France to be the country's premier geographer.

Edme Mentelle was probably the most prolific cabinet geographer of the late eighteenth century, having produced at least thirty-six works, totaling seventy-five volumes. On some of these works he collaborated with people such as Jean-Antoine Letronne or Pierre-Gregoire Chanlaire who either already were or were later to become famous in their own right. Mentelle also made maps and designed some ingenious globes. Named to the Institut de France at the same time as Buache de la Neuville, he was highly regarded and appreciated by his contemporaries, both as a geographer and as a teacher. He taught geography and history both privately and at the Collège royale militaire, the Lycée républican, the Lycée des arts, the École normale, and the École centrale du Panthéon and École centrale des Quatres-Nations. At least one famous geographer and academician, Letronne, was proud to claim Mentelle as his teacher and, in the early part of his career, even signed his articles, "A. Letronne, Élève de M. Mentelle."[8] He was widely known in geographical circles. The physical geographer Bory de Saint-Vincent described Mentelle and Voltaire together as "distinguished" modern scholars in his discussion of the possibility that the Canary Islands were once part of a larger Atlantic continent, perhaps associable with the Atlantis of the ancients.[9] Mentelle regarded himself as on par and to some extent, in terms of representation, dividing the field with d'Anville.[10] He was also known and appreciated among the Philosophes and Ideologues of his generation, including, in the Revolutionary years, Marquis Pierre-Simon de Laplace, Jean-Pierre Brissot de Warville, Antoine Laurent Lavoisier, Gaspard Monge, and Comte Joseph Louis Lagrange.[11] In these wider circles, it is clear that he was not regarded as an original creative mind but that he was respected as a good teacher and a solid scholar.

What geography, then, did these two geographers, arguably the best France had to offer in 1794–95, profess at the École normale? The geography taught by Buache de la Neuville and Mentelle, although apparently vast, was based on, and restricted to, description and lacked rationale, method, and explanatory power. What geography had to offer modern science, or even the education of children, was less clear at the end of the course than at its beginning. The poverty of this geography is apparent to the modern reader, but the criticism with which the classes were received suggests that it was also evident to contemporaries. The two geographers reacted differently to the criticism. However, neither understood its nature or its gravity.

In their lectures, Buache and Mentelle introduced geography as a vast subject—as vast as the universe—and daily growing in both detail and extent thanks to the tools provided explorers and astronomers by both the physical and pure sciences.[12] It consisted of the description of the earth as a planet, of the naming of its main physical features, and of the presentation of the nations of the world. Thus, they began their course by describing the place of the earth in the solar system, the effects of its rotation about its axis and its rotation about the sun on the seasons and the climates. They then undertook to describe the main physical features of the planet (largely the direction, extent, and height of the major mountain ranges and the courses of the major rivers), and closed with a political geography which consisted of a discussion of the extent of the geographic knowledge of the ancients and, as an example of the correct approach to the subject, a brief glance at the peoples, major historical events, finances, military tendencies, major towns, and the commerce of Russia.

Their description was not the rich description found in the accounts of many nineteenth-century explorers and travelers. Nor was it the erudite and at times almost literary description of Malte-Brun's universal geography (see chapter 3).[13] As though bound to the flat surface of a map, Mentelle and Buache restricted the meaning and purpose of description to the verbal depiction of "the entire surface of the globe, so as to know its forms and the great divisions."[14] Although Buache argued that a prior knowledge of how the globe functioned was essential to understanding the nature of the earth's divisions, when pressed on the nature of the ellipse formed by the earth as it circles the sun, Mentelle responded, "As a geographer . . . I would say nothing other than that the earth is distributed into parts and regions."[15] For both Buache and Mentelle, the earth was a fixed tableau of "the situation of peoples and the position of places" to be observed without comment or interpretation.[16] Thus, Mentelle's *La géographie enseignée par une methode nouvelle, ou Application de la synthèse à l'étude de la géographie,* published in 1795, consisted of the drawing of concentric circles around a central location on a map of France and the brief and superficial description of phenomena within each of those circles (towns, rivers, and the main kinds of production). In describing their approach to the natural world, and almost as an oversight, Mentelle and Buache even suggested that "description" lay more properly in the realm of another science:

> [Geography studies the products of the earth] not, it is true, in the way of a naturalist who describes them, nor as the chemist who analyzes them, but by indicating to each the places in which they must be looked for.[17]

Perhaps the clearest expression of the very surficial and uncritical nature of their conception of geography was their contention that history and geography could be easily and coherently summarized in some sort of combined map and chronological table: "a pictorial table which will at once reveal all of the peoples and all of the states that have existed since highest antiquity, their origin, their extent, their decadence. . . ."[18]

Mentelle and Buache offered no rationale for either their course or the geographic endeavor in general. In their view geography was knowledge. It was composed of irrefutable facts. How could that require a rationale? Their first aim in teaching geography was to demonstrate "the principal geographical truths."[19] Facts borrowed from the physical sciences had to come "proven true."[20] There was no question of hypotheses, arguments, interpretations, or even explanations:

> These are recognized truths, constant facts, upon which there will very soon be no doubt. Geography is happy to enounce these; mathematics and physics will give you their most satisfying demonstration.[21]

Invariably, when asked for an explanation of natural phenomena, Mentelle responded: "the question must be sent back to physics, which will provide you with an explanation. . . ."[22] Lacking hypotheses to separate the important and the relevant from meaningless detail, Buache and Mentelle found that they had no criteria to choose among the plethora of facts concerning the physical world around them. Rousseau had convinced Mentelle that excessive memorization was bad for children and they themselves found that they could not remember all the important facts of geography. The only way to proceed, then, was "to truly grasp the whole."[23] This, and perhaps also their training in small-scale cartography, led Buache and Mentelle to posit the essential sameness of phenomena the world over in precisely the period when the extraordinary variety and variations in natural phenomena, individuals, and societies were beginning to be perceived. After all, they argued,

> what is said about a mountain or the bed of a river can be said about the bed of another river or about another mountain as soon as there is conformity between these objects; same causes, same effects, with regard to the rainy season, with regard to the perenniality of the sources; with the same relationship between the nature of the terrain and the properties of its produce. Thus, a first lesson, well conceived and well developed, can put [the student] promptly in a position to be able to judge one country by another, especially if one does not lose sight of variations dependent on accidental causes.[24]

In the absence of a field dimension to their research, they were unlikely to understand or share Bernardin de Saint-Pierre's discovery of the extraordinary variety of landscapes and the fundamental differences between them.

Insofar as the geography of Buache and Mentelle represented a whole, it functioned on two widely separated scales in a diametrically opposed manner. This is clear in Mentelle's contention that while "physique" was to undertake explanation of the variety of human and animal species, and "morale" was to investigate the influence of climate on customs, it was geography's function to "indicate their generalities and their remarkable varieties."[25] In the geography taught in their course, generalization was restricted to the description of global- and sometimes regional-scale natural phenomena while the particular and the exceptional came into play when political geography or peoples or nations were discussed. Thus climate and physiography were discussed in only the most general terms, while particularities, such as the customs of the inhabitants of Kamchatka or the height of the grass growing in that region, dominated their political geography.[26] Interest in "man," the history of "man," and human institutions was recent, and evidently geographers had still not adequately grafted this new interest onto their traditional concern with location. This split between "man" and nature in geography, associated, as it was, with two radically different scales of conception and approaches, may also have made it impossible for geographers either to conceive of spatial systems or to consider the exploration of the interrelatedness of natural and human phenomena as a realm of inquiry proper to their field. Thus, although Mentelle and Buache repeated the commonly accepted view that mountains had an impact on climate and the customs of people, they did not discuss mountains as ecological systems to be explored and explained but as objects in space which mysteriously influenced the objects surrounding them.

The course was received very poorly in intellectual circles. The scholars attending the classes were irritated with the unwillingness or inability of the geographers to explain the natural phenomena they were describing. An intolerance for the frequent exclusion of questions on the grounds of intellectual territoriality soon developed. There was some attempt in the question sessions to engage the professors in their area of interest. Thus, there were many questions to Mentelle about the teaching of geography to young children, as that was the subject of the bulk of his publications. But, apart from a few general principles—that one should proceed from the familiar to the unknown, that one should avoid memorization, and that one should not teach above the level of the child—Mentelle appears not to have thought very deeply about the problems of teaching children geography. He was uncertain, for example, whether, when dealing with a complex phenomenon such as gravity and the reason why people living in the antipodes would not fall off the earth, children should be taught erroneous and incomplete explanations or whether they should be taught to assimilate correct facts without explanation. When pressed on this issue, he responded in exasperation "but in the first sessions, no one complained to Citizen Laplace that one could not discuss binomials and logarithms with a child."[27]

Devastating criticism came from an anonymous reviewer at the *Décade philosophique et littéraire* who after the first class taught by Buache and Mentelle complained that on the basis of that class he could not see the value of a geography course at the École normale, and if there were to be such a course it was evident that "géographes de profession" were not qualified to teach it. Considering the material to be covered by the professors, the reviewer asked: When the areas of overlap with the other fields were eliminated—mathematical geography with mathematics; physical geography with general "physique"; chemistry, natural history, and political geography with "the political history of nations"—what remained proper to the domain of geography? Nothing, he argued, but "details and nomenclature."[28]

> If one wanted a geography course in that school, it would be necessary to present there, as in the other courses, the advancement of the science, a review of methods, and the application of analysis to geography, which can just as easily submit to analysis as any of the other sciences. However, to arrive at that point, it seems to us that there was no point in seeking this from professional geographers. It is easy to sense the reasons why and, consequently, it is unnecessary to state them.[29]

Thus, although the geography taught at the École normale was condemned by the *Décade philosophique,* the review contained a sense of the possibility of another kind of geography, a reformed geography working at the forefront of science with rigorous methods, dominated by analysis.

It is clear that this assessment of the course taught by Mentelle and Buache held sway: when the curriculum was established and fixed for the écoles centrales on 3 brumaire, an IV, geography was not one of the subjects. According to a Ministerial report, which explained the exclusion of geography from secondary education in terms redolent of the assessment of the *Décade philosophique:*

> . . . geographic ideas fall into the realm of memory alone and belong in primary school. Physical geography can be taught in the écoles centrales by its professors of natural history and political geography by its professors of ancient languages or by the librarian. These professors only have to ensure that a few rudiments are presented to the students upon their entry into the school.[30]

The future mathematician, professor of the École polytechnique, member of the expedition to Egypt, and prefect of Isère Jean-Baptiste-Joseph Fourier provided a general impression of the École normale in a letter to his home province of Auxerre. He was critical of the whole venture but prepared to acknowledge intellectual quality even when delivery was far from ideal. However, he had little that was positive to say about the course taught by Buache and Mentelle. Catching Mentelle out at his habit of recycling old publications and courses, Fourier commented:

> Mentelle is known in Auxerre. His lessons are extremely familiar and have nothing worthy of that institution [the École normale]. He converses pass-

ably, as far as I can judge, as I never listen to him. Buache is a famous geographer who expresses himself very badly and occasionally has a scientific idea.[31]

In the end, it was the ministerial judgement that counted, as it effectively institutionalized a view of geography as unintellectual and elementary. The criticism elicited different responses from the two geographers. Mentelle was injured by it, particularly as the published review came from a journal and a group with which he was associated and for which he had respect. He understood the review primarily as a statement of his lack of qualification to teach either mathematical or physical geography and ironically he increasingly took refuge in nomenclature, listings of latitude and longitude, and especially in the proper method of teaching geography to children, with the result already described. Buache, who, by virtue of his long family association with geography and his longer association with institutions such as the Académie des sciences, had perhaps a greater stake in both geography and his own reputation, responded with anger. He accused the *Décade philosophique* of conspiracy and determinedly maintained the course of his lectures. His only major response to his critics appears to have been a multiplication of references to specialists. But it was also Buache who, looking over his shoulder at what the Ideologue Volney was teaching in his history course, noticed that Volney had demonstrated "the scope that geography can have, the details it includes, and the multiple perspectives from which one must consider a country in order to really understand the peoples who live there."[32] It was in Volney's lectures that Buache found the inspiration to look a little more closely at the mechanisms of climate, certainly not as well as Volney himself would have done, but there was in Buache a recognition of something of importance and of value to geography in the way in which Volney approached the subject. We will explore Volney's potential as a geographer more fully in chapter 6.

CASSINI IV AND THE DECLINE OF TERRAIN GEOGRAPHY

Jacques-Dominique Cassini, otherwise known as Cassini IV ([1748–1845] the son of César François Cassini de Thury [1714–1784] and the great grandson of Jacques-Dominique Cassini [1625–1712]), provides us with another and a particularly suggestive example of geography's loss of direction and status at the end of the eighteenth century. Some would regard Cassini's life path as highly particular and reflective of nothing so much as the man's distinctive personality.[33] This is as undoubtedly true for Cassini as for any other human being. Yet Cassini was also responding to forces—political, economic, and intellectual—which were far larger than the man himself. On the other extreme, some have seen Cassini IV as just one more late-eighteenth-century academician crushed by the events which so overturned and disrupted the values and way of life of his generation.[34] No one to date, to my knowledge, has attempted to link Cassini's life path and career decisions to his own sense of the nature and importance of "the sciences" (including geography) in the changing context of realities that were both political and intellectual. In short, no one has paid attention to what Cassini himself saw as the most important dimension of his public life: science or its antithesis. Neither has anyone sought to place his alienation into the intellectual context of the science of his day. Yet it scarcely seems possible to ignore either dimension when considering the career of a man who devoted his first twenty-eight years as an intellectual adult to "science" and then abandoned it at forty-six years of age for the remaining fifty-two years of his long life. He not only abandoned science but condemned it as part of a prevailing ideology which he accused of hubris, religious intolerance, and a lack of humanity both in terms of its disrespect for the humanities and because, in his view, it was the cool scientific rationalism of the Enlightenment which had served as the basis for the ready sacrifice of humans and human institutions to ideas during the Revolution. Given the chance to resume his scientific career after the Revolution, Cassini would have done so. But, after 1795 he was not given the chance because science (including geography, astronomy, physics, and the relations between them) had radically changed since the beginning of his active intellectual life. The changes had been, for the most part, gradual, but they were both methodological and conceptual and they were profound. Cassini was aware of them—which is not to say that he entirely understood them. Cassini was an exceptional man in that he was prepared to suffer the ridicule of his contemporaries and risk the opprobrium of subsequent generations in order to avoid living a lie—or several lies. He was like all other humans in that some of the reasons for his own alienation from the society and the scientific society within which he continued to live were beyond his understanding. Nevertheless, it is possible to trace in his life and career the alienation that many geographers must have felt as a result of the Revolutionary attack on both the Crown and the Church. His life and career further reveal that disciplinary dislocation emanated, in part, from the changing status and function of a geographer such as Cassini, essentially an astronomically trained terrain geographer who up to the nineteenth century might have expected to find himself at home in the

Academy of Sciences. And most tentatively, but perhaps most importantly, Cassini's example suggests that geography's loss of direction and status was in part born of the changing nature of astronomy and physics in the late eighteenth century.

Cassini IV was born at the Observatory in Paris in 1748. Cassini began his education at the Collège de Plessis, but finding the discipline needlessly harsh, he convinced his parents, with some difficulty, to transfer him to the Oratorian Collège de Juilly, where he flourished.[35] After completing his education at Juilly in 1765, he was privately tutored in physics, mathematics, and related subjects by Abbé Nollet and Antoine Mauduit and worked as an informal apprentice at the Observatory under the direction of Giovanni Maraldi and J.-B. Chappe d'Auteroche. From an early age he was equally drawn by the arts as by the sciences and, he tells us, had his name not been Cassini, he most certainly would have become either an artist or an "homme de lettres." His sense of responsibility to his family's name and to the tradition of science that they had practiced was both strong and sophisticated. There is every indication that he devoted the first forty-six years of his life to being a Cassini worthy of the name.

Cassini IV—A Geographer?

Ironically, just as the work of the Cassinis was declining in importance among astronomers, Cassini IV appears to have traded his father's identification with geography for an identification with astronomy. Indeed, he consistently referred to himself as an astronomer or even more comfortably as a scientist, or, in later years, as an ex-astronomer or an ex-scientist. Rarely, he also referred to himself as a geometer, but not once in his published works did he call himself a geographer.[36] This in spite of the fact that a good portion of his work was classically geographic: he played an important role in completing the 1:86,400 survey of France; he sought to extend that survey beyond the national borders; and he produced a number of other important cartographic works. From the perspective of the twentieth century and our conception of those two sciences, this simply does not make sense. But if we can understand what geography and astronomy were in that period, together with the changes which they were then undergoing, then this apparent anomaly vanishes.

We have seen in chapter 1 that a number of geographers at the beginning of the eighteenth century had strong ties to astronomy. Guillaume Delisle, for example was trained as both an astronomer and a geographer, and the geographers in the Academy of Sciences occupied astronomers' chairs until late in the eighteenth century.[37] Similarly, there are scholars whose primary affiliation was astronomy who, nevertheless, wrote some geographic works, Joseph-Jérôme de Lalande among them. What we now see as (at least) two separate disciplines were historically linked and only very gradually pulling apart in the eighteenth century. Fundamentally, both were subfields of cosmography and engaged in the description of the cosmos. Astronomy took as its major focus the heavens while geography principally concentrated on this veil of tears. But there was considerable overlap. How could a geographer seeking to correctly depict area and distance on the globe, and to whatever degree possible on maps, not be concerned with whether the earth was a prolate or an oblate spheroid? How could a geographer neglect what astronomy had to say about the effects of the movement of the sun and the moon on the earth? How, in turn, could astronomers ignore the important astronomical and meteorological field observations of nautical and terrain geographers?

The overlap between the two fields was apparently—but only apparently—growing in the eighteenth century. Certainly Cassini III saw it that way: geography was becoming increasingly dependent on astronomical observation for the determination of location. At the time that Cassini III composed his *Description géométrique de la France* there was even something of an ambiguity in the use of the term "géographe."[38] Did it more properly belong to Cassini III or to the engineers working on the survey? Cassini substantially reserved the title for himself, together with that of astronomer, while not in any way denying that the engineers were, indeed, practicing geography. In some ways, then, geography and astronomy were inseparable. On the other hand, as we have seen, already by the middle of the eighteenth century, astronomers willingly ignored the research and arguments of a cabinet geographer such as d'Anville. But the distance between the observational astronomer and the true field geographer was simply not that great. There is no doubt that there was a considerable distance between Newtonian and post-Newtonian theoretical astronomy and geography of any sort,[39] but the Cassinis were not theoretical astronomers—certainly not in the tradition of Newton. Arguably, then, through the eighteenth century one could choose to call oneself an astronomer without excluding geography from one's realm of interests or even from one's realm of primary interests.

Between Cassini III and Cassini IV there had been a shift. Cassini III reserved the title of geographer for himself. Both Cassinis saw geography as a great science at whose rebirth the Cassini family had served as midwife.[40] Yet Cassini IV, on the rare occasions when he used the word "geographer" at all, used it almost as a term of abuse aimed at mapmakers who did not engage in the fieldwork or quality of criticism employed by the Cassinis.[41] Yet if the Cassinis had reformed geography, who could better represent the field than the Cassinis? Cassini IV did not reject this role but he was more concerned to distance himself from non-astronomical geographers on the one hand and from engineers on the other than to assume the title of geographer.[42] Both Cassini III and Cassini IV saw themselves as the administrative and intellectual overseers of the map of France. For Cassini III, this made him an astronomer and geographer. For Cassini IV, concerned to distance himself from increasingly well-trained and competent engineers and from increasingly out-of-odor cabinet geographers and commercial mapmakers, the title astronomer seemed to work best.

Again, context makes this more comprehensible. Cassini IV's primary focus was on the Observatory: he was born there, he served his apprenticeship there and, as its second director, he spent many of the years up to 1793 turning it into a serious center of regular observation—which is not how it had started life. The mountains that had to be moved to achieve that end were not small. They included gaining government support for the funding of a constant and a sufficiently large personnel at the Observatory to undertake regular (but still not continuous) observations, rebuilding the Observatory so that it could house up-to-date instrumentation, and securing the most up-to-date instrumentation. This last need, in turn, entailed freeing a corps of highly skilled craftsmen from guild restrictions, training those craftsmen in astronomy, arranging for them the sort of apprenticeship that in the last half of the eighteenth century was really only available in London,[43] building a foundry attached to the Observatory, and securing a foundry master sufficiently skilled for the exigent requirements of astronomical instrumentation. And all this in a period of severe fiscal restraint and economic retrenchment on the part of the Crown.

Finally, as head of the Observatory, and as the principal agent of its reform, Cassini IV saw himself less as his father's son than as the fourth member of a distinguished family, stretching back to the great astronomer Jacques-Dominique Cassini (after whom, incidentally, he was named), to dominate the Observatory of Paris.[44] It was in this vein that he wrote a history of the Observatory which included his great grandfather's autobiography. Cassini I was considered one of the greatest of French astronomers and scientists, whereas in Cassini IV's day, Cassini III was merely regarded as a good astronomer and a great geographer. It is also important to add that the title "astronomer" carried more kudos than that of "geographer": as Chapin has noted, astronomy was one of the most established and successful sciences at the Academy of Sciences in the seventeenth and eighteenth centuries.[45] None of this need suggest that Cassini IV rejected his father's work. He willingly carried on as director of the association responsible for the survey of France in 1784 after his father's death. It should, however, be clear that Cassini IV had a number of converging reasons for calling himself an astronomer. We will return to the complex theme of Cassini's increasingly painful and troubled relations with astronomy later in this chapter.

The Nature of Cassini's Geography

While Cassini's description of himself might make us question his status as a "geographer," his activities leave us in little doubt. Cassini's involvement with the survey of France was considerable. From 1784 he was its director, he negotiated with the Associates funding the map and with the King, thanks to whose permission it was being produced, he oversaw the remaining fieldwork, and he negotiated and fought with the relevant provincial authorities who significantly hampered his work. He spent less time in the field than had his father, but by the time he took over the survey from his father, all of the engineers had been trained (the most important reason for Cassini III's occasional presence in the field) and most of the fieldwork was complete. At this stage in the survey's history, Cassini IV's involvement was largely confined to the Observatory, where the surveys were checked, compiled, engraved, and rechecked. But he was also engaged in extensive and tiresome negotiations with the Generality of Brittany, an intermediate agency, which according to an agreement with both the government and the Associates, who since 1756 owned the map of France, was responsible for paying for the survey undertaken in their territory. The Generality, according to Cassini's account well after the fact, had run out of money and, through the obstruction of the survey by a series of complaints about missing information or toponymy, hoped to put off payment indefinitely. In terms of payment for the map, it was a strategy which ultimately worked, thanks to the Revolution. However, thanks to Cassini IV, the obstruction was merely financial and administrative and the survey and office work went ahead in spite of those obstacles.

Cassini was also involved with a number of other mapping ventures distinct from but related to the survey of France. In 1775 he discussed with the Grand Duke of Tuscany the possibility of undertaking a survey modeled on the Cassini map of France in North Central Italy. This would have been a step in the direction of Cassini III's dream in which all of Europe was covered by a Cassini topographic survey. It was a serious offer of expertise, direction, and fine instrumentation. Cassini IV, whose enthusiasm for the idea is clear in the plan he composed for the Grand Duke, was making a very significant commitment of time away from astronomical observation. Cassini's biographer, Jean-François-Schlister Devic, tells us that he would have gone ahead but that the Grand Duke backed out after Cassini provided his account of the likely expenses. Also strongly related to the survey of France and to Cassini III's dreams for its expansion was Cassini IV's involvement in the operations to connect the Greenwich and Paris meridians through a network of triangles. Seconded by Pierre François André Méchain and Adrien Marie Legendre, it was Cassini who directed the operations, who maintained contact with his English colleagues, and who composed and published the report in 1790, once the operation on both sides of the Channel had been completed.

Beyond that, we know that Cassini IV had begun, just prior to the Revolution, an itinerary atlas for French travelers which not only indicated the route in the fashion of a strip map but also indicated the nature of the countryside as seen from the road. That project was never completed, although at least one plate was printed. A cartographic project that he did undertake and complete in the 1790s was the map depicting the new departmental divisions of France.[46]

Cassini once mentioned, in a love letter, a metaphorical lunar map[47]— but this was not the only evidence of his desire to map the heavens. In the archives of the Observatory of Paris there remains an incomplete manuscript atlas of the constellations—one clearly designed to be accessible and of use to those without astronomical training.[48] Each open face of the atlas includes on the left page a map of a particular constellation with an indication of the surrounding stars and constellations. While allowing readers to locate themselves in the heavens, so to speak, this method also avoids crowding the picture and, thus, confusing the reader. The right page contains a table listing the stars by number (the numbers correspond to those on the facing map). Information is systematically provided for each star, including its Bayer designation, its size, its right ascension, and its declination. In addition, Cassini also indicates, where appropriate, nebulae and double stars—providing both Herschel's and Flamsteed's numbers for them. At the end of the atlas there is an also incomplete "Situation respective" which provides a quick guide to the relative position of each constellation (relative to its nearest neighbors). The constellations appear in alphabetical order and Cassini clearly intended to map at least fifty constellations. Only twenty-nine plates are complete, with an additional six begun but not finished. This work was clearly begun while Cassini III was still alive as the elder Cassini mentioned that his son was working on this atlas in his *Description géométrique de la France* (1783).[49] There is no way of knowing why Cassini never completed and published this work. Perhaps the illness and death of his father, which meant that he had to take over Cassini III's responsibilities, left him with too little time.

Cassini IV's most scientifically significant work, both as an astronomer and as a geographer, was administrative and organizational rather than conceptual or even observational. If Cassini is remembered at all among astronomers and historians of astronomy, it is for his reform of the Observatory. But while this may seem far removed from geography, there are important ties between even this work and Cassini's activities as a geographer. The reform of the Observatory required the same combination of skills and knowledge that were needed to manage and direct the survey of France. Indeed his work in reforming the Observatory was similar to the type of work that Cassini III had had to engage in for the survey of France: for example, negotiating the continued existence of the survey with the King, recruiting Associates from the Academy of Sciences, training the engineers and commissioning and developing the instrumentation, establishing an engraving workshop and a press, etc. Cassini IV's first task was also to convince the King, his court, *and* the Academy of Sciences that rebuilding the Observatory, constructing and commissioning new instruments, establishing a foundry for this purpose, and recruiting and training astronomy students was truly necessary. Specialized scientific and even artistic knowledge was critical to his endeavor.[50] Cassini provided the architect Claude Perrault with a detailed set of design specifications for every working part of the Observatory.[51] His active involvement in the selection, design, and production of instruments was also critical to the establishment of a new Observatory. But Cassini was not merely an unusually good administrator. The reform and reconstruction of the Observatory (1784 to 1793) was not an aim in itself. From 1785, Cassini IV was using it for two scientific works of major importance: regular meteorological and astronomical observations which he both directed and shared with his students and (from 1784) the 1:86,400 survey of France.

His scientific credentials were also particularly necessary as the greatest opposition to the reforms came, for a variety of reasons related to intellectual territoriality, not from the King and court but from the Academy of Sciences and the old guild system. We have seen that in the context of the survey of Tuscany, Cassini sought to distance himself from "simple engineers." Nevertheless, in many regards, in his work at the Observatory he can be seen as one of science's democratizers—a movement with relevance to engineers in this and later periods.[52] He advocated the interests of educated but often middle-class engineers and artisans because he realized that the continued advancement of both astronomy and cartography demanded a closer link between artisans and scientists. In a sense, what Cassini IV, and in a quieter way Cassini III (in the training he provided his engineers), were working toward was a blurring of those heretofore sharp borders. Cassini IV went so far as to suggest that the Academy of Sciences might benefit from including highly skilled instrument makers among its members. This was not a suggestion welcome to many eighteenth-century scientists— particularly not to his fellow astronomer Joseph-Jérôme de Lalande. In the context of eighteenth-century observational astronomy, then, Cassini IV can be seen both as a radical and as an important innovator.

The rebuilding of the Observatory was only completed in 1793. His foundry manager left France for countries willing to better remunerate his skills. And the instruments that Cassini sought never arrived from the workshop of Ramsden—almost certainly due to the Revolution and the constant warring between Britain and France that ensued. So, in a sense, Cassini was never able to use the institution he helped to create. He did, however, produce a series of regular observations from 1784 to 1790.

If we were to compare Cassini's observations and records to those now regularly fed into the supercomputers at the National Oceanic and Atmospheric Administration, we would find them lacking indeed. In addition, other astronomers, even in his own time, kept sometimes more accurate and detailed records. Cassini's meteorological data was composed of extremely generalized barometric, thermometric, and wind data by month. Only days which differed outstandingly from the norm were mentioned and then, largely, descriptively. The astronomical observations were more extensive and included the observations of stars, constellations, planets, the moon, the eclipses of Jupiter's moons, and comets. But these observations, too, were far from exhaustive and functioned more as consistent and regular spot checks. It is important to remember, however, that Cassini IV was the first to insist upon, and to successfully fight for, regular observations in a national observatory supported by the state and by the primary scientific institution. He argued for this on the grounds that regular observations alone would provide theoreticians with sufficient data to check their hypotheses. He was convinced that it was only through the regular monitoring of nature that its regularity and the laws governing it could be discerned. Prior to the Revolution, he believed, above all, in those laws and in their importance. Talking about meteorological observations, he wrote in 1786,

If we consider only meteorological observations, which from one point of view, if we look at them as a whole without thinking of that multiplicity of accidental, momentary, and local causes which seem to influence the constitution of the atmosphere unceasingly and to occasion constant changes in temperature, then we might be tempted to believe that chance plays the biggest role in meteorological phenomena; that it is impossible to discern any certain rules; and that consequently there is no use, only simple curiosity, in looking at them. But the educated Philosopher who examines everything in depth before making pronouncements, does not think like this at all. The more he has observed nature, the more he penetrates into its secrets, the more he is persuaded that in the vast plan of this Universe there is no effect without cause, no event due to chance, no movement without law.[53]

As we will see later, however, his word choice in 1786 was radically different from the language he used after 1793. In particular, there is no direct reference to God in this passage and after 1789 he would use the word "philosopher" only as a term of derision.

As with many scientists or scholars, a large proportion of Cassini's work was never published and has therefore vanished from our view. In Cassini's case, thanks largely to the work of a biographer closely attached to the family and writing soon after his death, we can catch glimpses of this work. According to this biographer, Cassini not only oversaw but periodically took part in the engraving carried out at the Observatory. He was very attracted to this type of skilled artisanal artistry and, with his father, actively sought to improve the salaries and respect accorded for such fine and demanding work.[54] He did not confuse geographic engraving with artistic engraving, regarding the former as somewhat mechanical.[55] We also know that he made topographic relief models and that he was trained in this "under a highly expert master, under the author of the magnificent works of that type that one sees at the Invalides."[56] Cassini was particularly fond of sketching and drawing, and we also know that he passed his time in the prison during 1793 and 1794 sketching "paysages" (presumably from memory) and the facades of buildings and teaching his fellow inmates the principles of drawing. Little evidence of this has survived in his printed works. In addition, Cassini IV was a poet and appears to have composed for his friends and family from an early age, but he was very humble though unapologetic about his affinity for and talents in poetry. He, saw himself as a pleasing rimer, not a poetic genius.[57] This critical assessment of his poetic talent is fundamentally sound. Nevertheless there is both charm and originality in some of his poetry.

Cassini's Banishment from Science and Its Relevance to Geography

Cassini IV's assessment of himself as a poet could also be applied to his science. Like most people, Cassini was not a genius, and in the company of intellectual giants, he can be made to look tiny indeed. Such company was lacking in astronomy and physics in France through much of the eighteenth century but began to make itself felt toward the end of the century, most notably in the person of Pierre-Simon de Laplace. Since 1784 Cassini had held the positions of director of the Observatory and director of the map of France by virtue of his name and his dedication to that name. No doubt he would have remained in that position until his death or retirement but for the Revolution. It is arguable, however, that the potential of the position had outgrown him by the last decade of the eighteenth century.

Once the Revolution began, it was fairly inevitable that a man such as Cassini would fall afoul of the Revolutionaries. The Cassini name had a long association with French royalty and aristocracy. In addition, Cassini was a died-in-the-wool ancien régime scientist and royalist. For him the Church was an institution worthy of respect and utterly inseparable from the social and intellectual fabric of the country. The Academy of Sciences was an elite intellectual institution floating serenely above all possible political dispute. The seeking of royal patronage was the normal and natural course of events and had nothing whatsoever to do with politics. Stooping to political discussion, debate, and argument was, however, unbecoming to a scientist. A scientist might choose to serve the state or not, but he could in no way be mistaken for a state functionary. As Cassini had written to the Grand Duke of Tuscany in 1775, "I am responsible for my time to science and to myself."[58] His time was his own and his sense of his responsibilities as a scientist did not extend to inspecting prisons. These were his views and he made them clear to those who sought his opinion. Indeed, as secretary of the Academy in 1793, he was in a position to *have to* make them clear.

As a result of his views and due to the power and influence of his position, Cassini came under attack: he was successively deprived of membership, with all the other members, in the Academy of Sciences (9 August 1793); the directorship of the Observatory (21 August 1793); control and part ownership of the Cassini map of France and all of the equipment associated with its production (21 September 1793); the apartment in which he, his mother, and his family lived at the Observatory (22 September 1793); and his estate at Thury.[59] Finally on the 14th of February 1794 he was thrown into prison with a cousin, Mlle de Forceville, who happened to be visiting the family and who, his biographer tells us, for an ill-considered sarcastic remark uttered in prison, was ultimately guillotined.

For geography as an intellectual endeavor, the most serious of these events was the confiscation of the Cassini map. That it was Cassini who was deprived of his position as director of the Associates and of his personal investment of time, effort, and money was not the critical dimension. What did matter for geography was that the confiscation and the refusal of successive governments to restore the map to Cassini represented the nonrecognition of the map as the intellectual property of a member of the Academy of Sciences. There is no doubt that Cassini saw the map as his *intellectual* property:

> They took it away from me even before it was entirely finished and before I had added the final touches to it. This no other author has suffered before me. Is there a painter who has seen his painting seized before having put the final touches to it? What poet had his poem taken from him before completing the final scene? That fate was reserved for me.[60]

It is interesting to contrast this sense of personal ownership with the emphasis on "national interests" conveyed in Pierre Jacotin's ruling on the remuneration Cassini IV had the right to expect from the government.

Pierre Jacotin was typical of the "geographers" who in the course of the nineteenth century produced the new survey of France. Born of a family of small and poor landholders in 1765, he was recruited as a cadastral mapper by his uncle, Testevuide, who was one of the directors of the Corsica cadastral survey.[61] He worked on the cadastre for fifteen years, learning both office and field aspects of mapping in difficult terrain and in trying political circumstances. Soon after leaving Corsica to the British forces, Jacotin was assigned as a military geographer to the secret Napoleonic expedition to Egypt and Syria. After the death of his uncle in the Cairo uprising, Jacotin was made head of the ingénieurs-géographes and thus became responsible for all mapping operations in Egypt. In his time in Egypt, he had under his orders some of the first graduates of the prestigious École polytechnique, in particular the young Edme-François Jomard. It was for Jomard that he was to later write his critical description of the mapping of Egypt, his only published textual work. In the course of his military career, he served on the expedition to Egypt, in the attempted conquest of Syria, as office director and head of the engraving school at the Dépôt de la guerre, and as a head of the 1823 survey of Spain.[62] In 1817, Jacotin was one of six colonels named to the consultative committee charged with the conceptualization and design of the new 1:80,000 Carte de France. However, the monarchy, more concerned to demolish Napoleonic structures than to construct a new state, had disbanded the ingénieurs-géographes in 1817. This presented a considerable challenge to those charged with the direction of the new Carte de France. Jacotin served as head of the topographic wing of the Dépôt from 1817 to 1827 and as a director of the Dépôt itself from 1822 to his death in 1827. Jacotin was both typical, and among the most successful, of the many hundreds of ingénieurs-géographes who devoted their lives to mapping for the state in times of war and peace. After years of service in the field as one of a team of cartographers, he became a military cartographic bureaucrat virtually indistinguishable from the interests he served. He was typical of most of the ingénieurs-géographes in that he left little beyond his mapping work for scientific posterity. If his personality or insights marked his work, it is nowhere evident in what remains to us. His memoir detailing the decisions made in compiling the map of Egypt repeats the prevailing state ideology that it was France's mission and duty to conquer Egypt for science.[63] The memoir reveals the painstaking competence of its author, but there is nothing in it that is in any way intellectually personal. Jacotin's was a life of service with little or no sense of ownership over the product of his very considerable labor.

It was fitting, then, that it should be Jacotin whom the Minister of War chose to assess what the state owed the Cassini family for more than twenty years of the use of their map. It was also predictable that Jacotin would not (or perhaps could not) understand Cassini IV's sense of ownership. Indeed, he was offended by the concept. A map of France was "a national monument" above the interests and claims of individuals. Despite the fact that Jacotin recognized the Cassinis as the map's "former proprietors," and in spite of his recognition of the map's continued enormous value to France (given the slow pace at which its replacement was being made), he felt little was owed the Cassinis.[64] Jacotin calculated the family's due by estimating the change in value of the plates from 1790 to 1818 and further subtracting the costs of maintaining and touching up the plates incurred while the state had printed from them. Outraged at Jacotin's lack of understanding of the very personal and intellectual nature of the Cassinis' investment in the maps, Cassini IV fulminated, "Why didn't he just weigh the copper and compare its value then and now?"[65] Far from understanding the personal attachment of a scientist to the product of his thought and effort, Jacotin deemed Cassini IV's sense of his family's investment "exaggerated pretensions."

There is some ambivalence in modern historical accounts of the confiscation. This is because geographers and cartographers today recognize that the state confiscation of the map was a fundamentally important step in the establishment of regular, systematic, and well-funded topographic surveys.[66] For the community of geographers, however, the confiscation constituted a demotion. Henceforth large-scale mapping could not been seen as the result of intellectual activity on the level of art or science. It was, instead, a bureaucratic activity of primary interest to the state and controlled not by the Academy of Sciences or by the Observatory but by either the military or bureaucrats.

The government's action better represented the actual state of affairs in large-scale cartography than did Cassini's claim of authorship. Had the map indeed been the intellectual property of one man or even of a family, then the confiscation would have resulted in the death of the survey or at least its stuttering or stagnation. In fact, the map was completed without Cassini's help. The new survey rested on the knowledge and experience gained in the execution of the first and the second Cassini surveys and perhaps more importantly on the experience gained in surveys executed during the Revolutionary and Napoleonic wars by both the military and the ponts et chaussées. To his continuing chagrin, this survey was not to be executed under the direction of Cassini IV or even under that of any Cassini, but under the direction of the état-major, distantly overseen by a committee headed not by a geographer or even by a practical astronomer but by Laplace, the man who had successfully blocked Cassini's reentry into the Bureau de longitudes between 1799 and 1803.[67]

The fact was, that by about the middle of Cassini III's career, large-scale cartography had pupated into something altogether different from its original existence as the one-family-show of the first Cassini survey or even, to a degree, of the second survey. Cassini III sensed this. Cassini IV knew it. What other conclusion is possible if we take the latter's assessment of the future of large-scale geography at face value:

> It is not possible to deny France the honor of having effected a great reform in geography, in fact, of having made it an entirely new science by associating it so intimately with geometry and astronomy. Before that association, on what bases did geographers rely? On word of mouth, on the memories and drawings or accounts of voyages which were rarely in agreement. If travel accounts are sometimes dictated by conceit, ignorance has also sometimes drawn their itineraries. Let's pity the condition of geographers when they only have such materials to support their work and when they are obliged to stop there. The finest and most delicate criticism cannot always give shelter from the grossest errors. In order to escape them one must be endowed with that sort of instinct, with that sense of divination which seems to have guided the Delisles, the d'Anvilles, the Buaches, to whom we here fulfill our obligation by rendering the homage that is their due. But in praising their useful and

untiring work, we are happy for their successors who, more fortunately, will have less research to do in constructing maps infinitely more accurate than those. Thanks to the multiple voyages undertaken by educated men* all over the world; thanks to astronomy's, geometry's, and clock making's easy and rigorous methods for determining the position of all places, geographers will soon find that they have neither uncertainty, nor choice, nor need of a critical faculty in order to fix the principal positions of the four parts of the globe. The canvas will fill itself bit by bit as time passes, imitating the procedure that we followed for the production of the general map of France.[68]

A geography that had "neither uncertainty, nor choice, nor need of a critical faculty" was, by the most basic criteria, no longer a science.

Geography's demotion was complete when the last member of the great Cassini dynasty was refused reentry into the Bureau des longitudes, and was thus banished from active participation in post-Revolutionary science. He had been reinstated into the Institut and, later, the Academy of Sciences. But in pique at the apparent demotion that his proposed reinstitution as just another astronomer at the Bureau des longitudes represented in 1795, he had refused reentry. Once he had acquitted himself of the education of his sons, he found that he missed the life of the Observatory and reapplied to the Bureau des longitudes. His admission was blocked—not once, but twice—by Laplace. It is impossible to know precisely what Laplace had in mind when he voted against Cassini IV. But there are indications that two factors might have played an important role, both of which are relevant to geography: Cassini's strong attachment to both the Cassini tradition and the ancien régime, and the fact that Cassini was no longer respected as a scientist. These are, in turn, strongly related to Cassini's complaints against post-Revolutionary science. Cassini had three principal difficulties with science after the Revolution: its godlessness, its more overt handmaiden role, and its disrespect for the science of the Cassinis.

There is little indication of a man of deep religious conviction in Cassini's pre-Revolutionary writings. Nor is there any sign of disrespect or even criticism of the religious orders. Still, Cassini was a man of science and primarily devoted to science, not religion. Post-Revolutionary science, however, was another matter. This science claimed to be able to reshape society and, indeed, even to be able to revolutionize itself. There were no limits to its sense of its own power. It was not even beholden to God. This was incomprehensible to Cassini.

---

*The tireless and scholarly traveler Mr. Humboldt has on his own already determined more than 250 locations in South America. [Cassini's footnote]

. . . that the "géomètre," whose sublime study is truth, whose practice it is to reason clearly and to never lose the chain of consequences; that the physicist and the naturalist, who occupy themselves ceaselessly with the great operations of nature, gathering its riches, considering its beauties, penetrating its mysteries; that the astronomer, who spends his days and his nights contemplating the harmonies of the celestial bodies as they follow the immutable rules of their movement . . . that they can for one moment doubt the existence of a being superior to all, the author of such sublime works, that is what is inconceivable to me. I will say more, that is what seems absurd to me.[69]

What possible explanation was there for men of science who studied the cosmos, developed sophisticated mathematical theories about its workings and yet did not believe in God? In a fictitious discussion with a nonbeliever, Cassini had his opponent comment on this:

In effect, gentlemen scholars, I am beginning to believe that you are not all very honest with your systems, and I do not know if the inventors of hooked atoms [atomes crochus], vortexes, and other similar ideas, themselves believed in what they told us.[70]

To which Cassini responded:

To start with, no. But by repeating it they finally persuaded themselves, especially when contradiction and argument honed their pride to a fine point and provoked their stubbornness.[71]

Science without God was, then, a lie and dangerous intellectual hubris. But Cassini went beyond this in arguing, in direct opposition to the spirit behind the establishment of the École normale and the écoles centrales, that science, with or without God, was superfluous for the day-to-day running of society. It was mere froth, and society needed something of greater substance to keep it working smoothly.

It seems that those brilliant creators of the new [educational] establishments thought of public instruction from only one point of view, that of science. Thereby they separated instruction from education. One can only be shocked to see that at the birth of a French Republic they only thought to people the country with scientists instead of forming citizens. Mathematics, chemistry, natural history, drawing, these were the principal concerns of the écoles centrales. Languages and literature were merely secondary. Last, as an afterthought, came legislation and morality.[72]

For Cassini, the damage to society inflicted by the loss of religious education was irreparable.

For me never, it has to be admitted, no, never, will they be able to replace those valuable and respectable men who for many centuries preserved the repository of the humanities, who spread them throughout Europe, who upheld the utility and glory of our old schools, and who are the men to whom we are indebted for what we know.[73]

Many other geographers, long the recipients of Jesuit and Oratorian education, must have felt similarly. Nevertheless, this attitude did not win Cassini friends and respect among the now powerful in the world of science. De Lalande was happy to address his letters to his colleague mockingly "From Citizen Lalande to his Dear Friend in God and in Astronomy, Cassini," and Cassini was aware that he was the butt of jokes among his former colleagues.[74]

Geographers had long had a close relationship to royalty. Philippe Buache had served as royal tutor to three kings—Louis XVI, Louis XVIII, Charles X.[75] The most successful geographers benefited directly from royal privilege. The topographical engineers worked for the Crown as members of the military and in mapping expeditions beyond the borders of France or along the country's borders. But the best example of the importance of royal favor to geography was the Cassini family. Cassini I (Jacques-Dominique) was brought to France by Colbert for King Louis XIV in 1669 and thereafter the Cassini family benefited from four generations of royal favor and support. There was an inherent conservatism in the social structure of the ancien régime which meant that great success, once achieved, could secure the position of subsequent generations of a scientific family who demonstrated, not genius, but merely application and loyalty. Such was the case of the three subsequent Cassinis, Cassini III being the most creative and productive of the descendants of Jacques-Dominique (and also the most important for geography). In addition, it is arguable that geography, particularly its large-scale mapping branch, was, of all the sciences, the most dependent on Royal support. The Cassini map and much of the scientific work in which Cassini III and IV engaged in their lifetimes was inconceivable without just such support.

There is little doubt that Cassini IV owed his position in the Academy of Sciences, at the Observatory, and as the head of the Society of Associates charged with the production of the map of France very substantially to his name and the advantages and privileges that it had brought him. The old order was an order which had benefited him and which he had had every hope would benefit his sons and protect his daughters. The Revolution brought an abrupt end to that order of things, and Cassini had every reason for deeply regretting the loss. As long as the state had embodied an order understandable and, cynically, favorable to Cassini, he had not noticed the service nature of his scientific role which, indeed, was so light-handed as to leave him with the impression of intellectual liberty.[76]

> No one, I believe, had more enthusiasm for science and desire to support the glory of his name than I. No one, more than I, certainly, based his interests, his tranquillity, and his happiness on the study of the sciences and on the profession of scholar. Enclosed in the Observatory, I thought that there I was in a port sheltered from all storms, beyond the sphere of jealousies and intrigues that we call the world. I saw in the movement of the stars only the noble and sweet contemplation of the marvels of the universe and of the grandeur of the creator. In elevating my thoughts so high, I found the advantage of losing sight of, or forgetting, the earth, men, their mistakes and their miseries.[77]

The sovereignty of new and subsequent regimes was less established and consequently assurances of loyalty from men of science were more insistently and openly sought. Having served the King with numerous protestations of loyalty, Cassini consistently denied the right of the subsequent regimes to demand the loyalty of its scientists.

> Is it possible that they regard the exercise of science as a public office, and that the liberty of *thought* and of *knowledge* are somehow bound by these oaths?[78]

This, then, was someone who consistently stood apart from his fellows, who thought little of contemporary science and less of contemporary governments, and who said so. Was such a man likely to be welcome in a government bureau which served as the nerve center for knowledge of both France and the world?

Cassini's greatest difficulty with post-Revolutionary science (and post-Revolutionary science's greatest difficulty with Cassini) was the issue of the importance of the Cassini family. We have already seen that Cassini devoted his life to living up to the name of his family. His great grandfather had been an astronomer primarily remembered for his discovery and observation of the moons of Jupiter, which had, in turn, served geography by helping in the determination of longitude. Astronomy, for Cassini IV, in the image of Jacques-Dominique Cassini I, was a matter of observation: the facilitation and regularization of observation and the practical use to which observations could be put (e.g. geography and navigation). After four generations, however, astronomy was changing, thanks, in part, to the work of Newton (whom Cassini II had opposed in the debate about the shape of the earth; see chapter 1) but thanks, most recently and importantly, to the work of Laplace. The Cassinis had sought to trace and follow the placement and the movements of the planets and the constellations. Newton and Laplace wanted to *understand* the physical forces *lying behind* these movements. They wanted to be able to *explain* the impact of the movement of one planet upon another and upon the whole system. This required a highly mathematical and theoretical astronomy which the Cassinis seemed to reject in principle. Writing in general terms and referring to neither astronomy nor geography, Cassini described systems, a contemporary and slightly pejorative word for theory, in the following terms:

> With respect to the systematic approach [esprit de système], which is nothing but a more thought out and therefore more dangerous prejudice, it is important to protect oneself from that, as much as is possible. The observer most often becomes superfluous . . . to one preoccupied with a system. He no longer sees things as they are but only as they must be in order to favor the opinion he has previously adopted. Those [observations] which too directly oppose his system he rejects or ignores. A system must be a result; it must be the fruit of persistent observation and profound reflection.[79]

This might have been good advice for geography, which was in mid-tumble into its Buachian trap, but it was diametrically opposed to the direction of modern astronomy. Cassini IV was to some degree aware of this and was no more comfortable with modern astronomy than it was with him. Again, writing about his alienation from science to a colleague,

> How can I even recognize myself in this overturning of our old calculations, of our old measures, of those days which will have only 10 hours instead of 24; of those circles composed of 400 degrees. . . . Everything is changed; and I am a bit too old to lose the habit of my old methods and my old notions. . . . If Galileo, Newton, or Kepler fell from the sky into the middle of the Institut, they would not understand anything. . . . What would they say when they were told that such a profound conception, such a sublime language was the masterpiece of our own scholars who, in the middle of the Revolution did not want to be left behind and who gave themselves the honor of overturning everything, of changing everything without need, and for the sole pleasure of destruction or of saying things differently than those who had preceded them.[80]

This was much more than a distaste for uniform measures based on a regularity observable in nature (for which he had argued as early as 1775), or even of the Revolutionary passion for renaming and rationalizing the most banal aspects of life (although that clearly enraged him). Here Cassini was expressing discomfort at the ground shifting beneath his feet, although probably he would not have been able to identify precisely what was moving and why.

We get a final brief glimpse at this alienation between two sciences in Delambre's assessment of the Cassini family's contribution to astronomy in his *Histoire de l'astronomie au dix-huitième siècle* and in Cassini's sharp reaction to it. In his history, Delambre methodically reviewed the work of the Cassinis and found it seriously lacking in some important regards. He saw in it a lack of rigor, a lack of understanding of the meaning of the discoveries and work of scientists such as Kepler, Newton, and even Halley, and a tendency to draw pretty and satisfying Cartesian pictures rather than to seek explanations of anomalies in observations. He summed the problem up devastatingly:

> These various memoirs demonstrate well the consequences of family or sect prejudices which, in fact, it would be so easy to defend science from, given that science should only proceed from rigorous measurement. But, in general, the Cassini school calculates very little. Instead it brings everything down to graphic constructions, which even J. [Jacques] Cassini usually dispensed with. They treat all of this material in a manner whose only point is to retain consistency with the ideas of Dominique and to support Descartes's system.[81]

The Cartesian approach, which Delambre clearly regarded as unmathematical, nonexplanatory and simplistically descriptive, by virtue of its attempt to find geometric and graphic solutions to problems, was out of date and inadequate to the task at hand. There is every reason to believe that Delambre's view of the Cassinis when he wrote the book (sometime before his death in 1821) was similar to that of Laplace, whose *Celestial Mechanics* revolutionized astronomy. What this view of the Cassinis and their work represented was a definitive break on the part of modern astronomy with its old sister-in-cosmology, geography. Where did that leave geography, which was without question the endeavor which had most gained from the Cassini family's observational and graphic tendencies? Perhaps where it left Cassini: fuming in rage at the desecration of all he held dear.

With a different personality or under different circumstances, Cassini might have shuffled along, a respected old man of science like de Rossel, among the leaping and bounding scientific and mathematical gymnasts of his age. The Revolution and his reaction to it allowed for an abrupt break with the outmoded astronomy of the Cassinis. But, as I have argued, Cassini IV (and indeed Cassini III) was far more important to geography than ever he was to astronomy, and with Cassini's exile from science, and especially from the Bureau des longitudes, came geography's loss of direction and status. If those behind geography's greatest achievement did not merit the respect of other scientists, then where did geography stand?

CONCLUSION

We saw in chapter 1 that traditionally, and certainly by the eighteenth century, geography's major aim was the reflection of the unity and coherence of the world. Further, that this unity and coherence had been expressed and celebrated in two forms: text and cartography. By the end of the eighteenth century the mode of thought that supported this conception of geography was becoming outdated. On the one hand, huge strides had been made in the elaboration of the picture of the world, in part through the work of geographers. However, the picture revealed increasingly complex interactions between phenomena at a variety of scales and of various natures. How could description of either a textual or graphic sort possibly elucidate that complexity? The descriptive accounts of the world provided by Mentelle and Buache de Neuville to the young and critical minds at the École normale seemed pointless, unrelated to recent discoveries in any field, and uninteresting. Whether expressed by Mentelle and Buache or by Cassini IV, assertions as to the unity and coherence of the cosmos sounded more like the recitation of an article of faith than a point of investigation and therefore seemed to belong to another and more primitive mode of thought. Ironically, where geography could be said to have truly progressed under the influence of developments in mathematics, optics and physics, in the area of large-scale cartography, that progress alienated and cast geography's most famous highly achieving family out of the world of modern science. The new and prevailing direction of research seemed to be (in the fields most closely allied to large-scale mapping) away from the sort of observationally driven and graphos-based descriptive science practiced by the Cassinis to an intricate, mathematical explanation of forces, driven less by data than by elaborate hypotheses well tempered with data. In this context, the geographic map and even its more sophisticated brethren, the topographic map, began to seem outdated, or at the very least, no longer a scientific challenge.

·PART·TWO·

Reaction and Continuity

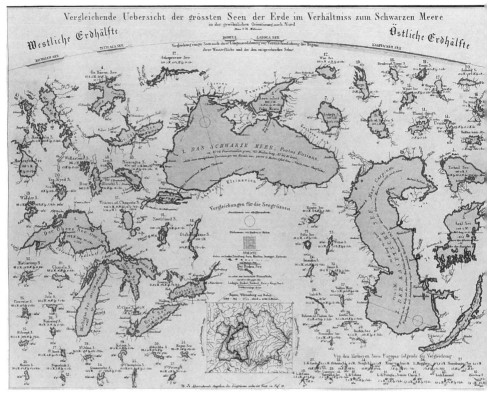

FIG. 3. "A comparison of the lakes of the world." A concern with relationships between phenomena and a graphic playfulness allowed Humboldt to compare phenomena in ways unimagined by most authors of universal geographies.

Universal Description

> It is not easy to arrive at a conception of a whole which is constructed
> from parts belonging to different dimensions. And not only nature,
> but also art, her transformed image, is such a whole. It is difficult
> enough, oneself, to survey this whole, whether nature or art, but still
> more difficult to help another to such a comprehensive view. —Paul
> Klee, *On Modern Art* (London, 1948), 15

The universal geography is one of geography's oldest genres. The oldest
surviving example was written by Strabo in the first century a.d. It is clear
from Strabo's description of his own work that the genre was already well
established two thousand years ago.[1] Since then, universal geographies have
been written in extraordinary number and under a variety of titles, from *The
Science of Geography*, to *A Complete System of the World*. What all these works
have in common is an avowed attempt to provide a multiscale description
of the earth from the cosmos to peoples. Most have described the place
of the earth within the astronomical (or cosmological) system; the physical
systems of the earth, including the continents, oceans, lakes, and rivers; and
the human world, including countries, nations, borders, peoples, societies,
governments, populations, etc. Their entertainment, educational, and even
administrative value was considerable in the era preceding widespread long-
distance travel and readily accessible dictionaries, encyclopedias, and atlases.
Until sometime in the eighteenth century, prior to the bibliographic explosion, universal geographies were a useful, if sometimes pedestrian, genre.[2]

They have varied significantly in physical and intellectual form, and
in content. Physically, they may be as small as pocketbooks or as large
as twenty-one-volume sets of barely liftable tomes. Sometimes a series of
sophisticated and interconnected essays, they have more often been little
more than lists of features in ascending scale and increasing detail. Most
have sought to provide exhaustive descriptions of "everything." But a very
few are philosophical treatises on the problems of describing "everything"
in a meaningful and coherent manner, or on taking a holistic approach to
the cosmos in all its detail and complexity. The solutions to the problem of
meaningful description and the principles of inclusion and exclusion have
also varied, creating a further diversity. In addition, the context—intellectual

and political—within which each universal geography was written played a powerful shaping role.

Given the long-standing popularity and diversity of the genre, it would be fair to say that there was nothing new or innovative about the concept of the universal geography in the early nineteenth century. Yet, it was through a return to the genre that Conrad Malte-Brun believed French geography would find its salvation. Born in Denmark in 1775, Malte-Brun came from a comfortably-off Danish family. His father was a military man and a government functionary who sent his son to the University of Copenhagen to study for the ministry. Conrad pursued a classical education emphasizing literature and history (in the course of which he was certainly introduced to geography and probably particularly to German geography). Ultimately he rejected his father's direction and began a literary career as a poet and publicist. In twentieth-century terms, from 1789 through the first decade and a half of the nineteenth century, Malte-Brun functioned as a literary, academic, and social critic. His political writings in favor of revolutionary reform in Denmark along the lines being pursued in France in 1789 led to his exile from Denmark. He was briefly readmitted in recognition of his merits as a poet, then exiled once again for his renewed writings critical of the government. He arrived in France in 1799 and, we are told by his biographers, resumed his political writings by criticizing Napoleon's usurpation of power on 18 brumaire. His biographers state that the fact that his political views were unacceptable to Napoleon subsequently forced him to retreat into the study of geography. It is probable, however, that he had well-developed geographical interests before 1799 and it is also clear that by 1806 Malte-Brun had a regular and sometimes openly political series of columns in the *Journal des débats* which were usually supportive (if not adulatory) of Napoleon. Although the combination strikes the modern thinker as peculiar, Malte-Brun was both a social and literary critic (with flexible political views) and a geographer who chose to examine and critique his times from the perspective of a field making an awkward transition into an era of modern theory-driven science.

Malte-Brun appears to have made his entry into French geography through the good offices of Edme Mentelle. It was through his participation in Mentelle's first universal geography that Malte-Brun became known as a geographer to the French public. The field that he encountered in his new home was one very different from what he had known in Denmark. Eighteenth-century French geography was much more focused on mapping and the resolution of locational problems. German geography was primarily concerned with synthetic historical description and particularly with the collation of geographic statistics descriptive of states. By the time Malte-Brun arrived in Paris, the decline—at least in status—of French cartographic geography, whether it involved large-scale field mapping or office compilation, was clear. Malte-Brun, trained in the classics, languages, literature, and history, and aware of sudden growth in the sciences, saw no reason to regret the loss of cartography, or even to pause to contemplate it. He took on the defense of geography, but the geography he defended was one in which cartography was relegated to the sidelines and from which theory was banished.

Geography can be described as the principal constant in Malte-Brun's professional life. Through the shifts in regime and the considerable gyrations in his political views (which earned him a place in the *Dictionnaire des girouettes*[3]), he retained a concern with geography and its place among the sciences. His identification with geography and his adopted role as its defender is clear in many of the articles he wrote as a journalist for the *Journal de l'empire*.[4] In articles having little to do with geography, he often found ways to bring it into the discussion. Thus, while reviewing a book on artificial memory aids by a M. de Feinaigle, he launched into a lengthy discussion of the role of erudition, memory, and philosophy in geography.[5] In an article defending the Napoleonic campaigns in Prussia, Saxony, and Poland, he opposed geography to all those who were gullible, had erroneous views, and were against Napoleon's wars.[6] More than anything else, Malte-Brun sought the esteem of other scholars for the kind of geography he wanted to impose in France: after all, did not "such a science offer as much of interest, or maybe even more, than zoology, botany, mineralogy, chemistry, and so many other sciences that today it is supposed to be desirable to teach everyone?"[7] He appears to have been particularly concerned to rescue geography from the reputation for incompetence and irrelevancy it gained in the mid-1790s from the courses taught at the École normale as a result of which "youth rejects it, scholars neglect it, and worldly people disdain it."[8]

Apart from his publicist's role at the *Journal de l'empire*, Malte-Brun responded to this sense of loss of direction and status in three ways. He established a journal, *Annales des voyages*, which he was convinced would demonstrate the utility of geography by collecting and evaluating all geographic research being undertaken in France and abroad. Toward the end of his life, he helped found the "Société de géographie" in order to give geography a particular direction and form. Finally, and perhaps most importantly, he composed a universal geography, his *Précis de la géographie universelle* (eight volumes).[9] This work can be seen as his life's work as he began it sometime after 1803 and did not succeed in completing it before his death in 1826. It was his most deliberate, sustained, and indeed monumental attempt to both shape geography and its image.

In writing and publishing his *Précis,* Malte-Brun saw himself as returning to the true path of geography, the most authentic and promising tradition possessed by the field. In his view, the opportunities and indeed the true historic role of geography lay in description, most particularly in accessible literary description, that is, in the universal geography. Malte-Brun, as we shall see, like so many of the geographers of the period, was fundamentally conservative. For him the way ahead lay in a well-tried approach adapted from the past to fit the needs of the present. Following Malte-Brun's own assessment, his biographers have tended to paint him as geography's savior, as "one of the greatest geographers of modern times" and the reformer of a field in which "until then the geographical treatises . . . were uncritical and tasteless compilations."[10] It is only relatively recently that a French geographer has commented somewhat mysteriously that Malte-Brun "without a doubt slowed geographic research in France more than he stimulated it."[11]

THE FUNDAMENTAL NATURE OF THE
UNIVERSAL GEOGRAPHY: STRABO'S APPROACH

Arguably, Strabo's *Geography* is the earliest surviving example of a universal geography. It is also a work which Malte-Brun had read, which he cited and discussed in several of his most important publications including the *Précis,* his *Annales de géographie,* and an article in the *Journal des débats,* and with which he identified. Indeed, in his *Journal des débats* article, he used Strabo as an ally in his battle against the use of theory in the geographical sciences and what he regarded as its excessive focus on measurement and cartography.[12] The political and disciplinary resonance between Malte-Brun and Strabo, even across two thousand years of history, is understandable—if a little surprising. Strabo, as a Greek and as a late Stoic, was fundamentally conservative and backward-looking to the glory of the Greek empire. For Strabo, the Romans amongst whom he lived, although militarily and strategically admirable, lacked the cultural depth of the Greeks. It was Greek philosophical thought that he admired, and he was concerned to protect the integrity of the Greek intellectual tradition, exemplified by the poetry of Homer, from all criticism—even that of later Greek writers.[13] A Dane living in exile in a French militaristic state and admiring of German-language culture may have found a kindred spirit in the politically astute cultural exile Strabo.

It is also clear, when one considers Strabo's preoccupations alongside those of later authors of universal geographies, that his definition of the genre and of geography itself captured the tone of universal geographies into the twentieth century. Strabo had a clear and somewhat rigid sense of what constituted geography and certainly of what belonged in a universal geography. Strabo argued that while geography had a significant "theoretical," or speculative, dimension, namely the many speculations on the nature, shape, and mathematical proportions of the physical world, it was practical geography—the utilitarian description of the earth—that defined the field. It was geography's link to action that gave it its superiority over both speculation and philosophy.[14] Further, it was practical geography that defined both his present endeavor, the universal geography, and the whole field. The highest ideal to which geography could aspire was the creation of an empire encompassing the entire habitable world:

> It seems to me excellent encouragement for the project at hand to say that geography is essentially oriented to the needs of politics. In effect, the scene of our actions is constituted by the earth and the sea that we inhabit: for small actions, small scenes; for grand actions, a large scene. The largest of all is the scene that we call the inhabited world. And that is the scene of the greatest actions. The greatest captains of war are thus those who can exercise their power over earth and sea, collecting peoples and cities together under a single empire, *controlled* by the same political structures. In these conditions it is clear that all of geography is oriented toward the practice of government: . . . It would be easier to take control of a country if we knew its dimensions, its relative location, and the original particularities of its climate and nature.[15]

This type of total and utilitarian description was the ideal. Strabo conceded, however, that even if one sovereign, well served by geographers, controlled all the world, it would still be impossible to describe all regions with an equal degree of fidelity. Ultimately, what was close would be described in more detail than what was far. This, then, was a legitimate reason to take a regional approach to the study of the known world.

Even within the regional approach, not everything that was close was of equal significance. Of special importance were borders and boundaries: the lines between phenomena. For Strabo, Eratosthenes's disdain for the delimitation of physical boundaries, such as those constitutive of peninsulas and islands, was peculiar.[16] But his uninterest in political borders was impossible.[17] Regional geography and particularly information on the borders between regions, even if incomplete or highly generalized,[18] was one of the most important elements of the universal geography because it enhanced the political utility of geography.[19]

There is a strong element of intellectual conservatism, typical of a late Stoic, in Strabo's *Geography* which placed scientific speculation beyond geography's bounds.[20] While knowledge of the inhabitable world might still be incomplete, Strabo did not see himself as a speculative voyager into the unknown and certainly not into the uninhabited unknown. The truly unknown was almost certainly of limited political value. It was important to focus on what was known to be important "and to enlarge at leisure upon those things which are known and important and also useful for action, memorable and pleasing." Like the making of a colossal statue, there was no point in focusing on insignificant details, "because this, too, is a colossal work."[21] Details, insignificant by virtue of their distance from political action, such as the configuration of the whole world,[22] or the action of celestial phenomena,[23] did not merit the kind of attention Eratosthenes and Posidonius had given them. This sort of speculation, together with the search for cause, was beyond the realm of geography. To question the origin of physical phenomena was philosophy, not geography.[24] It was better to simply believe the truth than to inquire into it.[25] The universal geography should serve everyday political action, not encourage occasional idle speculation.

The causes of physical phenomena, which required exploration of the essence of things, was certainly beyond the pale of geography. Nevertheless, what lay behind the distribution and variety of human existence on the earth was worthy of critical thought. These were not a consequence of a predetermined plan, but the product of historical contingency. Thus, the characteristics of a people—their knowledge, way of life, languages, and abilities—were the result not just of latitude but of many things: chance, habit, the qualities of the resources surrounding them, education, etc.[26] These a geographer could study.

If speculation and the search for cause belonged to philosophy, geography was still not a lowly activity like engineering or sea chart–making, which were so often practiced in ignorance of the basic principles of science.[27] No, geography was among the highest of endeavors; it was a science linked to poetry. The ancients had respected poetry and had used it to teach even adults "everything oriented to the social and the political and also historical information."[28] Prose, which was created to serve philosophy and history, was a fallen form of poetry which had cut learning off from the masses, from the women and the children. Geography was derived from the greatest of all poets, Homer.[29] Those, then, like Eratosthenes and Hipparchus who sought to banish poetry and fable from geography in favor of measurement and mathematics, threatened the very core of geography. In this sense,

Eratosthenes was a failed philosopher who had a "type of mind that made him always go just half-way."[30] The value of poetry, and by implication that of geography, could not be assessed the way one might assess the value of carpenters or blacksmiths. The value of poetry had everything to do with the quality of the man, the beauty of his expression, the veracity of his soul.[31]

Geography was closer to poetry in spirit, purpose, and form of thought than to "geometry." Hipparchus's concern with exact numbers and precise placement and configuration of lines was ungeographic and demonstrated his unfamiliarity with the concept of geographic approximation and how what was a sensible error varied with scale.[32] These were concepts central to geography which rendered it nontechnical and not exact. We see already, in this earliest of surviving universal geographies, a sharp line drawn between approximative geographic description and exact cartographic expression.

While Strabo argued that geography had to be distinguished from philosophy, geography's scope required the insight of a philosopher. Geography was related to philosophy in that its proper study required a holistic vision. Something that touched "such varied domains, which also touch political life and the practice of government, as well as the knowledge of celestial phenomena, the earth and the sea with everything they contain: living beings, plants, fruits, and also all the particularities that one can find in each country."[33] Such a huge subject could only be encompassed by a philosophically inclined individual with a holistic unifying vision.[34] From the divine to the human, all of creation fell within the descriptive and poetic realm of geography.

Most modern readers of Strabo's *Geography* are likely to be somewhat repelled and certainly puzzled by the amount of space Strabo devoted to a critique of his predecessors. Strabo's attention to his predecessors has a great deal to do with the establishment of authority. In his *Geography*, Strabo both constructs himself as a geographic authority and does so within the framework of a veritable hierarchy of authorities whom he discusses, critiques, and, in turn, legitimates. Strabo's claim to authority is partly based on his own personal knowledge of the world but more particularly on his ability to judge what has already been said.[35] Indeed, he maintains, we learn and have long learned far more from the sense of "hearing" (or by what has been said) than ever we learn from the sense of "sight."[36] Of course, not all voices are equal. Anyone discussed in this work, he claims, no matter how critically, is worthy.[37] Those not worthy are passed over in silence.[38] For Strabo, then, to write about an author was to dignify him and to admit him to the fraternity, to the path to truth, to the history of geography. Whom to cite was, then, among the most important decisions a geographer could make.

It will become clear as we explore a few of the most important universal geographies of the eighteenth and nineteenth centuries that both the tone and the form of the genre was already well established in Strabo's geography. Its dominant features, characteristic of the genre, included an emphasis on pragmatic, utilitarian description of the known world, accompanied by a decided impatience with theoretical and speculative discussion. The bulk of Strabo's geography was devoted to regional description with an emphasis on the limits and borders between phenomena. The description was not aimless but overtly linked to service to the politically powerful. Nevertheless, it was a literary, rather than a scientific, tradition of writing with little technical content. Indeed, although the entire work was structured around a map-like textual representation of reality, mapping itself and even the scientific foundations of mapping seem to have been regarded as part of another sort of endeavor. Strabo's geography embodied and explicitly stated the need for a holistic unifying vision of the known world. Finally, the genre spoke with an already established authority, the authority associated with tradition, the authority with which a scholarly tradition is emended.

THE DEGENERATION OF THE GENRE

After Strabo, universal geographies were written in Europe from at least the sixteenth century. As already described, the genre had considerable variety. Nevertheless, coherent and sustained geographic description required a unifying vision of the known world and all its inhabitants; a synthesis shaped by an argument about the nature of the world, its relationship to the larger cosmos, and the place of humans within it. Without some sort of synthesis or argument, there would be, and frequently was, a tendency to degenerate into lists of locations, names, features, and characteristics. Many authors of universal geographies misunderstood the challenge that the genre represented. Varenius, whose universal geography was much lauded by Newton,[39] managed to create a partial synthesis by banning most of human geography from his discussion of geography and by replacing the effort of synthesis with a form of intellectual cantonization of geography into physical geography, the classification of the earth by climates, and cartography. The eighteenth-century geographers Delisle, Expilly, and Mentelle, all well-known authors of universal geographies, both exaggerated the tendency toward a fragmented approach and abandoned epistemological reflection altogether. They seem to have had little sense of the increasing intellectual enormity of the task that the framework demanded. Instead, they understood their problem as physical and organizational rather than intellectual. They were not alone in this. If we trace the universal geographies produced from 1652 to 1990, we can see a marked trend (with some exceptions) toward expansion in volume with little corresponding concern for intellectual integration of the new material, culminating perhaps in Elisée Réclus's nineteen weighty volumes or Vidal de la Blache's twenty-three-tome universal geography, both of which have neither general introduction nor conclusion.

Universal geographies were certainly produced between Varenius's and those of Delisle, Expilly, and Mentelle. But the work of these latter three men is symptomatic of the state of the art in universal geographies by the eighteenth century. There are differences between the universal geographies produced by Delisle, Expilly, and Mentelle, born of the differences in the outlook, education, and experience of these men. They are nevertheless strikingly similar works. Certainly, reading them, one has a sense of a well-defined form, or genre. It is both surprising and significant that the works are so similar given the differences between the men. Delisle, as we saw in chapter 1, was one of the most respected of eighteenth-century geographers and made his reputation in cartography and to some extent astronomy. Mentelle, as we saw in chapter 2, made his career in geographic education, teaching at the École de Mezières, but in 1790 was considered, at least by some, as geography's best representative. Expilly was a popularizer in a period when geography was coming to be seen as a popularizing science.

The work of all men would have been known to Malte-Brun. Delisle's two-volume *Introduction à la géographie avec un traité de la sphère* (posthumously published in 1746) was highly regarded in its day, although the first volume (devoted to geography) was reviewed critically in the *Journal des savants*. The two works by Expilly examined here are his *La Polychrographie en six parties* (1756), which included a section on geography, together with his *Géographie manuel* (1757). Although there is duplication and overlap between these two works, they are worth looking at together in order to gain a picture of Expilly's sense of the relations between geography and the fields which, since at least the sixteenth century, geographers had seen as either a part of geography or closely related to it. Edme Mentelle produced a large number of universal geographies in the course of his lifetime—most at a primary- or secondary-school level. The universal geography that he produced in the early years of the nineteenth century with Malte-Brun is worthy of study but it is also important to look at what Mentelle was producing prior his contact with Malte-Brun as it is by this means that we gain a better sense of what Malte-Brun was trying to build on and to reform. It also gives us an idea of Mentelle's sense of geography prior to the disruptions of the French Revolution. Mentelle's *Cosmographie élémentaire, divisée en parties astronomique et géographique* (1781), although aimed at children, was the basis of most of his later universal geographies. In addition, it is worth looking at his *Géographie comparée ou Analyse de la géographie ancienne et moderne des peuples de tous les pays et de tous les âges* (1778–1784) because, although this latter work was a comparative geography rather than a cosmography or universal geography, Mentelle considered this work a more complete discussion of the nature of geography than he felt was appropriate in an elementary reader.

The universal geographies of Delisle, Expilly, and Mentelle share a number of characteristics, some of which were already clear in Strabo and to some extent in Varenius but which by the eighteenth century had become fixed and increasingly exaggerated. In particular, they all defined geography as primarily concerned with description and the conveying of certain knowledge. They regarded "scientific geography" (or research) as somehow outside the purview of universal geography and embodied in either cartography or the kind of painstaking research involved in the reconstruction of past geographies (which, from the time of Ortelius until roughly the early nineteenth century, was intimately connected with mapmaking). They all implicitly or explicitly rejected Varenius's exclusion of human geography from general or universal geography. Yet, at the same time, there is no sign in them of a quest for the unity of geography. Nor did they demonstrate epistemological sensitivity or concern. Instead, they based geography's value on its utility. Finally, there is something of an awareness apparent in these works, and particularly in Mentelle's works, that geography was falling out of favor, or at least open to some fairly severe criticism.

Each of the universal geographies retained a connection between geography and a number of the other sciences, and in particular between geography and astronomy. However, there was little in the way of open and clear discussion of the relationship between these fields. Nevertheless, the way in which these subjects were introduced and discussed in the texts suggests a great deal about these authors' conception of their own field and its relations with these other fields. The most striking example of this is to be found in Mentelle's *Cosmographie*. For Mentelle, part of the point of a universal geography was the presentation of the knowledge gathered in fields about which one simply could not afford to be ignorant. Thus, the first section of his cosmography was devoted to the explanation of

> the principal discoveries by the human mind about the System of the World. Those for which we are indebted to the great geometers of this century have been consigned to scholarly memoirs and are therefore impenetrable

to the curiosity of the greatest number of readers. As a result, in spite of the considerable progress made by astronomy, many otherwise well-read people think that they have a basis for doubting the most incontestable truths and consider unknown the causes of numerous phenomena . . . on which there is not the least doubt among those who devote themselves to geometry, physics, or astronomy. I therefore thought to render an essential service to the public in presenting the results of the best works in this genre, disencumbered of all of the apparatus of analysis that led to these sublime truths.[40]

Important to note in this quotation is the frequent use of words conveying certainty when referring to the hard sciences, the assignment to geography of the role of general education for the wider and less educated public, and, most importantly, the idea that the universal geography could play this role because it dispensed with calculation and analysis. For Mentelle, Delisle, and Expilly the universal geography was descriptive, restricted to the presentation of known facts and hence primarily concerned with the order and form appropriate to the clear understanding of those facts. Thus, Delisle defined the geography he was presenting as

the description, or at least the knowledge, of the different parts of the earth and the water in relation to their situation and their extent.[41]

Mentelle went into slightly more detail in describing its function:

Geography's purpose is 1st, the description of the surface of the globe, 2nd, pointing out its products, and 3rd, knowledge of the creatures which inhabit it.[42]

Here as elsewhere, the words Mentelle used most often when talking about the nature of what he was doing were "description," "indication," and "knowledge," although both he and Expilly also commonly used "divisions," and "location." The starkest example of this unreflective and unimaginative view of geography is expressed by Expilly:

Geography is the description of the earth. The earth is divided into four parts . . . .[43]

By implication, geography as presented in the universal geographies was a lower-order science, subaltern in a way in which Strabo had never imagined. Probably without meaning to, and while writing about recent developments in natural history and geology, Mentelle expressed this clearly:

The man of genius can look for some glory by offering daring conjectures on the first formation of the terrestrial globe and on the revolutions which it seems to have suffered. But the Geographer, whose primary talent is to make accounts of facts which are recognized as true, must aspire only to a reputation of simplicity, exactitude, and fidelity.[44]

What are the implications of this statement? This is geography: the brilliant and inspired need not apply? These statements were not merely slips of the tongue. These geographers did indeed restrict the geographic part of their universal geographies to division, subdivision, and explanations of relatively self-evident nomenclature (e.g. lake, river, island, peninsula, etc.). Expilly was the most extreme in this regard. His geography was a series of divisions and descriptions on an ever-decreasing scale of both size and interest. Precisely the kind of geography which 105 years earlier Le Père François had refused to write:

> My purpose was not to provide in this treatise a division of the terrestrial globe into all its parts, such as into all the Nations, nor of the Nations into all the Sovereignties and Realms, nor the realms into Provinces, nor these into Cities, nor the ecclesiastical state into Patriarchs, Primacies, Arch-Bishoprics, Bishoprics, and Parishes.[45]

Yet, by the eighteenth century, the universal geography seemed to have assumed this form.

Each of our three authors recognized and stated that there was another sort of geography—a scientific, academic geography—which was to be held in respect. We get a better sense of this from Delisle's editor than from Delisle himself. Commenting on the production of maps from the critical comparison of multiple sources, the editor wrote:

> All of that demands an immense detail, sufficient to try the most obstinate patience. . . . What a boring and tiring discussion! One must be born a geographer to engage in it.[46]

This implies, although perhaps Mentelle would not have put it so directly had he thought about it, that a tolerance for extraordinarily boring work was what principally distinguished the geographer from other scientists. Delisle's editor apparently felt no differently about the other related subfield devoted to the reconstruction of past geographies:

> This profound research . . . can only be the lot of scholars who are attached to this sort of study by either taste or necessity.[47]

Delisle himself did not separate maps from description, arguing "words are necessary on the one hand and so is application on the other," as without graphic application "we would easily forget the form and the size of countries, the distances between cities, the shape of the seas, the course of the rivers . . . ."[48] His successors had a different approach to cartography.

For Expilly, maps were evidence of science in geography and sufficient to establish his credentials and credibility as a geographer but they did not particularly enter into a universal geography or have any relation to its methodology.[49] Mentelle tended to regard cartography as something beyond, and somewhat unattached to, the universal geography—indeed as superior to it. In this vein he wrote about ancient geography in a passage in which he juxtaposed his own career and work in geography with that of d'Anville:

> The importance of his work did not really allow him to work for children and for the less educated. He engages in science for its own sake: happy are those who come and profit from his lessons![50]

In Mentelle's hierarchy of intellectual activities in geography, the universal geography (which he associated with the instruction of the relatively uneducated) was a step down from more scientific cartography. All geography, then, was not contained within the universal geography. A scientific geography, for the most part attached to cartography, also existed. However, our authors considered it relatively uninteresting and inaccessible to the uninitiated, a view that perhaps defended the necessity of the universal geography.

The primary justification for the universal geography was its utility. Not its intellectual utility, which had been Varenius's concern,[51] but its pragmatic utility. In Delisle's *Introduction à la géographie*, although he did not say precisely what he meant by utility, his use of the term implies a commercial or geopolitical meaning. Entirely in keeping with Strabo's (by then 1,800-year-old) views about the relevance of areas untouched by European interests, he commented on the description of the Americas:

> If the only people in that country were autochthonous, knowledge of it would not be very useful. But because a number of European countries extended their domain that far and established themselves there strongly, America must be considered almost a dependency of Europe and knowledge of it has become almost absolutely necessary.[52]

Lacking any other criteria, a vague sense of utility seems to have provided the primary criteria of inclusion and exclusion for Delisle's descriptions. In this vein, he regretted not being able to list all the rivers of the world in his book as he knew that this would be useful. The whole point of Expilly's *Polychrographie* and his *Géographie* was utility: they consisted mostly of descriptions and lists of phenomena, followed by tabular presentations of routes and distances from Paris, monies of the world, etc. (the whole presented in pocket format to be carried around while traveling). Expilly expressed his sense of utility more grandly and directly than Delisle as "the good of the Public."[53] Mentelle's sense of utility was directly tied to the educational mission he adopted for himself and for his universal geographies. The expression that best characterized his sense of the utility of the universal geography was, "an essential service to the Public."[54]

This emphasis on utility can be seen as a consequence of a lack of any coherent sense of the glue holding the disparate parts of a universal geography together. Or, otherwise expressed, of any sense of the fundamental unity of geography. Whether the authors realized it or not, the unity of geography was the premise of the genre within which they were writing. A universal geography behind which there was a conception of geography as a series of unrelated parts or as a collection of phenomena that one did not try to relate was intellectually problematic. There is no sign that Delisle considered this a problem at all. In this regard, he seems to have been epistemologically unreflective. Thus, in a phrase, he linked human and physical geography through the soil.

> The earth is that dry element which not only serves as the home of man but which also sustains houses and cities and which is clothed in trees, flowers and other things.[55]

He did not explore any of the implications of this link. Similarly, while he combined geography and astronomy in his book, he nevertheless, and without any explanation, considered that hydrography was a science distinct from geography.[56] Therefore seas, winds, and ocean voyages would not be discussed in any depth. Expilly was most blunt about the lack of unity in what others called cosmography or universal geography and what he chose to call "Polychrographie."

> Here "Polychrographie" means a multiplied description which has as its aim many distinct knowledges, which cannot be brought together under a single heading.[57]

Mentelle was both less blunt and more suggestive of the intellectual problem behind geography's apparent lack of unity. He wrote with some discomfort about the apparent difficulty that both geography and natural history encountered in trying to ensure that their students retained the key facts and details:

> . . . so many others, which are more methodological, such as mathematics for example, are learned with such benefit and are so well retained. . . . It is that with these, the facts have a connectivity which links them; the truths are related and seem to the mind to form a whole.[58]

About natural history, he went on to say that recently in their courses Bucquet and d'Aubenton "have shown . . . how the man of genius can link up the various branches of that study, and how to bring to it clarity, education, and interest." Geography, he felt, was still awaiting the "Savior" who would recognize, find, or build its potential:

> Geography may also have its own way of offering thought and judgment and a means of pleasing and interesting. It awaits only an able master who can present geography in all of its totality, with all the neatness of which it is capable. May my feeble efforts contribute to finding this![59]

With geography he did not dare to hope for a unity as "all the neatness of which it is capable" was less than the unity to be found in mathematics and more recently in natural history. He dreamed only of a way of rendering geography pleasing and interesting to reflection and judgment.

There was an increasing sense among the authors of the universal geographies that geography—and particularly the type of descriptive geography contained in the universal geographies—was open to criticism. Delisle mentioned the banality of some of his predecessors' tendency to define the obvious:

> I do not think it is necessary to explain here what a mountain, a valley, a forest, a copse, and other such things are, as some have done. These are common notions and those who have not studied geography know it as well as the best geographers.[60]

It was, as we have seen, Mentelle who had the most acute sense of geography's apparent inadequacy. He complained that, as late as 1760, he was one of the very first geographers to speak about the Copernican system. He pointed out that most treatises on the sphere produced by geographers were "almost copies of one another."[61] He was also keenly aware of a certain intellectual territoriality among the sciences which meant that there were "details that belong to another" and that geography's turf was far from evident.[62] What this demonstrates, particularly given the nature of the universal geographies produced by these authors, is not epistemological sensitivity but epistemological angst.

MALTE-BRUN RESUSCITATES THE GENRE

There is little doubt that Malte-Brun was aware that he was writing his *Précis de la géographie universelle* within an established tradition and that he was trying to deal directly with much of the criticism that had been leveled at geography toward the end of the eighteenth century.[63] Geography was superficial, it was badly written, it trespassed on the turf of other fields without significantly contributing to them and without necessarily understanding them and, well . . . it was boring. He referred obliquely to these criticisms throughout his *Précis* while discussing and explaining his methodology. Indeed, the whole work can be understood as an attempt to rejuvenate geography, to rescue its reputation, and to mark off some turf for the endeavor. That, of course, was not its only aim. Together with all of the universal geographies discussed here, Malte-Brun's was concerned also to popularize geography and to provide a geographic education for the educated and literate but not necessarily scientific. He claimed the public was his sole protector and that his aim was "to make geography loved and to spread the taste for it."[64] But it was both the scientific world as well as the public that he wanted to impress.

When one compares Malte-Brun's universal geography to that of Mentelle, or Expilly, or even to that of Varenius, one cannot but be impressed by the scope, ambition, and erudition of his work. The first volume of the *Précis* was a history of geographic thought and exploration (with some smatterings of historical geography) from Moses and Homer to 1809. His second volume introduced and explored all that could be generalized about mathematical, physical, and human geography. His subsequent volumes examined, continent by continent, region by region, country by country, and then province by province, every part of the earth. In these, he began by describing the general physical features of continents and regions, including the major features of their relief, the principal rivers and lakes, their typical vegetation, their climate, the animal life, and the level of civilization of the indigenous peoples. On the scale of the country and province, he looked again at physical features and then focused on the political structure, population, commerce, and way of life of the peoples.

The enormous scope of many universal geographies made a certain superficiality inevitable. Malte-Brun's *Précis* was no exception and Malte-Brun frequently had to end a section or a discussion with "but this task would take us too far. We will limit ourselves to a few features."[65] Nevertheless, in many regards Malte-Brun's work was, certainly within the context of French geography, exceptional. While Mentelle had mentioned the major works used for some portions of his geographies, Malte-Brun religiously cited the works and ideas that he used. His sources included French works but also English, German, Danish, and classical publications. He clearly understood the inner workings of many of the aspects of geography that he sought to explore, including the processes behind mapmaking. His coverage of the natural science literature was broad and rich and he had a keen sense of the key questions being asked in these areas. He maintained a critical and thoughtful approach to the research that he reviewed. And he rightly censored Philippe Buache and his successors for building vain systems and then clinging to them in the face of all evidence to the contrary. His descriptions were lively and entertainingly written and some of his characterizations of governments and peoples were strikingly prescient. It was a remarkable effort demanding considerable stamina and erudition. But was it a successful universal geography? Did it mark off turf that was particular to geography? Did it rescue geography from its critics, or throw the endeavor to them? And finally, what was its long-term impact on the field?

The tradition of geography with which Malte-Brun identified was closely linked to Strabo's geography and, like Strabo, somewhat hostile to the cartographic tradition in geography. Malte-Brun began by attacking the supposed antiquity of the mapping tradition, disengaging it from the descriptive tradition and suggesting the systematic and holistic vision necessary to the universal geography.

> The Egyptians could trace meridians; the regular flooding of the Nile may have rendered the art of topographic plan-making necessary. But this application of geometry does not presuppose geographic ideas in a people who had a horror of the sea and navigation. And the supposed map of Seostris is as problematic as the voyages attributed to that hero. . . . We have to admit that there is no geographic system worthy of attention prior to that of Moses.[66]

The distance Malte-Brun sought to put between geography and mapping was almost certainly due to the prevailing sense that large-scale cartography had become a technology more akin to surveying than to science. Ironically, as we will see, what he did retain from cartography was a remnant of a methodology already recognized as dated by most mathematically based cartographers. More importantly, however, Malte-Brun did not have an alternative conception capable of harnessing the historic scope of the field or able to respond to the challenges posed by recent developments in the sciences and social sciences. Rather, his conception of geography was vague, based on an attachment to the field's long-standing affiliation with history and the humanities, and hostile to the very foundations of modern science and social science.

Malte-Brun did frequently mention the globe in his *Précis*. Nevertheless, it was for him more of an instructional toy or gimmick than an instrument of any real scientific or conceptual value.[67] Cartography which had dominated geography's past was certainly worthy of mention, but it was absent from Malte-Brun's sense of geography's more recent history. Thus, while he recognized that maps had played an important role in geography's history, he substantially wrote the last three Cassinis out of that history.[68] Malte-Brun did not openly deny that mapmakers could be geographers, but he frequently referred to cartographers (who had always been known as "géographes") as "géographes-dessinateurs" or simply "dessinateurs." Despite the fact that he placed mathematical geography (or cartography) at the head of his discussion in volume 2 of *La Théorie général de la géographie,* he did not see large-scale cartography as an important part of the future of geography. In his view, there was simply nothing of interest left for a geographer to do with or in cartography.

> The elegance and exactitude that was so praised in the maps of the Cassinis has been attained by the Russians, the Danes, the Spaniards; but, making great strides, the French engineers surpass the Cassinis every day and leave little hope to those who would, in their turn, try to surpass them [the engineers].[69]

He recognized that other sciences were increasingly seeking to map out phenomena of interest to them, in the form of thematic maps. He did not, however, see this as an opportunity for geography or in any way related to a geographic conceptualization of the world or of particular problems.

> . . . finally, there are few objects which they have not tried to reduce to relations of locality in the form of maps. But the composition of these sorts of tables could not be submitted to constant rules other than those resulting from sciences foreign to geography.[70]

In his view, there was nothing particularly geographic about a thematic map. For Malte-Brun the globe was not a unifying image and cartography did not lie at the heart of the definition of geography.

There was something important in cartography for the future of geography, but this was neither mathematical nor graphic. What Malte-Brun took from and respected in cartography was a methodology that large-scale cartography itself was beginning to discard by the early nineteenth century, and which he used heavily in his universal geography. One could say that it was the central idea behind Malte-Brun's universal geography. This was the critical comparison of multiple sources to arrive at a true picture of the world. For Malte-Brun, as long as the geographer retained this approach, all of the sciences and all knowledge might serve as geography's raw material without leading to confusion between these sciences and geography itself. The critical comparison of bits and pieces of information related to the earth (drawn from any field), and compiled to create an engaging description of that earth, constituted the geographic method.[71] It was this method which set geography off from the sciences and confirmed its attachment to history, as history used much the same method to reconstruct and produce engaging descriptions of the past.[72] This explains, in part, one of the peculiar aspects of Malte-Brun's history of geography: his glorification of d'Anville, the critical comparison geographer par excellence. He drew a line at d'Anville and called all maps prior to his inaccurate and unaccountably attributed a revolution in longitudinal accuracy to d'Anville. He lauded d'Anville's toponymic accuracy and described it, inaccurately, as singular. D'Anville was a critical disciplinary touchstone for Malte-Brun, as in his view it was the critical comparison of sources which made a "dessinateur" or a "copiste" into a geographer. In fact, Malte-Brun's *Précis* can be seen as d'Anvillian geography in textual form. It is in this light that Malte-Brun's only reproach to d'Anville makes a great deal of sense:

> How wonderful if the healthiest critical faculty and a vast erudition had been combined with literary talent, which alone makes people passionate about a science![73]

Had d'Anville expressed his geographic synthesis in literary rather than cartographic form, Malte-Brun was sure that d'Anville's work would have won support and affection for geography.

Theory Rejected

As we will see, Malte-Brun's literary talents and broad knowledge base did revitalize the genre, but his approach to theory marginalized both it and, to the extent that the attitude was generalized, the geographic endeavor in the context of early-nineteenth-century science, which was increasingly theoretical. Malte-Brun was convinced of the intellectual value of physical geography but he was deeply troubled by the pretensions and conclusions of speculative theory, especially in the realm of the physical and natural sciences. His primary objection to such theory was religious and focused on the terror to which speculative theory exposed the human spirit.[74] Malte-Brun was prepared to admit that individual facts had to be preceded and organized by general principles and considered that a combination of such principles would amount to theory. Nevertheless, he categorically rejected explanatory theory in the human, plant and animal, and physical realms in favor of "the purely descriptive approach," "the only really scientific and instructive method."[75] Unlike the natural sciences, a geography that focused on the natural world, or what Malte-Brun referred to as physical geography, could not engage in classification and other subtle and rigorous methods because, "Mountains, valleys, waters, climates, physical regions appear to the eyes of a sincere friend of truth as very complex, very irregular, and easier to depict than to define."[76] In contrast to theory, which Malte-Brun saw as ungrounded, "general principles" were composed by the combination of facts.[77] Thus, while botany could focus on the classification of plants and on their individual detailed description, it was geography's role to combine the description of plants with landform and climate description. This would produce general principles proper to botanical geography. The precise nature of these principles is unclear: Malte-Brun stopped short of identifying botanical regions. Instead he felt that it was safer to subordinate botanical description to the more conventional political-regional structure of his *Précis*.[78] He was clearly only comfortable with a very limited degree of generalization from individual facts. This, in part, reflected his sense of the youth of the field of geography, as this type of work could only be "the work of centuries and of nations."[79]

The role of "general principles" in geography is least comprehensible in Malte-Brun's discussion of what would later be regarded as human geography: the languages, religions, social orders (family, civil society, political society [e.g., monarchy, democracy]) of the world. He warned the reader that in the case of the human world "those principles which, founded in the nature of our being, do not vary with capricious human will, are few in number."[80] Thus, about human matters there was little point in seeking rules or patterns. The generalizations he did engage in quickly landed him in circular reasoning. Thus, it was moral society (by which he meant religion) which "determines the circumscription of the states and empires that political geography is charged with describing."[81] But the "moral state" of a nation was "the result of all those political and social relationships [military force, class, religion, population, wealth . . .] that we have just shown."[82] In the end, Malte-Brun's general remarks about society and its evolution are so few and unconnected one to the other that they never move beyond a few thoughts or observations.

The term "general principles" was absent from Malte-Brun's discussion of physical geography but the message was the same, if more adamantly expressed. For Malte-Brun, physical geography focused on phenomena characterized by regularity and generality.[83] Phenomena with multiple and complex causes, such as rainbows, mirages, and other atmospheric manifestations, were by their very unpredictability not geographic.[84] Regularity and generality could only be determined by observation and description.

What could not be observed was beyond the realm of science. Thus, there really was no point in talking about the nature of the subterranean world and speculating on the existence of caverns: "The unknown, banished from the domain of the sciences, is today considered the exclusive patrimony of novelists."[85] Above all, Malte-Brun sought to distinguish and protect physical geography from the theories that were then driving geology: theories on the origin and evolution of the earth.

> Nothing stops the flight of human curiosity. In vain the earth, the waters, and the airs, in offering us a thousand insoluble difficulties, have reminded us of the impotence of our mind. We only know very imperfectly that which exists around us, and yet we dare to research how all began to exist! . . . what temerity! In the course of this work we have seen that physical geography cannot help but link together frequently occurring facts and to draw general conclusions from these. It is even sometimes forced to present facts in a hypothetical manner because the observers have provided their comments in this form. But physical geography neither adopts nor affirms anything that has not been proven by experience. Geological systems, on the contrary, have as their avowed aim the explanation of the course of unknown revolutions from monuments that are often equivocal. They allow themselves to supplement the silence of the facts with analogies and thus, from hypothesis to hypothesis, they decompose this vast body, as though it were a piece of metal that a chemist might have forged in his crucible. We will prove that that so-called science, or speculative geology, promises no certain results, from the moment it wanders from the path of physical geography.[86]

It was geology's attempt to explain the unknown and the unseen by analogy and hypothesis, in short, by theory, that Malte-Brun rejected consistently and unequivocally.

MALTE-BRUN AS THE NEW MODEL

Whatever its faults, Malte-Brun's universal geography was an important and influential book, certainly within geography. Indeed, it became the point of origin and the model for subsequent universal geographies. Writing in the 1830s, the editor of the publishing house Jules Renouard held up Malte-Brun's *Précis* as a model.

> Geographical treatises abound in France: after Mentelle and Pinkerton, each famous in their own time, came Malte-Brun;[87] and today after Malte-Brun and in large part with the help of the many varied documents gathered together in his *Précis*, a crowd of authors offer the public geographies, in every format and under the most seductive titles, which they claim are novel. . . .

May we repeat the universal complaint that he who gathers all these works into his library is still lacking a geography.[88]

Academic geographers used Malte-Brun's *Précis* to distinguish their work from the many popular world descriptions produced annually.[89] In particular, for some time the basic structure employed by Malte-Brun was followed: a long introductory volume entitled "general principles," followed by a detailed hierarchical description of the world entitled "descriptive part." A history of geography produced approximately fifty years after Malte-Brun's death still borrowed heavily from Malte-Brun[90] and, it appears, attributed a new descriptive geography, a new school of scholarly geography, to Malte-Brun and Carl Ritter.[91] It was Conrad Malte-Brun's admirers and successors at both the Société de géographie and the *Annales des voyages*—in particular Jean-Baptiste Marcellin baron de Bory de Saint-Vincent (who became an admirer later in life), Jean-Jacques-Nicolas Huot, Philippe François de La Renaudière, Adrien Balbi, Théophile Lavallée, Vivien de Saint-Martin, and the son of Conrad Malte-Brun, Victor—who guaranteed the successive republication of Malte-Brun's *Précis de la géographie universelle* well into the last decades of the nineteenth century.

It is unclear how much respect Malte-Brun's *Précis de la géographie universelle* garnered outside this circle of geographers. What did Malte-Brun's disassociation of the universal geography—and indeed of geography— from the creation of knowledge, from explanation, from cause, and from uncertain questing for understanding imply about the place of geography among the sciences? In an era of extensive and high-profile international exploration with the stirring of what has since become known as "discovery science," what place did a geography content to summarize have?[92] Malte-Brun's rejection of theory cast geography in the role of public repository of information on the earth, or handmaiden and gentle introduction to the more systematic sciences. It also left geography with a task—that of cataloging everything known about the earth—which was, as Malte-Brun correctly surmised, "vast enough"—indeed too vast. In addition, who, precisely, would this résumé serve? With general and specialist dictionaries and encyclopedias appearing at a smartly increasing rate and volume, and with specialist journals rapidly replacing the exchange of scholarly letters as a means of communication among scientists and scholars, what scholar or scientist would consult a once- or twice-removed synthesis of scientific information on the earth and its inhabitants composed by a nonspecialist?[93] In spite of his rigor, erudition, and attempt to appeal to a more educated audience, it is unclear whether Malte-Brun's work found academic or intellectual recognition beyond the community of geographers. The comments of a pair of anonymous critics, who were perhaps unusually and partisanly harsh, nevertheless give us a sense of one contemporary interpretation of Malte-Brun's work:

> It is a compilation with all of the force of that term. In any case, there are so many books devoted to that study that in the midst of that mass, and having already previously published a geography in sixteen volumes, M. Malte brun [*sic*] can only transport the materials from one place to another with only the few modifications that that type of work demands.[94]

Harsh though the assessment is, the *Précis de la géographie universelle* was indeed a compilation, like many others, whose principal originality lay in the nature and number of facts collected and the quality of the writing.

WAS A THEORY-RICH UNIVERSAL GEOGRAPHY POSSIBLE?

If we can agree that Malte-Brun's *Précis de la géographie universelle* was more concerned to catalog than to engage the major intellectual problems of his day, we are still left with several questions: was the intellectual softness of his *Précis* an inherent feature of the genre; did it instead reflect the limitations of the scholar; or was it a product of the interaction of the scholar, the genre, and the intellectual context? These are large questions, perhaps ultimately unanswerable. Nevertheless, if we can identify universal geographies which played a more vital intellectual role and isolate and contextualize the source of their originality and power, then at least part of the question may be answerable. Both Le Père Jean François's 1652 *La Science de la géographie* and Alexander von Humboldt's 1849 *Cosmos* were intellectually engaged and highly influential works in their own times. François described his work as a geography and it has the scope of a universal geography. Humboldt's *Cosmos* reached well beyond the relatively narrow contemporary conceptions of geography to encompass all of human understanding. Nor did Humboldt describe his *Cosmos* as a geography: he considered geography an endeavor preoccupied with enumeration.[95] However, he does seem to have linked the study of the cosmos with the study of physical geography. Talking about the enormous scope of his *Cosmos*, from the sidereal to the terrestrial and encompassing their empirical relations, Humboldt commented:

The hitherto undefined idea of a physical geography has thus, by an extended and perhaps too boldly imagined a plan, been comprehended, under the idea of a physical description of the universe, embracing all created things in the regions of space and the earth.[96]

Historians of geography have long treated the *Cosmos* as a work of geography, and particularly as a geographic manifesto.[97] There are two major reasons for this. Cynically, the disciplinary quest for "Great Predecessors" encourages uncritical integration of Humboldt's genius. Arguably too, Humboldt is impossible to ignore precisely because in his *Cosmos* he grappled with a problem still not entirely resolved by geographers today but which plagued geographers of the early nineteenth century: how to marry the new directions taken in theoretical and empirical science with a holistic approach to the earth. This is not to imply equivalence between the unity sought by Humboldt and that assumed by the composers of universal geographies. Humboldt's quest for unity was probably inspired by his association with the German Naturphilosophen, among whom he studied, whom he read, and with whom he corresponded and socialized.[98] Although in Humboldt's case this dimension was tempered by a strong commitment to empirical science, this quest for the unity of nature, including the organic, the inorganic, and the social, intellectual, and perceptual worlds, contradicted the mechanistic and classificatory spirit of Enlightenment science. Humboldt shared the sense expressed by the Romantics and the Naturphilosophen that something was lost in a purely mechanistic approach to the world. There was room for a metaphysics of nature, through a physics of nature.[99] The universal geographers discussed here, with the partial exception of Malte-Brun, simply assumed a unity to the cosmos expressed in its *order*. They did not seek to elucidate this order through observation or experimentation, or through empirical research of any kind, nor did they problematize it. They assumed that the order they imposed on their works would reflect the order of the cosmos, or at least that it would be useful. As the nineteenth century progressed, geographers had increasing difficulty imposing order, structure, and coherence on the diverse realms that had historically fallen within their purview. The fact that, in France in any case,[100] geographers did not turn to Humboldt for inspiration in this regard suggests the fundamental difference in outlook between the both scientific and metaphysical Humboldt and the enumerative descriptive geographers who throughout the nineteenth century seemed preoccupied with classification as a model of scientific research. To confuse Humboldt with these geographers is to fail to discern radically different intellectual traditions. Nevertheless, Humboldt's problem was similar to that faced by French geography as it moved from a context of classical science to modern science. Both Humboldt's *Cosmos* and Le Père François's *La Science de la géographie,* for all of their considerable differences in both form and content, constituted attempts to conceive of the cosmos as an integrated whole, and both authors saw this endeavor as fundamentally linked to, in François's case, geography or, in Humboldt's case, physical geography.

Le Père Jean François's La Science de la géographie

Jean François (1582–1668) was a Jesuit scholar and professor of philosophy and mathematics at the Jesuit college of La Flèche. He was a significant, if now largely forgotten, geographic scholar. In the course of his career, he wrote a treatise on geography, a book on practical mapmaking, a work on hydrography, and a treatise on the sphere. He is perhaps best remembered for having taught René Descartes.[101] The argument here is not that François's geography was modern or in some way correct, but that it was intellectually vital and coherent. His cosmography was Ptolemaic rather than Copernican. His thinking was in part teleological, in that he believed fundamentally that all phenomena reflected the will and nature of God and thus could be seen to have a reason for existing. In that vein, mountain chains had been formed to separate kingdoms, to function as asylums to peoples faced with powerful enemies, to form reservoirs of water in the form of snow, and to stop the clouds, "the vapors, and other smokes which the winds carry with them by making them thicken and resolve into rain."[102] Yet he understood what many nineteenth-century thinkers could not absorb: that it is important not to confuse types of explanation. That is, when talking about God, one may think in terms of miracles, but to say that a natural event is a miracle is not to *explain* it. Thus, concerning the possible miraculous nature of the variety and distribution of beings on the earth, he commented: "But it is not reasonable to put miracles into the effects which can be explained by natural means."[103]

A cursory glance will suggest a very alien—and certainly not modern—strategy behind the book. It begins as a philosophical treatise, examining first principles, the nature of cause, and the meaning of place, and then extends into relatively self-evident enumerations of physical features and national or regional characteristics. Between the two, it is easy for a modern reader to lose the thread of this work. Yet the author's strategy is straightforward and his concern for the reader stretches easily across the centuries to assure him or her that there is indeed a central thread and a strong argument. François undertakes five principal tasks in the book:

1. An explanation of the nature of geography and how it ought to be approached.
2. A discussion of what is necessary to know to study geography.
3. A suggestion of what may be profitably learned from cosmography and geometry.
4. An exploration of the meaning and importance of place in the four geographic divisions which he considers "traditional geography": the conceptual realm, or concepts and their cartographic expression; the natural realm; the civil or human realm; and the solar or celestial realm.
5. Finally, a description of how these various divisions can be depicted.

The most difficult and the most important part of the work is the first, his explanation of the nature of geography. The second part reflects what can be seen as a defining characteristic of universal geographies, a concern to teach and to make geography widely accessible. The third part discusses the properties of a sphere and some geometrical principles valuable in the study of the earth as a globe. The fourth part is what François considers "traditional geography," or the description of the different parts of the earth, which he brings to life through his reconsideration of the nature of geography.[104] The fifth part explains how to depict the earth's divisions cartographically together with a series of cartographic problems and their solutions. These parts do not necessarily correspond to chapters or separate sections of the book. Rather, they emerge as the book is read.

While this description provides a general sense of the form of François's discussion, it does not convey or explain the coherence and vitality of his geography, both now and for his contemporaries. This is because the coherence and vitality are derived not from the structure of the text but from four ideas which François held as self-evident and which infuse the work. The first was the view of geography as contemplative of the globe and thus concerned with the representation of place in all of its divine, natural, and human variety. The second was a strong sense of the connectedness—and the importance of this connectedness—of the various parts of geography, despite the considerable diversity present on the surface of the earth. The third was a conception of geography as a science concerned with explanation, reason, cause, and uncertainty—or the quest for understanding. The fourth was an understanding of the place of geography among the other sciences, such that what geography borrowed or bestowed could be clearly accounted for. The particular form of these ideas may only have been possible in François's time. It is arguable that they, or something very like them, hold the key to the coherence of any universalizing conception of geography.

François both began and ended his book with the contemplation of the globe and its derivative, the map. He opened telling the reader that he composed this work because he was asked by "one who commanded" to make him an "artificial terrestrial globe and a treatise on its properties:"

> I only thought, in order to obey and please him, to make a little book of Geography. But the fecundity of the subject insensibly carried me to compose an ample Cosmography in which I deduce the reasons and the noble effect of the Divine art and the adroitness and exact manner of the most important practices of human artistry.[105]

Unable to resist the richness of the subject, he found himself drawn from the globe itself into a study of the world as made by God and acted in by humans. But the point of his book remained, notwithstanding the explanation of that globe, the image of which he held before the reader throughout the book. François frequently played with the globe as a concept and abstraction on the one hand and as a reality created by God on the other—thus making the globe stand for the earth and the earth stand for the globe.[106] This forced the reader to consider the whole, while François described and explored its parts in conformity with one of the fundamental laws of geography (later cartography):

> We can only arrive at a total knowledge of the Globe through [the knowledge of] its parts of which it is composed and these can only be acquired through observations made on location.[107]

It was the globe and the map which allowed the joint and unified study of the general and the particular through the importance they gave to place (more on this later). Maps and globes were, however, more than abstractions. They were also a graphic form with great explanatory power. He explained this in terms reminiscent of the sensibilists of the late eighteenth century:

> The one who said this said it well: that speech is a speaking painting and that painting is a mute speech because it is through one or the other means that we express the objects of our thought. And, if on a number of subjects, the spoken word has an advantage over painting in making some truths more clearly understood, such as those which deal with spiritual things, it is assured that what touches upon quantity and the subject of mathematics, is better represented by what we have discussed than by word alone. And looking at just one figure which is the perfect image of some body will make us understand it better than would a long speech composed of a multitude of words and periods. The reason for this is drawn from two things: 1. from the object itself which being sensible wants to be represented according to its

own faculty [of sensibility] either immediately and by itself so that it can be understood perfectly or in a mediated manner by something similar to it but that is other than it, 2. on the part of our own faculties whose order is to pass from the senses to the imagination and from this to understanding. It is thus that a perfect image is presented to our senses to be carried to our two other faculties. And if there is some mistake, it comes from the first representation and not the following ones."[108]

Maps and globes were of value because as models they best reflected the nature of the earth and because they offered a more direct path to understanding through the imagination via the senses. At the heart of geography for François lay what was to become known as cartography: it was at once geography's conceptual focus, its method, and closely related to geography's unifying principles of movement and place.

François's sense of the unity of geography and the fundamental connectedness of all of its parts was born of careful thought. He described geography as traditionally concerned with the divisions to be found on the earth—natural, human and celestial—and their depiction. However, he argued that the subject of geography entailed more than that. Geography was composed not just of divisions but of universalities and particularities and it was these and the relations between them that gave geography scientific value.

Universalities included light, warmth, rarefaction, movement, and exchange, either physical or abstract. François provided a detailed discussion of the sources and nature of each of these. Universalities were to be found everywhere in the world and emanated, in the first instance, from the sun but also included human action across space. In one passage he gave the reader a clear idea of the actions of the universalities:

> Light which follows the sun makes the change between night and day: heat which follows light makes the changes of seasons and the benefits that the seasons bring. The rarefactions which follow the heat transform the water into a thousand "meteors" and transport them to a thousand different locations. The movement which comes from rarefaction causes a thousand transports of bodies from one location to another.[109]

Geography, as the study of such universalities, included the consideration of the sun's action on the earth,[110] seasonal variation, plant and animal distribution, meteorology, perhaps hydrography, and commerce and communication.[111]

Particularities were of three sorts: miraculous (or those associated with God acting as God, otherwise known as final causes), natural, and human. The first category truly belonged in the domain of philosophy or, because God created the universe, in the domain of cosmology. The latter two belonged in the domain of a "science humaine."[112] Every place on earth must be either divine, natural, or human and all human events were highly particular, that is, rooted in space and time. In order to understand the causes of these particularities and to see the relations between particularities and universalities, the best focus, in François's view, was place. History and geography met in "place" and place unified man, nature, and God.

This, then, was the science of geography. It found expression in divisions and cartography and relied on these for the organization of thought. These structures and methods were not, however, the whole and it was wholeness that was the aim of geography.[113] Wholeness did not constitute studying the whole of the cosmos. That was to study a particular particularity. Rather wholeness was to move constantly from the particular (such as a city) to the universal (such as its commercial relations with its neighbors) and between particularities (such as between a city and its emplacement), as a cartographer would, in search of the larger picture—in search of understanding and efficient (not final) cause.

The natural assumption of a nineteenth- or twentieth-century geographer, looking at François's preoccupation with cartography and his periodic tendency to enumerate, would be that he conceived of geography as primarily descriptive and concerned with the recording of known facts. Nothing could be further from the truth. Many times in his *La Science de la géographie* François rejected the role that universal geographies were to so readily assume in later centuries: that of recorder of known facts and enumerator of all possible divisions and subdivisions on the earth.[114] Divisions and subdivisions were important. However, geography's purpose was higher than subjects which "more properly belong to memory and the simplest operations of the mind rather than to discussion and reason." As Mentelle later discovered, and as François pointed out, memorization, which "amounts to expressing that which is already known and imprinting the image of things that are already conceived," is a child's science and requires no master to teach it. The real role of geography, for François, was twofold: to prepare the mind to explore—and to that end he saw his text as "freeing the imagination, speeding conceptualization, and firming up the memory"—and to engage in the creation of understanding "via the conquest of hidden truths."[115] Geography, then, was a questing, inquiring endeavor, unshy about open-ended questions. Indeed, his concern was to structure his book so as to encourage such questions. Thus, he decided to separate his discussion of natural phenomena from civil structures as, he felt, this would be most likely to enhance understanding of the relations between physical phenomena (which, he admitted, required a thorough grasp of natural history).[116] Neither did he hesitate to explore such questions himself. In physical geography, he wondered, for example, about where all the animals on faraway islands had come from. In human geography, he pondered the meaning and importance to history and civilization of the first Columbian voyage, and he explored the conditions which render a state flourishing; he even asked what makes a civil entity a unity (a question which in slightly more pointed form is still being asked the world over). He did designate some questions as beyond the purview of geography. Thus, fundamental or final causes (those related to God) were to be explored by philosophy. Similarly, it was preferable not "to dig and search through the bowels of the earth so as to reveal the metals, minerals, and diverse mixtures of substances that the elementary heat has formed in various locations."[117] That activity, a meritorious one too, more correctly belonged to the naturalists. François sought to concentrate instead on "the terrestrial surface which serves as home to men in order to show its diversity of base, matter, fruit, and figure which it has by virtue of nature and art."[118] That left geography more than enough to cope with.

François had a clear sense of the place of geography among the sciences. He described it as one of the subaltern sciences, that is a science which instead of studying principles and objects in themselves, studied the effects of principles and the relations between objects "on the globe of the earth as an artificial work."[119] The study of principles and the nature of objects themselves belonged to other sciences such as physics, astronomy, and geometry and could be studied and borrowed by geographers to better understand the world. But it was the effects of these on the surface of the earth (including all the creatures there living) which constituted geography.

With respect to the other subaltern sciences, François only mentioned history and natural history. Geography, in his view, provided the foundation for history. In this vein he held a very geographic conception of history. History was the account of events carried out in particular places and which were, in part, shaped by places.[120] In addition, history added depth to place and could stimulate memory of place. As such, history and geography were inseparable but radically different in their focus. Of natural history, he said relatively little except that it was subaltern to chemistry and that it was primarily concerned with the classification of plants and animals.

In what was to become a characteristic feature of universal geographies, François also engaged in a little geographic advocacy. Geography, he said, was a noble science because:

> It envisions the earth in a more noble and agreeable fashion than the neighboring sciences focused on the same globe. It [Geography] considers it [the

earth] as the source of all fecundity, as the nurse-mother of all animals, and as the home of Man.[121]

Geography's nobility, then, was a function of its holistic vision of the earth. According to François, it was also a pleasant and agreeable subject thanks to the passion and enjoyment evinced by traveling and discovering the world. It was natural to the human soul, he argued, to find pleasure in such activities. Finally it was an intellectually useful science: useful to history as already described; useful to cosmography, which, with geography, could begin to explore the differences that our globe has with others; and useful in and of itself as "knowledge acquired through very difficult observations."[122] Geography, focused on the map and composed of knowledge both local and universal, was a science at once noble, pleasant, and useful.

François's geography could not be used by geographers today to champion this or that modern school of thought or trend. It is clearly of a different age and assuming correspondences between his and many contemporary labels would carry considerable risk. Perhaps for that reason it is doomed to neglect—at least until geographers begin to rediscover the extraordinary richness of their past. It is striking, however, that in contrast to the vast majority of geographies, universal and otherwise, old and new, the reader who explores *La Science de la géographie* will find in it, more than 340 years after its publication, ideas that will excite and a conception of geography that still has life—despite the impossibility of its existence in the context of modern science and everyday life.

Alexander von Humboldt's Cosmos

Approximately 1,800 years after Strabo, 196 years after Le Père François, and a mere 38 years after the appearance of Malte-Brun's first volume, Alexander von Humboldt (1769–1859) began to publish his *Cosmos*. It was the culmination of a remarkable and extensive career which had spanned and far surpassed Malte-Brun's lifetime. Alexander von Humboldt was a scholar of unusual talent and breadth. Trained for the state bureaucracy, he pursued his interest in both nature and its study from an early age and quit the bureaucracy as soon as the death of his mother gave him the means to do so. In the course of his life he carried out research in subjects that we would today ascribe to linguistics, art history, botany, zoology, history of science, political history, geography, anthropology, languages, physics, meteorology, chemistry, astronomy, and optics. Trained in mining engineering and its associated sciences, which bred in him a respect for empirical research, throughout his life he taught himself the skills he needed to explore questions of interest to him, or he collaborated with scholars, such as Aimé Bonpland, Leopold von Buch, and Georg Forster, through whom he learnt the necessary skills, or he corresponded and exchanged ideas with those able to help him. In addition, for most of his active scientific life Humboldt was independently wealthy and funded his research without institutional or governmental direction. Unsurprisingly, then, Humboldt the polymath can be ascribed no single disciplinary affiliation.

In the course of his lifetime and his research, Humboldt witnessed a radical transformation in the nature of science. In the seventy-odd years beginning with his entry into the realm of empirical scientific research, the natural sciences multiplied, developed distinct methodologies, and carved nature into disciplinary realms. Humboldt played a significant role in the evolution of a number of those disciplines, particularly botany, plant geography, zoology, geology, geodesy, meteorology, and geophysics. After a lifetime of specialist scientific research, writing, and self-education in the natural sciences, Humboldt sought to place his work and thought within a larger intellectual frame. This is not to imply that Humboldt's belief in the unity of nature and his focus on interconnections appeared for the first time in his *Cosmos*. They are already apparent in his work on plant geography, which dates in its earliest form from 1793.[123] His *Cosmos* was in part an attempt to counter the prevailing intellectual fragmentation of research into nature. It was an attempt to return to a holistic approach to science without jettisoning the advances in nature study brought by the evolution of the systematic and theoretical natural sciences. To that end, in his *Cosmos* he sought to reconcile the metaphysical aims of Naturphilosophie which had always guided his research and the more immediate aims of empirical science, to integrate nature and thought about nature. Otherwise put, he sought to link the external natural world with human existence through the human creative imagination. The integration of scientific, humanistic, and artistic thought, even only that focussed on the observation of nature, was almost as problematic then as it is today. However, Humboldt fundamentally rejected the opposition and hierarchization of scientific, humanistic, and artistic endeavors. In the *Cosmos* he sought to reunify human creativity by presenting the current state and historical evolution of scientific, humanistic, and artistic perceptions and representations of nature.

The *Cosmos* differs so radically from the traditional universal geographies that it has never been regarded as belonging to the same genre. Yet it does bear considerable formal resemblance to the universal geographies of his predecessors, including Malte-Brun, and his successors. Like the majority of those publications, it was a monumental, multivolume work of some, in this case, 1,800 pages. Its avowed intent, in keeping with the aims of universal geographies since the time of Strabo, was the "description" of the universe[124] rather than the pursuit of abstract principles or engagement in "mysterious and insoluble problems of origin and existence."[125] Like the works of Delisle, Expilly, and Mentelle, it began by placing the earth within its astronomical context. Like Malte-Brun's *Précis*, it gave great importance and space to the historical evolution of current conceptions of the cosmos. Finally, Humboldt insisted that his *Cosmos* was not to be understood as "a mere encyclopaedic aggregation of the most important and general results that have been collected together from special branches of knowledge."[126] Such individual facts and their mere aggregation,

> are nothing more than the materials for a vast edifice, and their combination cannot constitute the physical history of the world, whose exalted part it is to show the simultaneous action and the connecting links of the forces which pervade the universe. The distribution of organic types in different climates and at different elevations—that is to say, the geography of plants and animals—differs as widely from botany and descriptive zoology as geology does from mineralogy, properly so called. The physical history of the universe must not, therefore, be confounded with the "Encyclopaedias of the Natural Sciences," as they have hitherto been compiled, and whose title is as vague as their limits are ill-defined. In the work before us, partial facts will be considered only in relation to the whole. The higher the point of view, the greater is the necessity for a systematic mode of treating the subject in language at once animated and picturesque.[127]

This, in theory, captures the nature of all universal geographies: historical, descriptive, integrative, and fundamentally spatial. In fact, few universal geographies achieved anything like this sort of synthesis, and the *Cosmos* was in this and other ways distinctly different from the classical universal geographies, and more similar to François's geography. Like François, Humboldt dismissed locational enumeration as of relatively little intellectual interest and, taking this further, even dismissed classification from the purview of his cosmology. Similarly, while Humboldt, like Malte-Brun, believed firmly in the value and power of description, like François, his commitment to theory and experimental science was considerable. Indeed, Humboldt was proposing the reform of contemporary physical geography, which amounted to a casting off of its traditional preoccupations in favor of a radically new rationale, methodology, and place among the "connaissances humaines." Humboldt certainly shared with Malte-Brun an appreciation of literary description, but he took a very critical approach to its role, power, and limitations in the new physical geography that was his *Cosmos*. Finally,

Humboldt and François shared a commitment to the unity of nature, born of religious conviction in François's case and of aesthetic philosophical conviction in Humboldt's case. Indeed, we can see Humboldt's *Cosmos* as the last work for over a century to treat nature, culture, science, and art as an integrated whole. Of all the attempts, Humboldt's was at once perhaps the greatest and the most impossible as he sought to encompass the cosmos from the animate to the inanimate world and from the telescopically far to the microscopically near within a single coherent vision informed by science, literature, and art.

As written, the *Cosmos* is a far more limited work. The first two volumes comprise a discursive introduction on the value and purpose of an inter-disciplinary and unifying study of the cosmos, an overview/general plan of the work, a brief history of the textual description of nature, a brief history of the graphic representation and modeling (as in gardens and gardening) of nature, and a more lengthy history of the scientific contemplation of the cosmos.[128] Volumes 3 and 4 are focussed on the celestial portion of the cosmos: stars and the theoretical and practical problems of their observation, and the region of the solar system. Volume 5 returns to earth to examine its physical nature from the size, form, and density of the earth to its internal heat, to its magnetic activity, to earthquakes, thermal springs, and finally volcanoes. That the *Cosmos* is an incomplete work is clear from a comparison of the subjects covered in the overview in volume 1 and the structure of the final work. For the most part, the published work follows the plan outlined in the overview but abruptly ends two-thirds of the way through the outlined material. Planned for in the overview but absent in the published work are a detailed discussion of paleontology, meteorology, climatology, organic life, motion in plants, the universality of animal life, the geography of plants and animals, floras of different countries, "man," races, and language. The *Cosmos* then, as written, is very different from Humboldt's intended *Cosmos* and the *Cosmos* discussed in his first two volumes. The intended *Cosmos* was to have the breadth typical of a universal geography and reach from the astronomical to the physical to the human. With Humboldt's commitment to both modern science and the humanities, there was reason to hope for even more: a truly integrative study of the cosmos with humanity—as perceiver, admirer, and imitator of nature—at its very center.

In his *Cosmos*, Humboldt drew a sharp distinction between geography and the new science advanced in the *Cosmos*, physical geography. The geography with which he associated place-name and physical feature enu-meration was, like many "branches of empirical knowledge," burdened by the weight of its past and in need of reform or replacement by a more modern conception.[129] Far from rejecting all intellectual boundaries and structures, Humboldt was just as concerned as Malte-Brun had been with issues of intellectual turf and status. Instead of defending an old field such as geography or one of the new systematic sciences, Humboldt, here again inspired by the Naturphilosophen,[130] was really proposing a radical reformulation of the fields recognized and named since antiquity to erect boundaries around his unifying study of nature. This new study would be devoted to interrelations and connections rather than to enumeration, classification,[131] or the character of phenomena "independent of geographical relations of space."[132] At the core of this new study was something akin to Carl Ritter's comparative method,[133] focussed not so much on particular physical features or human structures but on the relationship between human understanding of nature and nature itself.[134]

Humboldt believed as strongly in the importance of description to his new science as Malte-Brun, Strabo, or François had been convinced of its centrality to geography.[135] Good literary description precluded verbosity, excessive flourishes, or pastoral romance but emanated from culturally informed, attentive observation of nature. For Humboldt, then, good literary description was itself scientific and he did not oppose it to theory, explanation, or observation. Perhaps in response to the influence of the early positivists, he saw less opposition between description and theory than between pure empiricism and theoretical science. It was pure empiricism that was antithetical to the aims of his *Cosmos:* its unreflective and uncritical approach to nature would mislead.[136] The *Cosmos* embodied a philosophy of nature and nothing could replace a theoretical/philosophical approach to nature.[137] As for François, a good scientific description had to be informed by sound investigation of particularities in conjunction with contemplation of universalities (although clearly their definitions of universalities and particularities differed). Due, in part, to developments in the theoretical sciences since Malte-Brun's time, Humboldt was far less worried by the dangers of speculation than had been Malte-Brun. He held geology in high regard precisely for its "solid foundation of scientific deduction."[138] Speaking about contemplation of language, race, and descent, a decidedly more risky realm of speculation, he commented: "here, as in all domains of ideal speculation, the dangers of deception are closely linked to the rich and certain profit to be derived."[139]

The central concept behind Humboldt's *Cosmos* was the unity of nature—an argument also made by François. Most other authors of universal geographies took that unity as self-evident, or addressed it only through the structure of their publications. Humboldt sought to demonstrate the physical and ideational links between phenomena of all kinds. For Humboldt nature included the organic, inorganic, and ideal realms within both the celestial and the telluric realms.[140] He believed that the human spirit was naturally drawn to a unified vision of nature as a result of a tendency in the human imagination to search for "a certain secret analogy"[141] linking all things. Humboldt saw this tendency as reinforced by the very character of landscape, "since the character of the landscape, and of every imposing scene in nature, depends so materially upon the mutual relation of the ideas and sentiments simultaneously excited in the mind of the observer."[142] He harked back to periods during which separation between the human and physical realms had hardly existed: in the histories of the ancient Greeks[143] and in medieval travel accounts.[144] But he recognized that in the modern world a holistic vision that was not to be "vain illusion"[145] must be based on the solid foundation of the specialized observational and experimental sciences. This was because "the powerful effect of nature springs, as it were, from the connection and unity of the impressions and emotions produced; and we can only trace their different sources by analysing the individuality of objects, and the diversity of forces."[146] The unity of nature, then, lay as much in the human mind as in nature itself. Thus, to study nature was to study the human mind and to study art, science, and literature was to study nature. Humboldt saw little difference between the two foci: for the aim of both the human and physical sciences was the investigation and understanding of cause and process.[147] This is not to say that Humboldt recognized no differences between the sciences. Indeed he seems to have conceived of a hierarchy of sciences with the experimental sciences at the base, the science devoted to the description of the universe (and dependent on the experimental sciences) at the next layer up and something he called "higher speculative views"[148] at the pinnacle. Humboldt said little about the latter except that it dealt with questions beyond the realm of the experimental and descriptive sciences such as the origins of life. It is this unifying vision of both nature and its study which has so drawn geographers to Humboldt's *Cosmos* and which so contrasts with the development of the natural sciences until the rise of ecology. The unity posited by Humboldt and eloquently argued in his first two volumes is less evident in the last three volumes of the work and suffered as a result of the incomplete nature of the publication. Humboldt recognized that the very enormity of his proposed study of nature meant that no narrative no matter how long could ever be complete, but the *Cosmos* is certainly more incomplete than Humboldt had intended.[149]

CONCLUSION

A close reading of Humboldt's work, both his *Cosmos* and earlier work resulting from his expedition to the Americas, reveals the overwhelming influence of Romanticism and Naturphilosophie on his thought. Humboldt shared with the Naturphilosophen a belief in the unity of "man," nature, the intellectual endeavor to understand nature that is science, the spiritual endeavor to understand nature that is art, and the history of all of these. His *Cosmos* especially was an attempt to express and explore that unity. The nature and flavor of this unity is very different from the holism embodied in the universal geographies of Malte-Brun and his eighteenth-century predecessors. These geographers engaged in a descriptive geography analogous to descriptive natural history, though with a far less delimited sphere to describe and without the internal logic of Linnaeus's system or that of its successors. Concerned to assimilate established knowledge about the earth and to use it to arrive at as complete, as methodical, as systematic, and as accurate a description of the earth as possible, since the time of François, their work seems to have lost any higher or spiritual purpose. Humboldt's thought was leavened by metaphysics, by his almost religious belief in the unity of nature, culture, science, and art. In a way, Humboldt was very like the French geographers in that he sought to describe the cosmos and thus expressed his belief in its unity. Yet Humboldt's concept of unity was not an unstated assumption. It was a profound and frequently expressed belief that the cosmos was governed by natural laws acting in all spheres. This search for natural laws motivated his exploration of the cosmos through travel, observation, experimentation, specimen collection and comparison, mapping, study, and contemplation. Humboldt interests many modern geographers because his study of the cosmos suggests the possibility of a reconciliation between geography's holism on the one hand and the more specialized empirical sciences that have now held center stage for over 150 years on the other. It is interesting and significant that such a possibility did not occur to Humboldt's French geographic contemporaries, probably because his metaphysical expression of the unity of the cosmos, the essence of his approach, was so alien to them that they simply ignored it.

In many ways, then, Humboldt's *Cosmos* represents a major break with the tradition of universal geographies as written since the time of Strabo. Its stated aim was not political utility. Indeed, Humboldt was one of the few scholars of his age who could afford to substantially ignore such questions.[150] In contrast to the geographies of Strabo, Delisle, Expilly, Mentelle, and Malte-Brun, Humboldt's physical geography openly embraced the study of cause, the examination of the unknown, and the focus on detail typical of the empirical sciences. In an era in which description unlinked to either empirical research or theory had ceased to find favor, Humboldt sought to reintegrate description, empirical work, and theory about the cosmos. Although the contexts were radically different and François was not faced with the same volume of information or problems of interdisciplinary integration, their solution to the problem of writing a universal geography was remarkably similar. Both François and Humboldt took the unity of the globe or the cosmos as the fundamental point of demonstration. François achieved this by focussing on the globe, Humboldt by concentrating on the creations of the human mind. Both argued the importance of moving from a close study of the parts to an apprehension of the whole. In so doing, François mimicked the cartographic procedure while Humboldt sought to contextualize, catalog, and update a career of work in empirical research. Both warmly embraced the intellectual challenges of their age according to their lights, but each sat on the cusp of a new conception of the physical world which would date their work: for François it was the Copernican Revolution (already long discovered but still not absorbed), for Humboldt it was evolutionary thought. Both made strong statements about the role of their field of study in the "connaissances humaines." Neither work radically altered the way in which universal geographies were written, but both, for their originality and prescience, have increasingly drawn the attention of scholars and historians of geography. In particular, Humboldt's work, in spite of its very incomplete discussion of the organic realm, has served as a touchstone for late-twentieth-century ecologically inclined geographers. There is little doubt, however, that the most remarkable of the universal geographies discussed here was Le Père François's—for both its coherence and its elegance of vision.

It was, then, possible to write an intellectually engaged universal geography in the mid-seventeenth century without making it a life's work. It is hard, though, to imagine what conception of nature would have given Humboldt both the unity and detailed understanding he sought in all realms of nature and human activity. However, it is important to remember that our understanding is the product of the accumulation of generations of research into the specialist sciences, social sciences, and humanities. Consequently, the occasional unifying visions we periodically embrace, such as the Darwinian vision of nature and society and perhaps most recently the ecological view of nature and society, are always only partial. We readily accept that fact. Perhaps, we do not think in wholes anymore. Was, then, Humboldt's attempt doomed to failure or is that a flawed conceptualization of history? And what of the more traditional universal geographies? In a sense they served precisely their declared purpose: general education and the conveying of certain knowledge. However, they could do little in an age of increasingly rigorous theory-driven empirical research to win respect and attention from the larger scientific community. One traditional solution to geography's loss of direction and status was clearly a dead end or at the very least an intellectual backwater.

FIG. 4. "The monumental importance of Egypt." In their conquest of Egypt, French scholars and military leaders hoped to marry the art and architecture of ancient Egypt with French modern science. Capturing the monuments of Egypt, both by mapping them and by bringing some home, seemed the best way of integrating inscrutable ancient Egypt into France.

The Powerful Mapping Metaphor

Decipherment versus description: the debate poisoned the adolescent years of Egyptology. —Jean Lacouture, *Champollion, une vie de lumières* (1988), 325

The extraordinary mapping associated with the Revolutionary and Napoleonic wars effected a transformation in the status of the map over a relatively short period of time. In that period the large-scale map took major strides in its long descent from an instrument of kings to a common document. The implications for the broader society were profound but have never been studied. There is no question that increased access to and familiarity with maps and mapping enhanced military, local, and state administration. It may also have changed the way in which the state was conceptualized. Much less importantly for the larger society, but most importantly for geography, the increased access to maps also tended to banalize the map.[1] As has been described, this had consequences for the discipline, which for centuries had been closely and fruitfully associated with mapping. By the early nineteenth century a number of geographers were prepared to accept that the future of the intellectual endeavor of geography did not lie with the map. Many more geographers, however, were not prepared to abandon the map altogether. Instead, most sought to shift their affiliation to a more metaphorical map in which the aim of research was not so much the creation or production of a map per se, but addition to the map of the world in a more abstract sense. Geography's long flirtation in the course of the nineteenth century with exploration, the "filling in of the map," the history of discovery and exploration, and the growth of geographic knowledge is the clearest and most sustained manifestation of this new conception of geography. Beginning in earnest in the 1820s, this conception of the discipline rapidly took hold and for a time, at least, seemed to offer continuity, status, structure, and endless possibilities for action, contemplation, and growth. This generated a large literature surrounding the theme of the history of the exploration and mapping of the world by Europeans, to which most of the countries of Europe seem to have contributed. The most significant nineteenth-century geographers writing in this vein came from Spain, France, Italy, Portugal, Germany, and Britain, in approximately that order. Among the French scholars of note were Marie-Armand-Pascal d'Avezac de Castera-Macaya, Casimir Delamarre, Ludovic Drapeyron, Jules Théodore Ernest Hamy, Gabriel Marcel, Charles de la Roncière, and Baron Charles Athanase Walckenaer. Some of the most influential scholars working in this genre were not French but were deemed authorities by French geographers, including C. Raymond Beazley, Henry Harisse, Konrad Kretschmer, Joachim Lelewel, Martín Fernández Navarette, A. E. Nordenskiold, Sophus Ruge, and Viscount Manuel Francisco de Santarem. These scholars, through their interest in the history of exploration and its documentation in mapping, essentially launched the history of cartography as a viable subfield.[2] In the midst of this very general movement, a few geographers, especially those trained as large-scale mappers, clung more tenaciously to mapping as a more immediate aim, warning that divergence from the true path held great danger for geography. The most influential of these confirmed carto-geographers was Edme-François Jomard [1777–1862].

It was in the early 1820s that a group of eight colleagues gathered to found a geographical society. Only five or six of these men would have described themselves as geographers, including Edme-François Jomard, Conrad Malte-Brun, Jean-Denis Barbié de Bocage, Élisabeth-Paul-Édouard de Rossel, Jean-Antoine Letronne, and Baron Walckenaer. Malte-Brun, the lowest ranking of the eight, by virtue of being the only nonmember of the Académie des inscriptions, may, nevertheless, have been the most important founder. In 1807 he had founded a journal devoted to geographic research, the *Annales des voyages*. Arguably this journal, dedicated to the publication of geographic research and the results of voyages, served as the model for the Société de géographie. It was this group of eight, then, who founded the first geographical society in Europe and through the multidiscursivity and governmental contacts of their larger membership established the dominant focus and tone of French geography into the 1870s.

The rules governing the Society were outlined in the first volume of its *Bulletin*, published in 1822. According to the first article of the Society's stated aims, the Society was being founded to advance the progress of geographic knowledge, to encourage geographic voyages, to award prizes and perks for the best discoveries, to maintain a correspondence and contacts with voyagers and geographers, to publish their accounts, and to produce maps [les faire graver].[3] The making of maps was, thus, mentioned in the first article of the Society's regulations but the mention was muted. The Society was not interested in surveying or compiling maps or even in directing their compilation but merely in producing them. Most of the founders saw the Society as a sort of world repository for those interested in improving the metaphorical map of the world.

If anything, the relative maplessness of the Society increased with time. The *Bulletin* produced or reproduced few maps of its own and while it reviewed maps from time to time, this was a declining activity executed in a desultory fashion. Within a remarkably short period of time maps had become just one of the many tools that a geographer could use and by no means one of the most common or crucial. As secretary-general of the Society from 1827 to 1830, La Renaudière stridently advocated synthetic textual description over any other form of expression. In his view a rich description which took into account the nature of the terrain, the agricultural systems, the reasons for the wealth of the nation, and the characteristics of the races it contained would survive the ages, whereas, "It is in the nature of scientific works that they age rapidly. Statistics age even more quickly. Within a very short time of being written they are no longer accurate." His view of the geography of his predecessors, dominated by the cartographic approach, was unequivocal: "The description of terrain is a vulgar work if it is restricted only to the classification of observed facts."[4]

Perhaps most impressive are the instances where problems identified by the Society which were essentially locational and cartographic were not treated as though they were. A striking case of this is to be found in the 1830 issue of the Société de géographie's publication *Recueil de voyages et de mémoires*. This issue was devoted to a prizewinning essay on the orography of Europe (the description of land forms including mountain, valley, and river systems together with the provision of spot heights). The competition organizers had specified that competitors should present their entries in textual and tabular form. Consequently, the winning essay was composed of texts and tables of extraordinary complexity describing spatial information far more easily expressed and grasped in cartographic form. The examiners commented on the uselessness of the small-scale map that the author had included for those uncertain of the location of the Pyrenees and the Alps but did not lament the absence of large-scale cartography.[5]

The clearest expression of the alienation of cartography and cartographers from the Society is to be found in its membership trends. The engineers and graduates of the École polytechnique and the École des ponts et chaussées constituted a significant proportion of the Society's membership in the 1820s. By 1852 they had left the Society together with most of the mapmakers and engravers.[6] Through the 1830s the *Bulletin* received a number of letters from Coraboeuf, a geodesist (and a former member of the expedition to Egypt), on the mapping of the Pyrenees and the solution of a variety of geodetic and topographic mapping problems. However, such communications were rare and the Society does not appear to have followed the many mapping ventures of the nineteenth century at all closely. Pierre Jacotin had been a member of the Society since its inception. As described in chapter 2, he had had an illustrious cartographic career that had taken him to Corsica, Egypt, and Spain and had made him one of those responsible for the conceptualization and design of the new 1:80,000 Carte de France. When he died in 1827, he was described in the secretary-general's eulogy not as a important geographer but as one of those "useful scholars whose entire life was devoted to the exploration of a special branch of human knowledge."[7]

The two historians who have studied the history of the Société de géographie have done so from a social-contextual point of view.[8] This has meant that their histories, however commendable in other regards, have tended to ignore the very point of the Society, that is geography and its changing nature. Lejeune recounts an early dispute between the founding members of the Society which he traces by Jomard's sudden and fairly prolonged absence from the early meetings of the Society. Both he and Fierro attribute this apparent row to a difference of age between the men and Jomard's refusal to be marginalized as the youngest member of the group.[9] However, at the Society's founding, Jomard was already in his mid-forties and an established scholar by all possible criteria. Only Barbié du Bocage and Rossel were significantly older than he. Letronne was ten years younger. Malte-Brun was only two years older than Jomard. In fact, logically, age difference is unlikely to have been an issue, except in a case of profound difference of opinion. It is probable that if there was indeed a row or difference of view between the founders, with Jomard alone in his own camp, the cause was not an issue of age or rank but an issue of content or principle of some sort. The documents are silent on the subject, but the odds are, as I hope to suggest, that they revolved around the role that the map and mapping were to assume in the new geographical society.

Jomard is one of the most fascinating and yet disappointing geographers of the early nineteenth century. He lived eighty-five years and his experience spanned the ancien régime, the Revolution and its regimes, the First Empire, the Consulat, the Restoration (both soft and hard versions), the July Monarchy, the Second Republic, and the Second Empire. The early nineteenth century was not only a period of political and social experimentation in France, it was also a time of major developments in the life sciences, mathematics, physics, astronomy, and the then still nascent social sciences. Jomard lived in that context and took a major part in a wide variety of social and scientific activities. He was in the first class of students produced by the famous École polytechnique together with André Jean François Marie Brochant de Villiers, Louis Benjamin Francoeur, Etienne Louis Malus, Antoine-Léonard Chezy, Etienne Augustin de Wailly, Joseph Michel Dutens, Gilbert-Joseph-Gaspard Comte de Chabrol de Volvic, Tupinier, Pons Joseph Bernard, Louis Claire de Beaupoil Comte de Saint-Aulaire, Louis Poinsot, Jean Baptiste Biot, and Baron Charles Athanase Walckenaer.[10] The École polytechnique embraced both science and technology, and through its instructors and its students, in particular Auguste Comte and Barthélemy Prosper Enfantin, it ultimately elevated engineering and the engineering mentality to the status of an ideology to be embraced at home as a religion and exported abroad as enlightenment. Jomard was by no means the least of the products of this education. As a graduate of the school, he was among those selected by Monge and Napoleon for the expedition to Egypt, on which he accompanied Napoleon and 150 or so scholars. There he served as a geographer producing maps, exploring ruins, and writing reports on the state of this or that aspect of the country.

Upon his return from Egypt, he traveled to London, both to assist in the negotiations for the recuperation of the antiquities lost to General Sir J. Hely-Hutchinson, and to copy the Rosetta stone. He thus played an instrumental role in the decipherment of the hieroglyphs—but more about that later. He was then sent, as a topographical engineer, to the army of Bavaria. In 1803, as a result of the death of two other editors, Nicolas-Jacques Conté and Michel-Ange Lancret, he was recalled from the army to assume the title of secretary and then president of the Commission charged with editing and publishing the monumental twenty-two-volume *Description de l'Égypte*. He served as president of the Commission from 1807 to 1822, or from the age of thirty to forty-five. As already described, the expedition and Jomard's publication of it served to reawaken French interest in Egypt, in Egyptology, or otherwise put, in the place of ancient Egypt in the French intellectual (artistic, religious, and scientific) consciousness. The expedition and the *Description* reawakened interest in the entire Middle East, which became, and remains, one of France's principal global preoccupations.

A member of the Institut d'Égypte in his early twenties, and as a member of the Académie des inscriptions in his early forties, Jomard was an increasingly powerful and important man. His education in the École polytechnique and his experience of the Egyptian expedition had given him access to some of the best scientists and most powerful figures in France. His work in Egypt and with the Commission gave him an experience and insight that few contemporaries could boast. But above all, Jomard was every inch a geographer. Not only was he one of the principal founders of the Paris Geographical Society but he served as its president once, its vice-president twenty-six times, and as president of its central commission thirteen times.[11] Toward the end of his life, as geography turned more and more toward the study of peoples and social phenomena, he was also one of the founders of the Ethnographic Society of Paris, and served as its president the year of his death. As part of his work for the Paris Geographical Society, he became something of an official government-sponsored director of geographic research and exploration. In this capacity, he wrote letters to most of the explorers who sought any contact with the government of France in the first half of the nineteenth century. It was with, and often as a result of, Jomard's support that these explorers received financial assistance for their travels and found publishers and a network of intellectual contacts upon their return. Most geographers of the period had something of a preoccupation with education and in particular with primary education and the role that geography could play in forming the understanding of children—very much a continuing echo of the concerns of the Enlightenment and the Ideologues. Jomard spent much of his life fighting for the universalization of primary education. He also believed firmly in the benefit that French scientific and technical education could bestow on "less civilized" nations and advocated the establishment of French schools abroad and the transfer of foreign students to France for higher education—a tradition that continues in modified form today. Finally, as a geographer, he founded and headed the Département des cartes et plans, at the Royal Library in Paris. This establishment gave maps a status as historical records and national treasures on a par with books, medals, manuscripts, paintings, and prints, and thus led to the collection and preservation of an extraordinary number of cartographic treasures. Jomard, then, was an important geographer in his own time, and many of the effects and consequences of his efforts are still felt today—both in France and within the discipline of geography.

Jomard also closely identified himself with Egypt and Egyptology. One of his portraits described him as "Jomard l'orientaliste." Arguably, it was his Egyptian experience that gave him credibility in a far wider arena encompassing antiquities studies, mapping, and travel not just in Egypt or the Middle East but even in the Americas. He continued to publish on both ancient Egypt and matters concerning the relations between France and modern Egypt throughout his life. If the self-identification as an orientalist was strong, it was also acknowledged, if somewhat negatively, by contemporaries. The physical geographer Élie de Beaumont, author with Dufrésnoy of *La Carte géologique de la France,* referred somewhat disparagingly to him as "Jomard bey."[12] Jean-François Champollion, arguably *the* Egyptologist of the nineteenth century, may have better captured some of the spirit of the man. In response to Jomard's successful exclusion of Champollion's elder brother from the Académie des inscriptions, he described Jomard's tactics (and incidentally Jomard himself) as "the little hypogeomicroscopic ruse."[13] As we will see, this expression captures admirably Jomard's very particular and peculiarly geographic approach to Egyptology.

THE NATURE OF JOMARD'S GEOGRAPHIC IMAGINATION

Jomard represents the most conservative response to geography's crisis of identity. His understanding of geography started, culminated, and ended with maps. His first task as a working adult was the production of maps in Egypt and his last publication was a facsimile atlas of old maps: the *Monuments de la géographie.* Few of his many publications wandered very far from the subject of cartography. His understanding of cartography, however, went far beyond a graphic representation of topography. For Jomard, geography was not only the finished map but a research procedure. The geographic method entailed the ordered laying out and measurement of phenomena and a search for understanding through the spatial information and patterns that this created. This method could be applied to virtually everything in one way or another. For Jomard, that is what it meant to be a *modern* geographer. The modern geographer was not merely a mapmaker, someone who produced national, regional, or local surveys, but a scholar who applied the geographer's scientific methodologies to the problems of the day. Geographic method was to be applied to problems as wide ranging as the nature, chronology, and language of ancient Egypt, to the establishment of a way of studying the diverse peoples of the earth. His approach was modern, in Jomard's view, because it reflected the methodologies developed through some of the most respected sciences of his day, the natural sciences.

A search for understanding through the examination of spatial patterns has a decidedly modern ring to it. When we as social scientists talk about spatial distributions and spatial patterns, there is behind this discussion an awareness of, and a curiosity about, the complex and constructed nature of society and a conviction that exploring its spatial patterns can reveal formerly hidden dimensions of social existence. Therein lies the modernity of the concept of spatial analysis within the context of the social sciences.

For the most part, Jomard's spatial focus was far from modern. Indeed, it is characterized by the application of old geographic methodologies to new problems not readily subject to analysis by those means. There is in Jomard's writings one important exception to this general observation, his *Comparaison de plusieurs années d'observations faites sur la population française*.[14] As already described, Jomard was interested in primary education. In addition, as a former Polytechnicien who had worked with the ingénieurs-géographes (who had been ordered to collect such statistics), and as a colleague of both Jean-Baptiste Fourier and Comte Gilbert-Joseph-Gaspard Chabrol de Volvic, Jomard had some interest in social statistics. In 1827 he responded to a memoir written by Louis-François Benoiston de Châteauneuf in which the author had wondered aloud whether there might be a relationship between criminality or the incidence of criminal behavior and ignorance.[15] Jomard argued that the question raised by this author was unanswerable as so many other factors could be at play: domestic conflict, impropriety, the number of illegitimate children, the incidence of gambling, etc.[16] However, in 1832, based on statistics provided by the military and the judiciary, and those gathered on primary education, he concluded that, given the small number and longitudinally limited nature of the statistics collected to date, there was indeed some sign of a correlation between a lower incidence of criminality and primary education in both men and women.[17] As a geographer, and taking a leaf out the book of Charles Dupin, who had produced a map of France relating education and productivity in the different provinces of France,[18] he took the problem a step further and argued, characteristically, that the problem should be mapped. He proposed but did not execute a sort of a cartogram comparing by province population size, the number of delinquents, and the size of the population benefiting from primary education. The cartogram would retain approximative relative location only, and abandon both absolute location and topography. In a sense, the idea is strikingly modern, but if we think carefully about the proposed map and look at his description of the relationship of social statistics to geography we discover that Jomard had not really applied geography or a geographic methodology to a social problem. The map was not really part of the analysis. He was not proposing a spatial analysis of social behavior but the straightforward depiction of results on a map. The map was the form to which he was drawn, but geography was not part of the analysis. It was no more than a backdrop.

> With respect to the territorial surface, it is clear that it has no relationship to the moral and statistical problem which involves only population, that is those living and moving masses, as they are to be found in a variety of social

conditions, or at different stages of life. Still, geographic position should be considered from another point of view: that is the, proximity or distance of places is not without influence on people in the margins of regions.[19]

This is not to say that this exploration of the possibilities of statistical mapping was unimportant, but simply that its relationship to modern social scientific geography is limited. It was in this realm of social statistics that Jomard's mapping methodology has the most modern and intellectually convincing ring to it. It is worth looking critically at how he applied his mapping methodology to a variety of problems throughout his career.

DECIPHERING EGYPT

Much earlier in his career, when Napoleon, his forces, and the scholars selected to accompany Napoleon (including Jomard) invaded Egypt, in 1798, remarkably little was known about the language, culture, and history of ancient Egypt. The hieroglyphs were undecipherable and, naturally enough, there seemed to be little memory among the local peoples as to the meaning of the ruins amidst which they lived. The intellectuals of the expedition were overwhelmed by the monumental nature of the architecture, by the strangeness of the images, in short by the suggested wealth of the culture from which they found themselves barred. Unable to tell a 2,000-year-old building from 5,000-year-old remains, or a 1,600-year-old zodiacal representation from one of 5,000 years, they struggled to develop ways of interpreting and understanding what stood before them and of somehow fitting it into their knowledge of ancient history. Their sense of astonishment, awe, excitement, and anxious fear of missing something important are palpable in the sketches depicting their work to be found in the *Description de l'Égypte*. All the scholars tended to use the procedures and insight that their various educations had bestowed upon them. Fourier, for example, who was a mathematician, sought insight into the six zodiacal depictions he found on the ceiling of temples not far from Thebes through astronomical analysis.[20] He argued that one might be able to use ancient zodiacs as a calendar if they indicated clearly and incontestably the relationship of the signs with the solstices and the equinoxes. Sadly they did not.

Jomard principally sought to use the traditional cartographic methodologies of geography to break the Egyptian code: measurement and erudition. The temples and ruins of Egypt could be made comprehensible and reasonable if one could but understand the scientific knowledge that was their foundation. One might understand Egyptian science perhaps precisely through the science that was quintessentially, Jomard believed, theirs: geometry—or what he called "geometry," that is measurement and mapping.[21] Looking at the monuments surrounding him and increasingly convinced of the numerical and scientific sophistication of the ancient Egyptians and perhaps projecting his own fascination with "geometry," he became convinced that there was both a graphic and a numerical key to their interpretation. The regularity and monumentality of the pyramids alone suggested that. He was convinced that the Egyptians had been innately geometrical and that a rational system lay behind all of the ruins of Egypt. He further believed and argued that this system was decipherable and would give considerable insight into ancient Egypt and the influence of ancient Egypt on the modern world, if one but knew how to study it. So, together with many of the other members of the scientific commission, and particularly the other engineers and geographers, he laboriously measured and mapped the ruins of Egypt. This had a double function: it was integral to the analytic and descriptive methodology of a geographer, and, should analysis fail, Jomard believed that a topographic map, like any such map, would record *the truth* for analysis later perhaps from a different perspective.[22] In short, the engineers/geographers/scholars were also trying to capture and pocket ancient Egypt so that they could go home and think about it.

It was upon the basis of this measurement that Jomard later constructed an elaborate argument, supported by liberal interpretations of classical and some Arab sources (including Herodotus, Diodorus of Sicily, Strabo, Artemidorus of Ephesus in Strabo, Strabo and the periplus of the Erythrean Sea, Eratosthenes in Strabo, Hipparchus in Strabo, Aristides, the Antonin Itinerary, and Pliny among the ancient sources; and Abou-1-Farage, Abd el-Latyf, Mohalli, Joseph be Altiphasi, and Ebn Salamas among the Arab sources). The argument was that the ancient Egyptians had had a universal system of "natural" measurement not all that unlike the metric system of Revolutionary France, with the suggestion that it had even been perhaps more accurate than the modern French system.[23] This suggested that Egypt might be the cradle of Western and particularly French (via Greek and Roman) scientific culture.[24]

The modern reader of Jomard's "Mémoire sur le système métrique des anciens Égyptiens . . ." is immediately overwhelmed by the seemingly meaningless multiplication of measurements and detailed comparison of the writings of ancient sources and awed by the fastidiousness of the research. The memoir is not meaningless when taken on its own terms. Jomard fundamentally believed that insight into ancient Egypt would come from mapping its monuments. To that end, he employed both modern

methods of geographic research available to him: on-the-ground measurement with the best available instrumentation and the more venera-
ble and tried method, relatively critical consultation of the accounts of
the ancients.[25]

The results were extraordinarily erudite and admirable for their basis in
terrain measurement and field research, as Jomard was to frequently remind
his critics and colleagues, but nevertheless absurd. The methodology of
mapping simply did not match the problem, which at heart was the decipherment of the hieroglyphs. Jomard's memoir was published in 1817 and
was not vigorously attacked and discredited until 1823.[26] There is no sign
in Jomard's writings after 1823 that he ever accepted that his method and
approach had been in any way inappropriate to the problem. He published
an article in 1846 in which he discussed a new method of making copies
of ancient inscriptions (essentially by taking rubbings) and lamented the
fact that this technique for what still amounted to mapping inscriptions
had been unknown to him at the time of the expedition to Egypt. It began
to approach an admission that his maps and sketches of monuments were
not sufficiently accurate to serve the needs of modern Egyptology, but it
did not recognize that research into the hieroglyphs had taken a whole
new direction.[27]

CARTOGRAPHY AND "EXPLORATION AND DISCOVERY"

Much more appropriate to the problem at hand was his cartographic
approach to the direction of overseas, and particularly African, exploration
in the 1820s, the 1830s, and particularly the 1840s. Here Jomard played the
role of a director of research, both providing and requesting information.
In the case of instructions provided to M. Panet for his trip from Senegal
to Algeria, he began by creating a sort of a written map of the route to be
followed which was, in part, an annotation or criticism of existing maps.
According to this, and in addition to the route to be followed, M. Panet
was informed as to where and how he should select his guides; the type of
topographic information he should record; what he could expect in terms
of density of population in particular areas; what areas and peoples to avoid
as needlessly risky; how to dress and what demeanor to assume in particular
areas; places he should avoid tarrying for too long to avoid drawing suspicion
and hostility; what towns to visit and what he might expect to find there;
possible side trips to take; and, perhaps most importantly, the documents,

publications, and experts to consult before leaving. Here, clearly, a map was being filled in, and the role of the geographer, at once familiar with the state of the cartographic knowledge of the country and possessing sufficient erudition to be aware of the major publications in existence or research already carried out, was to direct future work. This geographic authority was recognized both by individual explorers who approached the Société de géographie (and sometimes Jomard himself for guidance and support) and by the government. In the case of M. Panet, it was the minister of the Marine who approached the Société de géographie for help in compiling the necessary instructions.

Jomard played this role with René Caillé,[28] Frédéric Cailliaud,[29] and a number of lesser explorers.[30] Often the principal aims of the expedition had already been laid out by either the explorer or the government. It was then Jomard's role to work through and around these. He unfailingly asked for material valuable for "constructing an exact map"—or a map as exact as possible. In the case of the travels of M. Prax through the northern Sahara, and no doubt inspired by the work of the Commission to Algeria, he asked Prax to gather linguistic information for the resolution of a geographic problem: he thought that perhaps inscriptions, and particularly Libyan inscriptions, collected throughout the region might give a clue as to the origins and past migrations of the various peoples of the northern Sahara. Here he was asking an explorer to carry out much the sort of work that he himself had carried out in Egypt. In this case, the aim was to resolve a different sort of geographic problem and one much more susceptible to solution through traditional geographic methods.

CARTOGRAPHIC ETHNOGRAPHY

Another area of research where we find Jomard unfailingly map-bound or, more correctly, bound by a preoccupation with the ordered laying out and measurement of phenomena and a search for understanding through the spatial information and patterns that this creates, is in his ethnographic work at the very end of his life. The results were no more happy than with the research he carried out in Egypt. Jomard was certainly aware of and responsive to the major intellectual concerns of his time, and from the late eighteenth century there had been a growing interest in the study of the various physical appearances of "man," the various languages of "man," and very gradually the social existence of humanity.[31] In this vein, Jomard broke the study of "ethnographie" into three major categories:

1. The study of "man" through his language.
2. The study of "man" through his physical condition.
3. The study of "man" through the works of his intelligence and industry.

Aware that something beyond both the variety of human language and physical anthropology was drawing the interest of scholars, Jomard proposed the study of objects produced by humans, or human artifacts.[32] The aim was not the interpretation of these artifacts so as to reveal the social organization or the lived reality of different peoples, but a means of mapping and fixing in place the different peoples of the world. Thus, what interested him about the third approach was that it had the potential to impose an order on the subject not unlike the order imposed in the natural sciences by classification. What he was after was "the order which puts each thing in its place and fixes it invariably"[33] and a "conducting thread of sorts which it will suffice, in a sense, to hold on to so as to not lose our way."[34] The study of human beings was clearly a complicated affair difficult to order and to map. What Jomard was proposing in his methodological classification of the products of extra-European industry (broadly understood) was an extension of maps and mapping to "include pretty much all of the objects that an observer will see while travelling."[35] Jomard was advocating a shift in the focus of the geographic imagination away from topography, physical geography, and location—"because today distance and space are nothing"—and towards the mapping of "man."[36]

Jomard divided the products of human industry into ten major categories: images, tools used for procuring food, clothing, objects to do with housing and building, domestic economy, defense, the arts and sciences (including commerce), music, customs and usages, and religion. A map composed of all of these products of human industry would necessarily have a distinctive form. It would be less a document than a museum. Indeed a "museum of geography and of travel" is just what Jomard had in mind.[37] Its creation, for all the difference in form, was no less a job for a geographer. All objects within it would have a double classification: "by subject" and "geographic." But for a geographer who, it could be said, had spent his most active years mapping monuments, the link between the map and the museum as the final product clearly lay in both the concept of the monument and in the process of research capable of elucidating monuments. In a passage which carries us back to Egypt and forward to his facsimile atlas, Jomard suggested the mapping methodology connecting all of his work:

It is thus through the thoughtful and tenacious study of the monuments of antiquity that one can begin to delve into the secrets of its architecture. It is even possible to say that all of science can be included, appreciated, and judged by its productions. This principle, which I consider a general principle, is above all applicable to ethnographic science.[38]

It was a mapping methodology precisely because what Jomard meant by a "thoughtful and tenacious study" was the erection of a descriptive system capable of putting every object definitively and unquestionably into its proper place.

MONUMENTS OF GEOGRAPHY

It was entirely fitting then, that Jomard's last work, the one he devoted to the history of geography, should have the title *Monuments de la géographie* and be devoted to the history of mapping.[39] This was not to be "a doctrinal treatise" or a "new general and critical history of geography."[40] What, asked Jomard, would be the point of reproducing what has already been said by a large number of predecessors?[41]

This enterprise, well executed, will be, in a way, a history of geography according to maps; a history written by itself, that is to say, by its graphic productions. It will thus be the most authentic and the most certain of histories.[42]

Just as the architecture of Egypt, graphically captured, had spoken to Jomard, and just as he anticipated that artifacts would provide all the insight necessary for a solid ethnography, maps would best tell the history of geography. All Jomard would have to do is to select and reproduce the best examples with fidelity. The unfolding map of human discovery would take shape in the minds of his readers as they gazed at his selection. The facsimile atlas, then, would serve to bring the history of geography to the attention of scholars who seemed to have forgotten its importance, and it would also democratize access to these rare documents.[43] Of course, Jomard, as a cartographer, was far from naive about maps and warned that maps had to be looked at critically: they could not be assumed to reflect the knowledge of the time in which they were produced, but merely the knowledge or understanding of a particular power or geographer.[44]

Jomard's atlas resembled an atlas factice, or gathering of maps. It was an unbound collection of individual, largely black and white sheets almost entirely lithographed by E. Rembiélinski and published chez Kaeppelin. It comprised a title page and table of contents and twenty-one cartographic representations in eighty sheets. The text was to be published separately but, in fact, appeared only posthumously. Six of the maps were sold individually. In the *Monuments* the images were presented chronologically but without any further systematic organization despite the fact that the collection included some works which were, strictly speaking, not maps. These included three celestial globes (two of which were Arabic or Kufic) and an astrolabe. The remaining maps were, for the most part, facsimiles of manuscript maps that represented the highlights of medieval cartography.

Jomard explained in his brief and posthumously published *Introduction* that he had chosen this period because an atlas composed of earlier or later maps would have been less interesting: maps produced after the sixteenth-century Ortelian reform were less strikingly different than modern maps and, as printed maps, were available in sufficiently large numbers to make their reproduction relatively less worthwhile; maps of antiquity and "oriental" maps would be too few to make a book or to allow conclusions about the history of discovery. His selection included the Matthieu Paris map; a medieval map of Lombardy; the Hereford world map; the Martin Behaim globe; the Juan de la Cosa map; and a Descelliers map of 1546. Some of these had been only recently discovered and their value was far from recognized even in Jomard's time.

Thanks to Jomard's dispute with Santarem over ownership of the concept of a facsimile atlas—about which more later—we know something of the procedures Jomard followed. The greatest part of the work, and it was work that went hand in hand with his functions as director of the Département des cartes et plans, was the identification, purchase, borrowing, or copying of maps around Europe. We know that one of the maps, the Juan de la Cosa map, belonged, at this stage, to a friend of Jomard's, the well-known geographer Baron Walckenaer. One of the maps, in a situation of possible conflict of interest, belonged to Jomard: the Descelliers world map of 1546.[45] In order to procure some of the others, he claimed to have traveled over a period of eleven years in England, Belgium, Italy, Germany, and Holland. Some maps were actually sent to him. Some of the most spectacular of the collection, such as the Hereford Mappaemundi or the Matthieu of Paris map were copied on the spot and the copies then sent to Jomard.

ONE MONUMENTAL BATTLE

While Jomard maintained some distance from the physical work on the atlas, it was a project close to his heart.[46] It positively resonated with the

principal preoccupations of his career as a geographer. Thus, when he wrote
in 1847—

that geography should tell its history through its own works, through the
variety of its graphic productions, . . . that idea does not belong to Mr. de
Santarem[47]

—he was speaking the truth. That was very particularly his idea and his
approach. It did not occur to him, however, or perhaps he did not care to
think, that someone else should arrive at the same, or a more complete,
conclusion by a different route. Jomard's *Monuments de la géographie* was
published sometime after 1847. The introduction to the work was published
posthumously in 1879, a considerable period after Jomard's death in 1862,
although it may have been written sometime in the 1840s. The Viscount
Santarem's *Atlas composé de mappemondes et de cartes hydrographiques et
historiques depuis le XIe jusqu'au XVIIe siècle* was published in France in
1842. This publication was followed in 1848, before the publication of either
Jomard's atlas or his *Introduction*, by a three-volume discussion and description, map by map, of the maps to be found in Santarem's facsimile atlas.[48]
While Jomard's reproductions were black and white, many of Santarem's
were in color. Further, a detailed comparison of a few sheets demonstrates
that at least some of Santarem's copies were truer to the original. Thus,
comparing Jomard's reproduction of the Turin Beatus with that of Santarem,
we find that Jomard's copyist has added a sophistication to the depictions
of the winds personified not to be found in the original. Getting a good
copy was clearly not a straightforward matter, as Santarem commented in
his discussion of this map:

Pasini produced a woodcut engraving of the copy of the world map to be
found in the manuscript of the Royal Library of Turin [see Pasini's catalogue
Codices Ms. Bibliothecae Regis Taurensis Athenoei digesti, etc. (Taurini, 1749),
vol. 2, p. 29], and we reproduced his engraving in our atlas with the utmost
fidelity. But when we were having it colored against the original, our scholarly
colleague at the Royal Academy of Sciences of Turin, Mr. Amédée Peyron,
had the extraordinary consideration to critically compare Pasini's engraving
with the original. He found important errors and he took the time to correct
them according to the manuscript. We are the beneficiary of his work on this
monument.[49]

In truth, neither reproduction is of sufficiently good quality to permit
informed work today, given present standards of reproduction and the
detailed level of analysis currently employed in the study of such documents.
Nevertheless, Santarem's reproduction is more correct than Jomard's. In
addition, Santarem labeled his maps more completely within his atlas

and certainly described them more completely within what amounted to his three-volume history of cartography. Finally Santarem's atlas was four or five times the size of Jomard's and, further, had been produced more systematically. That is, it was divided into sections devoted to maps of different types and periods.

In a vitriolic and public attack on Santarem in 1847, Jomard claimed to have been working on his own facsimile for many years.[50] Santarem, he argued, had stolen his idea and then had massively undercut his publication. There is no doubt that Jomard felt both indignation and deception—although why he waited until five years after the production of Santarem's atlas to express it is something of a mystery. Jomard's attack is full of self-contradiction and confusion. Carelessly and unwittingly, he claimed three separate dates for the conception of the facsimile atlas: 1829, 1830, and 1832. As proof of his priority, he rallied a considerable phalanx of scholars whom he claimed to have consulted. The list proves only that he was active as conservator of the map collection at the Bibliothèque nationale and that he might have been working for both himself and the library at the same time. He argued that most of the plates of his atlas had already been engraved in 1842. But they were not published before 1847 and his description of how many were already engraved and the number still awaiting engraving does not correspond to the final number produced. Further, Jomard claimed that he had had no financial assistance in the production of his atlas which, if one counts the costs of the copies from which the lithographer worked, is completely untrue.

Santarem's response to the attack was dignified but deadly. The proof of the quality of Santarem's research and thought was to be found in his atlas and in his three-volume analysis and discussion of the atlas—both of which works won accolades from, among others, Jean-Antoine Letronne and Carl Ritter. There is only one thing that can mark off scholarly terrain, Santarem pointed out, and that is scholarly publication, not the vague possession of an idea which one may or may not have shared with some friends. Further, what was Jomard really claiming? The concept of making facsimile maps? That idea had been around for some time and he duly provided a list of twenty-three map reproductions produced since 1730.[51] Was it the idea of a facsimile atlas capable of recounting the entire history of geography?

> He can relax: I am in no way trying to encompass the entire history of geography. That was never my idea. Reading the prodigious works of the most encyclopedic scholar of our day (M. de Humboldt) would have dissuaded me from that. And anyway, no matter how universal the knowl-

edge of a scholar, life would be too short to successfully complete such a project, that is in the manner in which I understand that it should be carried out. Mr. Jomard has nothing to fear in the way of competition from anyone.[52]

What really mattered, what really took time, effort, and scholarship, Santarem pointed out, was not the reproduction of maps but their study and analysis.[53]

In many ways, the fight with Santarem, in which Santarem engaged with considerable regret and distaste, calls to mind Jomard's more muted but no more successful battle with in the 1820s with Champollion, the scholar who without setting foot in Egypt or seeing a single monument in the flesh, so to speak, broke the code and first read the hieroglyphs of Egypt.[54] In that case too, Jomard was simply not prepared to admit that a scholar with an approach fundamentally different from his own (Santarem after all was less interested in the history of geography than in the history of the Portuguese discoveries) had achieved what he, the geographer Jomard, could not. This attitude reflected a marked sense of intellectual territoriality and possessiveness (well demonstrated in other publications primarily devoted to laying questionable claim to intellectual priority) and a peculiar attachment to a traditionally geographic methodology, the mapping of whatever problem arose.[55] Non-geographers had no attachment to this methodology. It was a matter, as Lacouture so aptly put it, of "Decipherment versus Description." It is, however, important to be fair to Jomard in this matter. Not so much out of a sense of solidarity, but because it casts light on the often bizarre way in which discoveries are made and how our collective thought and understanding takes shape. Jomard could not have hoped to decipher the hieroglyphs or to understand the civilization of ancient Egypt by mapping either the terrain or the monuments. That is utterly obvious to us today and it may even have been clear to some contemporaries. Due perhaps to his training and his received understanding of how to tackle unresolved problems, this was not clear to Jomard. So, in what seems to us utter futility, he undertook the mapping and cartographic analysis of whatever he deemed important and puzzling. Yet it was, in part, that mapping operation which drew the attention of a wide variety and a large number of French scholars (one is tempted to call it a critical mass of scholars), including Champollion. It was also those same maps or map-like sketches which allowed scholars such as Champollion to begin to study the hieroglyphs without setting foot in Egypt.[56] In a sense then, Jomard's impossible approach did lead scholars in the right direction, though not in the way he anticipated. Nor did it assign him the role he coveted.

CONCLUSION

As one of the most important and best known geographers of the first half of the nineteenth century, Jomard's conception of geography was important and had special weight. As we can see from an analysis of his most important publications, including his facsimile atlas, his conception of geography was intrinsically tied up with maps and mapping. As we saw, mapping was increasingly regarded, as the nineteenth century advanced, as a technology rather than as a science. Jomard, however, refused to abandon this conception of geography. As a result of his education and the many encounters with remarkable intellectuals and researchers that his position and long life afforded him, Jomard also sought to remain open and responsive to the major intellectual and social developments of his time. These included the growing interest in the study of humans and human societies, the increasingly important role accorded social statistics, the passion for and interest in pre-Greek societies, the advancement of French industrial society, and the development of universal primary education. Jomard struggled to render geography, as he understood it, useful in this new social world. For him, the essence of geography was not space, place, distribution, movement, and a propensity for spatial analysis, as it largely is today, but the map. For Jomard, the history of civilization could be summed up in that "monument" of Western thought, the map. It was through his facsimile atlas that he sought to bring the history of mapping to a wider audience. In that sense, we can think of Jomard as one of the earliest historians of cartography and one of the first academic map librarians. In addition, he sought to stretch the concept of the map and particularly the concept of mapping so that it could encompass the analysis of non-geographic problems. The results were often problematic and the attempt often placed Jomard in a difficult position vis-à-vis many of the other scholars less or differently bound by scholarly traditions. And while he did not always respond to valid if harshly phrased criticism in the most laudatory fashion—there is something enormously positive about the effort of extension. Perhaps, then, it is not so surprising that some measure of enlightenment and understanding came from it.

FIG. 5.   "Military geographers working in the field." Military geographers had to worry about housing, forage, their food supply, conducting their work under fire, as well as the quality and accuracy of their mapping. Uncertainty as to what was wanted at various levels of the chain of command added a further level of anxiety.

Geography as Handmaiden to Power

Geography, its first use is to wage war. —Yves Lacoste, 1976

Men in great place are thrice servants: servants of the sovereign or state, servants of fame, and servants of business. —Francis Bacon, *Essays* (1625), no. 11, "Of Great Place"

A state which dwarfs its men, in order that they may be more docile instruments in its hands even for beneficial purposes—will find that with small men no great thing can be accomplished. —John Stuart Mill, *On Liberty*, chapter 5

GEOGRAPHY AND POWER: AN HISTORIC ASSOCIATION

Geography has always had a close association with state power and co-ercion. Brian Harley has documented the presence of the interests of class, capital, and political power in maps as far back as the Renaissance.[1] Both he and others have commented on the use of maps as instruments of aggression and economic and political reorientation in the encounter between Western colonials and indigenous peoples.[2] The close link be-tween geography and the state at the height of European imperialism is now well documented.[3] Lesley Cormack has revealed the very political nature of the writings of famous Tudor British geographers such as Hak-luyt.[4] However, the association is even older than that. Christian Jacob has demonstrated the political cogency of the writing of ancient geogra-phers in Aristagoras of Miletus's attempt to hoodwink cartographically Cleomene, King of Sparta, into a disastrous war against Persia.[5] As we saw in chapter 3, the political was a major dimension of Strabo's writ-ing. In France, in the eighteenth century, geography was seen as critical to the education of the prince. Philippe Buache embodied this role as tutor to three kings of France: Louis XVI, Louis XVIII, and Charles X.[6] "The First Geographer to the King" was a title sought after by ge-ographers and one which significantly enhanced their chances of finan-cial success.[7] Topographic engineers were, from the middle of the eigh-teenth century, government employees, and some were sent on mapping expeditions to the Far East and South America.[8] The Cassini surveys and the mapping work of the topographical engineers in, for example, the Austrian Netherlands, constituted geography in the service of the state—to say nothing of their work in France.[9] The Jesuits can be seen as having employed the science of geography to open China to European interests, and in much more complicated ways they served the state by educating its elite and maintained an alliance of interests with political power.[10]

The Napoleonic era heralded a new intensity of cooperation between geography and political power. Never before had there been anything approaching the engagement of Napoleonic geographers in the rhetoric of state power or their unambiguous commitment to the imperialist aims of the state. Looking at the careers and language of key geographers, there is little question that geographers moved closer to the state and its aims. However, this new and stronger relationship between geography and the state was in part also formed of the state's recognition of the utility to war and population control of approaches perhaps most closely associated with geography: mapping, and information collection and ordering in general. The relationship seemed to offer nothing but positives to the geographers: security, status, social value, the opportunity to travel and conduct research beyond an individual's means. . . . In a time when there were few postsecondary teaching positions and those that existed entailed little remuneration,[11] such opportunities were not sneered at. In fact, in the Napoleonic era, apart from the academies devoted to sciences, inscriptions, or literature and the arts, there was no official source of funding for independent research not of immediate utility to the state. Geographers could either work for the government or engage in commerce. In the eighteenth century, then, geographers had functioned in one of four ways. They might be independent small businessmen who made their living from the commercial sale and production of maps and books on travel or geography, with or without the benefit of the title of First Geographer to the King. They might be teacher-scholars working either at the secondary level in Jesuit colleges (until these were abolished) or some of the newer specialized technical schools. They might be civilian mathematical terrain geographers, like the Cassinis or their assistants, who produced large-scale maps with the spasmodic support of the government. Finally, they might be military geographers who from the late eighteenth century increasingly took over the role of the scientists/bureaucrats. These field and office cartographers, thanks to specialist schools like the École des ponts et chaussées, the École militaire, and the Revolutionary École polytechnique, were some of the best educated applied scientists in Europe. The Revolution and the Napoleonic wars offered all of these geographers opportunities, though the most striking opportunities fell to those most useful to the state, and particularly to the military geographers.

In this chapter we will explore some of the sources of the new close relationship between geography and the state, including those generated by military innovations, by the rise of the modern citizen-based state, and perhaps more personally by the values and interests of Napoleon himself. No geographer alive and active in the Revolutionary and Napoleonic period—not even those engaged in erudite and arcane research—survived this era of extreme state aggression unscathed. However, the geographers most affected by the Napoleonic agenda were those prepared to commit their literary and scientific skills to the furthering of state interests, and especially those who served the state most directly as military men. Consequently, in this chapter we will examine the impact of this close relationship between the state and military geography both on the work carried out by military geographers and on their intellectual careers as scholars. The extraordinary value of military geography to the state, and the new need for information not just on position and landforms but on resources and society, meant that military geographers began to develop a more critical and rigorous descriptive geography. The Napoleonic wars also exposed these military geographers, many of whom were among the rising educated middle class, to regions, experiences, international scholars, and landscapes they might never have encountered in the course of a normal peaceful lifetime. Thus, the military and strong state interest in geography and the work of geographers brought, even in the Napoleonic period, some of the gains today associated with state support of higher education and science research. It is easy, today, as the mighty financial support of the state is gradually being replaced with "strings attached" private-sector funding, to be nostalgic about state support for intellectual endeavors. It is important to remember, however, that in the Napoleonic period, few of our modern university-associated structures had yet evolved to protect scientists and scholars from the capriciousness of state interest and uninterest. The consequences for the young intellectuals who were military geographers are hard to measure. Indeed, trying to assess the impact of over fifteen years of full participation in a militarized society on the mind and practices of an intellectual amounts to writing counterfactual history. Nevertheless, it is possible to trace in the career paths and writings of two military geographers, André de Férussac and Jean-Baptiste-Geneviève-Marcellin Bory de Saint-Vincent, who were both representative of military geography, and who also had exceptional intellectual talents, impacts that were far from positive. In the careers of these two men, then, it is possible to suggest, if not measure, both the costs and benefits of Napoleonic state involvement in geographic research.

SOURCES OF THE NEW RELATIONSHIP
BETWEEN GEOGRAPHY AND THE STATE

Military Innovation

The forces leading to the rise of state geography were both larger than the immediate circumstances of the Napoleonic wars and amplified by it. Toward the end of the eighteenth century, change in the broader society was creating a demand for a geographic research of a slightly different kind. The nature of warfare was changing in the last third of the eighteenth century. The Revolution was the source of a number of important innovations, but others were the culmination of technical and theoretical developments in eighteenth-century warfare whose significance only became apparent in the course of the Revolutionary and Napoleonic wars. The increasing emphasis placed on the rapid movement of troops in smaller units such as divisions and brigades and their reuniting for combat en masse, together with the development of lighter and more accurate artillery, gave cartography and road construction a new importance.[12] The rapid movement of smaller units required the careful guidance of troops through rough and irregular terrain. But as these new self-sufficient fighting units moved through or near enemy terrain, they could never be sure when they might encounter superior numbers of enemy forces. Thus, their movement had to be informed not merely by a cartographic knowledge of the terrain but by a strategic sense of the meaning of terrain features, human landscapes, and combinations of terrain features and human landscapes to both offensive and defensive operations. This required a knowledge of terrain and society that went well beyond traditional cartography and began to extend into what we would recognize today as social scientific preoccupations.[13]

It is generally agreed that the Revolution gave European warfare a strong ideological character, although not for the first or the last time. This ideology amounted to a collective illusion closely tied to a new nationalism which argued the coherence and superiority of the French social and political order, French civilization, and French science. This social and cultural superiority warranted the conquest and domination of neighboring peoples who, once conquered, would not only no longer pose a threat but would share in the benefits of French rule and participation in French science and technology while expanding the power of France and the historical depth of French civilization.[14] Thus, in the eighteenth century, war in Europe had been directed by aristocrats who identified at least as much with each other as with their sovereign's state, and had been fought by soldiers whose sentiments were treated as substantially irrelevant.[15] With the Revolution, the transborder identification between officers was to break down, to be replaced with a new emphasis on the nation and nationalism. In that shift, the hearts and loyalty of the rank and file became something to be wooed and won.

At the same time, France's financial difficulties and the government's need to defend the new social order required, at least in the early years of the Revolution, reliance on a conscripted army—an army that could only be kept in the field through ideological conviction.[16] It also required the mobilization of the entire society: from the recruitment of scientists, scholars, and artists for military expeditions, to the request that every basement owner collect saltpeter for the manufacture of gunpowder, to the Napoleonic militarization of the École polytechnique, the country's premier scientific educational institution.[17] The ideological tone was further raised when one of France's stated military aims became the export of Revolution: a more equitable system of government and a better way of life.[18] The French Revolutionary ideology, then, permitted and encouraged direct attack on the social system of the country being invaded—in particular on any aspect at variance with the French economic, governmental, or social system. This general atmosphere of social reform, which can also be seen as social intolerance, had, as we will see below, a significant effect on the activities and writings of geographers charged with studying the newly conquered realms.

The combined development of a more highly educated (and thus expensive) middle-class officer corps and the ideologization of the army had another effect ultimately of significance to geography. France's inability to pay its soldiers regularly and to keep them in clothing and shoes meant that the army had to live off the countryside it was invading. This led to what Geoffrey Best has described as "war addiction": an army on foreign soil could be kept functioning by predation of a more or less institutionalized sort. What began as a necessity evolved into a profitable venture.[19] In its least institutionalized form, it amounted to looting—an activity obliquely (and no doubt courageously) condemned by one geographer.[20] Institutionalized predation (which to my knowledge was never condemned by a geographer) entailed the assumption of the previous regime's taxfarming system, local conscription (which became essential as the French government encountered increasing resistance to conscription among its own people),[21] and a less pecuniary theft: the plundering of art works and historical monuments.[22] Effecting a conquest, then, entailed not only mapping phenomena of strategic importance but mapping and commenting on actual and potential wealth. The urgency of this activity and the precise definition of wealth was ultimately largely shaped by the economic war waged between France and England and the effects of Britain's blockade and France's continental system.

Political and Social Change

The nature of civil society and particularly the French government's sense of both its ability and right to control day-to-day and local activities also underwent radical, if gradual and incremental, revolution in the course of the eighteenth century. These changes were directionally similar to those taking place in the military realm—and they had just as significant an effect on geography. Whether examined from the perspective of the historical geography of France, the history of statistics, or the history of geography, it is clear that in the course of the eighteenth century the French Government developed Colbert's initial desire to have a general description of the kingdom for the purposes of administration into a commitment to increasingly centralized, regular, consistent, and exhaustive statistical and cartographic surveys of the realm. This commitment was not complete by the time of the Revolution and was not realized, in part due to the disruption caused by the Revolution and the Napoleonic wars, until well into the nineteenth century.

One of the characteristics of the state's growing awareness of its realm had also been an evolution in understanding of what constituted national wealth and, thus, what merited the attention of the government. From a primary preoccupation with territory and tax, a more sophisticated appreciation had evolved of the relationship between population (and population change), agriculture (and agricultural practices), industry, commerce (and its structural obstacles), and wages (and poverty). The effect of this was a growth in belief at all levels of government in the value of statistics, including population, industrial, agricultural, commercial, and, to a limited extent, social statistics. In the Revolutionary and Napoleonic period statistical inquiry became something of a universal preoccupation.[23] Geographers, whether the emphasis of their work was on graphic or textual description, could not fail to be strongly influenced by this general atmosphere and hence were not only convinced of the state's need and fundamental right to collect such statistics but, just as the state was taking over responsibility for the collection of statistics, increasingly saw the endeavor as an integral part of their mission.

Direct Pressure from Napoleon

Long-term sociohistorical forces were not alone in suggesting to geographers the appropriateness of a close marriage between geography and the modernizing and imperialist state. There was also a more immediate pressure from Napoleon himself. Napoleon's appreciation of maps and cartography is legendary.[24] What is less well known is that he had a clear sense of what descriptive, noncartographic geography could offer the state and, further, that he expressed those views in writing. It is equally likely that he conveyed this sense verbally to key geographers with whom he came into contact—although this cannot be proven absolutely.

Napoleon had little time for the humanities, which he felt had reached their pinnacle in classical times and were therefore unworthy of a place in institutions of higher learning. Anyone with a sound education could acquire all the insight such subjects had to offer by reading the classics for themselves. Napoleon reserved his respect for the sciences—in particular mathematics, physics, and chemistry. These formed the basis for engineering—perhaps the most useful of sciences. In fact, Napoleon shared the middle-class conviction (which developed and hardened through the nineteenth century) that it was the applied sciences rather than poetry, law, or philosophy which were responsible for Europe's global dominance.[25] Geography, a proto–applied science taught in special schools to graduates of the elite science school for engineers, the École polytechnique, benefited from this conception of the intellectual hierarchy. Napoleon saw geography (and to a degree history, by virtue of the critical approach necessary to its study) as scientific—or reasonably so. In his view, geography revolved around facts, not interpretations or style. Yet not all the necessary facts had been established and there were points of dispute that could involve scholars in advanced research. In addition, its domain was a changing and expanding one by virtue of exploration, the growth of human understanding, and the constantly changing political and physical nature of the world.[26]

Together with his minister of the interior, Champagny, Napoleon planned for the establishment of a school of geography and history at another elite institution of higher research and instruction in the pure sciences and the humanities, the Collège de France. But his description of this new school makes clear that he was less concerned with making geography part of the college's research and teaching agenda than in reforming the Collège de France with the sciences, including geography and history, into a useful appendage of the Great State. The school of geography was to be a central information bureau which would gather information on different parts of the world, trace and record any changes, make these known, and provide guidance to those who sought more information. The chairs of geography were to be four: maritime, European, extra-European, and commercial and statistical geography. This last chair was to be devoted to the study of the source of the power of states, in the cameralist tradition and in keeping with the primary concerns of contemporary statisticians/bureaucrats.[27] For reasons that remain unclear, Napoleon's school of geography and history at the Collège de France was never instituted. The very idea of it nevertheless suggests Napoleon's sense of geography's appropriate nature and role.

THE HANDMAIDEN ROLE: MILITARY GEOGRAPHY

The most valuable service to the state and to imperialism was that of Napoleon's military geographers, the ingénieurs-géographes.[28] These geographers were to be found wherever Napoleon's forces were engaged in combat, sometimes in advance of the army but most often either alongside or behind the army and employed in mapping operations designed to consolidate French control. It is hard to exaggerate the contribution made by the military geographers, to the conquest of territories invaded by Napoleon's forces. They mapped rivers, mountains, canals, towns, cities, roads and ways, monuments, and the movement of armies all over Europe from Italy, Piedmont, and Savoy to Elbe, Corsica, Bavaria, Austria, Russia, Poland, the Netherlands, Spain, and Egypt. They even carried out mapping missions on special assignment beyond the reach of the army in Persia, Algeria, Santo Domingo, Louisiana, and Greece. They must have produced tens of thousands of maps in the course of Napoleon's reign: a report by Alexandre Berthier written in October of 1802 stated that in the last war alone,[29] these geographers had produced 7,278 engraved maps, 207 manuscript maps, 51 atlases, and 600 descriptive memoirs.[30] For the expedition to Egypt (1798–1801), they sketched hundreds of manuscript maps and ultimately published an atlas of 50 topographic sheets together with innumerable town plans, maps of monuments, a hydrographic map of the country, and a map of the ancient geography of Egypt.[31] There are over 1,800 extant manuscript maps from the various Italian campaigns.

The Innovative Potential of Military Geography: A Critical and Rigorous Descriptive Geography

Military geography, or a corps of military personnel working together to create a body of useful geographic information, could be said to have existed since the establishment of the Dépôt de la guerre in 1743. Henri Berthaut traces the existence of a military mapping and information collection corps back to the late sixteenth century and the wars of Henry IV. However, these mapping personnel had a shifting status: being phased in and out of existence as need for their services waxed and waned; receiving new titles and uniforms; being subsumed within other specialist corps such as the Génie; or being attached in a more or less random way to the regular infantry. Their history prior to the middle of the eighteenth century is so complex and discontinuous that it is debatable whether such a corps could be said to have existed prior to the establishment of the Dépôt. With the establishment of the Dépôt, the mapping and geographic activities of the army acquired a nerve center and the ingénieurs-géographes gained direction, protection, and something of an identity. From 1743 the Dépôt grew in importance: reassigning the ingénieurs-géographes to colonial and frontier mapping operations in times of peace, and by 1769 acquiring a training function complete with professorships of mathematics and foreign languages. However, it was in the course of the Revolutionary and especially the Napoleonic wars that the Dépôt de la guerre was to acquire its full importance as a repository of geographic information, as a center of geographic research, and as a both formal and informal training center for military field geographers. Patrice Bret has described the evolution of the "Dépôt général de la guerre et de la géographie," as it was then known, from 1791 to 1830, with particular emphasis on the mathematical training of geographic personnel.[32] Over that period, training varied from informal classes held at the Dépôt with more or less well trained instructors, to the farming out of topographic education to bodies able or prepared to teach little more than the mathematical basis of mapping or merely the rudiments of non-geodetic mapping, to the establishment of advanced and comprehensive training programs designed to prepare the military geographer for a wide array of activities. What is clear from Bret's account is that, in the face of bureaucratic resistance and financial constraint, there was a strong movement in this period toward institutionalized, more rigorous, and far broader training of military geographers. The evidence for this is to be found both in the courses taught at the Dépôt de la guerre and in the journal published through the Dépôt from 1802. In particular, Director-General Etienne-Nicolas de

Calon sought to establish a school at the Dépôt de la guerre which was to draw on the talents, among others, of the physical geographer Nicolas Desmarest, the ancient geographer Pascal-François-Joseph Gosselin, and the astronomers Nicolas-Antoine Nouet and Perny. Calon's school eventually fell to both hostile lobbying from the rival specialist corps of the Génie and Calon's own political misfortune. However, a number of Calon's ideas were resurrected in 1801–1803 in the proposals of Napoleonic directors of the Dépôt de la guerre General and Count Antoine-François Andréossy and General Pascal Vallongue and again in 1809–1830 in the actual École d'application du Corps royal des ingénieurs-géographes. What all of these educational propositions and institutions had in common was a concern not just to train topographers and mapmakers but to produce officers who by virtue of their knowledge of mathematics, physics, mineralogy, or geology—and indeed proto-economics, history, military theory, and languages—were capable of interpreting all the countryside had to tell them.[33] Precisely what was taught under these large rubrics is not clear from the records of the Dépôt de la guerre. A sense of the nature and purpose of these training programs can be gained from both the journal produced at the Dépôt de la guerre under the impetus of Generals Vallongue, Andréossy, and Nicolas-Antoine Sanson, *Le Mémorial du Dépôt de la guerre,* and the topographic memoirs to be found in the war archives at Vincennes.[34]

The geography advocated in the pages of the *Mémorial* was far removed from traditional cabinet geography. In a very general sense, its principal aim was geographic synthesis, both graphic and textual, which would provide a solid enough understanding to give military decision-makers some forewarning of what they could not themselves know.[35] More precisely, this was a utilitarian and military geography whose principal function was to inform the decisions of senior officers. That did not mean supplying them with ready-made decisions, or even a list of alternatives. Rather the function of this geography was to give those officers the information necessary to understand the circumstances in which they found themselves, so as to be able to concentrate their forces rapidly in key strategic locations.[36] Their "forces" included not only their soldiers and weaponry but the terrain itself, the population of that region, the resources of the area (especially any surplus over subsistence), its fortifications, its institutions, political and religious attitudes, views on the war and its combatants, attitudes to government and authority, etc.[37] As a consequence, this geography encompassed many aspects of both human and physical geography. There was a sense at the Dépôt de la guerre that the syntheses produced by its geographers, while useful in periods of war, should also be useful to commerce, industry, and the advancement of general statistical knowledge, science, and the larger civil society.[38] The geographical engineers at the Dépôt were, thus, as involved in producing a general map of France, in the measurements by Méchain and Delambre for the arc of the meridian, in producing high-quality topographic surveys of the newly conquered territories, in producing maps for a cartographic history of the Napoleonic wars, and in work on the cadastre as they were in regular military reconnaissances.[39] Good geographic work could be equally valuable in military and civilian circumstances, and in the view of the officers of the Dépôt de la guerre it was its rigor, exactitude, coherence, comprehensiveness, and coverage that defined its value.

The military geographers and especially the "ingénieurs géographes" of the early nineteenth century are best known for their high-quality and large-scale maps of territories traversed by Napoleon's armies. In spite of the importance and value of these maps, another aspect of their work is more accessible and recognizable to modern geographers. In addition to writing cartographic memoirs like those composed by erudite office cartographers such as d'Anville (see chapter 1),[40] the military geographers can also be seen as innovators in a new genre. This genre, a refinement on descriptive geography, was the topographic memoir. Born of a new interest in population and society and part of a larger civilian administrative concern, the topographic memoir was designed to assist in the administration of the newly conquered territory.[41]

The topographic memoir was essentially a little physical, social, histori-cal, and military treatise on the region being mapped (generally at the scale of the canton). It was an innovation of the post-Revolutionary era.[42] The topographic memoir can be traced to the military memoir which came into being in the course of the eighteenth century. But the eighteenth-century military memoir was more immediately focussed on more narrowly defined military interests and did not extend beyond a description of the relief, major routes, and defensive aspects of the territory.[43] The topographic memoirs of the post-Revolutionary military geographers are far richer documents and bear witness to the influence of a new statistical curiosity and some of the geographic themes introduced by the ideologues (see chapter 6).[44] Neither were the questions they asked limited to the kind of question dealt with by the cadastral survey. The topographic memoirs dealt with agricultural techniques, the comfort or lack thereof of the homes, the nature of the social hierarchy, and the impact of government activity on local commerce and on the very existence of local communities. They described the professions, occupations, and skills of the population, often in tabular form. They detailed the staples, the commercial network, the educational system and its relationship to the state, and suggested the principal social, economic, and political problems of the community. When the military geographers could not understand what they considered an important aspect of the life of these communities (for example why the peasants of a region with a good system of canals were convinced that irrigation would render their fields sterile, or why the trees were cut or harvested using one method rather than another, or why there were not enough glass makers in a community clearly in need of more), they suggested possible avenues of research. Although they were clearly servants of the state, the military geographers were sometimes willing to represent the interests of the peasants against those of the state, or at least, to suggest to their superiors the limits of the legitimate authority of the state.[45]

Utilitarian to its very core, this geography had coherence, structure, and power. Malte-Brun and the other universal geographers of the early nineteenth century were trying to provide a complete description of the human world before understanding of the nature and structure of society had been developed. The result was a mass of information structured more by an arbitrary order of presentation than by any internal logic or argument. In the case of the geography practiced at the Dépôt de la guerre, it was military utility which functioned as the structuring agent and the principle of exclusion and inclusion.[46] For example, it was not all aspects of peasant life that interested the military geographers. They paid attention only to what might play a role in the relations between the state (frequently represented by the military) and the local society. To this end, they were instructed to ignore useless details and to avoid the literary style of "sentimental travelers" who describe everything but understand little.[47] This concern with concision and the relatively narrow focus of this work is what renders it accessible and even interesting to us. It offers more than a concatenation of the sources of information available on this or that part of the world. It provides a glimpse into what the government and the military considered important in the landscape. Indeed, with understanding of the perspective from which the topographical descriptions were written, the landscapes and societies described actually come alive. Their concern with clarity and concision also encouraged the military geographers to continue to make geographic "language"—especially their cartographic language and the language used to describe the physical world—more precise.[48] They had apparently hoped to extend this precision of language to the realm of human geography, but that was not seriously attempted before the 1830s. Concision and a purpose are good guides in most writing, and we can even see its effects when we compare a work on the history of cartography written by the military geographer Soulavie and another work on the same subject by a well-established civilian geographer, Barbié du Bocage. Soulavie focussed on the practical utility of the maps he was describing and consequently began to explore the relationship of the maps to the societies which had produced them.[49] By contrast, Barbié du Bocage's account reads much more like a list of the most important events in the history of the map.[50]

The utilitarian model with which they functioned did not obviously detract from the scientific spirit of their work. The military geographers trained at the Dépôt de la guerre were educated men, some of them graduates of the École polytechnique. It was expected that they would have the knowledge not only to carry out their work but to contemplate the meaning of the landscapes about them:

> The educated officer, who has reconnoitered and secured his position, can allow himself to contemplate the ravishing scene in which glory has placed him . . . he can ask himself what powerful hand formed these global skeletal structures [mountain ranges], and to what secondary but terrible forces one can attribute such a chaos of forms and substance. He can evoke the insight of the ancients, or the scholarship of our age, and be surprised to receive only dim and recent light on phenomena which are as incommensurably old as their effects are immense.[51]

Geology was an important and controversial subject in this period, characterized by a tendency to unsubstantiated speculation on the origins and history of the earth. As described in chapter 3, this evinced a very negative reaction from some cabinet geographers who, in trying to resist the radical change in worldview implied by a fathomable earth, rejected hypothesis, theory, and even explanation. In contrast, General Vallongue, adjunct to the head of the Dépôt de la guerre (Andréossy), looked at the question in greater depth. He suggested to the military geographers that there was an important distinction to be made between speculation and theory. Speculation was born of impatience in the face of the silent and mysterious earth. However, systems such as Buache's were part of a quest for order and coherence. That quest would be fruitless without careful observation of the countryside. Observations could in the long run overturn a theory, but a theory, even one as questionable as Buache's, could bring insight and coherence that would not come from observation alone. Thus, according to Vallongue, the horizons of an educated military man would be "enlarged by the light of theory" and he could curtail the influence of "optical illusions" "by practicing observation."[52] He advised the military geographer to know the various geological systems and hypotheses and to explore their predictive potential on the present state of the topography. It was already fairly clear that Buache's system would help little in the prediction of mountain forms and systems, but it might be of considerable use in the interpretation of rivers and their basins, which had such a large role in the shaping of topography. Philosophical questions about origins, he suggested, should be left to philosophers and scholars such as Pallas, Humboldt, and Ramond.[53] Vallongue's views were shared by Chevalier Allent who argued that the function of theory was to free topography from both arbitrary guessing and empiricism.[54]

The purpose of military geography, then, was to produce spatial syntheses, informed by broad reading and theory, which would at once enhance interpretation of the landscape and bring into relief the natural and human resources available for exploitation. The method by which this was to be achieved was equally well laid out in the *Mémorial*. Good-quality geography had to be based on archival and field research and on questions put to appropriately chosen local inhabitants. As life is the art of the possible, especially in conditions of war, the military geographers were instructed to begin their inquiries in the archives and libraries of the regions of interest. A number of engineers were sent on extensive missions in search of geographic documents in and covering the regions of interest to the government. They did not stop at government offices and libraries but visited local scholars and checked the stocks of book and map sellers.[55] Military geographers were rarely prepared to accept such sources as reliable—even after a careful study of the sources of the documents. For the military geographer and his superiors, the essence of the military approach lay in field research, or the observation and measurement of all aspects of the landscape deemed important. In order to understand a landscape, a geographer needed more than a plane table, a repeating circle, and a compass. On the ground, the information sought was not always readily accessible to either the eye or the hand. To know, for example, the relationship between riverbed fluctuation and rainfall—failing long-term records—the officer would have to examine the bed of the river and the terrain nearby. Were there damaged bridges to be seen? Would it be possible to determine how they were destroyed?[56] Similarly, if the aim was to determine if the population had recently dropped, the geographer had to consider the local use of space. Were there deserted houses or abandoned fields?[57] Did there appear to be a disequilibrium between the sexes or a tendency to either precocious or late marriage? If the geographer wanted to calculate the maximum possible production of a particular region, he must consider the nature of the soil, notice if the fields were under- or over-watered, estimate the number and state of the barns and stables, and roughly itemize the nature of the implements used and the general upkeep of the fields.[58] What the geographer could observe, calculate, and figure out for himself within a reasonable period of time would be limited. For some details it would be necessary to question the local inhabitants, and in his mémoire in the *Mémorial du Dépôt de la guerre* Chevalier Allent laid down the procedures to be followed in that eventuality. The geographer should begin with the authorities, then consult the educated people of the region, and then seek the perspective of older residents. The geographers were instructed to avoid both creating an atmosphere of authority and appearing too interested in the question being posed. They were never to use force or intimidation. It would be much more fruitful to try to simply engage people in a conversation in which the details sought were mixed with trivialities and subjects of interest to the interlocutors.[59] This kind of questioning necessitated very particular abilities and talents, which Allent listed. The geographer had to be able to acquire rapidly a certain sympathy for local beliefs and customs. He needed a prodigious memory and a structured way of thinking. He would have to be able to learn new languages quickly and well and to be able to cope with dialects and accents. He would need fluent Latin to be able to converse with religious officials in any region. Finally, he would have to have an easy personality which would allow him to appear comfortable in any social environment.[60] Once acquired, the information had to be verified, ideally by the critical comparison of testimony and where possible by direct observation of the phenomena in question. If there was the least doubt about the reliability of an important piece of information, it had to be rejected. A lack of information was always preferable to misinformation.

The geographers at the Dépôt de la guerre were known as "ingénieurs-géographes" or more generally "ingénieurs-militaires." When the value of their military knowledge was deemed greater than their value as technical mappers, they might be seconded to the État major (general headquarters) and become simply known as "officiers de l'état-major" (officers of the general staff). After the Napoleonic period, when the pressure of war was removed, the ingénieurs-géographes were merged with the officers of the État major and, together with those officers, they were given responsibility for the production of the 1:80,000 map of France. In a sense then, in the peace which reigned after 1815, they returned to their original function as highly trained cartographers. In the Napoleonic period, and especially from 1793 to approximately 1809, they were equally interested in what we would today describe as physical and human geography. It is true that the ingénieurs-géographes, in particular, often described themselves simply as "topographers." Yet they were instilled with a pride of profession, which encompassed the attributes and title of "geographer." This was perhaps best expressed in a contemporary report on the courses taught to the ingénieurs-géographes: "with the qualification of the geographer attached to that of the engineer, all of the earth is our domain . . . ."[61] In response to the prevailing passion for natural history and the rising interest in "man" and society, the military geographers sometimes claimed that while geography might be limited to indicating and fixing location,[62] it was topography that offered a new future. Topography, linked to statistics, could function as a sort of intermediary between the geographers and the earth and could guide the work of geologists, commercial agents, and public figures in their work for the state.[63] Interestingly, while the "ingénieurs-géographes" were prepared to distinguish topography from geography, they were less willing to differentiate topography from the broader category of military engineer. While a military engineer might work in different services in the course of his career, the study and practice of "topography" would be useful in all services.[64]

The topography, or geography, practiced by military geographers, then, was an analysis of the physical and human landscape in terms of its potential role in times of war or in a post-conquest peace. It was equally concerned with the morphology of landscape, its forests, marshes, and climate, and the role of all of these in the lives and fortunes of the inhabitants. It was a geography which was interested in the impact of industrialization on the countryside and on the way of life of the peasants. It paid attention to commercial centers, transportation networks, and administrative structures, especially those related to taxes and conscription. The study of population was largely restricted to the description of more or less verifiable characteristics such as height, origins, diseases, customs, and attachment to religion. It was acceptable to comment on their character but without giving such assessments too much importance.[65] There was even a little room for history (long since linked to descriptive geography) in this "topography." They were aware of making history and they asked themselves a question redolent of a remark made by Napoleon himself: if topography is better than cabinet geography, wouldn't a history on the ground, so to speak, be better than armchair histories?[66] One ingénieur-géographe who had a brilliant career as a general staff officer, Jean-Jacques Pelet, spent a good portion of the post-Napoleonic part of his career writing just that kind of experience- and field-informed history.[67]

But Power Corrupts

There are dangers in close association with state power. There are also grave dangers in the adoption of utility as the primary organizing principle of research. This could be well illustrated with the work carried out by military geographers in Egypt and even with the work of the military engineer Chabrol de Volvic and his geographical colleagues in Italy.[68] However, here the focus will be on the careers of two military geographers and the impact of close association with power on their lives and thought. Trained in the Napoleonic period, these military geographers remained active in both military administration and geographic research well after the end of the Napoleonic era. They were André de Férussac and Jean-Baptiste-Geneviève-Marcellin Bory de Saint-Vincent. They devoted the better part of their academic careers to natural geography. Both geographers took the geographic training and understanding they had gained during the wars and applied it to a series of nonmilitary problems. Perhaps due to the very practical nature of some of the problems they were trying to solve, their research was sometimes innovative. Yet, arguably, their close association with state power entailed unacceptable costs. The clearest costs were those associated with the Napoleonic wars: physical danger and political disfavor. These were borne to some extent by both de Férussac and Bory de Saint-Vincent. The militarization of geography temporarily provided purpose and an organizing principle for geographic research. What it did not do was to move the work of these geographers beyond description. How to identify a place for geography amidst the rising ventures of geology and statistics, or how to give geography some explanatory power, was never resolved by these geographers. Given the nature and number of problems that modern governments encounter, and in the absence of science councils or academies designed to protect intellectuals, political support for research is typically capricious and of short duration. Most significant research requires sustained attention. The work of both of these geographers was limited by both official estimations of what was valuable and a lack of resources. This meant that they lacked the freedom and opportunity to pursue often excellent ideas. There are signs too of a lack of an intellectual community through which to explore and sort out their ideas. Both primarily defined by their significant military careers and, thus, in a sense beneficiaries of a strong nonacademic affiliation, they lacked intellectual affiliation or status. In quest of these, both scholars allowed their research to wander across fields far too vast to permit them ever to make a significant mark. Finally, and perhaps most seriously, close association with the state's interests in the form of imperialism twisted and distorted their judgement in ways that they, themselves, could not have recognized. This is clearer in the career and writings of Bory de Saint-Vincent than in those of de Férussac, whose childhood and adolescent experiences had taught him to maintain a stance of skeptical watchfulness in the face of state power.

Life Trajectories and Career Paths: Military Power and Geographic Imagination

André de Férussac's childhood was spent in a period of extraordinary social upheaval. Born just two and a half years before the Revolution, political and military events substantially shaped the first twenty-nine years of his life. When André was just four, his father, a minor nobleman, an officer of the rank of captain, and an amateur naturalist, fled Revolutionary France to join the royalist army of the Prince de Condé. Consequently, the young de Férussac at an early and impressionable age experienced jail with his mother at the hands of Revolutionaries. Indeed, his family was not reunited until Napoleon declared amnesty for all émigrés in 1801. The new Napoleonic regime may have temporarily reunited the d'Audebard de Férussacs, but the Napoleonic wars were to send father and son, the father an officer and the son first a soldier and then an officer, to the far corners of Europe. André joined Napoleon's new light infantry corps (vélites) at the age of seventeen (in 1803) and was engaged in active military service, including fighting in the battles of Jena, Austerlitz, Heilsberg, Friedland, Eylau, and the siege of Saragossa, until he was seriously injured while on a military intelligence (mapping) mission in the Spanish Sierra Ronda in 1810. De Férussac's biographers agree that it was that injury, a bullet through the shoulder or upper chest, that ultimately killed the younger de Férussac twenty-four years later at the age of fifty.[69] After receiving this injury, André de Férussac never again saw active duty but continued to serve in the military as an aide-de-camp to General Darricaud, as a chef de bataillon d'état-major (1817), as a military advisor to Gouvion de Saint-Cyr (1818), as an instructor in the newly established École de l'état-major (1819), and as a military advisor, with intermittent periods of convalescence and secondment, until 1828. Through the remaining years of Napoleonic rule, he was only intermittently able to serve and not able to serve in a manner likely to gain him status or glory. In the most immediate sense, then, participation in the Napoleonic wars provided de Férussac with a rich array of experiences but cost him his health.

De Férussac also had a brief political/administrative career. In 1814 he published an account of the siege of Saragossa and a "Coup d'oeil sur Andalousie" in Malte-Brun's *Annales des voyages* which was noticed by Napoleon.[70] Soon after, he was named sub-prefect of Oleron, Basses-Pyrénées, just in time to find himself in a position of responsibility at Napoleon's fall from power. During the hundred days, and assuming a position of skeptical caution, he served France, having refused to swear allegiance to either Napoleon or his prefect, as sub-prefect of Compeigne. He quickly relinquished the position to his predecessor on Napoleon's final fall from power. Clearly uncomfortable with a political role in a regime he regarded as recklessly violent, de Férussac never again sought a position of political executive responsibility, although he did serve as an elected member of the Chamber of Deputies representing the Tarn-et-Garonne in the constitutional monarchy of Louis Philippe from 1830–1832.[71]

The military experience of his adolescence both offered de Férussac some opportunities and cut him off from other perhaps more important opportunities. Prior to becoming a soldier, de Férussac was strongly influenced by his father's interest in natural history and particularly in the study and classification of terrestrial and freshwater snails (mollusks). Together, they produced work which attracted the attention and support of some of the most powerful figures in early-nineteenth-century natural science. It was at the age of eighteen that de Férussac presented his first study before the Academy of Sciences. The younger de Férussac had a passion for study which, perhaps as a result of his military experience, extended well beyond natural history. As he served as a foot soldier and then as an officer, he studied the countryside, history, military history, "statistiques," and natural history of the regions through which he passed. Sometimes he gathered together enough material to deliver a paper,[72] or to write a small book.[73] Sometimes he lost the material he had collected to a militarily decreed change in venue.[74] How he systematically collected shells, their descriptions, and information on the geological context in which they were found under such conditions is mystifying, but according to his Preface to *Histoire naturelle*,[75] both he and his father did so. Whenever he was in Paris, both during and after the Napoleonic period, de Férussac presented papers on his research and findings in Paris's major intellectual centers: the Société philomatique, the Institut, the Académie celtique, and the Académie des sciences. They were well received. What de Férussac could not do while serving the military was to devote concentrated attention to any particular field of study. So, instead of focussing on what in the absence of war would probably have occupied his younger years, the natural history of mollusks, de Férussac's scholarly activity was dispersed across a wide variety of subjects. His work on mollusks was highly original, in part because in it he combined, without perhaps fully recognizing it, an interest in natural history classification and a spatial sense. However, apart from a brief period in the 1820s, he did not give his mollusk studies sufficient attention to fully develop and test his ideas. The period he devoted to concentrated natural history research fell in the three years between the end of the Napoleonic wars and his attempt to identify a meaningful terrain for peacetime geography. This project was immediately followed by another even more ambitious one designed to cope with one of geography's overwhelming problems: the vast increase in information and scholarship that characterized the early nineteenth century. Before we consider either his thinking about the nature of geography or his solution to the bibliographic explosion, we will examine the significance of his natural history research. Natural history, arguably, preceded his interest in geography and reflects the interests and potential of the unmilitarized de Férussac.

At first glance, de Férussac's joint work with his father on the description and classification of freshwater and terrestrial mollusks seems squarely within the tradition of natural history.[76] In a way, so it was. The two de Férussacs were the first to achieve an extensive classification of mollusks based not just on their shells but—in keeping with the ideas of Cuvier and Lamarck—on the form and anatomy of the animal living in the shell. To that end, father and son spent years collecting and studying both living and fossilized specimens, from the microscopic to the prosaic. This was no easy feat given the extraordinary variety of these creatures and the considerable variation between individuals. Father and son traveled to the collections of foreign natural historians and André de Férussac arranged that all diplomatic missions sent by the Naval Ministry and the Ministry of Foreign Affairs collect and bring back the mollusks they found in the territories they visited.[77] Their work was admired and lauded by both Cuvier and Lamarck.[78]

The de Férussacs's interest in mollusks was not purely classificatory. In his own description of his contribution to science, André de Férussac described his natural history work as the application of specialized study to "determine the laws of distribution of species on the globe."[79] The ubiquity of mollusks and the extraordinary number of fossilized survivals suggested the possibility of asking larger questions than those concerning interior form. Already prior to 1815 the two de Férussacs wondered at the cause of apparent large-scale disappearances of particular mollusk species in the geologic record. Mollusks, they felt at the time, were probably not as sensitive to temperature change as larger species. How extensive were these disappearances? Were they seeing the impact of local causes, such as huge landslides, or had there been global-scale catastrophes in the past? Was it reasonable to think that some species had actually changed form? But why would some species change form while others remained the same? Surely it was ridiculous to argue, as had M. Brard, that all mollusks had come "from a single fluid" [from the oceans?] and that this had determined "the laws of distribution of species on the globe," because "it would be necessary to assume that their respiratory system, their behavior, and their eating habits have also entirely changed; and then, why would only shelled animals have preserved their first forms?" Why, that was as likely as arguing that "lizards once inhabited rivers because today the Nile nourishes crocodiles." No, given the state of the research in 1814 and how little was known beyond Europe, the de Férussacs preferred to conclude: "today we have learned to doubt: that is the best proof of progress in our knowledge."[80]

By the mid-1820s André de Férussac was prepared to draw much broader and more significant conclusions from his work. Describing the implication of his mollusk research on both the geography of plant and animal distribution and on sedimentary geology, he concluded: (1) families, genera, and species increase significantly in number from ancient to more recent times; (2) they become more and more similar to living families, genera, and species; (3) families, genera, and species of both fossils and living mollusks "follow the increase of latitude and the decline of elevation, apart from some anomalies which have to do with the local conditions [the law of stations]"; (4) some species seem to have been progressively destroyed "because they were deprived of the living conditions which they had previously had"; (5) given that the greatest resemblance between living and fossil species exists in southern regions, "it is permissible to conclude that the lowering of temperature (really the average temperature) was the most important condition of life lacking for those species which no longer exist today"; and finally (6) looking at the series of fossils, there is no clear break suggesting that life was ever (once or many times) renewed over the whole earth. "On the contrary, we see in them proof of a successive and gradual change."[81] From this he concluded, still more generally:

It seems, then, that we can conclude from the preceding that (1) analogy of station and of destination, that is, the conditions of life and the role life fulfills, is the general law that has presided over the distribution of life over the globe; and (2) the changes that life has undergone on the earth's surface have been gradual. Life has not been renewed. Races have not been modified. However, insofar as the conditions of existence have changed or new conditions have formed, new species have replaced those which could no longer exist and which no longer had a role to fulfill. This continues up to the period when, for each portion of the earth successively, an equilibrium between influencing causes is established.[82]

In this very carefully worded semi-Lamarckian conclusion, de Férussac put his finger on the crux of the question that remained to be answered: *how did new species replace those which could no longer exist?*

De Férussac's work on mollusks, then, was far from ordinary. Given its focus on distribution and earth history, it was geographic in a Humboldtian sense. Based on his work on mollusks and the applications of that work to the history of life on earth and indeed to the history of the earth, he applied for an open chair in zoology at the Academy of Sciences. He was refused. De Férussac was declined a chair four times before he finally gave up trying.[83] One problem was that his work fell between proto-disciplinary stools. Although by the 1820s he had published voluminously, a substantial portion of his writing fell well outside of the purview of zoology. He could not compete in zoology, narrowly defined, with the likes of de Savigny, de Blainville, and Frédéric Cuvier. In one of his later applications, he sought to argue the relevance of all of his mollusk work to zoology. Referring to an earlier submitted dossier, he tried to frame himself as a zoologist pure and simple:

> As the works presented in the second part of this Notice are "zoological applications" of the study of living and fossil mollusks, they are not "geological works," per se. The papers mentioned in this notice are only presented as applied works in which the geological consequences of purely zoological facts are demonstrated.[84]

De Férussac was hurt and confused by rejection on the basis of disciplinary criteria. In a sad letter acknowledging his final failure to gain admission, he wrote uncomprehendingly:

> I dare to hope for a recognition that would be, for me, the most precious recompense for a lifetime entirely devoted to the progress of science.[85]

Unfortunately for de Férussac, the nineteenth-century Academy of Sciences was less interested in recognizing contributions to science in general than to the advancement of particular sciences. The second difficulty was that de Férussac did not entirely finish any major piece of work. There was always something else, and generally something central to his project, that he had intended to do. A significant proportion of his zoological work was published posthumously by scientists who had to complete, supplement, and correct de Férussac's research.[86] The reasons for this are almost certainly complex. De Férussac had a family to support but he had also adopted a pattern of activity in the course of the Napoleonic wars which, to gain admission to the Academy of Sciences, he would have to break. Each time he applied to the Academy, he was told to concentrate on his work on the classification of mollusks. Yet if he was to truly engage in the work he was doing for the military, essentially military geography and administration, then he could not so focus. Thus, he found himself in a professional dilemma: remain active as the kind of broad intellectual valued by the military and never gain access to the Academy of Sciences, or consider the military nothing more than an employer, and focus on natural history research. Perhaps, it was his attachment to geography and geographically framed questions which made him choose the former. Or perhaps the constant interruptions of his military life in the troubled Napoleonic period had left him unable to treat natural history as anything but "a personal taste," a "consolation," and an escape from "deplorable circumstances" to "the urbanity and generosity of scholars."[87]

De Férussac's life as an officer is inextricably tied to his identity as a geographer. De Férussac entered the military the year that Chevalier Pierre-André-Joseph Allent published his "Essai sur les reconnaissances militaires." It was precisely in this period that the Dépôt de la guerre was creating an active center of geographic field research through its publication, the *Mémorial*,[88] and through the activities of the military geographers. The geography developed and practiced through the Dépôt had a significant impact on de Férussac's understanding of the subject and his commitment to that kind of research and thinking. De Férussac seems to have fully embraced the values and approaches advanced by Chevalier Allent and his colleagues in the first decade of the nineteenth century. In his book recounting the siege of Saragossa and describing Andalusia, de Férussac argued, in terms redolent of the words of General Allent, that the military offered him, and many like him, a vista on unfamiliar societies available to very few travelers:

> The circumstances encountered by a military man are very advantageous for the acquisition of knowledge. He finds himself in contact with people of all classes. He integrates himself, so to speak, into families and lives with them. In that way he becomes closer to the private individual. He sees him in situations that make it easier to understand his character than is evident in the rash pronouncements of men of the world or in the accounts of travelers who cannot see as he [the military man] can. The military man leaves the main roads, crisscrosses the countryside, and can appreciate its fertility, its produce, its various aspects. By virtue of having to live off the land, he can better judge its resources.[89]

Further echoing General Allent, he argued that ability to read enemy or conquered countryside was not innate. It required careful study, observation, and reflection. It was also absolutely essential, said de Férussac, to the successful conquest and integration of enemy territory.[90] There is little doubt that de Férussac drew his inspiration directly from the writings of General Allent and the spirit of research reigning among the ingénieurs-géographes from roughly 1793 to 1817. Indeed, through de Férussac and other young officers of his ilk, the influence of Allent and his colleagues at the Dépôt de la guerre extended into the second and third decades of the nineteenth century.

De Férussac, with a wife and three children to support, continued to work as an officer of the état-major on the committee responsible for designing and setting up the corps' training school while at same time designing an outline of a course and textbook on geography and statistics as it might be taught to officers of the état-major for the then Minister of War Gouvion de Saint-Cyr;[91] as Chef du Bureau de la statistique étrangère at the Dépôt de la guerre for many years; and then as Chef de la Division de statistique at the Ministère du commerce. In 1822 he shifted his considerable geographic imagination and the bulk of his attention to his work on his *Bulletin*.

One finds in de Férussac's writings an abiding concern with the contextualization of knowledge. All of his works in natural history, geology, or mollusk geography discussed the role of predecessors and contemporaries in great depth. Indeed, in one of his works on terrestrial and fluvial mollusks, he promised his readers a complete analytical bibliography of all of the works of his predecessors and claimed to have already spent six months compiling it.[92] Interestingly, he described this sort of review as "the topographic study of known observations."[93] His research into mollusks had also taught him that collaboration multiplied the amount of material one could assimilate and broadened one's understanding. Nor was he satisfied merely to design a treatise on geography. Instead, he sought to understand geography within the context of "human knowledge." He even contextualized that concern within the tradition of writing about "human knowledge." De Férussac believed strongly in the scientific endeavor and he saw science as something which built on the work, insight, and writings of predecessors. Speaking of science, and the role of theory and observation in its advancement, he commented:

> The consequences deduced from noted facts and known as systems or theories are verified or rejected, confirmed or extended by the results of experiments and new discoveries. Thus, truths that have the force of law are established to govern and assuage the mind of man. And when one thinks that often it is an isolated fact (sometimes apparently insignificant, or whose relationships to

[other facts] one cannot perceive but which become luminous for the scholar who knows how to discover them) that can structure a whole system, or destroy the strongest theory—then it is that one recognizes how important prompt and complete communication of results is for the positive sciences.[94]

It was with that vision of science and with that conviction of the utility of such a tool that de Férussac founded his *Bulletin*.

The aim of the *Bulletin général et universel des annonces et des nouvelles scientifiques* was to provide an ongoing bibliography of all new work being conducted in all of the sciences. The *Bulletin* was to be complete and international, covering not just books but journals and journal articles. To that end publishers and journal editors were asked to send works to the office of the *Bulletin* on the understanding that the publication would either be returned or sold on their behalf with the proceeds returned to the press or editor. As many of the works as possible would be reviewed by scholars recruited and listed on the masthead of the *Bulletin*. Most, though not all, reviews were signed. Reviews were to err on the side of being informational and to avoid all "esprit de clique." For the most part the *Bulletin* fully lived up to these intentions. It was divided into eight principal subject areas and the monthly publications were gathered and published as subject-centered volumes:

1. Mathematics, astronomy, physics, and chemistry.[95]
2. The natural and geological sciences.[96]
3. The medical sciences.[97]
4. The agricultural and economic sciences.[98]
5. The technological sciences.[99]
6. The geographical sciences, political economy, and voyages.[100]
7. The historical sciences, antiquities, and philology.[101]
8. The military sciences.[102]

Five years into the project, de Férussac estimated that 80,000 reviews had been written by somewhere around 300 collaborators. On average, only about a third of the works listed were reviewed. De Férussac explained in response to critics that not all works were worth reviewing and that, in a sense, the very decision to review or to not review necessarily amounted to a judgment.[103] De Férussac seems to have run the *Bulletin* from an office with the help of two to three assistants. Each of the eight fields had its own special editor or group of editors with a tenure of anywhere from six months to ten years. According to the published record, there were years when some fields had no special editor. Presumably, in those periods, the task reverted to de Férussac himself. His office quickly became one of the most interesting and accessible libraries in Paris and, consequently, an important gathering place for scholars.[104]

Despite France's relatively early start in book collection and bibliography, de Férussac's *Bulletin*, or the concept of a universal, all-inclusive, and critical bibliography of books and articles ordered by subject, was well in advance of its time.[105] But it was nonetheless a very timely response to the explosion of research, international collaboration, publication, and particularly publication in specialized journals which characterized the nineteenth century. In a sense, it was even more than an attempt to control information. It was an attempt to come to terms with the new structure and geography of nineteenth-century research and there is every evidence that de Férussac saw it in those terms:

> Thus, centers of activity and propagation established themselves over the whole of the earth and divide it into a multitude of distinct circles unequal in extent and influence. The sphere of activity of these circles was determined by the combination of political circumstances, geographic position, the ease or difficulty of contact, etc.
>
> Seen in general terms, this multitude of particular circles . . . can be related to larger perimeters upon which they are to some extent dependent for reciprocal ties, such as community of origin, language, values, and customs, which constitute nationality. These large circles of culture and propagation can be classed according to the combined favorable circumstances which determine the greater or lesser extent of their relative influence.[106]

De Férussac was not only interested in the great centers of research in France, Germany, and England but collected information from Denmark, Norway, Sweden, Switzerland, Holland, Belgium, Russia, Poland, the Baltic states, the southern Russian states, Finland, Spain, Portugal, the United States, . . . In short, he was trying, within the means at his disposal and within a Eurocentric mode of thought, to create a map of world research in the sciences.[107] That was not his only aim. De Férussac was also convinced that his country was making significant mistakes in the way it was managing science. He lamented in particular the decline of France's great public institutions and pointed to the lack of support for pure research in France. Comparing Germany and France on this score, he commented with respect to scholars in Germany:

> They do not solicit political or administrative employment, or financial burdens. Rarely, and only in the case of men privileged by high ability, do governments, which, besides, are pleased to award [these men] honorific titles, divert them from the careers to which they have devoted themselves to throw them into an unknown world for which they have had no apprenticeship.[108]

After the Revolutionary and Napoleonic periods, most scholars in France, including André de Férussac, made their living as either government functionaries or private businessmen. This, de Férussac was convinced, perhaps as a result of the experience of his own career, would ultimately retard French science.[109] One of the aims of his *Bulletin* was to bring French scientists, busy and preoccupied with earning a living, into touch with new developments at home and abroad and generally to raise the profile of science in France.

By all accounts, the *Bulletin* was an enormous intellectual success. For ten years, de Férussac succeeded in mobilizing an extraordinary number of academics, many of whom worked very hard for him.[110] According to René Taton, he inspired sufficient trust to obtain the regular delivery of 500 international and national journals.[111] The *Bulletin* was influential and widely circulated in its day and there is no doubt of its value as a snapshot of science in that decade.[112] However, it was not a financial success. De Férussac simply could not keep up with the costs of producing the *Bulletin*. To help finance it, in 1828 he founded the "Société anonyme du Bulletin universel pour la propagation des connaissances scientifiques et industrielles." The Society was placed under the protection of the Dauphin, then that of the King of France, the Grand Duke of Baden, the King of Württemberg, and the Emperor of Brazil. Still, by 1831 the Society was failing and in February 1832 the Chamber of Deputies finally refused to bail the Society out. By 1835, de Férussac was in a bad way. Having retired from the military in 1828 to devote himself full-time to the *Bulletin,* and having invested very heavily in a scholarly venture, he was bankrupt and—at the age of forty-nine—found that he had to request some sort of duty with the Corps royal d'état-major.[113] In a sense, then, the military was kinder to de Férussac than was the gradually cantonizing world of nineteenth-century science. For twenty-five years, the military educated him, housed him, fed him, and gave him status.[114] Throughout, and in the end, it provided him with an income of last resort.

It is perhaps possible to generalize that de Férussac could not realize the various intellectual projects in which he believed because he did not have adequate support to do so. In geography, in natural history, and in what was to become information science, he had ideas of considerable value and a willingness to undertake the enormous amount of work their realization demanded. Born to a noble family without resources in the midst of Revolutionary France and drawn into the military both by family tradition and the urgency of national need, de Férussac did not have the means to become a man of science. What he achieved using his own resources and as he worked for the military is truly remarkable. The fact is that the academic infrastructure of the time was inadequate to support an intellectual of de Férussac's limited means but unconventional and challenging imagination. As is clear from his proposed innovative course at the École de l'état-major, that imagination took him beyond universal descriptive geography as conceived by Malte-Brun or Malte-Brun's admirer and intellectual successor, Adrien Balbi.[115] What de Férussac proposed in that work was a new social geography based on statistical field research at the level of individuals and communities. His free-ranging imagination also put him beyond the pale of the sciences emerging from natural history. Trained as a geographer, keen to promote geography, in the end it was his lack of affiliation beyond the military, his lack of a discipline, that limited de Férussac. In the end it was the military and his experiences of the Napoleonic wars that claimed him. For that reason, de Férussac, like many military geographers, has never found a place in the history of geography.

The life trajectories of de Férussac and Bory de Saint-Vincent bear remarkable similarity. In a sense, the political and military storms which characterized their age were so overwhelmingly powerful as to wash away at least the surface layer of individual differences. In addition, as geographers and scholars, the men came from the same place and followed very much the same course, from officer of the état-major, to a brief political career, to postwar advisor to the état-major, to a long attempt to realize a scientific career. For all their political and social differences, de Férussac and Bory de Saint-Vincent made similar decisions in their scientific careers. In terms of a subject focus, both were trained as geographers with the skills array of field mappers with training in descriptive statistics collection. Both retained a commitment to that type of geography which they carried into their later careers as peacetime officers of the état-major. Yet prior to the Napoleonic wars both had been taught to identify themselves, and good science, with natural history. The military geography they were taught and practiced in no way altered that prejudice. Thus, while de Férussac sought to gain admission to the Academy of Sciences as a zoologist, Bory de Saint-Vincent founded, edited, and contributed to a dictionary of natural history comprising seven volumes. Both men were struck by the research explosion of the early nineteenth century and sought in their own ways to control and increase access to it. While de Férussac better understood the magnitude and implications of the information explosion, Bory de Saint-Vincent took a more typically early-nineteenth-century approach: he established a journal devoted to his conception of the physical sciences. That they were functioning in significantly overlapping realms is clear from their collaboration on each other's major life projects: Bory de Saint-Vincent was an editor and contributor to the *Bulletin* and was also reviewed in the work; de Férussac contributed to Bory de Saint-Vincent's *Dictionnaire classique d'histoire naturelle* and was originally slated as one of its principal contributors.[116] In a sense, then, they represent the strongest and best of the military/engineering geographers who took a predominant interest in "natural geography" as opposed to geodesy or large-scale topographic mapping or state administration.[117] While the similarities in their career paths reflect similar origins and interests, the differences stem from differences in personality, fortune, and age at the onset of war.

Bory de Saint-Vincent had a stellar military career, a troubled political career, and what was deemed in his own time a successful scientific career. Six years older than de Férussac, and despite the latter's precocity, Bory de Saint-Vincent was in a better position to establish his scientific reputation before entering active military service. Thus, when de Férussac was just beginning to botanize, Bory de Saint-Vincent was a young but respected natural historian on the Baudin expedition to the South Pacific. Although he was forced by ill-health to abandon the expedition in Mauritius, he was able to publish two major works relating to that expedition after his return to France.[118] As a result, both his military career and his scientific forays while with the army were more socially supported and facilitated. Thus, in a sense, de Férussac started his military career far closer to the bottom of the rank structure and was more subject to the whims of war. Both men saw a great deal of action and took significant risks. As officers, both were known for their originality, intelligence, and bravery. Both served in the état-major and engaged in extensive mapping operations, especially in Spain. Bory de Saint-Vincent had an ideal état-major training, including experience in infantry, in the cavalry, in a variety of état-majors, and at the Dépôt de la guerre. Finally, both used every possible opportunity to observe, analyze, and comment on the natural and human structures about them. It was de Férussac who received an injury that ended his active military career while Bory de Saint-Vincent continued to rise in the military establishment for the next two decades.

Politically the men were further apart. While de Férussac was at best uninterested in political matters and often repulsed by them, Bory de Saint-Vincent was politically engaged. De Férussac never forgave the Revolution its damage to his family, or Napoleon his harm to France. Whereas Bory de Saint-Vincent came to believe in the social benefits of the Revolution and saw Napoleon as both the preserver of its best aspects and one of the enlightened monarchs of the nineteenth century.[119] Bory de Saint-Vincent's Bonapartism is perhaps clearest in his work on the Morea expedition, in which one senses his joy at rediscovering many of his colleagues from the Napoleonic wars.[120] De Férussac was prepared to take on minor administrative roles and even to play a part in representative government when asked. Bory de Saint-Vincent, for a time, sought a political career for which he paid with exile from 1815 to 1820. His political activities included participation in the political satire of the *Nain jaune*, which in 1815 took on the task of critiquing the self-censored press, the regime, and any powerful individuals too slavishly supportive of the new regime (Malte-Brun in particular). They also included the less critical but perhaps even more tainting support for the military achievements and political ideas of Field Marshal Soult. Bory de Saint-Vincent had entered General Soult's état-major for the first time in 1808 and served with him later in Andalusia when the field marshal essentially functioned with the power of viceroy in Spain. In particular, he publicly declared support for Field Marshal Soult's views on the radical land and social reform necessary to make Spain a prosperous and cohesive country.[121] Once exiled, he further antagonized the authorities by declaring himself privileged to share exile with a man such as Field Marshal Soult.[122] His other major political contribution was a very brief stint as deputy for the Lot-et-Garonne to the House of Representatives. This was interrupted by the edict ordering him into exile. Bory de Saint-Vincent's political commitment cost him several years of peaceful scientific and professional activity. He did manage to carry out a limited amount of research and writing and more editorial work in exile, in particular researching and writing his essay *Description du plateau de Saint-Pierre* and founding his *Annales générales des sciences physiques*. Further, he made a number of valuable scientific contacts while in exile, in particular forging a bond between himself and Alexander von Humboldt.[123] The exile clearly limited his ability to carry out research but may have added to his political and social profile. Arguably, military aggression had a more limiting impact on de Férussac's life and social opportunities but, perhaps because it benefited him far more, it had a far deeper impact on the nature of Bory de Saint-Vincent's thinking and research.

Bory de Saint-Vincent's research can be seen as marked in two significant ways by the military. In the first place, because his activities were substantially directed by state interests, from his mapping work in Spain to his participation in both the Morea and Algeria expeditions, he developed a tendency to jump from subject to subject and from approach to approach as the interests of the moment dictated. Secondly, state ideology fundamentally warped his thinking. Each had a profound effect on his work, though Bory de Saint-Vincent may have imagined himself, apart from his largely self-inflicted moments of penury, to have acted as a free and independent agent.

A lack of sustained intellectual focus seems to have been an even stronger tendency in Bory de Saint-Vincent than in de Férussac. Perhaps consequently, Bory de Saint-Vincent played a more typically early nineteenth-century and less modern role than did de Férussac in both natural history and geography. With his focus structured by his alphabetical *Dictionnaire* or by regional descriptive works such as his *Essais sur les isles fortunées*, Bory de Saint-Vincent's contributions were of a more general nature. While de Férussac's conclusions emanated from his mollusk research and gained an intellectual rigor from the limitations imposed by mollusks as evidence, Bory de Saint-Vincent speculated freely on the polygenic origins of the human species, the role of the environment and/or climate in shaping plants, animals, and humans, and the links between geological and human history. His speculations were not idle or ill-informed. Bory de Saint-Vincent carried out botanizing research in the Canary Islands, Mauritius and Réunion, Spain, Maestricht, Greece, and Algeria. However, while he carried out significant botanical work, as Jean-Marc Drouin demonstrates,[124] and sometimes directed the research of other scholars, his own research contributions were usually more general than rigorous.[125] Bory de Saint-Vincent produced a number of classically geographic works in the course of his life. As an officer of the état-major in Spain, he authored a number of reconnaissance maps, some of which survived.[126] He also shared responsibility some fifteen years later for the production of a new map of Spain.[127] In addition, he wrote the account of the first year of the Baudin expedition[128] and reported on, wrote up, and directed the physical sciences division (including geography) of the expedition to Morea of 1831–1838.[129] He headed the scientific expedition to Algeria and wrote four works in association with that expedition.[130] He also wrote, and was well known for, two major regional works: an essay on the Canary Islands[131] and a physical and human geography of Spain.[132] Other works of less evidently geographic importance were an essay on the Plateau of Maestricht, his "Géographie, sous les rapports de l'histoire naturelle," in his *Dictionnaire classique d'histoire naturelle,* and a work he planned but which seems to have been completed substantially without him, a statistical atlas of France.[133] In spite of this quantity and diversity of writing, which also included numerous articles for the Comptes rendus of the Academy of Sciences and encyclopedia entries, Bory de Saint-Vincent's inspiration was of a more literary and philosophical nature and he is better remembered for his polygenic arguments about the origin of "man" than his botanizing or his geography per se.[134]

Despite this considerable geographical writing, apart from his geography of Spain, there is little indication in Bory's writings of a self-directed scientific mission. Nor is his conception of geography or its role in scientific thought clear, despite his strong identification with the subject. In contrast to de Férussac, who seems to have written from the margins and who cited geographers largely to critique their approach, Bory de Saint-Vincent had a broad and engaged consciousness of the field. He cited geographers heavily in all of his works. Certainly, he was not always flattering. Malte-Brun came in for his full share of criticism from Bory de Saint-Vincent: for favoring foreign, while denigrating French, authors; for republishing what was already well known; for inanely slashing a man Bory de Saint-Vincent deemed one of the greatest spirits of the Enlightenment and the soul of the French Revolution, Voltaire; for repeating erroneous information on explored parts of the world; . . . Yet, after Malte-Brun's death and after the divisive and treacherous political atmosphere of 1814–1815 had been forgotten, he was prepared to acknowledge Malte-Brun's importance as one of the primary geographers of his era and he composed a biographical notice on him for the *Revue encyclopédique.*[135] Beyond Malte-Brun, who, by virtue of his *Annales* (in which Bory de Saint-Vincent published) and his *Précis,* was one of the highest profile geographers of the period, Bory de Saint-Vincent knew of and cited many geographers. From the disciplinary touchstones, including Strabo, Ptolemy, Willem Blaeu, Jean-Baptiste-Bourguignon d'Anville, and Nicolas Desmarest, to the contemporaries he admired or simply acknowledged, including Jean Baptiste Benoît Eyriés, Jacques Nicolas Bellin, Pierre Lapie, William Bowles, Isidoro de Antillon, François Raimond Joseph de Pons,[136] José Lanz, Silvestre-François Lacroix, John Pinkerton, Edme Mentelle, William Guthrie, Charles Athanase Walckenaer, Seco,[137] Jean-Jacques-Germain Pelet, Jean-Denis Barbié du Bocage, and Émile Le Puillon de Boblaye, Bory de Saint-Vincent cited, discussed, and corresponded with geographers all over Europe. His writings, then, convey a sense of belonging to the field of geography and a centeredness not evident in de Férussac's work.

Yet, it is only in his geography of Spain that there is anything resembling a thoughtful treatment of the subject and one of its major dilemmas: the problem of the unity of nature and how that unity might be explored scientifically. In his geography of Spain, Bory de Saint-Vincent seemed to advocate an historical (or evolutionary) regional geography. This geography, he felt, could integrate humanity and nature. Not particularly drawn to geology,[138] Bory de Saint-Vincent was nevertheless fascinated by the natural landscape that he had experienced as an officer in Spain. He was interested in it per se and as a backdrop to and function of the activities of nations and individuals. For Bory de Saint-Vincent, nature and humanity were dynamic. Consequently, the landscape had to be understood in such terms. What principally drew his curiosity were the processes by which landscapes or human societies had arrived at their present form. He was convinced that they had done so through a complex process of interaction: involving individuals, populations, races, societies interacting with nature, cataclysmic change, and local conditions—to produce the sort of distinctive regional geographies to be found in Spain: "the physical physiognomy of the Peninsula, if one may be permitted to express onself thus . . ."[139] Perhaps echoing the ideas of Augustin Pyramus de Candolle, Alexander von Humboldt, Arthur Young, and Girault-Soulavie, he identified these distinctive regions with natural regions.[140] He then pointed out the greater heat of the peninsula, the prevalence of grape cultivation, and some typical vegetation.[141] This, it was clear, had, over time, and through interaction with "human behaviors," created distinctive peoples:

> The inhabitants of this side [of the Pyrenees] have some customs in common despite the fact that long-established political divisions have considerably influenced their character and have divided them into distinct peoples between whom deeply rooted enmities appear to prevail. These customs probably derive from the general influence of place.[142]

These peoples were not predeterminately as they were. Bory de Saint-Vincent expressed a strong sense of the historical contingency of multiple human migrations from the Greeks, to the Carthaginians, to the Romans, to the Arabs and the Jews, which had created in Spain a distinctive mixture.[143] Nevertheless, there was between this diverse population and the varied, yet unified, landscapes in which they lived a profound bond. A bond so important that any work on Spain, as Bory de Saint-Vincent explained in his introduction to Bigland's *Histoire d'Espagne,* should be accompanied by a map depicting the peninsula's physical features.[144] For Bory de Saint-Vincent, to understand the geography of Spain was to textually disarticulate both the human and natural body of the country, to recombine each into a physical and human map, and then to ponder the unity of the two cartographic visions. To a degree his thinking was like Humboldt's in that he believed the link between man and nature lay in the characteristic landscape of the natural region which bound together local physical conditions, way of life, landholding patterns, and political conditions.[145] The geography that Bory de Saint-Vincent practiced in his work on Spain—a geography "guided by the relations of natural history," or "natural geography"—was a unifying subject focused on "the entire history of all the bodies, whether simple or organized, of which the planet on which we dwell is composed, and everything which gives an idea of its physiognomy." The fundamental unity of the physical world lay in the complex, multiple, and confounding causes of its development.[146] It was a geography deeply influenced by the field observation skills of the ingénieurs-géographes and suggestive of an early sensitivity to the way space and place can be defined by social behavior. In contrast to Alexander von Humboldt's contributions to natural geography (see chapter 7), and largely because his research directions were more shaped by national interests than by an intellectual agenda, Bory de Saint-Vincent never developed the ideas he had begun to articulate in his work on Spain.

While Bory de Saint-Vincent wrote many works of geography, and while his citation patterns suggest a mind thoroughly immersed in the literature of geographers, he was not committed to a geographic mode of thought. This is perhaps clearest in his essay *L'Homme,* published just three years after his *Guide du voyageur en Espagne.* In this work, Bory de Saint-Vincent abandoned the subtle interaction of human movement in space, historical accumulation of behaviors and values (or culture), and interaction with the environment in order to conceive of the study of "man" in biological terms. In order to make a contribution to the current debate about the place of "man" in Linnaeus's classification (perhaps a debate well beyond the purview of geography but one which another geographer, the Baron Walckenaer, also could not resist), he argued the primacy of race and the unimportance of environment in human history. That this contradicted his earlier more subtle view of the geography of Spain does not seem to have occurred to him. Undisciplined in his thinking, encouraged by the pressures of the military life to jump from subject to subject and from approach to approach as the interests of the moment dictated, he easily shed any commitment to his earlier geographical argument.

The second, related, and perhaps most significant influence of military aggression on Bory de Saint-Vincent was that it colored and twisted his thinking to its very roots. A post-Enlightenment thinker with a strong commitment to social reform, he nevertheless was fully committed to French imperialism within and beyond Europe. In contrast to de Férussac, who developed pacifistic tendencies, he embraced much of the ideology later used to support European colonialism in the late nineteenth century, including the "French man's burden" and the "other's" obligation to "share" resources and "donate" labor. This is clear throughout his career from his participation in the Baudin expedition to his advocacy of the colonial cause in Morea and especially Algeria.[147] The nature of his advocacy is clearest in an early disquisition on the distinction between conquerors and barbarians. Among the former he included the French colonialists and among the latter the Spanish empire.

> Conquerors are sometimes humane and enlightened. They do not always run across the theater of their exploits with feet of steel. Some have even been useful to the people they have forced into submission by spreading, in their impetuous charges, the sciences and the light of the nations which have followed their flags. Seostris, Alexander, Caesar, and most recently our record of great deeds have shown us that in spreading war over the earth we do not always leave behind us an execrated name. . . . Conquerors do not destroy peoples; they liberally share with them their knowledge and their customs. They respect the opinions and customs of the defeated. It is not unheard of that they study them. They do not destroy the monuments; often they repair them and transmit them to posterity. Finally, far from reducing those who have experienced battle to desperation, conquerors allow them to benefit from a liberty that they did not know how to defend themselves. Barbarians, on the contrary, who have no superiority to their victims beyond that of force, the virtue of ferocity alone, annihilate everything that dares to resist them.[148]

Here, in this apology for French colonialism, we see a man whose thought was profoundly marked by the need to justify state violence. It is possible to see his polygenetic view of the origins of the human species—the argument that humans are composed of distinct races that do not share a common origin—as providing the ultimate justification for European exploitation *through science.* Mary Louise Pratt may consider that Humboldt, by virtue of his participation in the construction of a European conception of the world, contributed to Western imperialism. Yet, for Humboldt, the ultimate aim of science, as the highest expression of civilization, was to better the condition of all peoples. For Bory de Saint Vincent, science, an expression of state power, overrode "philanthropy."

> We recognize that it would be consoling to philanthropy if one could make all men understand that, no matter what species they might belong to, they must love each other as members of the same family, and that they must not butcher each other or sell each other [into slavery]. But the quest for truth does not allow for such considerations.[149]

In this statement, unimaginable emanating from Humboldt, science in the guise of truth is placed above the interests of humanity. Yet at the same time, Bory de Saint-Vincent was a great believer in the human will and in civilization and culture as the realization of such will. Thus, while he toyed and tinkered with the ideas of Montesquieu throughout his career,[150] his commitment to the Revolution would not allow Nature dominance over human will, culture, and history.

But the possible events, however one imagines them, which take from man all that surrounds him and all that is not within him, cannot, as long as he still has his life, take away his views. Those views, born of the way education has modified his mind, have aged with him. He has grown so used to them that he cannot change them. He can only abandon them with the greatest difficulty. And as his customs are the consequence of his views, and as views can be transmitted, customs are perpetuated among the descendants of a people, even when they are dispersed and no longer form a body.[151]

How could a man with such subtle social understanding draw simple distinctions between conquest and barbarity, equating each with different European imperialisms? In fact, here, and in many of Bory de Saint-Vincent's writings, there is evidence of a profound contradiction. It was a contradiction born of his role as servant and soldier of the state and his identity as a free-thinking intellectual, heir to the Enlightenment.

In spite of this, Bory de Saint-Vincent would be easy to romanticize. An adventurer in every sense, he spent much of his life either in active service in the military or on high-profile scientific expeditions. Whereas de Férussac wrote with the studied dispassion and distance of "science," Bory de Saint-Vincent peppered his works with personal observations. Perhaps consequently, whereas de Férussac's descriptions of Spain are more likely to induce somnolence in the modern reader, Bory de Saint-Vincent's accounts of Spain and his travels elsewhere are alive with lived experience. His descriptions of the street scenes of Tenerife engage the reader's olfactory senses.[152] His voyage accounts make such frequent reference to food that the reader is rarely able to forget either the author's or her/his own daily needs and their possible implications while travelling.[153] Finally, he had, and managed to convey, an almost tactile sense of horror. His sinister description of the valley of the Duero after the rains captures, in the modern mind, some of the horror of the battlefields of World War I.[154] He forces the reader to identify with the monks who died, lost and entombed in the silence of the Maestricht caverns.[155] And he captures both the cruelty and beauty of nature in his brief description of the life choices of the exocet (Exocoetidae fam., a flying fish).[156] Again in contrast to de Férussac, and indeed to most of his contemporaries, women make repeated appearances in his writings, always framed according to his lights and values, but nevertheless there. Consequently, he conveys at once a sense of his own interaction with women (which was somewhat predatory, at times compassionate and at times disparaging) and provides insight into the role and status of women in both his own society and the societies he visited.[157] Although he landed in debtor's prison and spent approximately three years there, Bory de Saint-Vincent was richer, luckier, and more flamboyant than de Férussac.[158] His lifelong friend and associate Dr. Dufour captured admirably the personality and flamboyance which found expression in Bory de Saint-Vincent's writings:

> Small in stature and tilting a little to one side, yet believing himself to stand straight, pale and colorless with a lively and always changing expression, gay, playful, impassioned by music and humming all tunes well, infinitely natural and with a remarkable conversational facility without being chatty, an exquisite grace in story and anecdote telling, very amiable and trying very hard to appear to be so, a friend of the world and all ostentation, educated but skimming over many sciences while entering deeply into little, often giving money away ostentatiously, but habitually without a cent, ambitious for titles which he sometimes usurped, writing well off the cuff and on the run, but sometimes injuring orthography, although married, living as a bachelor, creating mistresses and debts everywhere, a very individual life, lived for today![159]

Perhaps inevitably, given his close association with state violence, there was an ugly side to his nature. This is perhaps clearest in his own—one hopes exaggerated—description of his evenings as a soldier in the area of Austerlitz (Slakov, Czech Republic).

> It feels good, after a hard day, covered in mud and sweat, in spite of the snow and ice, and after having come under fire and left friends behind in it, to go back to headquarters, which for eight days we have situated in the homes of the rich boyars and there to lord over it, as though one were in one's own home, to help oneself to stag or pheasant or venison for dinner, accompanied by Hungarian wines, and then to caress some of the pretty girls of the region without anyone being able to object, to borrow a horse and money that one will never return and being well mounted when one leaves, to regret only not having done more and not having put it all to flames, after breaking the limbs of the proprietors.[160]

We have in Bory de Saint-Vincent a man who passed through over fifteen years of intermittent war and state aggression apparently unscathed but deeply marked by the experience.

There is no question that Bory de Saint-Vincent was an active and sometimes inspired scholar. If he lacked de Férussac's originality and idealism, he devoted most of his spare time to research and scholarly writing. It is also clear that he had a large and sustained commitment to geography. Yet, in spite of his productivity and his involvement in three of the major scientific exploratory expeditions of his age, Bory de Saint-Vincent's writings do not add up to either a major contribution to geography or, indeed, to a significant contribution to any modern field. While his work is often clearly geographic, sometimes by its subject matter and sometimes in his approach, it would be difficult to point to any one of his works and argue with confidence that it defines the nature of his geography. His ideas and thinking were deeply marked by state aggression but he was almost certainly unaware of that influence. With his interests largely directed by his assigned role either in the military or on government-sponsored expeditions, his research was scattered across what became a wide array of disciplines. From botany to geography to anthropology. In the end, he was easily forgotten in all.

CONCLUSION

Geography and power have an ancient association which derives, in part, from the utility of textual and graphic description of the surface of the earth and all of its resources to both the politically and commercially powerful. An equally important part of the bond between geographers and power has been the gradually increasing cost of geographic description, in particular the detailed graphic depiction of extensive regions of the earth's surface. Although mapping clearly involved a considerable investment of time, and skilled personnel, even textual description entailed considerable cost: in terms of the informal or formal collection of information, its storage in libraries and archives, its analysis and synthesis, and finally its production in coherent and usable form. Thus, it was the value of geography to the prince and the cost of geographic work that constantly renewed and refreshed the relationship between the state, and more exceptionally, commercial power, and the geographer.

That old and established association began a new phase of intensification in the Revolutionary and Napoleonic periods. This was due to changes in the nature of warfare, including the increased size and mobility of armies, advancements in artillery and theories of supply, and a shift to more ideological warfare. Larger social and intellectual forces also played a role in that intensification. Toward the end of the eighteenth century, most of the states of Europe were moving toward an understanding of national wealth which included population, agriculture, industry, commerce, and even, to a degree, the social order. This national wealth had to be managed and, in a sense, farmed. Management according to rational principles, it was coming to be understood, required information on all of these aspects of national wealth, and understanding required not just tables of indigestible and incommensurable figures, but graphs and maps of the sort ultimately produced by the officers of the état-major. The very collection of such information would in turn act as a stimulus to the development of statistical language and analysis and a far more statistically sophisticated understanding of society and national wealth. In addition to these larger social forces, which were far larger than Napoleonic France, there was the particular stimulus to geography provided by Napoleon and his sense of what was intellectually and socially valuable. Napoleon's personal interest in not only maps but geography as a repository of spatial information brought geography closer to the state than it had been even in the ancien régime, in spite of ancien régime geography's attention to the measurement of the earth and the search for the solution to the longitudes.

The participation of geographers in state aggression took two principal forms during the Revolutionary and especially the Napoleonic period. These were practical assistance in military conquest and propaganda designed to further the acceptance of state policies. Some geographers also took part in the local administration of conquered territories and even in informally advising the government in the development of its foreign and colonial policies. It is not surprising, then, that, in the course of the Revolutionary and Napoleonic wars, military geographers acquired a new administrative importance and began to develop a new more critical and more purposeful descriptive geography. This geography incorporated the collection and organization of any geographic information that might be valuable to decision-makers, including many aspects of both human and physical geography. In fact, we can see in the topographic memoirs composed by these geographers a number of important innovations, including a new statistical curiosity, an enhanced interest in social structure and the way of life of local peasants and townspeople, a greater appreciation of the use being made of local resources, and attention to commercial networks. In addition, the need that these geographers had, as both geographers and soldiers, to anticipate what could not be seen in the landscape, made them less hostile to theory than the "pure" descriptive geographers such as Malte-Brun and William Guthrie. However, the immediate and unacceptably high cost of misinformation made them intolerant of ill-informed speculation. Consequently, they developed and began to refine a method of field observation and inquiry which is strikingly modern in its sophistication and its awareness of the situatedness of the informants.

Although this military geography was significantly more innovative than the descriptive geography practiced at the Academy of Sciences well into the 1830s, its potential was never realized. Memoirs were written, perhaps used, and then archived. Composed to solve problems of immediate interest to the military, they did not evolve into a body of material accessible to the larger public. Nor did most of the military geographers go on to become proto-anthropologists or regional planners. With the research directed by immediate need, work was often abandoned after considerable investment. There was little room for the perfectionism needed for fine cartographic work. And most geographers must have had a limited sense of ownership over their work. When the Napoleonic regime finally fell, there was more concern to dismantle the Napoleonic war machine than there was concern to capitalize on its innovations and achievements. As a result, the skill and training of many military geographers was lost, and, only a few years after the end of the Napoleonic wars, the military administrators of the Carte de la France, founded in 1817, had great difficulty finding and recruiting the talent needed for high-quality large-scale mapping.

This lost opportunity reflects perhaps inherent limitations in the nature of state or military support for geography, or for that matter science in general: in particular, a relatively short attention span and reluctance to invest in long-term projects whose utility, or political value, is not immediate. The careers of the two highly successful military geographers André de Férussac and Bory de Saint-Vincent further reveal both the personal and professional dangers of close association with state power. Standing above the crowd, even among the scientific corps, these geographers risked their physical well-being and exposed themselves to political danger. André de Férussac lost the first gamble, and Bory de Saint-Vincent, the second. Nor did either of these geographers succeed after the Napoleonic period in securing the freedom and support to carry out research *they* deemed of intellectual merit. In a sense Bory de Saint-Vincent gained the most by following the political winds and participating in a number of official state scientific missions. De Férussac persisted in trying to convince either the military or the government of the need to support his own research, in each case unsuccessfully. The originality of his ideas, his very critical approach to descriptive geography, his enormous capacity for work—all suggest the potential of this creative thinker, had he received adequate support. Trained by the military and protected by military employment, neither scholar had the relative freedom enjoyed by Humboldt for much of his career to pursue his thoughts with concentration and coherence. It seems that the lives they lived in the course of the Napoleonic wars taught them to undertake multiple and varied projects which could be picked up and dropped as circumstances dictated. This pattern of research and writing meant, in Bory de Saint-Vincent's case, that subjects were rarely fully developed or pursued consistently to their logical conclusion. As a result, both men found themselves a little beyond the academic pale and deprived of the intellectual community capable of raising the standards of their work through criticism and discussion.

Bory de Saint-Vincent benefited personally far more from his close association with the state and, in particular, from a successful military career. But the very success of that career meant that the state, and its military arm, virtually shaped his intellectual life. In spite of considerable activity and prodigious research and publication, and in contrast to de Férussac and Alexander von Humboldt, he does not appear to have had a personal scientific mission. He went where opportunity and the state directed him and asked the questions that seemed to be uppermost in people's minds. Thus, while, like de Férussac and Alexander von Humboldt, he published in a wide variety of fields that we would scarcely consider part of the geographic mainstream today, unlike these two scholars, there is no line of enquiry linking all of these works. In the case of both de Férussac and, as we shall see in chapter 7, Humboldt, the coherence came from a commitment to geographic enquiry, or from following one, or a series of, geographic problems through observation and theory to where they might lead. In a sense, Bory de Saint-Vincent's work was more traditionally geographic: his work was descriptive, often cartographic, and his citation pattern was much more that of a geographer. Description offered no obstacles to his intellectual wandering and he traveled from literary description to botanical description to regional description, with little if anything to link these efforts in his own or his readers' minds. If that was his natural tendency, a life of close association with state power exacerbated that tendency. Thus, the rich geographic ideas to be found in his discussion of the human and physical landscapes of Spain were never developed in his work on Morea or Algeria. In a sense then, the work of Bory de Saint-Vincent was undisciplined. Not so much because he lacked attachment to a field of study, as there are many signs that he took such affiliation more seriously than did either de Férussac or Humboldt, but because the field to which he was affiliated could provide little direction in the developing world of the modern theoretical and explanatory sciences. The gap was filled by another form of discipline—military and state power—which gave his work a structure and meaning related less to the ongoing intellectual discussion we have come to regard as "science" than to the rise of imperial power.

In a sense, this latter is an absurd statement as, since *at least* the time of Napoleon, and certainly since the World War II, science and the state have been intertwined and interdependent. While this is true, and while any sense of "science" apart from political power is naive, the relationship between science and the state has been a negotiated one. In that negotiation, turf, control, assessment, and value have been, and continue to be, fiercely—although not always successfully—defended by scholars. In this defense, there are many defensive walls. One of these is the discipline. Weakness in that wall, among others, opens scholars to influences beyond the normal realm of academic discourse, sometimes with nefarious consequences and sometimes with extraordinary benefits. Since the Nazi era brought the ugly ideologies of imperialism home to Europe, Western academia may feel it has developed the skepticism and caution to deal with state power. It now remains to be seen whether it can use some of that understanding to defend itself from the even larger and far less "responsible" economic powers developing beyond the state. In the time of Napoleon, and his immediate successors, scholars had little defense from state power, much to gain from the association, and only a weak sense of its danger. As I have tried to show in this chapter, the careers and minds of some of the best geographers were both informed and derailed by state power.

·PART·THREE·

Innovation at the Margins

FIG. 6. "Paris at the time of Chabrol de Volvic's administration." Although intended as a caricature, this image of Paris, with its foundations in industry and the police, captures something of the spirit of the city in the 1820s.

Explaining the Social Realm

Volney, more closely listened to, would have provided the example of a sociologist, an economist, and a geographer: a lucid traveler freed of all metaphysics. —Gaulmier, *L'Idéologue Volney 1757–1820* (1951), 345

INTRODUCTION

It was from the relationship of "science" with state power that the social sciences were to emerge: from the desire to manage production, commerce, transportation, and most especially population (including labor and minds), both at home, in the cities and the countryside, and abroad, in the colonies and the spaces beyond the colonies. The social sciences evolved as one of the principal tools of the secular Western state. However, also integral to the social sciences, from their very first stirrings in Enlightenment thought, was social criticism, a critical watchdog function over the power and activities of the state. Thus, the same social sciences that produced phrenology, ultimately also produced Frantz Fanon. It isn't that phrenology was wrong and Fanon was right, but that phrenology was social science turned on the other (with little consciousness of the interests of the self), while *Les Damnés de la terre* was the same critical gaze turned back on Western state power.

As we saw in the previous chapter, social scientific geography received an early impetus from the attempts of the Revolutionary and Napoleonic states to control their own and newly conquered territories and peoples. Engineers, military geographers, and officers began to gather information of a nature and in ways we now associate with social science. We have seen that that inquiry was not sustained. That the state was not yet prepared or able to maintain that level of investment in personnel, training, and information collection beyond military or political exigency. We have also seen that the close association with state power had costs. Some of which were clear to contemporaries and some of which were apparently not. There were other signs of the stirrings of social science in the Revolutionary and Napoleonic periods in the activities of a few scholars in the Class of Moral and Political Sciences, particularly in the "analysis of sensation and ideas" section and from the early work of the "Académie celtique." Yet, as

Martin Staum has so clearly shown, a new social scientific geography did not emerge from the work of geographers in these institutions. Those who called themselves geographers and who arguably were well placed to explore the possibilities of a social geography, either because they participated in voyages of exploration or because they had at their fingertips the best libraries and map collections in the world, showed little inclination to do so.[1] Suggestions of the possibilities of the subject came from the fringes of the community. It came from scholars who saw in questions that married the physical and the human and that spatialized the social an inherent interest and a political value that had barely begun to be explored. This emerged from their remarkable imaginations, sometimes stimulated by contact with geographers. Their imaginations were more deeply fed by the ideas of their predecessors, the men who, in the eighteenth century, had begun to formulate questions focussed on society and to develop methodologies suited to those questions. They were also nourished and sustained by state interest in, and support for, their activities. This nascent social scientific geography substantially depended on such state support, and when it was withdrawn, this geography deflated and collapsed.

In this chapter, we will look at the writings and activities of three men whose work, collectively, covers the period from the late 1780s to the late 1830s. This is a very early phase of social science. Most historians of social science consider it a period of proto–social science. Early-nineteenth-century scholars who thought critically about social problems were rare. Those who thought about them with anything approaching spatial sensitivity can be counted on the fingers of one hand. Thus, it would not be easy to replace either of the two innovators in this chapter with another contemporary scholar. They are not representative of many others who were undertaking similar research. They do, however, represent a growing interest in the kinds of questions they asked and a growing sense that the way such questions were asked was important. The two innovators discussed in this chapter are Constantin-François Volney, who began to develop a geographically sensitive approach to the study of other peoples and their societies; and Gilbert-Joseph-Gaspard, comte de Chabrol de Volvic, who had an unusually developed spatial sense of the state and used statistics as a powerful tool of social analysis and administration. The third figure in the trilogy does represent the field of geography relatively well. He was a cabinet geographer who had occasion to travel and who sought to apply the essentials of statistics to geography. He also proposed, at least seventy years before the subfield came into existence, a new and altogether different approach to the study of the "other," or ethnography: linguistic geography. It is clear from the work of Volney and Chabrol de Volvic, who worked with, influenced, and were influenced by geographers, that there were some significant signs of innovation on the very margins of what was then understood as geography. Mainstream geographers such as Adrien Balbi, that is men who had a strong self-identification as geographers and were recognized as such by their contemporaries, had great difficulty incorporating these innovations into their research. Balbi failed to notice both the relevance of Volney's ideas about the study of foreign societies and Volney's argument and demonstration of the importance of fieldwork to social inquiry. Nor did he fully understand the concepts behind the statistical terminology which he did, in fact, adopt. This was less the result of one man's limitations than a consequence of his identification with a field which possessed a strong attachment to a particular publication genre, scope, and method of study. Nevertheless, in the 1820s and '30s, innovation was in the air, and there are signs in the work of Balbi of a dawning interest in social phenomena, reflected in his attention to demographics; the economic use of territory; the role of infrastructure in facilitating commerce, agriculture, and industry; and in his abandonment of characterizations of "peoples" in favor of analysis of the sources of their behavior and attitudes.

VOLNEY AND HIS SOCIAL AND
SCIENTIFIC GEOGRAPHY

Constantin-François Volney is a complex historical figure who, in spite of a number of excellent articles, a good historical biography, and an exceptional intellectual biography, is still vague in detail and sometimes in outline. Born into aristocracy in 1757 and trained by the Oratorians, he eventually studied medicine, which he practiced, as a philosopher rather than as a healer, much of his life. Very much an intellectual, he studied Hebrew and Arabic early in his career and retained a strong interest in language and a conviction that language might provide the only possible access to dying or dead civilizations. He wrote a number of important and highly influential works, particularly in the sixteen years from 1787 to 1803, but then seems to have abandoned the intellectual project in which he had been engaged. His ideas were still considered sufficiently influential and persuasive to warrant proscription in the 1820s.

For the historian of social science, one of the most striking aspects of Volney's life and intellectual career is his ambiguous relationship to political power. Volney perhaps best embodies the social scientist: thinking for himself, at times in open radical opposition to state power, and utterly determined to thwart its excesses and guide its more positive impulses. Yet thoroughly implicated by it: paid by the state, beholden to the state, stimulated by and serving its interests and silenced by its opposition. As his biographer, Sibenaler, has pointed out, it is likely that funding for Volney's expedition to Egypt from 1783–1785 came from the French royal government. It is possible that he traveled to that country as a spy for the French and against the Turkish empire and its allies. However, he did not restrict his attention to phenomena of interest to the government. In Egypt, Syria, Corsica, and the United States, he had his own ideological axes to grind which had less to do with political service than with social criticism. Nevertheless, his activities in Brittany in 1787 were political and it is likely that so too were his activities in Corsica in the early 1790s where he appears to have been sent to explore the possibility of using Corsica as a French fruit and bread basket. He hoped that his trip to the United States (1795–1798) would be paid for by the government and perhaps lead to a diplomatic post, but in the end Volney himself bore the cost of that three-year exploration.[2] His links to Napoleon are well known and there is evidence that he assisted in the coup d'état that brought Napoleon to power. Volney's biography tells two tales: one of a political figure usually on the margins, yet playing an important (and well remunerated) role in the major events of his time, from the Revolution, to the invasion of Egypt, to the rise of Napoleon; and the other of a quiet intellectual, pacifist, and disappointed idealist who abandoned the political scene and the concerns of social science for more erudite study when his hero, Napoleon, proved not to be the enlightened representative of the people that Volney had believed him to be. The picture of the man, no matter how often one reads his biographers or Volney himself, always retains the two-dimensionality and fuzziness of a stereograph, before the eyes have reconciled the two images into one. Yet if one brings to the image an understanding that, especially in the realm of social science, the interests of the scholar and the state are almost invariably inextricable, the image of the man and his work begin to acquire definition.

Volney's interest in what we today would describe as geography resulted from the confluence of a number of important intellectual influences and from the nature of his own political and social commitment. Volney, as his chosen name, derived from the words Voltaire and Ferney (Voltaire's estate), would suggest, was heir to the political, social, and philosophical thought of the Encyclopédistes. He shared this inheritance, together with the epithet "ideologue," with a number of the other prominent intellectuals of the

1790s, many of whom he came to know through the salons of Paul Henri Dietrich baron d'Holbach and Claude Adrien Helvétius. The ideologues, including Antoine Louis Claude Comte Destutt de Tracy, Georges Cabanis, Joseph Marie Baron de Gérando, Pierre Louis Comte Roederer, Joseph Lakanal, were a group of intellectuals who survived the Revolution without abandoning the Enlightenment belief in rationalism; the perfectibility of "man" through education and good government; the unity of all peoples; liberty, equality, and justice; and the power of observationally based science to reveal the nature and advance the well-being of "man." Volney can be seen as part of this group by virtue of his close relations with these intellectuals, the ideas he shared with them, and Napoleon's own inclusion of Volney within this group of "dangerous" intellectuals. With Cabanis, Volney inherited from John Locke, and from the study of medicine, a view of the human being as a creature who acquires knowledge and understanding through the world acting upon his senses. This invited study into two principal realms: human physiology and the action of the outside world upon the human senses. While Cabanis and, eventually, Destutt de Tracy focussed principally on the first realm, Volney devoted his attention to the second: the world—both social and natural—acting upon "man." His approach to that world was also strongly shaped by his intellectual predecessors. Apart from his name, he inherited from Voltaire firm skepticism, a fascination for history, detestation of religion, and a belief in rationalist science. He shared with Condorcet his belief in progress and the perfectibility of "man" and his interest in studying "man" as a social creature using a scientific methodology based on probability. From Condillac, Volney learned a scientific method based on observation and both the importance and subtlety of language. He shared with Helvétius the conviction that all people are alike, in that all individuals, families, peoples, and governments are comparable. From Jean François de Galaup Comte de La Pérouse, Charles Marie de La Condamine, and Louis Antoine Comte de Bougainville, and the numerous other scientific voyagers of the eighteenth century, he acquired a belief in the value of scientific travel which also conformed to the sensibilist's belief in the importance of "seeing" for oneself. Consequently, for Volney, the travel relation was not novelesque diversion, but a way in which the lessons of history could be learnt across cultures without the sharp pain of bitter experience. Volney's anthropological curiosity about other human societies, their governments, ways of life, commerce, and laws was one he shared with Dégerando and Cabanis. There was, also, a certain continuity with earlier eighteenth-century use of the "other" as a cover/foil for social criticism and this newer anthropology. That is, although Volney and his colleagues

in the second class of the Institut and at the Académie celtique were
genuinely interested in the societies on which they turned the beginnings
of an anthropological curiosity, this curiosity was not pure, innocent, and
unlinked to criticism of French society, as is clear in both Volney's *Ruines*
and in Napoleon's intolerance for the speculations of the "ideologues." In
short, Volney's ideas about nature, government, natural law, hygiene, and
language were deeply rooted in the thought of both the Revolutionary and
immediately post-Revolutionary period and in the writings and ideals of
the Enlightenment.

Among the ideologues at the end of the eighteenth century, Volney
was not a prodigious writer.[3] His relatively few works can be divided into
four major categories, although some of his works fall into two or more
categories: what has been seen as travel description, his historical writings,
his philosophical and pragmatic explorations into the nature of language,
and his philosophical/political works. The first and, for geographers, the
most significant category of his publications has begun to receive serious
attention from historians, although geographers continue to neglect it.[4]
These writings represent either partial or complete attempts to explore the
relations between "man," nature, and society with a view to improving
society. In them Volney expressed a clear and consistent conception of
"man" and society, described them as the product of the interaction of social
and natural conditions, and called for, and took the first steps toward, an
integrated and rigorous study of "man," nature, and society. In so doing he
laid the groundwork for a new social scientific geography.

Volney's geographical writings include his *Voyage en Syrie et en Égypte*,
published in 1787 and which went through eight French and three foreign
editions;[5] *Tableau du climat et du sol des États-Unis*, published in 1803 with
five French editions and two foreign editions;[6] *Questions de statistique à
l'usage des voyageurs*, actually written some twenty years before its publication
in 1813;[7] and "État physique de la Corse," which was not published before
his death.[8] *Voyage en Syrie* was the first and among contemporaries, perhaps
apart from his *Ruines*, the most famous of his major publications. In it he was
concerned to measure the relative impact of climate,[9] soil, and government
on the various peoples inhabiting these countries. He had intended his
works on America and Corsica to explore the same general question but
in neither case did he complete his analysis.[10] His *Questions de statistique*,
although designed to elicit field notes and not memoirs, embodied his
approach to geography and can be seen as the skeletal plan of a complete
geographical treatise.[11]

In all of his geographic writings, we find humanity and society at the center of concern, shaped both by nature and the human environment and at the same time influencing both the natural and the human environment. It was people and their relationship with both the environment and with other peoples that interested Volney. While among his philosophic and "ideologue" colleagues this interest resulted in an almost exclusive focus on human physiology and humanity and society,[12] Volney was at least as interested in the human and natural environment—in humanity's social and physical context—as he was in the nature of humanity.[13] Volney's approach to the realm of geography was rigorous, and in his study of Corsica, Egypt, and the United States he demonstrated to anyone who cared to follow his lead a methodology appropriate to such a science. It entailed the use of three elements: hypothesis, field research, and criticism—the correct use of which he both described and demonstrated.

A Geographic Methodology

The statement of hypothesis or the posing of key questions was a matter of importance to Volney. For Volney there was no real distinction between a hypothesis and a good question: behind any good question lay a hypothesis. A question asked without a sense of what would issue from it was a child's, not a scientist's, question. His *Voyage en Syrie* was structured around the question of whether the primary influence on humanity was nature or government.[14] Within the overall structure of the work he asked more specific and limited questions: was Bedouin society healthy?[15] had the Nile Delta undergone radical change in the course of history?[16] what was the motor force behind winds? what was the relationship between climate, soil, and vegetation?[17] etc. In contrast to the questions posed by Buache de la Neuville and Mentelle, these were questions for which he did not have a ready answer. Their value lay in their very state of irresolution. In Volney's view, both the human and natural sciences were at a stage where, while these questions had to be asked (if only to structure research into the nature of humanity, society, and nature), they could not and should not be answered with any certainty. One could make observations and generalizations, but the latter had to remain open to contradiction: "certainty is the doctrine of error or fallacy, and the abiding weapon of tyranny."[18] Frequently, in the course of his own argument he contradicted his hypotheses, thereby suggesting the possibility of an alternative explanation. Thus, a question was a tool for the gathering of facts and for the structuring of thought. Description for description's sake was much less than what he was engaged in.

After a clear and concise statement of hypothesis, the second important element in Volney's method was the use of field research. This was a dimension entirely absent from the geographies of Edme Mentelle, Jean-Nicolas Buache de la Neuville, William Guthrie, Conrad Malte-Brun, and a whole host of cabinet geographers. In fact, Mentelle's primary-school geography book, which was substantially devoted to France, was based on the answers to questionnaires he had sent the administration of each département. The inadequacy of this approach even for a primary-school book is clear from his indignant publication of the extensive list of those who either did not respond or who responded inadequately.[19] In contrast, fieldwork offered the possibility of both framing and answering one's own questions, of seeing for oneself, and of recognizing patterns not evident to others because their assumptions had been different.[20] The nature of the terrain of Egypt, the climate of the United States, the farming possibilities of Corsica, and the nature of Ottoman government were not to be learned in books but where and as they were lived and experienced.[21] This was why plagiarism was such a grave error.[22] It could only lead to misunderstanding and the formation of ill-informed stereotypes.

In a sense, Volney can be understood to have shared the eighteenth-century philosophic view that all peoples were essentially the same in that they shared the same basic needs and aspirations, that all societies had the same functions and were subject to the same corrupting forces, and that the geography of the world formed part of an enormous and complex yet single natural system.[23] That Volney's purview was neither restricted to the thin surface and untheorized space which were the preoccupation of cabinet geographers nor the universal geographies of the early nineteenth century is particularly clear in Volney's approach to the human/cultural realm. Although all peoples were basically the same, Volney recognized that they appeared very different, had varied ways of life, and often espoused apparently diametrically opposed value systems. How could this be explained? In Volney's view, the differences between peoples were substantially due to the circumstances in which they lived, either in terms of the natural environment, or in terms of the human/political environment.[24] These differences and their causes could not be explored without field research. There was no point in relying on the secondhand generalizations of a Buffon or a Montesquieu.[25] The only way to understand a people, their customs, and their institutions was to immerse oneself in the context which had given birth to them.

> For such a study one must communicate with the people that one wishes to
> understand, one must adopt their circumstances in order to feel the forces

that are acting upon them, and the attachments that result: one must live in their country, understand their language, practice their customs . . . .[26]

From the outside, much might seem irrational, bizarre, and even capricious. From the inside (and in a healthy society)[27] the human realm and its place in the universe would be ordered. Here then lay the difficulty. All peoples were indeed alike, but they lived within different contexts and their context limited their ability to perceive the similarities they shared with other people in other contexts. The field researcher would have to make a special effort to leave his culture behind:

> Not only must one fight the prejudice that one encounters, one must defeat those that one carries: the heart is partial, custom is powerful, facts are insidious, and illusion easy.[28]

Once his field research was complete, he would return to his own culture and discover the difficulty of representing the "other." Because, upon his return, his pen would be guided by the views and interests of those around him:

> Soon there arises between his listeners and himself an emulation and a commerce according to which he returns in astonishment what they pay him in admiration.[29]

The problem was a thorny one and one which, since the writings of Clifford Geertz, has caused open anguish in modern anthropology and ethnogeography. Fieldwork, as conceived by Volney, represented at least a partial solution to the problem. In the field the sources were multiplied. It might be possible, for example, to use verbal testimony and to truly interrogate the source. For some kinds of information there was no better means of obtaining data.[30] For others the earth was a more honest book than most of those published—once one had learnt how to read it. In order to remain true to one's sources, whether human or natural, it was most important not only to collect data on location but also to write it up on location. There is evidence, both in the structure of his presentation in his *Voyage en Syrie* and in the testimony of his posthumous editor, that Volney composed notes in argument form on location and then wrote directly from these once back in France.

The third element in Volney's methodology, criticism, is implicit in the first two and has therefore already been indirectly addressed. He stressed repeatedly the need to multiply sources, to greet them with apprehension and uncertainty, to question them, and to test them against each other. In this process, skepticism and one's sense of what was reasonable and probable based on one's experience of the world was the best, and really the only, guide. Doubt and uncertainty were part of the intellectual stance appropriate to any intellectual. Referring to his own *Les Ruines*, he commented:

> In general, it exudes a spirit of doubt and uncertainty which seems most appropriate to the limited nature of human understanding and most likely to advance that understanding in that it leaves a door open to new truths. In contrast, the spirit of certainty and fixed beliefs restricts us to received ideas, ties us to chance, . . . to the yoke of error and untruth.[31]

Thus, when many of his contemporaries still gravitated to the authority of the ancients for the resolution of problems beyond the realm of even modern science, Volney asked of those scholars, "but what are the facts on which they [the ancients] based their opinions? . . ."[32] He attempted to thus arm his readers and repeatedly invited them to criticize an author's assumptions and conclusions. He himself made that possible by unfailingly explaining the sources of his information and the process of his reasoning. Hence, although his diagnosis of what ailed the French Illinois settlement of Vincennes at the time of his visit was uncertain and vague, his description of the features of the problem and his insight into the possible relationship between the problem and differences in the local French and English methods of measuring space and time is so complete as to make its source, the meeting of two radically different colonial and economic systems which used space and time differently and incompatibly, jump out from the page at the modern reader. A critical approach was essential to Volney's method: without it, his hypotheses would be no better founded than the vague and ridiculous generalizations that he so little respected in the writings of both Montesquieu and Rousseau.

To the modern social scientist, trained in quantitative techniques and steeped in modern philosophy and epistemology, Volney's methodology appears simple and straightforward. In fact, many of the questions he was asking were difficult and often irresolvable. The real problem for contemporaries was that the resolution of the vast majority of his questions required extensive and coordinated research in widely separated parts of the globe. On numerous occasions Volney emphasized the insufficiency of the efforts of single travelers moving across terrain and gathering information for only short periods of time. Consequently, he repeatedly called for interdisciplinary group work and government support and direction on subjects as varied as the winds and topography in Africa, the Middle East, and North America;[33] the Kurdish language (in order to compile a dictionary); the gradual movement of Niagara Falls; past earthquake and volcanic activity; and the languages of the North American Indians (with a view to tracing—in the absence of written works and monuments, and given the changeability of oral traditions—the origins of these peoples).

Geographical Problems Posed by Volney

Volney was not alone among the ideologues in having a geographic imagination. Indeed, he shared with most of them the conviction that the study of humanity within its geographic and social milieu should be one of the principal preoccupations of modern science.[34] He was different from most of his contemporaries in his preparedness to himself explore these themes in distant and disparate parts of the globe. Volney's particular geographic imagination stemmed from his willingness and ability, perhaps best metaphorically expressed in his Piranesque[35] *Les Ruines*, to shift his focus from the minutiae of daily life, legislation, and nature to their larger context and interactions. He tried in his geographic writings to harness the power of the spirit with whom his character conversed in *Les Ruines* to view human existence from a distance of several thousand miles above the earth and yet to see all its details with extraordinary clarity of vision.[36] The scale of his interests, then, ranged from Cabanis's attention to the physical influences on the human senses, to nature's interaction with the "society," so recently discovered and explored by the "philosophes" and the "ideologues." Society was not fixed but had evolved and would continue to do so under the influence of both human will and nature. For Volney, nature itself was as open to investigation as the physiological individual and society.[37] New questions, however, needed new information. And this Volney sought in his travels to Egypt, Syria, Corsica, and the United States.[38]

In his geographic writings, then, Volney focused on the relations between the individual, nature, and society. Human beings, both individually and in society, were ruled by nature. Not in any immediate brutal sense. As a sentient being, whether alone or in society, the human being was always part of nature and would seek to fulfill basic natural needs:

> . . . self-love, the desire for well-being, and aversion to pain! Here are the essential and primordial laws imposed by nature itself on man.[39]

These were very basic laws, which if ignored, as they frequently were by the composers of laws, would pervert individuals and societies. Understanding a society, then, required bearing in mind these basic laws of nature; knowing and perhaps experiencing the physical conditions, climate, soil, etc., of its existence; and penetrating the social forces acting upon both individuals and groups within the society. Multiple-factored explanations which took into account both the natural and the human environment were always called for. The best example of this approach is his analysis of the life-style of the Bedouin. The Bedouin lived, he argued, a "way of life which revolts us,"[40] but in spite of their wandering life they appeared to have maintained their traditions across great expanses of both space and time:

> to this unity of character preserved over considerable periods of time is combined a similar continuity over vast expanses of space, that is, that the most distant tribes do infinitely resemble one another, it would certainly be interesting to examine the circumstances which accompany such a particular social state.[41]

Volney concluded that theirs was a healthy society because their way of life was an entirely reasonable and informed response to the poor quality of the soil and a hostile human environment. The poor soil did not allow them to settle but did permit them to live off a constantly moving flock. The political instability and repression prevalent in the Turkish empire constantly replenished their ranks and removed all temptation to settle on agricultural land. When both physical and social factors were taken into account, their society was comprehensible.

A less successful analysis was the one he applied to the North American Indians. Although, in his view, nature offered them the possibility of a settled agricultural life, they nevertheless lived a nomadic existence. Volney suggested that this nomadic life in the midst of plenty was anomalous and in part explained the degeneracy (violence, alcoholism, etc.) of their society. In trying to understand North American indigenous society, Volney did not, as he himself advocated, rise above ignorance and separate out cause and effect.[42] Instead of assuming, as he had with the Bedouin, that the North American Indian was reacting to a combination of forces, Volney pointed to a poorly defined combination of social and physical influences, which he called their "way of life," as cause.[43] By this he meant their hunting and gathering means of livelihood. This he was convinced was backward or wrong in a context of plenty. In a sense, Volney was taking a far more critical and nuanced approach to the question than Cornelius de Pauw,[44] who saw the entire geography of America as corrupting, or Jean-Jacques Rousseau,[45] who considered the "savage," by virtue of his distance from "society," noble and pure. But to point to their way of life as wrong in the context of plenty did not explain the "wrongness." Volney was not in a position to grapple honestly with that question. He was less interested in understanding indigenous societies on their own terms than in understanding their place in a hierarchy of cultures and in justifying ongoing Euro-American conquest:

> What it is important to know is whether the Savages reasonably have the right to refuse land to those agricultural peoples who would not have enough of it to live off.[46]

Volney's perspective on indigenous America was encumbered by interest and prejudice fueled by a fundamental belief in property as one of the foundations of society. His perspective became a major obstacle to insight because Volney failed to follow his own method in his analysis of the societies of indigenous North America. Volney, in keeping with his methodological strictures, had lived among the Bedouin and had sought to see their society from within. In the case of the Great Plains Indians, however, he relied far too uncritically on Euro-American perceptions. Indeed, although he had considered it, he decided not to go and live among the "Indians" on the advice of Europeans who described their society as structureless, anarchic, and dangerous.[47] This, Volney had warned in the context of the Bedouin, is how all societies appear from the outside. His American trip, although restricted to a relatively small portion of the then United States of America, was nevertheless extensive and exhausting. He spent three years in the country and traveled from Philadelphia to Washington, to Monticello, twice. From there in 1796 he ascended the Ohio River and then struck out for the French Illinois territory, which was in that period in decline under the demographic and political pressure of the great American push over the Alleghenies. Instead of pushing south along the Mississippi, which had been his original plan, Volney struck north to the lower Great Lakes. Traveling via Cincinnati, Detroit, and Buffalo, he crossed Lake Erie, visited Niagara Falls, and then descended New York State via the Hudson River to finally arrive in Philadelphia after a trip of seven months. Conducted on horseback, foot, and by boat, it must have been a significant physical challenge. It was in many ways a far greater challenge than his more narrow itinerary through lower Egypt and the coast of "Syria." In the United States, Volney found a multiplicity of geographies and societies well beyond his ability, as a solitary traveler, to assimilate, never mind to render analytically. It is perhaps for that reason that he chose to restrict his book, which he initially planned to devote to both the human and the physical geography of the United States, to the physical condition of the country.[48]

Where Volney took the time to live among those he was studying, he found sympathy for them, no matter how inimical their philosophy was to his own. One of the best examples of this was his response to the Coptic monks of the Mar-Hanna monastery. On his voyage through Syria, Volney and his interpreter stopped and stayed in that monastery for eight months. There he read, studied Arabic, and conversed with the monks. Volney was hostile to religion and not just to Christian religion but to all of the religions of the world. He regarded them as complex systems for the propagation of injustice and ignorance.[49] Faced with the peaceful and measured life of the monks of Mar-Hanna, Volney could only marvel at the society's ability to wrestle the calm necessary to contemplative erudition from a countryside ruled by tyranny and the chaos of misery and want. Characteristically, he refused to see the life of the monks as separate and distinguishable from the surrounding society. Indeed, he considered that it was precisely the arbitrary tyranny which ruled the lives of peasants, shopkeepers, or soldiers in Syria which, despite its rigorous rule and restrictive life, kept the monastery well stocked with novices.[50]

While clearly not a simple determinist, and repulsed by the climatic arguments of Montesquieu, Volney nevertheless attributed great importance to the impact of nature on both individuals and society. He divided nature into two components: climate (in the study of which he saw himself as a primary researcher) and soil (for which he tended to rely much more heavily on experts). As a doctor and as a sensibilist, he saw climate as playing a major role in an individual's sense of well-being and susceptibility to disease.[51] This was the source of his considerable interest in climatology. However, Volney was more interested in the individual in society than in the physiological individual. Consequently, in his *Questions de statistiques* he also emphasized the importance of climate to agriculture, commerce, and shipping. Climate, for Volney, was not a fixed presence but a complex and changing global system in which the action of the sun on the surface of the earth, topography, and winds interacted to create both micro and macro climates. He deemed climate as complex as humans and human society, and in order to understand its role in influencing both, it had to be an object of study in its own right, as it manifested itself and as it was transformed from place to place. Thus, he explored the importance of temperature variation over short periods of time and the dangers of averaging out temperature data.[52] He developed a concept very close to our concept of pressure to explain wind and humidity change.[53] He examined the impact of bodies of water on local and global climate as a result of wind movement across terrain.[54] Finally, he produced a clear and almost perfect description of alternating coastal winds, and pondered the causes of desertification.[55]

Climate was most closely linked to humanity through disease but again not as a single causal factor. Volney rarely claimed to be sure of the cause of any particular disease and he often combined likely causes. In the Middle East he did find a correlation between the presence of water and what we now know as River Blindness;[56] in the United States he noted a correlation between a marshy or polluted water supply, urbanization, and the incidence of yellow fever;[57] and he tracked the seasonal movement of the plague from Constantinople to Egypt.[58] His approach to disease was both subtle and supple, and, without the knowledge of germs, viruses, and bacteria which would have allowed him to understand its cause, he nevertheless often managed to suggest the methods appropriate to its control or elimination. What lay at the success of his analysis was his tendency both in questions of climate and disease to look at the spatial distribution of the phenomena in conjunction with the distribution of any possible causal factors.

The relationship between nature and humanity was not unidirectional. Nature was clearly the dominant force but humanity had some power to alter nature—generally not for the better. In reference to Egypt he commented on the disastrous impact that a redirecting of the Nile into the Red Sea long before it reached Syene would have had on the land of the Nile and commented acidly:

> Looking at the uses to which man puts his power, should we reproach nature for not having accorded man more power?[59]

A less incidental series of observations concerned his study of the impact of deforestation on climate. He was convinced that the deforestation taking place as part of the process of agricultural settlement in North America had two measurable small-scale impacts: it led to the local swelling of rivers and it led to less rainfall in the region.[60] But he was equally convinced that continental-scale climatic change was attributable to local and even minor deforestation through the impact of the absence of trees on winds and the sun's warming of the earth.[61]

In his geographical writings, then, we find the individual and society at the center of concern, shaped both by nature and the human environment of government and legislation and at the same time influencing both the natural and the human environment. For Volney, the social field represented a complex system which he barely felt able to describe and about which he held few certainties. Rather, its contemplation left him with a plethora of complex questions. Nevertheless, the way he structured and approached the subject, and the many strictures he laid down for those concerned to explore similar questions, would have suggested to anyone who cared to follow his lead the scientific methodology appropriate to such a study.

The Baton Dropped

Volney was not a geographer and never referred to himself as such. He occasionally described his writings on Egypt and Syria as geography and it is clear from these references that he included under this rubric the study of both physical and human phenomena. Volney saw the geography that he practiced as a science. This did not mean that it was banned from the study of the individual and the larger society. It meant that in contrast to history, in which facts could not be verified but only suggested by analogy because historical facts subsist "only as apparitions in the magic mirror of human understanding where they are submitted to the most bizarre projections,"[62] geography was a science where one could go and see for oneself. It was a science where one could test one's hypotheses. Volney's references to geographers were largely restricted to those concerned with maps, place names, or the determination of either location, boundaries, or area.[63] Volney was not very interested in these questions (they occupy only three percent of his *Questions de statistiques* and he rarely sought to resolve them himself). Nor did he particularly value description undriven by theory, whether in written or cartographic form.

Volney and his geographical writings represented both a major challenge and an opportunity to the geographers of his era. There is little indication that prior to 1795 geographers had recognized Volney's writings as related to their work. His *Voyage en Syrie* was known to and respected by the geographers on the expedition to Egypt, but they used it as a source of accurate information rather than as a research and methodological guide. It is clear that, through the courses at the École normale, contact was made between Volney and mainstream geographers. We also know that he had contacts with the military geographers at the Dépôt de la guerre in the course of his career. He had met Pierre Jacotin and the team working on the Corsica cadastre while in Corsica, and a decade later knew or remembered them well enough to intercede on behalf of one of the engineers with Napoleon.[64] His advice was sought in the determination of the system of transcription most appropriate to the topographic map of Egypt and he was brought into the Dépôt to review and evaluate the cartographic use of that system.[65] And we know that he corresponded with both Andréossy, who received recognition from Napoleon in Egypt for his mapping operations in the region of Lake Menzaleh, and one of the founders of the French geographical society,

Edme-François Jomard.[66] But it is equally clear from the subsequent history of French geographic thought that little resulted from these contacts. It is impossible to say with certainty whether Volney's influence might have led to reform in geography had the tide not turned, had Napoleon limited his power and aggression, and had he not eliminated the Class of Moral and Political Sciences of the Institut, part of which was the very embodiment of ideologue philosophy.[67] Certainly, Volney was already worried by the course of political events when he left France for America in 1796.[68] However, he remained engaged and involved in both political activity and social research up to the 1801 papal concordat. At that point, his respect for Napoleon and his belief in the possibility of profound social change dissipated with Napoleon's militarization of the country, the muzzling of the press, the amnesty of aristocrats, and Napoleon's lack of respect for the rule of law, but most especially with Napoleon's rapprochement with the clergy. After 1801 Volney largely abstained from active political life and seems to have withdrawn to a more erudite and restrictive form of study.[69] From mid-1801, he maintained an attitude of open hostility to Napoleon and his policies, while Napoleon continued to reward Volney's achievements as a younger man with honors, and sizeable salaries and pensions. Thus, under the late Empire and the early Restoration, which was no more favorable to critical social inquiry, Volney found himself a disillusioned but wealthy man, ill-inclined to undertake the kind of social research that his ideology had inspired him to pursue so vigorously in Egypt and Syria.

SUGGESTIONS OF A SOCIAL SCIENTIFIC GEOGRAPHY IN THE WORK OF CHABROL DE VOLVIC

If in Constantin-François Volney we find the first stirrings of ideas and approaches reminiscent of social scientific geography, in Gilbert-Joseph-Gaspard, comte de Chabrol de Volvic we find a man who practiced the social scientific geography of the military engineers and brought it to bear in some of the most vital areas of the French realm. Trained as a military engineer and then reeducated at the École polytechnique, Chabrol de Volvic graduated in the same year as Edme-François Jomard. His was a remarkable career, conducted in the very corridors of power, yet reflective of a highly imaginative and independent mind. Assigned to the expedition to Egypt as an officer of the ponts et chaussées, he was one of the approximately

150 half-militarized scholars whose duty it was to integrate Egypt, ancient and modern, into modern France. Men such as Chabrol de Volvic, that is engineers, geographers, and cartographers, were the key men in this operation. In fact, in Egypt, and subsequently, Chabrol de Volvic's skills were repeatedly called upon to assist in the integration of recently conquered territories into the French realm. In all of these operations, Chabrol de Volvic worked closely with geographers.[70] In Egypt, we know that he worked on supplying Alexandria with a regular and permanent source of water; exploring, mapping, and commenting on antiquities; and assisting in determining the extent of the population, the cultivable land, and the number of villages in Egypt for the purposes of general administration and in order to establish a cadastre to regularize the system of taxation.

The primary objective of the Egyptian cadastre was to wrestle control of the taxation system from the Copts so as to cut out the middlemen and render taxation both more efficient (and arguably, perhaps, more fair) and more lucrative for the French rulers of Egypt. The Ottoman-dominated Mameluke system of government, which had been in existence in Egypt since the sixteenth century, had always suffered periods of disruption. As a result, Egypt had developed a Mafia-style system of government which emphasized complex relations of protection and fealty (based on an ongoing process of negotiation) between different groups: the fellaheen (the Egyptian peasantry), the settled Arabs, the Bedouin, the Mamelukes, the Pasha and his entourage, the foreign traders, and the Copts. Over the course of the expedition to Egypt, the French moved back and forth between wanting all or some of the middlemen to assist in the government of the country, to wishing to eliminate them altogether. The French conception of the ideal Egyptian society clarified over time and is evident in the writings of the members of the expedition, including Chabrol de Volvic. Egyptian society would be composed of rulers and producers. There would be a small French elite directing the country, a small governing class composed principally of former Mamelukes, and beneath them peasants, merchants, and industrialists. There was no room for semi-settled Arabs, for Coptic officials, and least of all for the Bedouin, whom Volney had so respected.

The Muslim system of landownership of the Iltizam, the Talaq, and the Waqf, which was at the heart and soul of relations between the Pasha, the Mameluke, the Sheik, and the peasants, was seen as an impediment to the rational development of agriculture and industry for export. Prior to his departure for France, Napoleon had attempted to institute private property by decree (or the confiscation of all lands not registered as private).[71] The result had been a revolt in which, not at all incidentally, the office of the chief geographer/topographer on the expedition was attacked and the geographer, himself, killed. The cadastre, which became the primary responsibility of the expedition's cartographers/geographers, was an attempt to introduce private property (or the registration of land ownership) slowly, systematically, and irremediably. The essays written by the officials with geographic curiosities and affiliations such as Edme-François Jomard, General Antoine François Comte Andréossy, Du Bois Aymé, Michel-Ange Lancret, Louis Costaz, and Chabrol de Volvic on the population, agriculture, industry, commerce, and landownership system of Egypt were all directed toward this end.[72]

In a work dominated by attention to the remains of ancient Egyptian civilization and mapping as an approach, the young Chabrol de Volvic's "Essai sur les moeurs des habitans modernes de l'Égypte," in *Description de l'Égypte*, stands a little apart. This essay was strongly influenced by Volney's own voyage to Egypt, sharing both its interest in peoples and the sources of their customs and the conviction that poor government lay at the root of the country's most serious problems.[73] The other major influence in this essay is traceable to the geographical thinking emanating from the Dépôt de la guerre. Chabrol de Volvic went to great pains to provide population numbers for the distinctive peoples of Egypt, from the Copts (for whom he had considerable sympathy), to the fellaheen, to the Mamelukes, to the Arabs, both settled and nomadic (for whom he had considerably less sympathy). This sort of statistical geography was, in Chabrol de Volvic's view, a contribution to science, so valuable that it represented the best of the work carried out by the French in Egypt.[74] Chabrol de Volvic was not the only geographer paying attention to population statistics. He appreciated and acknowledged Jomard's "Mémoire sur la population comparée de l'Égypte ancienne et moderne" in the same collection. Chabrol de Volvic's essay expressed most clearly the view, then prevailing at the Dépôt de la guerre, that it was population statistics that would allow the government to assess the resources of Egypt, to compare production to consumption, to compare tax to revenue, and to regulate the public economy. While, at this stage, Chabrol de Volvic lacked some of the sympathy of Volney (together with the opportunity to live amongst the people on their own terms), there was in his work a nascent anthropological interest which, in Jomard's writing, was subordinated to numerology and the political value of his statistics. Yet, Chabrol de Volvic was unable to structure his considerable interest in local customs, and his essay often disintegrates into a series of mini-essays on particular customs. In these mini-essays, little treasures of observation are to be found on the street life of Cairo, the lack of physical violence among the urban dwellers even in the midst of personal conflict, and the role and behaviors of women in society.

After his return from Egypt, where he had come to the attention of Napoleon, Chabrol de Volvic was named sub-prefect of Pontivy. He was a good choice for the job. It was Napoleon's idea to raise a new town on that site, briefly known as Napoléonville, which would open that part of Brittany to imperial control. As Paolo Morachiello and Georges Teyssot have pointed out, the foundation of Napoléonville was part of a new vision of national/imperial territory, born out of physiocratic principles, which sought to integrate regions through a system of internal connections and communications and the creation of a clear state presence. The connections and communications included the canalization of previously non-navigable rivers, the construction of state roads, and the development of port facilities. The creation of a clear state presence might be achieved with the establishment of nodes of state power far from the heart of power but at the intersection of vital communication routes. These nodes, or towns, by their structure and architecture would speak clearly of the presence of national power. This was a most modern conceptualization of state control and entailed a form of internal colonization whereby centers of resistance to the centralized state would be integrated into the national system.[75]

> At the bottom of this policy of communication lies the concept of the State as fiscal and formal unit, no longer a set of persons or classes, but a territory made compact and impenetrable by a central authority. Such a territorial concept heightens the role of the state's technical bodies and therefore of the engineers entrusted with carrying out the policy.[76]

Chabrol de Volvic, the engineer and, increasingly, the urban and regional planner, oversaw the designs for the official state buildings: the new city hall, school, justice building, and prison. It was these buildings, above all, that made the state authority in the region clear. He also drew up the overall plan for the town, which was to house 6,000 people. In his activities in Napoléonville, Chabrol de Volvic clearly served as an instrument of the state. Yet even at this early stage we have a sense of a man with his own, and quite modern, convictions. The original plans for Napoléonville by Jean-Baptiste Pichot, engineer in chief of the département of Morbihan, sought to regulate the building style, location, and use of not just the public but also the private buildings in the town. Chabrol de Volvic's plan, which was finally adopted in May 1805, provided the communication infrastructure and the form and position of the public buildings but left residents largely free to build and use their land as they wished. Here we see the beginnings of the symbiotic relationship between free enterprise and the state that was to characterize Chabrol de Volvic's administration of Paris and which so differed from ancien régime attempts to control both the population and commercial growth of Paris. It is also interesting to note here the attachment to the ideal in Pichot's thinking, which almost amounted to a greater belief in the reality of his maps and plans than in street reality. This reflects well the fixation with numerically based representation, exemplified by descriptive geometry, so evident among the geographers and engineers in Egypt.[77] It is interesting and significant that Chabrol de Volvic was already, by 1805, beginning to move away from that vision toward a more nuanced sense of the complexity of society. His success in his role of sub-prefect of Napoléonville again brought him to Napoleon's attention. As a result, in 1806 he was named prefect of Montenotte.

If Chabrol de Volvic's role was, politically, relatively minor in Egypt and perhaps second rank in Napoléonville, as prefect from 1806 to 1812, he embodied the state in the new French province of Montenotte. Montenotte was a French administrative creation that ran counter to the traditional trade and administrative patterns of Northern Italy. Liguria, formerly under Genoa, and Piedmont, formerly under Sardinia, were combined by Napoleon to create an extension of France into Northern Italy. In Montenotte, it was Chabrol de Volvic's assigned duty to redirect these two independent economies into a littoral-Apennines symbiosis. In this new economic geography, the trees and charcoal of the hills could feed the iron production and ceramics industries of the coast, while both agricultural and industrial produce would be exported (via a heavy tariff) either directly to France by road or to France or elsewhere by sea. Montenotte was to produce and export primary products and to import manufactured goods from France.[78] To that end, Chabrol de Volvic proposed the construction of a road from Nice to Genoa and a major Mediterranean port at Spezzia.[79] This was a direct practical translation of Napoleon's desire to augment France's maritime power relative to "the oppressor of the seas," Britain.[80] However, the first step in any such reform, as the civilian and military engineers, and increasingly the French government itself, had come to recognize, was the collection of information on the actual or potential wealth of the countryside in terms of its natural resources, population (including calculations concerning the average strength of workers in particular regions and industries), agriculture (olives, silk, fruit trees, grain, etc.), industry, and commerce.[81] In what remains a valuable geographical study, his *Statistique des provinces de Savone, d'Oneille, d'Acqui, et de partie de la province de Mondovi, formant l'ancien département de Montenotte*, Chabrol de Volvic used the research carried out by the military geographers working in that area together with his own observations to create a remarkably detailed picture of the society and economy of the region. In this study, Chabrol made few grand statements about the benefits of French government in Italy. His administration had brought significant agricultural reform, including reforestation to counter erosion and rebuild an exhausted resource, the extension of olive cultivation, experimentation with sugar beet (soon to destroy the wealth of the West Indies) and cotton production, and the establishment of an organized system of markets according to which supply and prices could be monitored and hoarding or monopolistic behavior curtailed. He sought to introduce or reintroduce local industries, including pottery production in Savona and lignite mining in Cadibona. He even proposed the introduction of iron smelting works which were to work with raw material from Noli and the Island of Elba. He also brought modern urban reform to the town of Savona in the form of sewers, a fire brigade, reliable drinking water, small pox vaccination, public baths, and the removal of the cemeteries to the outskirts of the city.[82] In the book in which he recounted his activities and which he published a decade later, Chabrol de Volvic was not concerned to stand back and assess the pros and cons of French rule in the region. Rather, he assumed the whole imposed structure to be sound and positive and a simple matter of good administration. Here, as in Egypt, it was clear that what was needed was "a benevolent authority"[83] to overcome the difficulties presented by the climate and physical geography of the region and by "the old customs and above all . . . institutions of the country."[84] Never before had this people had the benefit of an administration that, with firmness, held itself above all local interests ("a multitude of little passions in a constant state of agitation"[85]) and moved the region forward in a single direction. It was such single-minded vision and an engineer's sense of the landscape that would be needed for Chabrol de Volvic's pet project, in fact not realized until the twentieth century, of canalizing the Northern Italian waterways into an east-west navigable water route from the Gulf of Genoa to the Adriatic. For Chabrol de Volvic rational French administration, infrastructural reform, and grand projects such as the great canal had "spread ease and comfort among almost all the classes and particularly among the people."[86]

Chabrol de Volvic, an active and imaginative administrator, seems to have won Napoleon's approval especially for the delicate way in which he handled Pope Pius VII. The Pope had been in a long struggle with Napoleon and the French government over the delimitation of church and state powers. What had appeared an amicable enough compromise in 1801 with the concordat, disintegrated into troubled relations by 1802 with the French government's rejection of papal jurisdiction in France, and into open conflict by 1808. In 1809 Pius was taken prisoner and exiled from Rome. Napoleon sent him to Savona, where he was added to Chabrol de Volvic's responsibilities. The situation was a delicate one. France was a Catholic country, historically so and declaredly so in Napoleon's concordat with the Pope. So too, of course, were the newly conquered Italian territories. Napoleon was fully aware that the attachment of the people to their religion was a very real one. Chabrol de Volvic's task was to function as the Pope's jailor and to see to it that he did not succeed in communicating secretly with his clergy, that neither the Pope nor his clergy published or fomented the publication of anything in any way politically sensitive, and that acts and pronouncements of loyalty to the Pope were kept to a minimum among the general population.[87] It was a difficult and unpleasant task but Chabrol de Volvic handled his prisoner with a respect that was remembered by all sides and with the firmness and rigour necessary to avoid political disaster. It was partially in recognition of this achievement that Napoleon named him prefect of the Seine, in December of 1812.

As prefect of the département of the Seine, Chabrol de Volvic proved himself such an imaginative and efficient administrator that, in a time when senior administrators changed with each regime, he was deemed indispensable. In the very troubled days of the end of Napoleon's rule in 1814, and then again during the first restoration, he was able to keep the city functioning smoothly, in spite of the presence of considerable numbers of foreign troops. Principally for that reason, he remained prefect of Paris from 1812 to 1830, with a brief hiatus during Napoleon's hundred days in Paris in 1815. The Restoration was a period of rapid urban and economic growth which was felt in Paris as crowding, noise, congestion, delay, and a climbing incidence of road accidents. As prefect, Chabrol de Volvic was responsible for a number of important innovations in the city. Most importantly, he embraced an analytic view of the city according to which information on urban activities was gathered and analyzed to produce an overall picture of the city's functioning. Thus it was that he developed and implemented an urban census for Paris. This census was also an attempt to see the larger picture through the rendering of details. What Chabrol de Volvic needed to capture in this census was not a static picture of Paris but movement and change in its population, their activities, and their use of space.

Chabrol de Volvic did not see Paris as an isolated island. As Nicholas Papayanis has pointed out, he adopted a modern view of the city as a center of production and exchange, key to the prosperity of rural France.[88] In that sense there is a real consistency between Chabrol de Volvic's work in

Napoléonville, in Montenotte, and in Paris. All three regions had a role to play in the circulation, economic, and territorial systems of the modern French state, a role it was possible to manage and enhance. In order to function to its full capacity as an economic center, in Chabrol de Volvic's view, Paris needed an advanced circulation system capable of moving people, goods, water, and sewage. To that end, it was under his administration that the canals of Saint-Martin and Saint-Denis were dug and that a variety of improvements in the city's commercial and physical structure were proposed and begun, including the building of warehouses and slaughterhouses, the widening of key commercial streets (although less was effected than proposed), the general distribution of the water of the Ourcq canal to all of the public buildings in the city, and the introduction of a system of sewers to take the used water away. The appearance of the city was also of importance to the prefect, though he deemed beautification secondary to the city's efficient movement of people and goods. Nevertheless, his beautification projects included the construction of the exterior colonnade of the stock exchange, some bridges, a few churches, a number of fountains and markets, and the proposed construction of sidewalks on the boulevards. Arguably the construction of fountains, markets, and sidewalks had as much to do with the city's efficiency as it did with its beauty. Paris's efficient beauty, which allows easy movement through the center of the city without sacrificing vistas, open public spaces, and a whole series of amenities from hospitals to monuments, may only have been realized in the course of the nineteenth century, but it was set in motion by Chabrol de Volvic.

The enlightened nature of Chabrol de Volvic's administration is clearest in the first four volumes of what became, from 1821 to 1860, the annual report of the prefecture of the Seine: *Recherches statistiques sur la ville de Paris et le département de la Seine.* It was a major innovation in administration and represented some of the most advanced statistical analysis then taking place in France. It was composed with the help of the mathematician Jean-Baptiste Fourier, who may have been the editor of the first five volumes of the *Recherches statistiques.*[89] In addition, Fourier was responsible for two theoretical essays on error in measurement and probability, and one on the theory of population statistics, in volumes 3 (1826), 4 (1829), and 5 (1833) of the collection. Chabrol de Volvic was the impulse behind the work and wrote on policy applications of the statistical results. Two geographers, Edme-François Jomard and the Baron Walckenaer, the latter as secretary-general of the prefecture of the Seine, were also involved in its production.[90] Also clear in this work was Chabrol de Volvic's commitment to the sort of statistics gathering to which he must have first been introduced as an engineer, either at the École polytechnique or in Egypt among his geographic colleagues. In this first volume, Chabrol de Volvic included an extract of his report to the minister of the interior in 1818 in which he described how he had tailored the 1817 census to the needs and realities of urban life. Full-time census workers were hired and were given financial incentives to exclude no one. A hierarchy of workers was established so that statistics collection was independently verified on a daily basis and so that complications could be passed up the chain instead of causing delays in collection. Speed was of the utmost importance and Chabrol de Volvic congratulated himself for having accounted for 700,000 people in forty days. Far more was sought than generalized information on the number of people living in Paris. The census-takers inquired into not just the number of people in each house, but their gender, marital status, age, profession, . . . As a result, Chabrol de Volvic and his colleagues were able to produce tables that began to reflect life in Paris—though Chabrol de Volvic himself commented that no statistic could truly capture the diversity of activities in the city.[91]

These statistics were for Chabrol de Volvic much more than a collection of facts. They represented a commitment to social science: to rigorous observation; to the elaboration of method; and to the identification of relationships between social phenomena, and, where possible, their expression in constant laws. We find here, in his words, the same sort of commitment to social research to be found in Volney's writings:

> We chose the table as the form of expression that was most concise and that would make comparison easy. The style of dissertation and conjecture is, in general, opposed to the real progress of statistics, which is above all a science of observation. Tables have the advantage of excluding useless discussion and of bringing all research to its principal aim, that is the methodical enumeration of facts.
>
> In truth, it would be very important to know the relationship between these facts; but the study of causes is slow, difficult, and uncertain. It is rich in error when it is not the product of wisdom, knowledge, and long meditation.[92]

Chabrol de Volvic hoped to find social laws in an early form of the population pyramid, to which he referred as "the law of population," in the constancy of large numbers, and in the similarity of pattern that he discerned in both natural systems and in population and population movement statistics. Indeed, he referred to population statistics as "the natural history of man." His commitment to tables, as opposed to maps, in no way detracts from the very geographic nature of the research he was carrying out. Maps would have required a completeness of statistic that was simply not available or feasible in the first two decades of the nineteenth century. As Bernard Lepetit has pointed out, in this realm too (in the realm of the social) from roughly the 1790s to the 1830s, the relatively thin description of the topographic map, produced for example by the ponts et chaussées for the construction and improvement of the roads of France, was increasingly superseded by tabular statistics. These statistics enhanced the possibilities of analyzing multiple variables, which might include not just location but the condition of the road, the timing of the work to be effected, the time from completion of any or all road segments, the cost per toise for repair or construction, the importance or the usefulness of the route, . . . Further, figures did not bind the data to a scale, nor would distances and measures have to be recalculated for each use (as they would have to be for analysis of data in map form).[93] In short, statistics in tabular form enhanced analysis. Geographic information was deemed important. However, given the map reproduction technology of the period and the difficulty of obtaining complete and consistent data, even geographic information might more easily and fruitfully be represented in tabular form. What Chabrol de Volvic and his team could do, and did do, was to collect some statistics, in particular those on age, gender, and profession, by arrondissement. So that many of his tables contained an interesting geographic element.[94] In a way, too, Chabrol de Volvic expressed his way of seeing and understanding the world, or administrative knowledge as opposed to the "specialist" sciences, in much the way that geographers of the early nineteenth century were wont to describe their field:

> There is research that is devoted to the assiduous study of nature and the arts, but which does not fall within the competence of administration. [Administration] embraces objects in their totality and must limit itself to a general knowledge of the facts.[95]

The accuracy appropriate to the specialist sciences and arts was simply not necessary for administration, nor, as far as Adrien Balbi, Barbié du Bocage, and S. Berthelot were concerned, was such accuracy necessary to geography.[96]

The first volume (1821) was composed of approximately 120 pages of introductory and explanatory text and more than 60 tables. In these tables, Chabrol de Volvic and his colleagues explored physical conditions important to life in Paris since 1789, such as the monthly average maximum and minimum temperatures, humidity, "conditions" (sleet, rain, etc.), wind direction, and total rainfall. In a separate table, they recorded the water level of the Seine for every day of the year. This was followed by tables describing the population of Paris in terms of the total number of houses, households, individuals, and collective groups in each quartier in Paris. Individual tables displayed a generalized age, gender, and marriage distribution for Paris by age (with five-year intervals); and total population by age in of each of the twelve arrondissements. Also included was a diagram showing which age groups predominated in all of the arrondissements, and a table showing the growth in the extent of Paris since Caesar (with some sub-tables: on, for example, the nature of the terrain in contemporary Paris). Other tables explained the house numbering system and provided statistics on lighting and street sweeping. A subsequent set of tables provided information on birth, death, and marriage in 1816, 1817, 1818, and 1819, with details on the cause of death and even on the means by which suicide (more common among men than women, and more common among married people than singles) was effected. The next and perhaps most interesting set of tables described the infrastructure of public help in the city of Paris, including the number of hospices and hospitals and their male and female and married and unmarried inhabitants, and also the profession they had held before entering. Also measured were the number and rate of indigents in each arrondissement, the number of people helped by charitable assistance in their homes, and the type and number of goods given out to the poor. A final set of tables in this group sought to compare assistance for the poor in Paris in periods since the 1790s. Other sets of tables looked at agricultural production in Paris; consumption of all sorts of goods; public instruction (very much a pet interest of Jomard and also, incidentally, of Volney); receipts of major theatres and minor spectacles; and the number of registered and unregistered vehicles and drivers in Paris. Chabrol de Volvic did not consider all of this information of equal importance, but he was correct in his assessment: that statistical analysis of Paris provided an understanding of the city that volumes of description could not have achieved.

Subsequent volumes introduced an historical essay on the population of Paris since 1670; tables on property values; tables on taxes paid by Paris to the state; a memoir on the straightening and enlargement of the roads of Paris; some one-time tables addressing for example the number and location of fires in the city since 1794; a memoir by Daubanton on the means of paying for Chabrol de Volvic's proposed changes to the structure of the city; and continuing attention to the population, industry, commerce, civil institutions, consumption, and finance of Paris, with increasing emphasis on a comparable core of statistics.

In the work and career of Chabrol de Volvic we see amongst the happiest of alliances between the state and the beginnings of social scientific thought. We also see a man who devoted his career to the management, structuring, and organization of territory.[97] Chabrol de Volvic did not identify himself as a geographer. In a sense, he functioned at the margins of the field. He worked closely with geographers throughout his career and he applied, and sometimes developed, concepts and techniques, particularly in the realm of statistical geography, but to some extent in a nascent anthropology or ethnography, that had found early expression with the military geographers and in the writings of Volney. By virtue of his own administrative skill, his fiscal order and responsibility, and his usefulness to the state and because, although he had served Napoleon loyally, his origins and sympathies were more naturally royalist, he managed to keep his balance as the political sands shifted in the 1814–1815 period. He was less successful in 1830. Although his attitude to urban growth had been enlightened relative to the ancien régime's tendency to control and curtail, Paris did have a significant and growing population and housing problem. Chabrol de Volvic had sought to facilitate the movement of goods, people, and both fresh and used water, but he had done little about the crowding in many of the poorer central areas of the city. Paris was also vulnerable to financial scares and cycles of unemployment, and had experienced significant fluctuations in both the supply and price of basic foods in 1811–1812, 1815–1816, and 1827–1829. Chabrol de Volvic was not altogether insensitive to how these problems were experienced by the very poor. He probably did not fully understand the degree of crowding in the city's center. He did try to cope with supplying the city's poor with adequate bread in 1815–1816 with a proposed differential bread tax. He may have engaged in excessive borrowing to pay for some of the larger public works projects of the 1820s, and he acknowledged that perhaps some of these, with more experience, might have been implemented more cheaply. To a large degree many of these problems were far larger than the prefect or, indeed, Paris.[98] Yet, when the misery became intolerable in the streets of Paris, shortages, high prices, and unemployment were seen as his responsibility. In addition, Chabrol de Volvic's administrative style was born and developed under autocratic rule, first that of Napoleon and then that of the restored monarchy. The structures, procedures, and mood under the constitutional monarchy, not to mention the very different role of a relatively uncensored press, changed the prefect's position and stature. The cumulative effect was that Chabrol de Volvic no longer seemed the man for the job. Indeed, the job itself underwent some modification after his departure in 1830.[99]

ADRIEN BALBI: THE GEOGRAPHER'S TAKE ON
SOCIAL STATISTICS AND ETHNOGRAPHY

Chabrol de Volvic's approach to social statistics must have been among the most sophisticated in early-1820s France. His *Recherches statistiques sur la ville de Paris et le département de la Seine* were compiled and designed with the assistance of one of the foremost mathematicians of the day, Jean-Baptiste Fourier. The statistics were presented with a view to solving a series of complex socio-spatial problems. Those problems manifested themselves at the very heart of the French territory, where the threat of unrest made it impossible for any regime to ignore them. Consequently, there was considerable infrastructural and financial support for the investigations Chabrol de Volvic directed. Although he worked with geographers throughout his career and was unquestionably influenced by them at an early stage of his intellectual formation, Chabrol de Volvic was not a geographer. It is therefore interesting to contrast this powerful and highly trained individual's use of statistics with the way mainstream geographers understood and used them. Apart from André de Férussac and his short foray into the writing of geographical texts, the geographer most interested in using statistics in his research and writing was Adrien Balbi.

Adrien Balbi was born in Venice in 1782. A professor of mathematics, physics, and mathematical and physical geography in 1815, he left Italy in 1819 for personal reasons. He spent approximately two years in Portugal and then came to France, where he worked for fourteen years. His geography was deeply influenced by the writings and reputation of Conrad Malte-Brun, whom he seems to have regarded as the world's premier geographer even before leaving Italy.[100] Contemporaries, however, saw Malte-Brun and Balbi as different kinds of geographers. In his biography of Malte-Brun for the *Biographie universelle*, La Renaudière, writing in approximately 1843, described Malte-Brun as the founder of a Romantic school of geography, just as, in his view, Ritter had founded the philosophic school of German geography, and Balbi had founded the positive school of geography.[101] Certainly, in some of his correspondence with Adrien Balbi, Malte-Brun emphasized his skepticism with regard to statistics and the use of a scientific model of research.[102] This skepticism was entirely in keeping with Malte-Brun's more literary approach to geography.[103]

Adrien Balbi's interest in statistics and the limitations of his approach to statistics are nowhere clearer than in two of his works: his *Essai statistique sur le royaume de Portugal* and his *Essai statistique sur les bibliothèques de Vienne*.[104] The first work, published in 1822, might be regarded as the product of a younger scholar, in spite of the fact that he had already published two universal geographies (in a number of versions and editions) and one geography of Europe. The statistical study of the libraries of Vienna was published in 1835, three years after he had published his magnum opus: a one-volume universal geography, the 1,500-page micro-print *Abrégé de géographie rédigé sur un nouveau plan*. Balbi's two-volume essay on Portugal was a detailed study of the country, the purpose of which was to make Portugal, long-since located and mapped, known to Europe on a different level. It was Balbi's contention that the rest of Europe functioned with a knowledge of Portugal that was caricatural and highly prejudicial. It was his intention to start from the facts and rebuild Portugal in the European mind. To that end he provided a full description of the country, from its history to its natural conditions and products, its government structure, and its commerce, industry, agriculture, and literature. The one-volume essay on the libraries of Vienna was, in scale, a far less ambitious work. Yet, here again, it was Balbi's aim to make those libraries and their relative importance within European civilization known to the educated people of Europe. In both works, both the final aim and the approach was description. In addition to these two works, Balbi was also the author of an ethnographic atlas of the world, one of the very first such productions, and an extensive introduction to that work.[105] Although his ethnographic works had relatively little to do with statistics, they showed the same structural tendencies as his more statistical works and cast light on his thinking and approach.

There are, in Balbi's work, the early glimmerings of a social scientific geography. He was certainly aware of a broad array of statistical work, including Sébastien Bottin's commercial statistics and Chabrol de Volvic's research on Paris.[106] However, it is conceptually that Balbi's work has the traces of more modern social concerns. On the simplest level, the desire to move away from stereotypical characterizations is important and certainly more sustained than in many geographical texts of the period. Balbi rejected the idea that Portuguese agriculture was weak because the Portuguese were lazy.[107] The reasons for Portugal's stagnation in agriculture were historical and structural, having to do, as Volney might have argued, with bad government, but also related to the impact of colonial agriculture, a lack of investment in roads and communications, the excessively controlling role of the Church in Portugal, the landholding system, and low population. Yet, elsewhere in his text, although he was clearly uncomfortable with such assertions, he was still prepared to engage in the "character of the people" discourse so typical of the eighteenth century.[108] Dotted through the text is evidence of a modern social curiosity. Balbi asked the more traditional eighteenth-century question: What makes the wealth of the State?[109] He also asked: What constitutes poverty and what impoverishes a peasantry?[110] Why might a territory as rich and diverse as Portugal suffer depopulation, while the Ohio Valley's demographic growth seemed limitless?[111] What are the costs and benefits to the people, the countryside, and government of a commercial association like the "Compagnie générale de vins du Haut-Douro?"[112] What impact might an ancient legal code, based on the inevitably conflicting and now out-of-date principles derived from Roman, Barbaro-Roman, Medieval Christian, Jewish, Islamic, and more modern Portuguese law have on justice and the efficient functioning of the society?[113] In addition, Balbi had flashes of spatial intuition. He could not understand why, given Portugal's considerable annual wheat crop, a goodly portion of which it was in a position to export, the country nevertheless regularly imported a substantial quantity of wheat. It was his suspicion, and not something he sought to prove, that the wheat was required to feed the population of Lisbon, essentially inaccessible by road from the rest of Portugal.[114] The intuition was limited and remained untested, but it suggests the begin-nings of an understanding, so well demonstrated in Chabrol de Volvic's work, of the interrelated dynamics of economic production, exchange, and transportation infrastructure. The very idea, inherent in Balbi's study of Portugal, that a society could be broken down into measurable elements, or manifestations, was modern and reflected the first important step to social inquiry. In taking that step, government was both an important source of statistical information, which Balbi hunted down relentlessly, and the possible source of reform as varied as the use of import control to protect local agriculture and tree planting to provide physical protection from coastal sand dunes.[115] It is important to note that Balbi did not problematize the identification of the elements of society or even *how* they should be measured. Still, he was far from simple-minded and realized just how much insight could be gained from a well-chosen statistic, such as the number of volumes in a library.[116]

Far more innovative than his description of either Portugal or the libraries of Vienna was his ethnographic atlas and supporting text. In this work Balbi sought to gather together all current and acceptable past research on what was becoming known as linguistics to produce a geographic synthesis of the languages of the world. Behind this was an argument which, as we shall see, Balbi found difficult to sustain, that ethnography was or should be synonymous with linguistics.[117] That is, that peoples should be classed not by race but by language. The idea was a fascinating and highly original one. Indeed the modern reader may regret that Balbi did not take this altogether original idea and devote his life to its development. Within it, perhaps, was the possibility of a more egalitarian, less prejudicial, and less hard (as in sharp dividing lines) approach to the study of the "other" than was ultimately developed in nineteenth-century race-based anthropology.[118] Balbi argued that, in such a study, geography, history, ethnography, and linguistics would be all be one, not separate cantonized fields.[119] Further, language study could extend geography's history into prehistory and provide a sense of the movements in space of the ancient peoples of the earth.[120] Implicit in his thinking was the view that languages and dialects could be mapped.[121] Implicitly, then, he was beginning to approach the concept of mapping movement and change: an important element of the thematic map. The innovative nature of this idea of combining geography and linguistics, which was very much his own idea, also forced him into beginning to rethink some key concepts.[122] What is a nation? How is it defined? Does the nature of the definition change what is included within the borders of the nation?[123] What is the relationship between language, origin, and race?[124] These kinds of questions, it could be argued, were the beginnings of a stirring interest in culture and cultural geography, not as a static picture but incorporating the concept of change. If the root idea was highly innovative, its development was not. So much so that the atlas seems to have quickly fallen into oblivion. What was lacking in the publication was not research and hard work, but an approach that had contemporary scientific resonance. Balbi's approach to both ethnography and statistics was limited by his attachment to description. Balbi was either not able, or not willing, to engage in the hard thinking necessary to bring such an idea to life. As a result, his work in both social statistics and linguistic ethnography lacked the sophisticated analysis so evident in Chabrol de Volvic's *Recherches statistiques sur la ville de Paris*.

One frequently has the impression, reading Balbi, that for him, statistics were data, numbers, lists, and, at a stretch, tables. Statistics was less about probability, analysis, and the graphic display of quantitative information, than about information per se.[125] It was a comfortable sister field to geography because, to Balbi in any case, it appeared to be another form of description, and one respected by contemporary science. Both his work on Portugal and on the Vienna libraries was structured around the presentation of very close to raw data and facts which he described variously as truth, fact, and positive fact.[126] The extremely numerous lists, numbers, and tables in his work on Portugal were often taken wholesale from his sources, and their manner of collection, or indeed the criteria behind their collection, was frequently left unexplained. Balbi seems to have regarded method of data collection and criteria for selection of information as secondary to data collection. Thus, it is only on page 60 of his essay on the libraries of Vienna and the world, after countless lists of statistics, that Balbi explains according to what principles volumes in libraries should be counted. The reader is left with the distinct impression that most of the sources used by Balbi did not follow such criteria.

The word "list" best captures the nature of Balbi's presentation of statistical information. It is clear that some of his official sources provided Balbi with information in tabular form, which he then imported into his text. However, tables designed by Balbi lacked the essential concision, comparability, and graphicacy of statistics. On the simplest level, this is reflected in the endless titles to his tables, which in themselves required unwarranted attention. Discursive text also found its way into almost all of his own tables, obscuring the figures and dates. More importantly, the structure of his "tables" was problematic in that dates were frequently incommensurable; categories often contained several types of information, leaving what was being compared unclear; statistics that might be compared were presented under different categories and separated by pages; and sometimes how key statistics were calculated would vary from statistic to statistic within the same category.[127] The author delivered the data. It was up to the reader to restructure it for analysis and to engage in the analysis him- or herself. This meant that Balbi sometimes offered his readers a table, generally taken directly from one of his sources, packed full of highly suggestive information, which, however, remained unanalyzed. An excellent demonstration of this is to be found in the three tables giving the annual production of wine, oil, and fruits in Portugal from 1795–1820. The patterns of production were very different between the products and highly variable temporally. These tables had to be graphed to be comprehensible and, most importantly, they had to be analyzed and contextualized. Balbi did neither. There are no graphs of any sort in any of these works by Balbi. Instead, there is a similarity in the tabular structure of all of Balbi's works which seems to derive from the outline structure of most universal geographies:

Government		Modern geography
Justice		Physical geography
Juntas	OR	Lakes and Rivers
Positions		Tagus R.
Individuals		Elga R.

This, for Balbi, was classification: it was this order that lay at the heart of statistics which lent scientificity to description.[128] With such an order inherent in the very structure of his book, inherent in his description, why would his statistics necessarily need an argument-based structure?

Beyond fact, and classification, Balbi understood his science to be providing comparison. He gave tables of temperatures the world over to be compared with temperatures in Portugal, he provided population statistics for Portugal and then compared the highest density area to the lowest density area and calculated what the population of the lowest density area would be if it had the highest density . . . ; he even calculated and compared the number of words on a parchment roll to the number of words in a volume of text to arrive at comparative levels of civilization represented by the Bibliothèque nationale of Paris and the Alexandria Library.[129] He calculated the lengths of all the rivers in Portugal, presumably just to provide the statistic should anyone wish to compare the lengths of rivers in Germany and Portugal, for example.[130] The height of this might well have been his plan to count the total number of plant species in all of the botanical collections in Europe. Fortunately, he consulted Alexander von Humboldt prior to undertaking this research.

> It was my intention to also compose a table of the number of different plant species cultivated in the principal botanical gardens of Europe. However, the difficulty of procuring exact information and of reconciling such different estimates by travelers and geographers led us to consult Alexander von Humboldt. This scholar, more than any other, could guide us as to how to draft it correctly. We renounced the project based on his advice and after the remarks he addressed to us on the subject.[131]

One can almost hear von Humboldt's groan between the lines of his gentle response.

The strongest indication that Balbi did not understand the fundamentally analytical nature of statistics and its link to graphic expression is that he appears to have completely abandoned geography's most spatially analytical tool, the map, which he replaced with significantly less powerful lists and textual layouts. These were of the genre already described above and well illustrated in his "Tableau hydrographique des principaux fleuves du Portugal," which described the river system of Portugal in the following terms:

> Minho.
> Lima.
> Cavade.
> Ave.
> Douro.
>   Agueda. g.

Coa. g.
Sabor. d.
Tua. d.
Tavora. g.
Tamega. d.
Vouga.
Mondego.
　　Dào. d.
　　Alva. g.
　　Ciera. g.
Lis.
Alcoa.
Tage.
　　Elga. d.
　　Sever. g.
　　Ponsel. d.
　　Zezere. d.
　　Sorraya. g.
　　Cunha. g.[132]

Where "g" meant that the river flowed to the left and "d" meant that it flowed to the right. Such a diagram had none of the analytic potential and mnemonic power in the declarative description that is a topographic map, never mind the analytical and explanatory power of the thematic map (see chapter 7). Not a single map is to be found in his description of Portugal. Neither did he choose to explore the spatial structure of any of the statistics he presented in either his work on libraries or his description of institutions of higher learning in Portugal and Europe, or indeed in his *Atlas ethnographique du globe*. Incredibly, this atlas also contained not a single map. There were three types of plate in the atlas: six plates that described the ethnographic divisions of the world; thirty plates that described the families of languages, groups of languages, and languages within the ethnographic divisions; and five plates that presented and allowed the comparison of twenty-six words in the vocabularies of the five parts of the world. The "ethnographic" plates provided a classification of the languages of the world in a tabular structure based, largely, on geographic location. The classification-of-languages plates described each language textually, with the text organized into table-like structures.[133] The vocabularies allowed the reader to quickly assess the number of world languages known, how well those languages were known, and the degree of similarity between them based on the twenty-six words chosen by Balbi (with the advice of Jean Pierre Abel Remusat [1788–1832], the famous French Sinologist).

Thus, although the idea of one or many thematic maps of the languages of the world was implicit in his title and in the idea of the project, Balbi was a considerable distance from understanding the sort of analysis, and perhaps even the type of data, such a product would have required. Abel Remusat, who was a friend and supporter, pointed out the two major weaknesses of the atlas in a friendly review in the *Journal des savants*.[134] As Malte-Brun had warned Balbi, languages simply did not lend themselves to classification, certainly not the natural classification that was now the only acceptable form of classification in science. Abel Remusat expressed the problem in the following terms:

> There is always something vague, and necessarily there is something a little arbitrary, about these classifications imitated from the natural sciences. In their use one is deprived of the support that nature itself provides to the study of organized bodies, which is the only solid and invariable part of nomenclatures: the succession or physical descent of individuals of the same species.[135]

Beyond that problem, however, was one perhaps even more severe. Tellingly, Remusat observed that "the type of presentation adopted by this author makes analysis impossible."[136] The data presented by Balbi was presented in a form that allowed only the continued collection of data according to the same plan. It in no way furthered or enhanced linguistic, or for that matter geographic, analysis.

Balbi's method was the same in all of his works: it was description, the geographer's method. Unsurprisingly, then, he ran into many of the same problems with description encountered by the authors of universal geographies. These included the problem of establishing for contemporaries what in his work could be deemed original and, thus, his own contribution; how to establish criteria for inclusion and exclusion; and the place, if any, for analysis (see chapter 3).

Balbi collected an enormous amount of information for all of his publications but especially for his work on Portugal. In the case of Portugal, in contrast to the information he needed for his universal geographies, he was able to collect only a small portion of that information from already published authorities. European knowledge of Portugal was such that most of the information he needed had to be collected from authorities in Portugal, particularly from government officials and commercial agents. To that end, he wrote to them requesting information. One has a sense from his many parenthetical references to such letters that Balbi maintained a dauntingly large correspondence designed primarily to procure information. He also maintained a correspondence with the "celebrities" of his day, especially people like Alexander von Humboldt, to whom he referred as "that sublime talent," or his heroes in geography, Barbié du Bocage and especially Conrad Malte-Brun, to whom he referred as "a model of eloquence."[137] He sought advice from these people, which it is clear he did not always take, on the plan of his book and what research would be worthwhile or feasible. His books, then, were very much based on the "authority" of others. Unfortunately exactly who those others were and precisely what they were responsible for in Balbi's book was not always clear. While Balbi used footnotes extensively, and they, and the text, suggested the sources he had consulted in a global sense, they were not used to separate out his own from his sources' observations. Arguably, this was not yet part of the academic etiquette of the early nineteenth century. However, for geographers such as Balbi it did create some problems. Their work was so derived, and was seen to be so, that it seems to have encouraged plagiarism and pirate editions. Balbi took up eight pages in complaint over these problems in his *Statistical Essay on the Libraries of Vienna and the World*.[138] His complaints had been expressed by geographers since at least the early eighteenth century. The problem went to the heart of the geographic method. Balbi, in sharp contrast to Volney, but like his cabinet geographer colleagues, did not engage in fieldwork. He also does not seem to have had very much sense of the value of being there and seeing for oneself. Yet, on the one hand, he felt that not having spent a sufficient number of years living among the Portuguese limited the number of comments he could make about them and their way of life. On the other, although he was in Portugal and on the spot, he spent little time traveling through the country to assess, for example, the impact the wine company was having on the Haut-Douro.[139] Neither does he seem to have valued eye-witness accounts over those of compilers. Thus, for example, in his *Introduction à l'Atlas ethnographique du globe,* he saw his source for Carib vocabulary and grammar not as Father Raymond Breton (1609–1679), the Dominican missionary who collected information on the language before it died out, but as Conrad Malte-Brun, who could have had no other source than Father Breton.[140]

This may seem a twentieth-century quibble, but excessive borrowing from sometimes inappropriate sources limited the coherence and value of Balbi's work and may explain its neglect by the next generation of linguistic geographers.[141] Ironically, while his linguistic research was perhaps the most original part of Balbi's work, massive borrowings and an inability to rework these according to an informed argument of his own also make this work among his most derived. This peculiar combination of originality of idea but derived data and argument led Balbi into contradiction. Such contradictions occurred in his essay on Portugal but are most noticeable in his *Introduction* à *l'Atlas ethnographique du globe*. In his essay on Portugal, although he wanted to absolve the Church of blame for the underdevelopment of the country, many of his sources took aim at the Church, and Balbi found himself unable to maintain a consistent line of argument on that issue.[142] Similarly, his sources seem to have differed on whether agriculture or commerce was the basis of a healthy and growing society. So, then, did Balbi.[143] In his ethnographic essay the problems were more severe. One of his authorities deemed language unalterable and the traces of its influence on other peoples and non-native speakers ineffaceable. Balbi agreed and argued this forcefully. Elsewhere, Balbi claimed that there were three things which could alter and altogether destroy a language, the first being conquest by another people.[144] In a contradiction that ran to the very heart of his work, Balbi used a monogenetic racial argument (all humans are of one racial origin) to structure his *Atlas* but adopted a polygenetic racial argument (the human race comes from perhaps five distinct origins) in his *Introduction à l'Atlas*.[145] This was a major debate of the early nineteenth century that exercised the minds of many scientific and religious thinkers. It was not a subtle difference of opinion but a highly contested issue. Those of Balbi's sources who maintained the monogenetic argument, primarily Blumenbach, Cuvier, and von Humboldt, argued that researchers would not be able to use language to trace a people to its origins ("souche"); whereas the polygenetic argument, primarily supported by Desmoulins, was accompanied by the argument that language could be used to trace a people to its "souche." These were incompatible arguments with implications absolutely central to his *Atlas:* What was the relationship between language, or language and culture, and race? Was race fixed and language not? Or were both fluid? And what about the character of a people? Was that racial and fixed? Or was it cultural and fluid?[146] Balbi argued all sides, depending on the source he was consulting, apparently unaware, if not of the contradictions, then certainly of their importance. There are other contradictions in his work, about the importance and variability of pronunciation, about whether or not the highly educated are a good source of information on a language as it is spoken, and about the distinction between a language and a dialect.[147] Contradiction was not the only negative result of his critical compilation technique, it also led to a lack of follow through. With each shift to a new source, often designed to provide an example from another part of the world, the line of argument maintained by the previous source was dropped. This created a staccato effect, a little like reading a list of ingredients, with little flow and not much to capture and hold the attention of the reader.

Balbi, every inch a geographer, was primarily concerned with description.[148] To finally and factually "forever present the present state of the earth" was his unrealizable ideal.[149] Where his responsibility for complete description ended was as unclear to him as it was to all universal geographers, and perhaps, given his embrace of statistics, which he saw as description on a larger scale, it was even more unclear for Balbi. He was responsible for giving all significant spot heights in Portugal, for outlining the finest details of government structure, for listing all the works published on Portugal between 1800 and 1822, for recounting the number of threads of cotton in a private collection of industrial raw materials in Vienna.[150] Balbi was aware that this burden of description functioned as a constraint. He could not make his own calculations for his Portugal essay because the sheer volume of material he had to consult and order was overwhelming.[151] The amount of work his 1,500-page *Abrégé de géographie* demanded made it impossible for him to complete his comparison of world libraries.[152]

This volume and detail of information seems to have absolved him of any commitment to analysis. After the collation and presentation of two volumes of detailed information on Portugal, Balbi did not feel that he had the time or energy to conclude:

> Unfortunate and unanticipated circumstances, having removed the leisure and peace of mind so necessary to properly develop a sense of all that an enlightened government could undertake to make Portugal altogether flourishing . . . The reader will see from looking at the elements in this section the procedure we followed, and those elements will lead those, even those unfamiliar with these matters, to guess what we might have said to them.[153]

The elements to which Balbi was referring were global figures comparing the population, revenue, debt, and size of the army and navy in Portugal with those of numerous countries of Europe. His point, then, was that he had provided the important part, the figures. It was up to the reader to analyze the information. Periodically the reader senses that Balbi was fully aware of, but resigned to, the impossibility of the tasks he set himself. The research he was engaged in really demanded the knowledge and time of a team of scholars.[154]

CONCLUSION

If the limitations of Adrien Balbi's work were apparent to some contemporaries, they are far more apparent to the modern reader. This is why it is so important to be fair in our assessment of his work. Balbi may not have fully understood statistics, and his ethnographic work may have been far too derived. This limited the impact of his writing and meant that he could do little to reform geography, though he had hopes of doing just that. Still, his work was influenced by the kind of statistical questions and procedures being employed by Chabrol de Volvic. Volney's influence is less apparent, although he did refer to Volney's work on transcription and oriental languages. In addition, Balbi did live in the period that witnessed the rising of ethnography from the ashes of the Class of Moral and Political Sciences and he sought to play a role in defining that field. The contrast between the geography practiced during the expedition to Egypt, so tightly focussed on description and mapping, and the kind of research practiced by the few geographers on the expedition to Algeria, which showed more concern with commerce, society, and the transportation/communication network, was in formation in Balbi's work. Society was coming to the attention of geographers as well as nascent economists, sociologists, and ethnographers. The awakening was slow and this slowness had a great deal to do with a sense of the kinds of questions geographers asked and how they asked them. Nevertheless, there was an awakening.

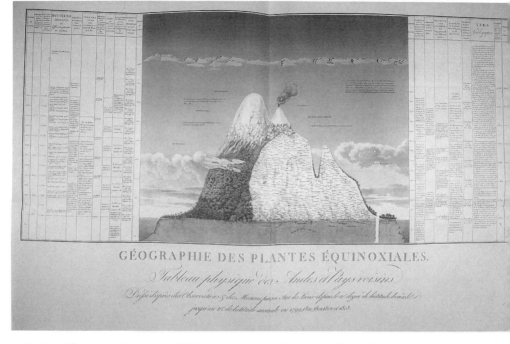

Fɪɢ. 7. "The geography of plants." This imaginative graphic captures Humboldt's concern to compare the results of instrument-based empirical work in the field to arrive at an overall understanding of the interactions of physical phenomena.

Innovation in Natural Geography

With respect to Humboldt, his interest in the historical occupation
of planet earth appears in a number of his works . . . . A tradition
was thus begun and one of the first links [in that chain] would be
the book by Lucien Febvre and Bataillon: *La Terre et l'évolution
humaine* (1922). However, only minds of an exceptional breadth
seem able to encompass the unity of geographic problems. One could
point out that neither Daunou, nor Humboldt, nor Lucien Lefebvre
was a geographer in the restrictive sense of the term. Clearly one
must conclude that the specialist is, by vocation, incapable of taking
distance from the object of his study. Without knowing what he is
looking for, he cannot know exactly what he has found. As paradoxical
as this must seem, even the meaning of geographical truth does not
seem to have been clearly established. Perhaps the mere existence
of geographers is not sufficient proof of the existence of geography,
whose status, unity, and autonomy remain contested, while those most
concerned do not trouble themselves to ask the question profoundly.
—Georges Gusdorf, *Les Sciences humaines et la pensée occidentale,*
vol. 1, *La Révolution Galiléenne* (1969), 366–67

Meditation, the word is not too strong to describe the intensity
and the continuity of Humboldtian thought. —Charles Minguet,
*Alexandre de Humboldt, historien et géographe de l'Amérique espagnole,
1799–1804* (1969), 64

INTRODUCTION: GEOGRAPHY, NATURAL
GEOGRAPHY, AND NATURAL HISTORY

Some of the most imaginative of French geographers of the early nineteenth
century worked not in physical geography but in something describable as
"natural geography." As I have argued in chapter 1, eighteenth-century
geography was description, either graphic or textual. As such, its function
was the description of *all* of nature. It was, then, in keeping with the
geographic tradition that geographers might continue to describe all of
nature, from geology, to plants, to human societies. This is what the two

natural geographers discussed in chapter 5, and particularly Bory de Saint-Vincent, did do. With most of the life and earth sciences specializing at an increasing rate in the first half of the nineteenth century, this served to marginalize the natural geographers, who failed to make a mark in any of these new emerging sciences. Nevertheless, a number of synthetic and analytic concepts, later associated with physical geography, emerged from the late eighteenth and early nineteenth centuries. These include the idea that landscapes and natural regions are worthy objects and scales of analysis; the insight that interrelationships between realms (for example plants and rocks, or climate and soil) are as revealing of the functioning of the natural world as is the study of the classification of rocks, or plants, or climates; and the realization that comparison of like phenomena across space can reveal much about the past and present nature of the earth. These ideas emerged as a result of work and developments in field observation, mathematization, the use of probability, theory-driven search for cause, and the development of more sophisticated descriptive, analytical, and graphic forms of expression. Although we see these ideas as fundamentally geographic today, they did not, and could not, emerge from French geography as it was then practiced. Nor were they embraced by geographers as central to their concerns until a much later period. Instead, these ideas were the result of the work and discussions of an innovative group of scholars who had begun to move beyond classificatory natural history to more systematic research. The most remarkable of these, in large part because he encompassed both holism (redolent of eighteenth-century French geography's concern with universal description but fundamentally inspired by nineteenth-century German Naturphilosophie) and the systematic nature of the emerging modern earth and life sciences, was Alexander von Humboldt. Although von Humboldt had extensive contacts in the world of science and art, he worked in substantial isolation from geographers, who, although they read him and sometimes corresponded with him, did not see that in his work there was a focus and a series of approaches which had the potential to fit their field like no other. It would be wrong to condemn French geographers as small minded. They were working within discursive and structural constraints. Geography was not the free-ranging and intellectually exciting research carried out by von Humboldt. It was a field tied by tradition to textual and graphic description of the world and everything in it. It was a field that had been inoculated against theory by both geology's flights of fantasy and Buache's attempts to "predict" the structure of the earth. French geographers were not engaged in what we today understand by physical geography, or anything like it. The term existed but contemporaries used it for the depiction of the nature-generated rather than the human-made lines on geographic-scale maps. Certainly, the term then carried little of its late-nineteenth- and early-twentieth-century identification with geomorphology as the study of landforms, and their formation and decay. Instead there was a descriptive geography of the natural world, both organic and inorganic, which tended to surrender its identity to botany, zoology, or geology whenever questions of cause were broached. Its absence is made clear by Augustin Pyramus de Candolle's call for its creation in 1820.[1] It would, then, be more appropriate to describe the work of geographers of this period, so broadly focussed on the natural realm, as "natural geography."

There is another reason such terminology is more appropriate. The geographers of this period were far more influenced by the classificatory model of natural history than by the kind of research being carried out by Georges Cuvier, von Humboldt, and de Candolle—no matter how much lip service they paid these men. Even more than the model of physics, with its cause-and-effect relationships and its increasingly mathematical mode of expression and analysis, the traditional approaches of natural history initially seemed ideal for taming the variety, diversity, and particularity of the physical and social worlds. After all, natural history had succeeded in putting order into the plant and animal kingdom. This had been achieved through the slow and careful collection of specimens and sketches, their comparison, and their naming and placement (and renaming and replacement . . .) within a genealogical-tree-like structure of similarities. In the most profound sense, human knowledge itself ("les connaissances humaines") was conceived of in such terms, by many geographers and non-geographers, and trees and schemata of knowledge, incorporating similar exercises of placement and renaming, grew in popularity and number through the nineteenth century.[2] We have already seen evidence of the influence of this conception of science in early ethnography or ethnogeography (see chapter 4). There are also strong signs of it in the regional geography of the late nineteenth century.[3] The influence of that model of descriptive, incremental, and classification-driven science was, however, strongest in the nascent earth sciences of the early nineteenth century. In the earth sciences the tendency was perhaps accentuated by the confusion, flawed research, and theological furor that seemed to have been the principal product of geology's eighteenth-century foray into questions of cause, origin, and theory.

While the early social and earth sciences were looking to natural history and its preoccupation with classification as a model of science, as Foucault and others have already pointed out, natural history itself was going through something of a change in the early nineteenth century.[4] In part, this emerged from a growing sense of the importance of the relationship between plants and animals and their habitat and the discovery that European climates, at least, had changed over time (in some cases, leaving a fossil record of plants and animals unsuited to the modern climate of Europe). The sense of the importance of habitat combined with the concepts of local climate change, extinction, and transformation, led to an exploration of the functional relationships between the internal structure of plants and animals and the natural environment. Thus, even as eighteenth-century natural history was used as a model for some sciences, the study of the plant and animal worlds was undergoing modification that eventually led to the birth of modern biology. A part of that modification involved a shift away from classification and description toward a more explanatory and hypothesis-driven exploration of the functions and mechanisms of the natural world. It was a slow transformation in which aspects of the older approach to the natural world subsisted alongside radical new ways of thinking. Indeed, arguably, the vast majority of the work carried out in natural history in the first few decades of the century still amounted to description and classification.

Some of those who declared an interest in natural geography sought to break away from elements of the patterns set by eighteenth-century natural history and indeed those set by geography. Alexander von Humboldt was in some ways a traditionalist and in others a remarkable innovator. In his *Géographie des plantes*, in particular, he provides a strong contemporary sense of the nature of much early-nineteenth-century natural history research, together with the nature of the innovation he proposed:

> The research carried out by botanists is generally focussed on a very small portion of that science. Botanists are almost entirely occupied with the discovery of new plant species, the study of their exterior structure, the features which distinguish them, and the analogies which unite them into classes and families. This knowledge of the form of organized beings is without a doubt the principal basis of descriptive natural history. It must be seen as indispensable to the advancement of the sciences that deal with the medicinal properties of plants, their cultivation, and their industrial application. But if this science is to be worthy of the attention of a large number of botanists, if it is to have a philosophical dimension, it is no less important to consider the geography of plants. This science, which now exists only in name, nevertheless forms an important part of general physics. This is the science that studies plants in terms of their local associations in different climates.[5]

If he was impatient of traditional natural history and its endless description and classification, he was equally so of traditional geographic approaches.

Thus, he rejected the restriction of approach to description that was advocated by Conrad Malte-Brun, John Pinkerton, Adrien Balbi, and many other geographers. He also did not share the prevailing tendency among nonmilitary geographers to prefer scholarly authority over field observation.[6] Nor did he see the topographic or geographic map as the ultimate aim of all research. In ignoring these geographic prejudices, Humboldt and a few other innovating scientists contributed, often knowingly, to the ongoing reformulation of the life and earth sciences. Their attempts to find alternative approaches are recorded in their published research, but also in the neologisms with which they sought to baptize their new interests, including "botanical geography or geography of plants" as opposed to "general phytology,"[7] "geognosy"[8] as opposed to "geology" or "physical geography,"[9] "oryctognosy" as opposed to "mineralogy,"[10] and a series of other terms such as "the natural history of the epochs of the physical world,"[11] "vegetal statistics,"[12] "the physics of the earth,"[13] "geogeny,"[14] "geogony," "geognomy," "ontogony," "ontography," "ontonomy,"[15] "the physiognomy of the earth,"[16] "comparative orography and hydrography,"[17] . . . This proliferation of terminology is a surficial manifestation of a profound attempt to reconceptualize the natural world, and control and direct its study in new directions. Arguably, Alexander von Humboldt was the most eclectic of these innovators, if he was not always the most original.[18] However, if we set Humboldt into the context of geography as it was written and practiced in the early nineteenth century, he was radically different and an innovator in two ways: in the nature of the science he practiced and the way he practiced it (the focus of this chapter), and in the particular flavor of the holism, or quest for unity, that he brought to the study of nature (discussed both in this chapter and in chapter 3).

Humboldt's quest for unity was less an argument or a theory in search of proof, than a fundamental presupposition and a motivating force guiding his research. At the end of the day, he was looking for "knowledge of the laws of nature"[19] which played a fundamental (but not necessarily determinative) role in all realms. "The isochronism of widely extended formations" that he observed in Europe and the Americas and "their admirable order of succession"[20] fed and consolidated this belief and drove what historians of geography have described as his comparative method. As he himself declared, while he had devoted a significant moment in his life to observation, sketching, mapping unknown regions, in short, to the collection of facts and the examination of details, he infinitely preferred the study of the links and relationships between phenomena such as the geography of plants, the migration of social plants, and the relationship between these and climate, soil, human activity, etc.[21] He even sought to connect and study the interrelationships between, for example, his study of the geography of plants and his study of landforms, each of which was in itself a study of interrelationships.[22] Humboldt was convinced that there was an analogy to be drawn from the relationship between form, structure, and function that Cuvier had found in his anatomical studies of organisms that would apply to the kind of interior forces Humboldt was seeking in the physical world.[23] The invigorating tension in Humboldt's work, then, was the very dilemma faced by nineteenth- and twentieth-century geography: how to reconcile the study of the physical and social realms; and how to make compatible a unified vision and approach to the world with the detailed, observation-based, theory-driven search for explanation and cause. In all of this innovation Humboldt was trying to retain a holistic vision of the cosmos while taking full advantage of the insight to be gained from empirical, experimental, and theoretical science. In a sense, he was trying to reconcile an emerging nineteenth-century science already shaped by discursive structures with the less defined and delimited Enlightenment science.

Enlightenment science is not easy to characterize. In scientific circles, the Enlightenment in France was characterized by a rational enthusiasm or the sense that, with the right questions, most puzzles could be elucidated; by a belief in progress, both human and natural; by a passion for exploration, observation, and the collection of information (and specimens), and the sense that the collection and classification of observations would in itself lead to enlightenment; and by an Aristotelian cosmological legacy that suggested that all in the Cosmos was linked by a universal chain of being (with or without angels and God at its apex). Bernard Smith has discussed the artistic vision that accompanied Enlightenment science as an artistic neoclassicism which regarded unity of mood and expression as more important than fidelity and detail of depiction. Smith argues that Humboldt sought a new unity—an ecological unity—which could reconcile the increasing development of an empirical science tending toward the disruption of unity and the holism of an artistic vision and expression. This artistic unity was not neoclassical. This art did not entail the sacrificing of nature's detail, roughness, and particularity to some higher sense of unity. It was artistry—and unity—derived from, and based on, an analytical and penetrating approach to the depiction of landforms and landscapes.[24]

My argument runs parallel to that of Bernard Smith. Smith argues that Humboldt provided the "typical landscape" (or a landscape designed to capture the look of an ecological system), then evolving among landscape artists such as Hodges and Webber, with "theoretical justification." Smith based his argument largely on Humboldt's writings about the look of nature, landscape, landscape painting, and even his few but emphatic comments about photography. In this chapter, I want to explore that "theoretical justification," or the hard thinking behind Humboldt's attitude to representation. Mine is perhaps closer to the focus of Michael Dettelbach's concern.[25] Instead of concentrating on landscape painting, I have chosen to explore the innovative ideas and approaches developed by Humboldt and their link to representation. Otherwise put, I am interested in the intellectual structure of the unity Humboldt was trying to discern and its relationship, less to landscape painting, than to his experimentation with maps, proto-thematic mapping, graphs, and diagrams. It is my contention that in his scientific graphics he was trying to develop or adapt from the work of others a language—or a way of seeing—that would encourage both conceptual depth and rigor and holistic vision. I am not saying that Humboldt was successful in this endeavor, but that the endeavor was coherent and powerful.

Having explored Humboldt's quest for the unity of both nature and science[26] primarily through the *Cosmos* in chapter 3, here we will consider in depth the strongly innovative nature of Humboldt's physical science, particularly in the context of the research of his contemporary "natural geographers": because, in spite of his fundamental commitment to the unity of the sciences, Humboldt was among the early-nineteenth-century scientists, including Cuvier and Marquis Pierre Simon de Laplace, who led the general movement away from predominantly descriptive and classificatory sciences, the very sciences which fully incorporated a unified vision of nature. As should be clear from the discussion of their work in chapter 5, there is no such unifying vision of nature, or even evidence of a sustained struggle to attain a unified vision of nature, in the work of either André de Férussac or Jean-Baptiste-Geneviève-Marcellin Bory de Saint-Vincent. De Férussac spread his attention across a number of questions focussed on natural history, military history, geography, statistics, and the bibliographic managing of science. His work in natural history on terrestrial and freshwater snails was the most synthetic of his scientific work. There he sought to combine classificatory natural history with the study of geological formation and the distribution of species. However, his effort in this direction was not sustained. His work in geography was separate from that in natural history and was related to his role as a general staff officer. De Férussac brought breadth of experience to geography and some knowledge of the fields most closely related to geography, including a spatially sensitive natural history, geology, and statistics. He sought to reestablish geography relative to those burgeoning fields. Failing adequate support, he abandoned that effort and never returned to it. His work on the *Bulletin* (fully discussed in chapter 5) was again distinct from his other two endeavors, although there is the coherence and consistency in all three projects that comes from a single enlightened imagination. Still, it seems clear that de Férussac saw these three projects as distinct, part of different phases of his life, unified only by his authorship. Bory de Saint-Vincent's career reflects well the dilemma faced by geographers and, in its breadth, closely resembles that of Charles Athanase Walckenaer (see chapter 8). He felt competent to participate in a wide variety of types of research from speculation on the origins of "man," to zoology and botany. He wrote military histories and travel accounts, and made maps. An authority on nothing, but a significant participant in most of the major scientific state expeditions in his lifetime, there is no unifying vision in his work. One has the impression of someone always acting, responding to every invitation to write or lead, but rarely thinking hard, rarely, and never sustainedly, trying to put all the pieces together. When seen in relation to the cabinet geographers of the early nineteenth century, these militarily trained natural geographers were highly innovative. They did not follow their cabinet colleagues who rejected theory, shied away from explanation, and described a wider and wider purview of knowledge. Instead they responded to the changing nature of science and the rapid growth of ways to study the aspects of nature by getting their hands dirty in the field. This presented two problems: there was little limit to the kind of research they might do, and there was nothing holding their research together or giving it coherence. They were far closer in spirit and content to the work and thought of Humboldt than to the cabinet geographers, but, by virtue of their relatively unconsidered and unrigorous approach to the study of the world around them, they lack much of the power to be found in Humboldt's thought and writing.

HUMBOLDT'S INNOVATION IN GEOGRAPHIC RESEARCH

Theory, Explanation, and Cause

At a time when many—though by no means all—geographers, repulsed by geology's excesses and Buache's stunningly misguided generalizations, readily condemned "theory," Humboldt embraced theory, explanation, and cause wholeheartedly. Equally disturbed by some of geology's more ridiculous pronouncements, he eschewed the investigation of origin, or more particularly, the investigation of ultimate origin.[27] Nor did he in any way accept the Buachian claim to be able to predict the mountainous structure of the globe from the hydrographic system.[28] Theory in scientific hydrography, as opposed to the sort of divinatory hydrography advocated by Buache and his followers, had to be built from the ground up.

> It is from the intimate knowledge of the influence exerted by the inequalities of the surface, the melting of snow, periodical rains and tides, on the swiftness, the sinuosities, the contradictions, the bifurcations and the form of the mouths of the Danube, of the Nile, of the Ganges, and of the Amazon, that we form a general theory of rivers, or rather a *system of empirical laws,* that includes all that is common and analogous, in local and partial phenomena.[29]

In contrast to the focus on identity and analogy of form that characterized traditional natural history, theory, for Humboldt, entailed the search for analogies, parallels, and equivalents.[30] Empirical laws, or extraction of what was analogous in physical phenomena, and the combination of those laws or analogies, was theory. And the elaboration of theory was the point of science.

Identification of analogies and empirical laws demanded understanding of cause. In his introduction to the concept of the use of contours as isothermal lines, Humboldt cautioned that if the various causes of local temperature were not understood, then attempts at generalization and averaging might well eliminate the most important factors determining the distribution and development of organic life "as outside and disruptive circumstances."[31] One must, Humboldt argued, be very careful in averaging out phenomena which "depend by their nature on locality, the constitution of the soil, the disposition vis-à-vis the sun's rays of the surface of the globe."[32] Above all, when one generalizes temperature, one must bear in mind its multiple causes and their relative weight.[33] Attention to cause required a careful balance between, on the one side, excessive empiricism and, on the other, neglect of the facts. Thus, while explaining the theoretical importance of his work, Humboldt wrote against pure empiricism.[34] While convinced that the sciences could not be profitably structured or driven by empiricism, he was equally certain that "it would harm the advancement of science to try to rise to the level of general ideas while neglecting the knowledge of the facts themselves."[35]

Humboldt's work was well laced with theory, analogy, generalization, and exploration of cause. It is difficult to privilege any one of the many theories he developed over others. Perhaps, however, the most pervasive theory in his work, which was common to all of his theories, was the argument that location, conceived of in three dimensions (latitude, longitude, and altitude; or, in geognostic terms, position and superposition), was the key to understanding the natural world. Location, as he demonstrated in his famous thematic diagram showing plant distribution as related to altitude, embodies the particularities of temperature, the chemical composition of the air, the presence or absence of particular rock formations (which themselves are a function of the historical action of similar forces), intensity of light, humidity, refraction, but also location in terms of position relative to bodies of water and continents . . . etc.[36] Thus, Humboldtian location was a concept far more complex and rich than location in the traditional geographic sense. Other scientists found his approach to location sufficiently convincing and powerful to build on. De Candolle's theoretical argument that plant geographers should expect to find analogous species in similar habitats can be seen as an elaboration of some of Humboldt's arguments about the role of location in the natural world.[37]

Movement, Change, and Distribution

One of the most pronounced differences between Humboldt's research and most of the work carried out by geographers and most natural historians in France in his day was his attention to the dynamic and changing elements of the landscape. Geography's attachment to description, and particularly to the map, made movement and change precisely the dimension excluded from depiction and thus from possible analysis. In natural history, classification was still something of a craze in the late eighteenth and early nineteenth centuries and this directed the field's focus to the static and, as already discussed, to identity. Humboldt was dissatisfied with this approach to the natural world. He expressed his frustration with this perspective eloquently in a letter to Schiller in 1794:

> To date, the history of nature has been studied in such a way as to retain only differences of form in the study of the physiognomy of plants. This study of the physiognomy of plants and animals has made the teaching of characteristics and identification such a sacred science that our botanical science can only seem an object of meditation for men given to speculation. But you sense, as do I, that something a little higher must be sought; that there is something that we must rediscover. We must follow Aristotle and Pliny, who included in their description of Nature both the aesthetic sense and the artistic education of man. Those ancients certainly had a larger vision than our miserable archivists [Registratoren] of Nature. These are, to my mind, the objects which seem worthy of attention and which are almost never taken up: the general harmony of form; the problem of whether there was an original plant form, which is now to be found in thousands of gradations; the spreading of these forms over the surface of the earth; the diverse impressions of joy and melancholy that the world of plants evokes in sensitive men; the contrast

between the massive rocks, dead, immobile (and even the trunks of trees which seem inorganic) and the living carpet of vegetation which, in a sense, delicately covers the skeleton with a more tender flesh; the history and the geography of plants, that is, the historical description of the general extension of plants over the surface of the earth which is an unstudied part of the general history of the world; the investigation of the oldest primitive vegetation to be found in those funerary monuments (petrifactions, fossils, carbon minerals, coals); the progressive habitability of the surface of the earth; the migrations and journey of plants—social plants and isolated plants—with the use of maps in this; which are the plants that have followed certain peoples?; a general history of agriculture; a comparison of cultivated plants and domesticated animals, the origin of two degenerations; which plants more or less strictly or more or less liberally abide by the law of symmetrical form; the return of domesticated plants to a wild state (American plants as well as Persian ones—wild plants from the Tage to the Oby); the general confusion in plant geography caused by colonization. I worry about these questions constantly but the noise being made around me on this subject makes it impossible for me to abandon myself to it in a systematic fashion. I find that I have expressed myself as one demented. Still, I hope that you fully share my sentiments.[38]

In their study of plant geography, both Humboldt and de Candolle, whose work on plant geography was a refinement of Humboldt's, had to resist strong tendencies to static classification. De Candolle's emphasis on the importance of the "station," or the special conditions under which each species would grow, entailed far more than the addition of a line of description at the bottom of a classification. In fact, he argued against simple-minded or any attempt at a *final* classification of stations. What he was trying to get at were the interactions between the exterior forces acting on plants and between competing plants, and the influence of these and other factors (such as differing reproduction modes, and thresholds of tolerance to sub-ideal conditions) on both plant structure and distribution. A static picture of plant distribution was not the final aim of this research: understanding the interaction of forces was.

Arguably, however, Humboldt's interest in the migration of plants (and animals and humans), their colonization of particular territories, and their tendency to form relatively uniform masses of vegetation, or not, etc., was part of a larger interest in movement, change, and indeed exchange. This extended beyond the plant world to the movement or tilt of rocks (whose real extent and significance he could not quite believe) and well beyond the natural world to movement and exchange in precious metals. In *The Fluctuations of Gold*, Humboldt studied the dynamic forces behind the distribution of precious metals and their impact on prices with his accustomed sensitivity to the complexity of interactions.[39]

There was another significant difference between Humboldt and de Candolle's curiosity and that of most geographers and natural historians of the early nineteenth century. Humboldt, especially, was as interested in distribution as in location. His plant geography, and that of de Candolle, necessarily focussed on the distribution of organic life in relation to the distribution of influencing factors. Humboldt's "géognosie" had a primary focus on the location, distribution, and comparison of rocks. His isoline work sought to provide the means to depict the distribution of temperature around the globe. He even extended his interest in plant distribution to fish distribution (under water at different levels of oxygen) and to parasites within the fish or animal (to different levels of oxygen in different organs).[40] Finally, his work on the fluctuation of prices was all about the global distribution of precious metals. In developing these ideas, he frequently demonstrated a strong sense of the spatial and dynamic or fluid nature (at least over time) of phenomena.[41]

Multidiscursivity, Mathematization, and Observation

Accuracy, numeracy, and measurement were major preoccupations of late-eighteenth-century geographic and, more generally, scientific thought. On first glance, Humboldt seems to have participated in this. If, however, we compare the overwhelming preoccupation with measuring and mapping "anything and everything" displayed by geographers on the expedition to Egypt with Humboldt's own description of his mapping and astronomical operations in the Americas, there is a decided difference in flavor.[42] Humboldt's concern was not so much accuracy and somehow capturing on paper and in number the true America. Instead, his aim was to provide information in a form which would enhance data comparability and data analysis from multiple points of view. He was attempting an abstraction. The purpose of this abstraction was not possession and recreation but enhanced analysis. Accuracy, then, did not have the same importance. Thus, Humboldt explained that the careful reader of his geographic "Map of New Spain" would see that it did not quite agree with his "Map of the Route from Mexico to Durango." This, he explained, was simply the result of a combination of the different sources used in the two maps and the process of reconciliation employed in compiling each.[43] So, for Humboldt, if the discrepancy were known and its source understood, then there was nothing further to worry about. The absolute truth of the picture was not the issue.

What measurement, numbers, and statistics did allow was a certain transferability of data and analysis across fields: a true interdiscursiveness.

For Humboldt, ever concerned with the unity of nature and the sciences, this was power indeed. His memoir on isolines was designed precisely to bring together and compare temperature measurements so that "theory can draw on the corrections provided by the diverse elements considered."[44] The quality of the data, especially the number of observations and the quality of instrumentation used, was important, but some of these problems could be minimized through the use of probability theory.[45] What was more important was having information on temperature in a form which could equally well be used to elaborate and test theory in plant geography, in economics, and even in geognosy.

For Humboldt, the isoline and more generally a numerically based graphic scale was analogous to and perhaps a continuation of the revolution brought about by the thermometer in the seventeenth century.[46] The thermometer for the first time allowed tracking exactly what was happening on the surface of the earth, comparing yesterday with today, comparing here with there, and recording information. Data in comparable numerical form was the basis from which the study of phenomena as complex as, say, fossil remains (historic plant and animal geography) could begin to be useful to the elucidation of geologic chronology.

> To compare formations with relation to fossils, is to compare the *Floras* and *Faunas* of various countries at various periods; it is to solve a problem so much the more complicated, as it is modified at once by space and time.[47]

Without numbers and comparable statistics, how could the processes behind geologic formations be studied at all?

Some areas of inquiry were more open to mathematization. "The details of natural history are foreign to statistics," Humboldt observed. But natural geography, by virtue of its ability to convert natural history's data to number and statistic, could substantially contribute to forming an exact idea of the territorial wealth of a state.[48] In this sense, through comparable numerical data, it could be a truly unifying field capable of linking nature and society.

Of course, for Humboldt, behind number lay observation. His voyage through South and Central America was all about observation and even occasionally experimentation.[49] In the account of his travels to South America, he pointed out that while the eighteenth century had been characterized by extraordinary developments in instrumentation, there had not been a growth in observations sufficient to take advantage of those improvements.[50] He was prepared to use the observations of others, as indeed, given the scope and scale of his focus, he had to.[51] Nevertheless, it was his own observations which gave him the ability to sort through and judge that great mass of facts.

Scale, Landscape, and the Natural Region

One of the most fascinating features of Humboldt's work was his sensitivity to scale. He had a strong tendency, particularly in his work on the geography of plants, to move up and down the scale from a micro scale discussion of a particular plant, to the analysis of local climate and conditions, to observations and generalizations on a continental scale. He was, however, acutely aware of the dangers of this sort of elision in scale, particularly in the realms of analysis and evidence.[52] Successful comparison across large areas demanded careful attention to local conditions and mechanisms.

Humboldt did have a favorite scale of analysis. This could be described as geographic, ecological, or landscape-based. It was this scale of analysis which he felt had the power not just to interest but to move the human spirit. He argued that it was as valid as the exploration of any other scale of existence.

> The configuration of great mountain masses, the great diversity of the contours of the high summits, situated like the lower lands in the midst of the agitations of the atmospheric ocean, are amongst the elements that constitute what we might call *the physiognomy of nature*. The look of the mountains contributes no less than their form, their size, and the grouping of plants, nor less than the different species of animals, the nuance of the celestial vault, and the intensity of reflected light in determining the character of a landscape and the general impression made upon man by the different zones of the earth.[53]

Perhaps more than anything else, Humboldt was interested in landscapes and their characteristic look. What, for example, was responsible for the very different look of the landscape in temperate versus tropical zones? In answering this question, and in his attention to landscape in general, he never lost sight of cause and process or of the action of process over time. Thus, for example, he described the topography of the Valley of Mexico, which he considered ideal for trigonometric mapping, in the following terms: "the vast plains of Zelaya and of Salamanca, united like the surface of the waters which seem to have covered the soil for many centuries . . ."[54] But it was the beauty of the landscape—of the human-scale landscape—of the view of the volcanos from Mexico City—to which Humboldt was most susceptible, and only a geographic-scale focus could bring that beauty home.[55]

Humboldt's focus on landscape closely approximates another concept that became important in the course of the nineteenth and twentieth centuries: the natural region. Bory de Saint-Vincent seems to have been gravitating to a similar concept in his *Guide du voyageur en Espagne*. However, whereas Bory de Saint-Vincent only touched upon the idea, Alexander von Humboldt explored the concept of the natural region and the landscape scale of analysis in a number of his texts, maps, and landscape depictions. The idea of the natural region has been traced by Paul Claval and Michel Chevalier to Girault-Soulavie, the provincial prelate who, during the first half of his highly checkered career, wrote *Histoire naturelle de la France méridionale* (1780–1784) and the unfinished *Histoire du Vivarais* (1779–1788).[56] Girault-Soulavie and Humboldt, separated by age, politics, nationality, and religion, nevertheless shared the same preoccupation with the vegetation characteristic of particular altitudes, climates, soil types . . . in short, natural regions. In his 1820 encyclopedia article on plant geography, in which he provided something of a historiography of the field, de Candolle linked the work of Girault-Soulavie and Humboldt.[57] Girault-Soulavie had developed some of the key basic ideas behind plant geography, albeit in an impressionistic, almost literary way, while Humboldt had laid down the essential elements of the science. However, both de Candolle and Humboldt, more concerned with the processes behind plant distribution than with the delimitation of regions, can only be said to have given the concept a nudge. The natural region, as an idea, was instead picked up, developed, and given a more geological emphasis in the 1820s and '30s by both Omalius d'Halloy and Coquebert de Montbret as the "physical region."[58]

Interior Structure and Function

If, rather than describing Humboldt's interests in our own disciplinary terms, dividing the natural from the human sciences, we see them with his unifying vision, we could say that Humboldt had a primary interest in the "way of life" of plants, animals, humans, and even, in a sense, of rocks. Today, we would distinguish these as an interest in the way of life of plants, animals, and humans and a focus on the interactions between physical phenomena. Yet, the curiosity is fundamentally the same in both cases. Humboldt wanted to understand how things worked and how they worked together to create what they are to the casual or admiring human gaze. Thus the differences in "morals or customs" noted by travel writers were worthy of attention but only if the narrative writer could remove him- or herself from the center of the stage to focus on the relations between the people under consideration. Humboldt did not doubt that it was of such interactions between individuals that the larger society was composed.[59] The failure of most travel writers to enter into this sort of social depth frustrated Humboldt. Indeed, he had so much trouble with the overbearing flow of the travel narrative that he found himself, in the end, unable to write one.[60]

Humboldt's interest in the interactions behind social characteristics is parallelled by his attention to the aspects of animal physiology that might influence animal behaviour. This sort of inquiry required dissection, and Humboldt averred that, as animal anatomy was not his principal focus, his occasional work in dissection might lack refinement. Nevertheless, he saw the dissections he carried out in South America as a minor contribution to Cuvier's larger efforts to get beneath the surface in understanding the relationship between physiological structure and animal behavior and to relate these to classification.[61] But here again, Humboldt found his dual aims of systematic anatomical study and his effort to acquire a sort of cartographic overview of the territory through which he was moving (perhaps analogous to the travel narrative) in conflict.[62]

As Humboldt sought to understand how human societies worked and animals behaved, he strove to understand plant geography. Thus, for both de Candolle and Humboldt, understanding the external forces acting on plants provided only part of the explanation. Humboldt was primarily concerned to refocus the work of natural historians on the geographic forces creating plant communities. However, he was aware that plant communities and landscapes were the result of the interaction of external forces and internal structures and constraints. Whether the external layers of a particular plant were composed of carboniferous or resinous material might have everything to do with its survival in cold conditions. Survival in different levels of light might be determined by the plant's particular root or leaf structure. And survival in different conditions of wetness would have everything to do with the plant's relative sponginess, its number of pores, etc. Indeed, de Candolle was convinced that the precise conditions under which a species would geminate could only be understood at the level of the excitability of its fiber or tissue.[63] Moving beyond the surface layer—which had been the principal preoccupation of natural historians for generations—was, once again, essential to understanding the way of life of plants.

Humboldt was similarly concerned with exploring complex internal interactions in his geognostical studies. His aim was to move beyond the classification of rock types to acquire understanding of the geognostical structure of the globe. Indeed, he equated the kinds of questions he was asking in terms of geology, stratigraphy, and landform to the types of questions asked by Cuvier of the interior structures of animals.[64] Rocks, of course, do not have a way of life. Still, the course of life or nature, over time, had helped to create the earth's geognostical structure. In Humboldt's view, then, his study of geognosy went beyond Dolomieu and Saussure's surficial topography to Werner's preoccupation with the interior structure of particular regions.[65] Humboldt's aim, as in his studies of society, animal behavior and plant geography, was not to focus on the particularity of phenomena, but to understand the complex interactions between particular phenomena that had created, in the case of geognosy, the physical region.[66]

A New Kind of Map

Humboldt did not "invent" the thematic map. A small number of such works predate Humboldt by hundreds of years.[67] He did, however, develop and explain some of the basic forms of expression used thenceforth in such maps. He also thought through and demonstrated the relevance of these maps to the new science he was practicing.[68] Denis Wood has argued that there is no such thing as a thematic map per se; that all maps have themes and, he might have added, that all maps embody a theory (perhaps scientific) about the nature of the world.[69]

Nevertheless, even those relatively little informed about the history of mapping would find most thematic maps fundamentally different in nature from, say, a topographic map. What then is the difference and why is it important? There are as many assumptions about the nature of reality and what is important to depict in it behind the topographic map as behind the thematic map. Nevertheless, there is a different *sense* of certainty and reality surrounding the two maps. The stated aim of the topographic map is mimesis. It is designed to replace reality with a more useful and simplified picture of commonly accepted salient aspects of reality. In that sense, grammatically speaking, it is a simple declarative sentence: one very extended descriptive statement. Recognition of the importance of this sort of statement, almost classification, to state administration was one of the great products of the eighteenth century. It was in the eighteenth century that codification of the language of this description took place and began to spread throughout Europe.

The thematic map has a logical structure more akin to an hypothetical argument: if we take this (the basic topographic information on the map) to be more or less true, and we use it to give a spatial structure to this (hierarchically structured data not generally found on topographic maps) and perhaps also this (other hierarchically structured data), then we have an interesting pattern not generally visible either on the topographic map or in the landscape. The thematic map, then, embodies an argument about invisible phenomena, or phenomena undergoing change, or about interactions between phenomena. It does not function as a record but as an analytical device, set aside when the question at hand is resolved. Arguably, then, although everything is a mental construct, there are different kinds of mental constructs.

Another difference between the topographic map and the thematic map is its place in the division of labor. The topographic map was, and is still, made by map specialists; people primarily concerned with the correct, consistent, and complete expression of an elaborate descriptive topographic statement: people who came to be called cartographers. The thematic map was, and is, more commonly designed and often made by the scholars and scientists who are specialists in the hierarchically structured data: social scientists, earth scientists, etc. Their concern is less with the language of description than with the logic of the argument, and any innovations in expression flow from the nature of the argument.

All this might be little more than an interesting definitional aside except that the development and elaboration of the thematic map corresponds to a shift in the nature of social and natural science. It is not coincidental that the thematic map came into its own in the early nineteenth century. It is also not coincidental that it received special attention from Alexander von Humboldt, whose work and preoccupations exemplified the shift from descriptive science to theoretically driven explanatory science focussed on the observation and analysis of change, distribution, interactions, and the interior functioning of natural and social phenomena.

By 1807 Humboldt had a knowledge of maps which could rival that of any topographical engineer, commercial cartographer, or academic geographer. Not only did he collect astronomic, geodetic, and itinerary data for many of his own maps and plans, he understood the critical decisions behind the compilation of that data. Thus, in his *Atlas géographique et physique du Royaume de la Nouvelle-Espagne,* he was able to explain his maps, his sources, and his decisions in a way that few before or since d'Anville had managed. Indeed, Humboldt had a compilation cartographer's sense of what parts of the world had been mapped to what scales and what degree of accuracy. He was aware of just how few places, even in the heart of Europe, had been well determined.[70] It was his knowledge of far more than maps that gave his understanding of maps themselves even greater depth than is to be found in any of d'Anville's works. For example, his sensitivity to landscape form and his sense of its relatedness to hydrology, geology, and history gave him a more critical attitude to relief depiction. Thus, he complained about the contemporary maps of the Valley of Mexico:

> In spite of how interesting this country is on three levels, in terms of its history, its geology, and its hydraulic architecture, there is not a single map in existence whose contemplation will give birth to a real sense of the valley's form.[71]

How, then, should such form be depicted? In practice, he generally chose hachures. However, his understanding of just how little was known about landscape and topography in most parts of the world made him cautious in advocating the use of either hachures or contours. Thus, at a time when many cartographers were arguing that contours by virtue of their greater accuracy ought to replace the more traditional hachure method of relief depiction in topographic mapping, Humboldt argued that pictorial hill symbols might more accurately reflect the state of current knowledge.[72] Humboldt was far from naive about traditional topographic maps and their limitations for the interpretation of the phenomena he sought to study.

It was from this base of knowledge, both about maps and about physical and social processes, that Humboldt began to experiment with graphic representation. He was not alone in his experimentation but was conscious of developing something analogous to some of the explanatory graphic tools being developed by people like Girault-Soulavie, August Crome, Jean-Étienne Guettard, J. L. Dupain-Triel, and, most importantly, William Playfair. While his graphic experimentation was intellectually playful, it was also purposive. Humboldt was seeking a more analytical spatial language which would allow the almost intuitive transfer of understanding from one graphic genre to another and from one specialist body of knowledge to another. Humboldt was trying to find a language capable of expressing his vision of the unity of nature with the newly found rigor of the systematic sciences. To that end, he experimented with isolines, distribution maps, flow maps, a map of error, proportional squares, something he called "pasigraphy," and a multidimensional pictorial graph. These eloquent graphic arguments about dynamic relationships in space, distributions, and interactions—which often revealed patterns not visible to the naked eye and created a new systematic, rather than geographic, time/space—seemed to Humboldt to suggest that his intuition about the unity of nature was sound.

Isolines (or lines drawn between measures of equal values of a particular phenomenon [height, temperature, degree of cloud cover, etc.]) were not a recent innovation. Nevertheless, although they had been developed by Halley over a century earlier to show patterns of magnetic declination, Humboldt pointed out that their true potential had not been realized. Humboldt rescued isolines from relative obscurity precisely for their ability to show patterns not visible to the eye or patterns so complex as to be obscure to the senses. Isolines were a method by which disparate numerical data could be brought together and rendered accessible to theoretical interpretation.[73] With such a rigorous yet simplifying mode of graphic representation, great volumes of data could be managed, rendering it possible to even consider "the influence of local causes of perturbations."[74] Indeed, the use of isolines, by revealing new patterns and enhancing comparability of data, would cause scientists to reflect back on the data they were collecting and to reexamine its rigor and value. Once this sort of care had been employed in the collection and interpretation of the data, Humboldt predicted that it might even be possible to adequately theorize solar action upon the earth and to calculate the distribution of heat received from the sun around the globe.

The greatest value of graphic representation, and of isolines more particularly, was its remarkable ability to reveal relationships between and the relative importance of, for example, latitude and continentality; altitude and wind patterns; or humidity, heat, light, and pressure.[75] Geographic maps and their power to reveal geographic patterns in locational data were analogous to thematic maps and their ability to reveal heretofore invisible patterns in systematic data. The space and especially the time displayed by these maps was a little different than geographic space and the relatively thin temporal code embodied in most maps. The space was systematic, structured so as to reveal the relationships between physical phenomena.[76] The time, instead of expressing, for example, France in 1756, might describe temperature variation over months, decades, or even hundreds of years. Given all this, one would expect Humboldt to have demonstrated the cartographic power of his isolines with a world map. As Arthur Robinson and Helen Wallis have explained, we do not know precisely why Humboldt did not include the map that was ultimately published in the *Annales de chimie et de physique* in his article on isothermes. It may have had more to do with the publication trials and tribulations of the *Mémoires . . . d'Arcueil*.[77] Certainly, Humboldt knew that at times the space of the map can be problematic and intellectually constraining. Thus, he may have considered that his 1813 essay "Des lignes isothermes et la distribution de la chaleur sur le globe" was well illustrated with a table structured by "isothermal bands" of 5 degrees Celsius. In fact, this was entirely in keeping with one of the principal aims of his essay, which was to unshackle the concept of temperature from latitude. He created a table, entirely relatable to a map, but whose space was defined not by latitude and longitude but by bands of average annual temperature.[78] This amounted to giving "systematic space" priority of expression over "geographic space."

Maps, tables, and graphs were all very well, but intelligent use of isotherms had to be informed by knowledge of physical and natural systems. Relationships might only be apparent, or unimportant, if unsupported by data and theory from, for example, plant geography, physics, chemistry, etc. Isolines, then, offered the possibility of truly capitalizing on numerical data while also increasing the likelihood of more intuitive insight.[79]

In some cases, Humboldt's experimentation with thematic mapping was more speculative than real. Often because the data was not available to undertake or even mock up the cartography. Such was the case with a proposed botanical map of the regions inhabited by single species (or social plants). Humboldt speculated that this map would almost certainly reveal such plants to have posed significant obstacles to human settlement and the movement of armies. They would, he thought, be seen to have formed barriers to human movement as significant as the mountains and the seas.

Standard topographic or geographic maps have limited means of depicting flow and movement. Humboldt produced at least one graphic designed to show the flow of metals between Europe and North America (see figure 8). The map, really a pseudo-thematic map, carries relatively little information. It is a map of the world centered on the Pacific. It shows only, as its title suggests, the routes by which precious metals were moved between the continents. However, Humboldt placed the map alongside four graphs (see figure 9) showing the volume of precious metals extracted from mines in the Americas since 1500; the amount of gold and silver extracted from Mexican mines since 1700; the proportion of gold and silver mined in different parts of Central and South America; and the proportion of silver produced by America, Europe, and Asia. It is but a tiny step from this map and its associated graphs to a thematic map, such as those produced by C. J. Minard, showing volume by thickness of arrow. It is important to note that in separating out map and graph, Humboldt was able to fully develop both the dimensions of space and time. He thus suggested the depth and importance of the relationship between the Americas and Europe.[80] Although his graphic is relatively informationally thin, Humboldt was fully aware of the interesting insight that might result from the mapping of human activities of all sorts and even from the comparison of these results with the results of investigations into the spatial patterns of the physical world.

On a somewhat more sardonic note, Humboldt explored the alternative spaces created by contemporary and early geographers in the south-central region of North America in a map of error. On this map, entitled "Map of the False Positions for Mexico, Acapulco, Veracruz and the peak of Orizaba," Humboldt drew the outline of Mexico and placed these locations according to Arrowsmith, d'Anville, Covens, Harris, and the Connaissance du temps of 1804, among others. Many of these authorities effectively placed Acapulco well out in the Pacific, and Mexico City almost anywhere in Mexico, including in the Gulf of Mexico or close to the Pacific coast. This map, based on Tobias Mayer's *Mapa critica Germaniae,* was designed

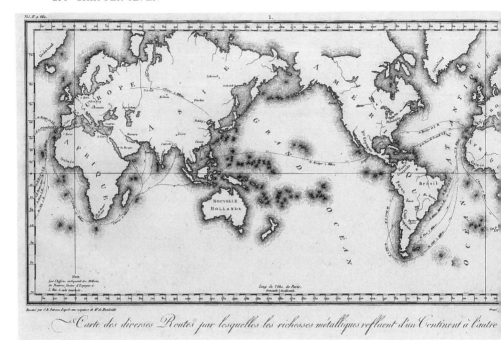

FIG. 8.   Alexander von Humboldt, "Cartes des diverse routes par lesquelles les richesses métalliques refluent continent à l'autre."

to show "just how imperfect the published maps of Mexico had been,"[81] but it also nicely portrays the imaginative geography practiced by many of Humboldt's contemporaries.

In one of Humboldt's more playful graphics he compared the relative territorial extent of Spain and the Spanish colonial possessions by means of proportional squares (see figure 10).[82] Together with this graphic is another which compares four dimensions: the comparative population size of European versus colonial territories and the comparative territorial extent of these same areas. This representation is highly suggestive of questions about the nature of colonialism that would only be explored in the twentieth century. Humboldt commented about this depiction:

> The figures combined on this plate demonstrate what is said below on the extraordinary disproportion observable between the extent of the colonies and the area of the European metropolis. The inequality of the territorial division of New Spain has been rendered apparent through a depiction of the intendancies by concentric squares. This graphic method is analogous to that first ingeniously used by Mr. Playfair in his commercial and political atlas and in his statistical maps of Europe. Without attributing too much importance to these sketches, I cannot either regard them as mere

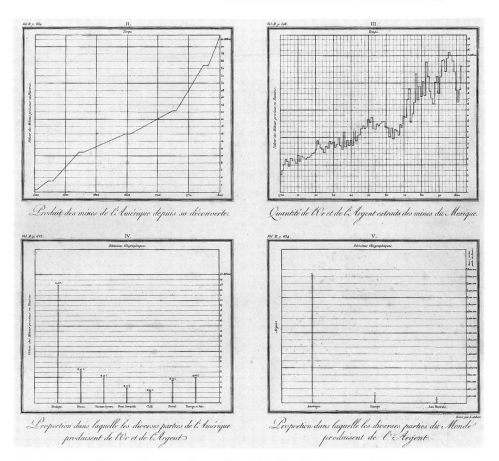

FIG. 9. A set of four graphs: Alexander von Humboldt, "Produit des mines de l'Amérique depuis sa découverte; Quantité de l'or et de l'argent extraits des mines du Mexique; Proportion dans laquelle les diverses parties de l'Amérique produisent de l'or et de l'argent; Proportion dans laquelle les diverses parties du Monde produisent de l'argent."

intellectual games unrelated to science. It is true that Playfair showed the growth of the English national debt which had a strong resemblance to the peak of Tenerife. But physicists have long used similar figures to show the rise and fall of the barometer and the average monthly temperature. It would be ridiculous to try to express by curved lines moral ideas, the prosperity of peoples, or the decadence of their literature. But anything that has to do with extent or quantity can be represented geometrically. Statistical projections, which speak to the senses without tiring the intellect, have the advantage of bringing attention to a large number of important facts.[83]

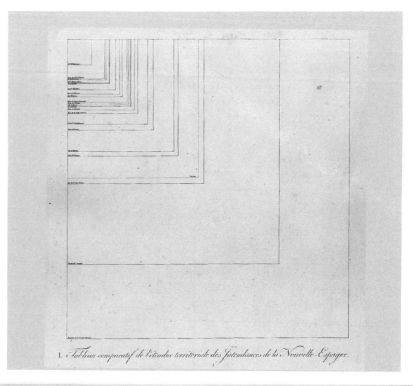

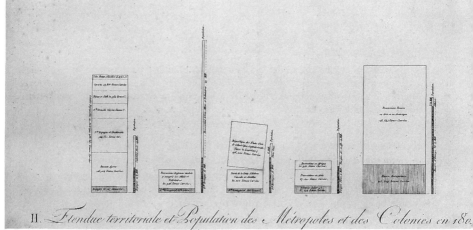

FIG. 10.   Alexander von Humboldt, "Tableau comparatif de l'étendue territoriale des Intendances de la Nouvelle-Espagne" (I), together with "Etendue territoriale et Population des Métropoles et des Colonies en 1804" (II).

Humboldt was not to know that it was precisely the uses that he deemed absurd for such representations that would become the most persuasive in both social science and the public realm.

Humboldt was more interested in natural than social phenomena and far more interested in landforms and landscapes than he was in geological structures. However, he was aware of research and innovation in all of these areas. He had noticed the stratigraphic diagrams being developed in geological studies and became convinced that they could be modified to enhance the study of geognosy. By geognosy Humboldt meant the comparative study of the superposition of rocks and rock types around the globe. The purpose of this study was not the detailed mapping of bed stratigraphy but the acquisition of a sense of common form, common development, and the ways in which landscapes had altered as a result of local conditions. Thus, he was not interested in the sort of chemical and mineral detail pursued by geologists but sought a more generalized picture which would allow him to immediately apprehend similar or identical patterns of superposition in widely separated areas. To this end he advocated two techniques of data representation, one graphic ("imitative or figured") and the other "algorithmic." The graphic technique Humboldt proposed looked very much like the geologist's stratigraphic diagram except that it covered a far larger area, paid more attention to the landforms in question, and generalized the formations into parallelograms with perhaps some additional stippling to suggest "the relations of composition and structure" which so preoccupied geologists.[84] Humboldt's "Esquisse géognostique des formations entre la Vallée de Mexico, Moran et Totonilco," drawn in 1803 and engraved in 1833, seems to be the only example of this graphic pasigraphy (see figure 11). It bears ample sign of inadequate information on the structures beneath the surface topography for much of the area depicted.[85] Humboldt also produced profiles designed to reveal and accentuate the topography of Central Mexico (see figure 12).[86] He was looking for a particular type of information, about the succession and relative age of rocks. He believed he could decipher these by "fixing the attention on the most general relations of *relative position, alternation,* and the *suppression* of certain terms of the series":

> The whole geognosy of positions being a problem of *series,* or the simple or periodical succession of *certain terms,* the various superimposed formations may be expressed by general characters, for instance, by the letters of the alphabet. . . . The more we make abstraction of the value of signs (of the composition and structure of the rocks), the better we seize, by the conciseness of a language in some degree algebraic, the most complicated relations of

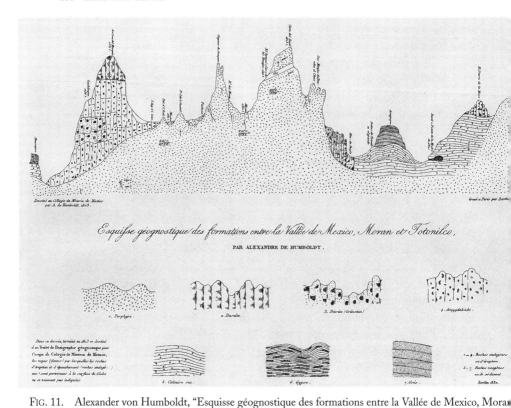

FIG. 11.   Alexander von Humboldt, "Esquisse géognostique des formations entre la Vallée de Mexico, Moran Totonilco."

position, and the periodical return of formations. The signs "α" "B" and "Υ" will no longer represent granite, gneiss, and mica-slate; red sandstone, zechstein, and variegated sandstone; chalk, tertiary sandstone with lignite, and Parisian limestone; they will only be the terms of a series, simple abstractions of the mind.[87]

Beyond looking for patterns in the structures beneath the earth and between continents, Humboldt was looking for modes of representation that could generate ideas capable of bridging the gap between different realms:

In this *geognostical essay*, as well as in my researches on the *isothermal lines*, on the *geography of plants*, and on the laws which have been observed in the *distribution of organic bodies*, I have endeavoured, at the same time that I presented the detail of the phenomena, to generalize the ideas respecting them, and to connect them with the great questions in natural philosophy. I have dwelt chiefly on the phenomena of *alternation*, of *oscillation*, and of *local suppression*, and in those which result from the *passage* of one formation to another in consequence of *interior development*. These subjects are not mere

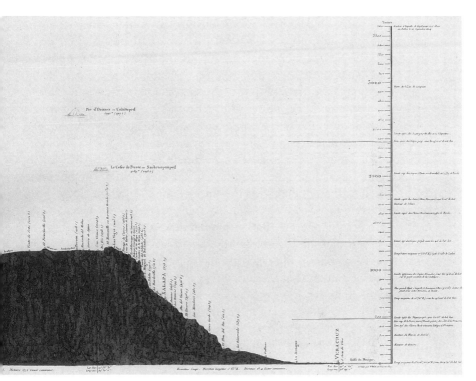

FIG. 12.   Alexander von Humboldt, "Tableau physique de la Nouvelle-Espagne. Profile du Chemin d'Acapulco à Mexico, et de Mexico à Veracruz."

theoretical speculations; far from being useless, they lead us to the knowledge of the laws of nature.[88]

If the final aim was the decipherment of the laws of nature and his method was generalization to emphasize a few key concepts, he also sought to make the link between his geognostical and botanical work on a more banal level. Thus, his landform diagrams included information on temperature, the snow line, etc., that was designed to provide measurement-based scalar links to "the large picture attached to my *Géographie des plantes.*"[89]

The most influential of Humboldt's experiments into graphic expression was without question the remarkable multidimensional pictorial graph reproduced in his *Essai sur la géographie des plantes* and entitled "Géographie des plantes équinoxiales. Tableau physique des Andes et Pays voisins" (see figure 7, at the beginning of this chapter). Humboldt's overriding concern in this essay, and in his "Tableau," was both to introduce a new way of looking at the world of plants to his natural history colleagues and to the larger public and to demonstrate how interconnected all of nature's forces and aspects really were. In a sense, then, his *Essai sur la géographie des plantes,* more than any of his other works, was a concise version of the *Cosmos.* Arguably, too, his "Tableau" was an even more concise version of the *Essai.* In spite of the length of the five-volume *Cosmos,* Humboldt believed in the power of concise and precise expression which might convince skeptics of the importance of looking for interconnections between different realms and different lines of enquiry.[90] The "Tableau" presented a view of three peaks of a mountain chain, one of which was colored with vegetation while the other two carried the names of the plants to be found at that altitude. On the sides of the illustration there was a scale marked in toises and in meters. Columns parallel to the meters/toises columns indicated minimum and maximum temperatures at given heights; the chemical composition of the air at different heights (oxygen, hydrogen, carbon); the lower limits of perpetual snow at different latitudes; the typical animals to be found at each elevation; the temperature at which water boils (at different elevations); "Vues géologiques" in which he observed that rock type was independent of elevation but that in any given area there is an order of superposition, inclination, and direction of beds which is determined by a "système de forces particulier." In that same column he then described the standard superpositioning and pointed out what was geologically particular about the equatorial regions: depth of the beds, the elevation of the post-granitic formations, etc. Subsequent parallel columns described: the intensity of light at different elevations; the humidity level at different elevations; how blue the sky looked measured in cyanometers at different elevations; the gravitational force at different elevations; the kinds of agriculture engaged in at various elevations; the incidence of electrical phenomena at various elevations; measured heights in different parts of the world; the distance at which the mountains would be visible from the sea, making abstraction of refraction; and the refraction at a given height and at 0 degrees of temperature.

In the text of his *Essai,* Humboldt then tried to draw the reader to his own conclusions by discussing linkages between phenomena that he had been unable to bring out in the graphic. These included the link between plant geography and geology to be found in fossil remains of plants (which added the dimension of geologic time);[91] the link between plant geography and the political and moral history of "man" forged by the human need for food (which added the dimension of human material history);[92] and the link between nature and "man's" spiritual being to be found in the imitative arts (which added the dimension of intellectual history).[93] These linkages raised fundamental scientific questions both about the history of the earth and about human history on the earth—precisely the sorts of questions Humboldt wanted his readers to ask.

At the end of this remarkable essay, which illustrated his even more remarkable "Tableau," Humboldt, one of the greatest early-nineteenth-century exponents of field research, spoke of the power of study through books and art. Broaching a theme strikingly reminiscent of those later developed by Foucault in his "Fantasia of the Library,"[94] he commented:

> That is how, without a doubt, enlightenment and civilization most influence our individual happiness. They [books and art (or text and graphic)] make us live at one and the same time in the present and in the past. They gather around us all that nature has produced in the diverse climates and put us into contact with all of the peoples of the earth. Sustained by the discoveries already made, we can leap into the future, and with a premonition of the consequences of phenomena, we can establish the laws of nature. It is through this research that we open the way to intellectual delight, to a moral liberty which fortifies us against the blows of destiny and which no exterior power can undermine.[95]

This, then, was the purpose of his oeuvre: to place "man" in a position from which he could observe nature, himself, and the workings of the entire cosmos.

CONCLUSION

This chapter has sought to capture the essence of Humboldt's "meditation" on the cosmos over a lifetime of exploration, research, experimentation, discussion, and thought. When we examine his work and contrast it with the natural history or geography of the Enlightenment, its innovative nature jumps out at us. The contrast with the type of questions being asked by contemporary cabinet geographers—still focussed on the flat and static surface of the map, or on literary description derived from second-, third-, or fourth-hand results—is stark. The military geographers, represented in this work by de Férussac and Bory de Saint-Vincent, were interested in theory, explanation, and cause; explored change and distribution, especially plant and mollusc distribution; and, in the case of Bory de Saint-Vincent, toyed with the concept of the natural region. In contrast to Humboldt, however, they took their cue from the growth of the life and earth sciences developing around them and essentially abandoned the unity that lay at the heart of the geographic endeavor.

Standing astride a major shift in the nature of the modern earth and life sciences away from description and toward explanation, Humboldt embraced theory, its explanatory power, and its exploration of immediate cause. Not satisfied with the study of the unchanging and the static, his curiosity turned to movement, change, and distribution. Stretching beneath the surface, he strove to understand the interior structures and functions of phenomena, whether organic, inorganic, or social. He advocated and pointed the way to a new multidiscursivity through the universal language of mathematics. He played with concepts of scale to try to arrive at a unit of analysis that would allow him to both analyze and synthesize.

Yet, Humboldt was also trying to retain something of the holism of traditional geography and, for that matter, of Enlightenment science, which saw the cosmos as a whole that could be elucidated by description. In geography that description took two forms: textual description and cartographic description, which sought to place "everything" within an x,y coordinate system. In natural history the evolving descriptive system was classification. Both implied a whole which could be gradually filled in according to a usefully rigid conception of nature. Both required critical observation and data collection. Humboldt's work was innovative in his attempt to break through and go beyond the classification systems of the "miserable archivists of Nature" to see and combine what these cataloguing systems could not perceive. In plant geography that meant sacrificing classification based on superficial characteristics to a combined analysis of the interaction of interior structures and location-based exterior influences, a physics of nature. In mapping it meant challenging and testing the veracity of maps and their classic forms of expression, going beneath and above the surface of the map to explore location-based cross-sections and elevations and reformulating the map from a descriptive statement to an argument infused with theory. All of these were part of a larger effort to see nature at once in great detail, using the insights from experimental physics, anatomy, geodetic and astronomical observation . . . and at the same time with a holism capable of integrating organic, inorganic, and human nature in all of its diversity and complexity.

The rise of thematic mapping coincides with the development of the new technologies of graphic reproduction beginning with lithography in the 1820s. Undoubtedly thematic mapping was well served by these technologies, which permitted the use of color, and made the maps cheaper, easier to draw, and easier to integrate into texts. Likewise, the establishment of regular statistical surveys led to an accumulation of data which began to facilitate the graphic display of geographic information by mid-century. It was also the kinds of questions that scientists were beginning to ask of the natural and human world that stimulated the development of this new graphic form of expression. Thematic mapping came to be in precisely the period that witnessed a shift from a more descriptive to a more explanatory mode of investigation. That is far from coincidental: the thematic map enhanced analysis while permitting synthesis, both geographic and not. In his experimentation with maps, graphs, profiles, etc., Humboldt was looking for ways to express what was seminal in his work. He was looking for a language at once highly descriptive but also analytical: capable of moving beyond geographic space to assume the spaces of scientific theory; capable of comparing like things in very different places; and capable of revealing new linkages. Humboldt's experimentation with graphic expression is interesting because it is suggestive of a shift in both the nature and language of science.

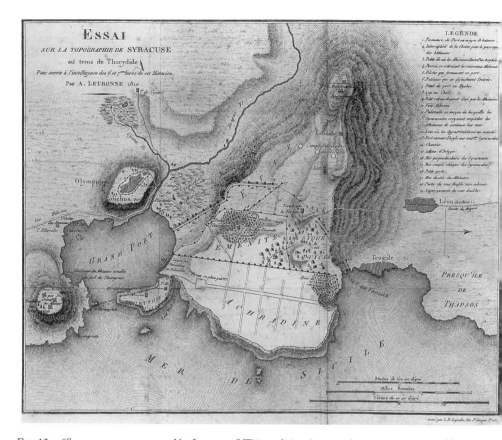

FIG. 13. "Syracuse as reconstructed by Letronne." This traditional geography, a reconstruction of Syracuse at the time of Thucydides, was the only such work carried out by Letronne. He moved away from asking traditionally geographic questions very early in his career.

Tough-Minded Historical Geography

Far from being a simple technique, epigraphy is the royal access to the most solemn or official, but also the most intimate, in all civilizations: laws, administration, religion, and so many details of material and everyday life. —J. Leclant, "Allocution d'ouverture" (1990), 314

Of all of the types of geography practiced in the early nineteenth century, historical geography is unquestionably the least accessible to moderns. This geography, preoccupied with the recreation of past geographies from ancient texts (sometimes with the help of travel accounts), was the immediate descendant of the geography of d'Anville and remained well within d'Anville's conception of geography. After all, once geographers, cartographers, geodesists, and astronomers were able to map and verify information on location, cabinet geographers such as Jean-Denis Barbié du Bocage, d'Anville's only student, could only under special circumstances sell their services as modern mapmakers: for historical texts or in conditions where geographic information was desperately sought and unavailable by any other means.[1] There was still room for commercial mapmakers such as Pierre Lapie, who in the course of his life made an extraordinary number of high-quality maps, largely from secondary sources.[2] In spite of the high regard in which his work was held, Lapie was considered neither "scientific" nor "erudite." Therefore he could not hope to gain access to one of France's scientific academies.[3] Perhaps it was because erudition brought greater kudos than modern mapping that Barbié du Bocage spent his academic career mapping ancient Greece and Rome, recreating the world as it must have been known to Herodotus, Strabo, Eratosthenes, etc., and writing a descriptive geography of the ancient world. Other kinds of early-nineteenth-century geography fall easily under the heading of historical, including reproductions and discussions of medieval or Renaissance maps; reproductions, translations, or new editions of the works of ancient geographers; and perhaps least appealing of all to us, analyses of the measurement systems of the ancients. This historical geography was "démodé" even in its own time, as dry and dusty (even musty) as it is possible for classical studies to be, untheoretical, unreflective of modern preoccupations and interests, and, because it was so narrowly focussed on the map and a cartographic understanding of the ancient world, superficial. If this adequately described the work of all historical geographers of the early nineteenth century, it would scarcely warrant more than the seven or eight lines accorded it by Robin Butlin.[4] There was, however, one geographer, Jean-Antoine Letronne, who stood apart from his colleagues, both in terms of the substance of his work and in terms of its reception among geographers. The quality and imagination of his work and its puzzling neglect by contemporary geographers alone makes it worthy of our attention.

THE MAINSTREAM IN HISTORICAL GEOGRAPHY

To best understand the ways in which Letronne's work was different and innovative, it is necessary to pause and—at least briefly—look at the mainstream. The mainstream in historical geography was principally occupied by five men. Barbié du Bocage was known to his contemporaries as d'Anville's only student, though, at one point in his career, he preferred to deny that.[5] Pascal-François-Joseph Gosselin wrote the *Rapports à l'Empereur* on ancient geography and thus, in a sense, defined the field.[6] The Baron Charles Athanase Walckenaer may be considered an historical geographer for a number of works, including an edition of Dicuil's geography, a general text on ancient and medieval geography, an essay on an itinerary dating from the Crusades, and his history of the Gauls.[7] Edme-François Jomard was certainly working within the tradition of historical geography in much of his writing on Egyptian antiquities and in his work on the history of cartography (both of which have been discussed in chapter 4). Finally, and in contrast to his innovation in natural geography, Alexander von Humboldt's work on the history of geography may also be considered part of mainstream historical geography.[8]

Jean-Denis Barbié du Bocage's Cartographic Erudition

It is virtually impossible to be engaged by the writing and thought of Barbié du Bocage. A member of the Académie des inscriptions, and respected in his own time for, as one of his necrologists, the Baron Walckenaer put it, "his fondness for the most difficult, the most useful, and the least appreciated of all of the branches of erudition,"[9] he devoted his life to establishing the geographic (fairly narrowly cartographic) background of ancient events for works by historians, travelers, and collectors. His best-known work, in fact the only work for which he was widely known, was the atlas for the Abbé Jean-Jacques Barthélemy's enormously popular *Voyage du jeune Anacharsis en Grèce.*[10] This work is in the genre of what we would today call "creative nonfiction," or "narrative nonfiction." It is the story of an invented character, the Scythian Anacharsis, who visits Greece a few years before the birth of Alexander the Great. There he travels widely and meets and discusses with all of the great men of the period including Epaminondas, Phocion, Xenophon, Plato, Aristotle, and Demosthenes. Barthélemy tells his readers that he chose to present this work in the form of a travel account because "everything is in action in a voyage and one is permitted details not permitted historians."[11] The main character is fictitious, but Barthélemy used all of the sources at his disposal to describe the geography, culture, polity, architecture, and thought of fourth-century b.c. Greece as correctly as possible. The book became what, in the context of its time, can only be described as a best-seller. It was accompanied by Barbié du Bocage's atlas and Barbié du Bocage thus benefited from the glamour and fame won by Barthélemy.[12]

Barbié du Bocage's atlas was composed of thirty-one plates, twenty-six of which were either geographic maps or plans, while the remaining plates were landscapes, perspectives on monuments, or reproductions of medals. Preceding the plates was a description by Barbié du Bocage of the decisions that had gone into the making of the maps, in particular the large geographic map from which all the others were drawn. They were based on maps made on location by M. Foucherot, a civil engineer, who had drawn them for M. Choiseul-Gouffier on the latter's trip to Greece from 1776 to 1782. Additional information was provided by maps held in the Ministry of Foreign Affairs, the Bibliothèque du roi, the private collection of the inheritors of d'Anville, and some detailed descriptions of Portolans found in Nicolas Fréret's papers. None of these documents would have been available to the public. After this very general description of his sources, the memoir entered into the minutiae of how he calculated ancient measures, his choice of projection, how he adjusted this to fit the newest theories about the shape of the earth (although it can have made no perceptible difference), the ways in which he fixed information to points of latitude and longitude, how he reconciled differences between sources, etc. In short, it was an essay a little like the Pompidou centre: with the plumbing on the outside. The maps themselves were well engraved and handsomely presented. Each was keyed to the relevant section of the text and Barbié du Bocage assured his readers that they had been constructed based on discussion with Barthélemy.

His other cartographic memoirs were much of a piece with this work. His essay on a map he composed for Mr. de Sainte-Croix's *Examen des historiens d'Alexandre,* which shows the route taken by Alexander the Great during his expedition to the Indus and back to Egypt, perhaps best reflects the strengths and weaknesses of Barbié du Bocage's work.[13] Sainte-Croix, although a reasonably popular academic, had nowhere near the impact of Barthélemy. Indeed, Sainte-Croix was later criticized both for sloppy research and his uninformed attribution of a number of antireligious tracts to the academician Nicolas Fréret.[14] More serious than a more limited respect from collaboration with Sainte-Croix were the limitations of Barbié du Bocage's memoirs themselves. Sainte-Croix's book, written on the assigned topic for the Académie des inscriptions' prize for 1770, was devoted to Alexander the Great's life and military activities as recounted by contemporary and subsequent ancient and medieval writers.[15] It sought to discern all that could be known, especially about Alexander's expedition to the east and about his moral character, through the critical comparison of these sources. The work's structure and logic was dictated by chronology and the agreement or disagreement between the accounts of the historians. It is clear that Sainte-Croix's history included geography and he devoted significant subsections and chapters of his book to geographic questions. He gave much attention to place names and sometimes discussed the terrain where it was relevant to battles or troop movements. He considered what the ancient historians had had to say about the speed of Alexander the Great's "marches." This had been studied by a number of famous geographers. Sainte-Croix summed their arguments up and presented his own views on the matter.[16] He dealt with other traditionally geographic themes, including the divisions of Alexander's empire, the geographic information, erroneous or valuable, brought back by Alexander from his expedition and disseminated by his historians, and the peoples in and around the regions invaded by Alexander. Finally, Sainte-Croix made clear that there were descriptions of landscape in his historians' writings—but he did not exploit them to create a sense of the space through which Alexander traveled. This did leave a space for Barbié du Bocage to fill. Rather than filling it, Barbié du Bocage devoted his monograph to a description of the sources he had used, a cartobibliography of all previous maps, and a vague discussion of the limitations of modern geographical knowledge. Nowhere did he attempt to engage the reader in Alexander's venture, his aims, the enormity and difficulty of his travels, or even to explore the reality behind the map. All of that was assumed or deemed unproblematic, which is what makes his essay so dry for us. In a sense, his memoir was nothing more than an oversized footnote to an historian's publication, which, in its avoidance of the historian's territory, perhaps showed too much respect to the historian. Who would take the time to read such a geographic essay? Someone conducting research on the area, who needed to be sure of a location, might look at it. Someone else concerned to make the best possible map of the area might read it with care, although given the rate at which the world was being explored and mapped, both the memoir and the map would become quickly dated. No one would read it for interest as it does not flow: there is no argument, no culmination, and, consequently, no conclusion. There was nothing objectively wrong with this sort of memoir. It was just a very limited intellectual product with a limited audience.

The only other type of work written by Barbié du Bocage was an essay on ancient geography for Pinkerton, Walckenaer, and Eyriés's universal geography.[17] What we see again in this publication is a determination to stay on the surface of problems. In the face of important new discoveries about the ancient civilizations of the Middle East, Barbié du Bocage decided to restrict his study to the classical civilizations of Greece and Rome. His essay is a description of the world as it would have been known to them, which, however, frequently fails to differentiate between Greek and Roman knowledge or knowledge of different periods. The essay has a regional structure. Apart from a single introductory paragraph at the beginning of each region which very briefly describes the people, their state of civilization, and a few additional tidbits about the people of that region, it is a list of places. The logic of the map so thoroughly dominates this essay that it is almost unreadable as a text.

Pascal-François-Joseph Gosselin's Relentless Erudition

The work of Gosselin is, if anything, even less accessible to us than that of Barbié du Bocage. Gosselin's biographers leave us with an impression of a fastidious man, who in order to concentrate on his work substantially cut himself off from the larger world, in which he had, in any case, relatively little interest.[18] A respected member of the Académie des inscriptions, he was cited with appreciation and occasionally gentle criticism by geographers, scholars, and classicists, including Alexander von Humboldt, Jean-Antoine Letronne, Conrad Malte-Brun, Bory de Saint-Vincent, Viscount M. F. de Santarem, Ernest Carette, Jarry de Mancy and Edme-François Jomard.[19] Gosselin devoted his life to what Martin Staum has aptly called "relentless erudition" and the resolution of a single question.[20] A question that we today would describe (and that some of his contemporaries did describe) as badly framed. Gosselin wanted to know the basis of the measurement system used by the ancient Greeks. Inspired at once by geography's traditional preoccupation with maps and the determination of location by d'Anvillian critical comparison techniques and by the Revolutionary adoption of a metric system, he spent years in calculation, devising methods to reconcile irreconcilable ancient sources.[21] He was convinced that the Greeks had had a single standard measure which had been used across the entire empire. This, he argued, had been lost by subsequent civilizations. No one had yet noticed the perfection of the Greek measurement system as a result of the corrupted translations and copies of the ancient texts with which scholars had to work. Gosselin was sure that he could devise a mathematical system which would reveal the truth and wipe away all corruptions. In his 1813 work, Gosselin went so far as to suggest that the Greeks had had a better and more complete knowledge of the entire Mediterranean region and, indeed, of astronomy and land measurement than was possessed by modern Europeans.[22] This reflected both an exaggerated respect for the ancient Greeks and profound ignorance of the last 150 years of European astronomy, physics, and cartography. By 1817, a careful reading of Strabo (forced upon him in 1801 by Napoleon's request that he edit a new French translation of the work) had made him nuance his argument. He explained that in his earlier work, he had been concerned not to go beyond the bounds of geography, and had therefore restricted his study to the stade and its relationship to a measurement of the circumference of the earth.[23] In fact, there had been, he thought, a number of different sorts of measurement systems in the Greek empire. The stade, however, was the foundation of the system used by Greek astronomers and cartographers. Further, in spite of the proliferation of stades of different lengths evident in the Greek texts, they were all derived from a single stade and a single measurement of an arc of the meridian.

There seems to have been a conspiracy of approbation—or, at least of peaceful coexistence—while Gosselin was alive. Such that most criticism of his elaborate geometric arguments remained muted or oblique. In an article written in 1817 but published in 1822 Letronne seemed to accept Gosselin's views or, more correctly, he put them aside as he disputed the originality and legitimacy of Eratosthenes' supposed determination of an arc of the meridian. Nevertheless, the trend of Letronne's argument, that the estimates of the size of the earth by both Eratosthenes and Posidonius were based on generally accepted measures, rather than on measured distances, ran away from Gosselin's sense of the precision of the ancients.[24] In 1823 Letronne published remarks very critical of Gosselin's "metrical" arguments. Nevertheless, he managed to formulate his criticism in such a way as to be perfectly clear about what he was criticizing, while leaving considerable ambiguity as to whether Gosselin was among those being criticized or not.[25] It must have been particularly difficult for Letronne, as the second assigned translator for Gosselin's edition of Strabo, to be openly critical. Even after Gosselin's death, and even while criticizing his work and conclusions, Letronne never failed to attach respect to his mention of Gosselin's name. By 1833, three years after Gosselin's death, Adrien Balbi (who was not a member of the Académie des inscriptions) announced that Gosselin had "disfigured" the writings of the Greeks with his "systems."[26] Still, in 1839 Walckenaer felt it necessary to praise Gosselin for his rigor even while describing the furthest extension of his arguments as erroneous, unsustainable, and unnecessary.[27] As late as 1885, scholars were still referring to Gosselin's "broad and deep science."[28] Although Gosselin's work was respected by some of his contemporaries, if not by others, Malte-Brun was reasonably sure that it was this sort of erudition that gave geography a bad name.[29] It is altogether possible.

Baron Charles Athanase Walckenaer's Wanderings

Baron Walckenaer is a far more engaging figure than either Barbié du Bocage or Gosselin. Born eighteen years prior to the Revolution, he was orphaned and taken under the wing of a wealthy uncle, Duclos-Dufresnoy. Under his uncle's protection, he was among the most privileged of the aristocracy. He received an excellent education, culminating with a degree at Oxford, where he met Banks and Solander. The Revolutionary menace in Paris obliged him to cancel a proposed voyage of exploration to Spain, Africa, and Asia. During the Revolutionary wars he served as an officer in military transport. While serving, and much like Bory de Saint-Vincent and de Férussac, he traveled the countryside, taking notes on the nature and productions of the soil, on the landscape, on the monuments and local customs, and collecting both plants and insects. One such venture landed him in trouble: mapping off the coast near Poitou, he was stopped by the French military and found to have English equipment. On the basis of his equipment and his social class, he was arrested and imprisoned as a spy. By a combination of flight and petition, he managed to survive the Terror. After the Terror, with his considerable inheritance indefinitely confiscated, he entered the École de ponts et chaussées and then transferred to the newly established École polytechnique, graduating in the same year as Jomard and Chabrol de Volvic. He regained his fortune soon after graduating and took up the free life of an independently wealthy scholar.

Unsurprisingly, Walckenaer was, certainly after the violence of the Revolution, a royalist, a Catholic, and a traditionalist. This conservative tendency, to which his financial freedom allowed him to give full reign, appears in his work and in his network of friends and associates. He collected maps and built up a huge library which he shared with his colleagues and associates.[30] It was also his financial independence which allowed him to turn down offers of employment, to avoid or delay publication of many of the works that he wrote in the course of his life, and to choose his fields of study as his taste, friendships, and ambition dictated. Consequently, he dabbled in a number of areas: from natural history, with a special focus on insects,[31] to the writing of popular fiction and popular science,[32] to the history of France,[33] to the subject to which he finally committed and with which he is now identified, geography. He published a number of works in geography deemed important in his day, from an updated and improved edition of Pinkerton's geography,[34] to a series of essays on the geography of ancient Gaul,[35] to a first published edition of Dicuil's *Liber de mensura orbis terrae*,[36] to an analysis of the anonymous "Itinerary from Bordeaux to Jerusalem,"[37] to an essay resuscitating and restoring the reputation of an eighteenth-century classical geographer,[38] to a speech on geography delivered to the Société de géographie de Paris.[39]

Several things are clear from this pattern of publication. Walckenaer, in his approach to geography, was both traditional and respectful of the established authorities. His wealth may have given him the freedom to study and behave as he liked, but it was to Walckenaer's taste to work within the protection of established tradition. Thus, he tipped his hat to Gosselin and considered his metrical calculations at the heart and soul of the geographic endeavor. Similarly, he liked and respected the type of cartographic reconstruction of past geographies practiced by Barbié du Bocage and contributed to it through his collection and study of ancient maps, ancient or medieval itineraries, and medieval geographies. His identification with geography was clear particularly in his lament over a perceived decline in the popularity of geography before the Société de géographie and in his article on the eighteenth-century historical geographer who published extensively on geographic themes, Nicolas Fréret. In this article, Walckenaer recounted to the Académie des inscriptions, and to the larger world, the way in which Voltaire's coterie had hijacked the name and reputation of this relatively obscure late-seventeenth- and early-eighteenth-century scholar. Walckenaer explained that Naigeon had used Fréret's name for the publication of fictitious tenth-century antireligious tracts, arguing that these were the unpublished works for which the pre-Revolutionary hero Fréret had been imprisoned in 1714. In fact, Fréret had been briefly arrested and genteelly detained for his Jansenism and an ill-timed essay on the origins of the French nation.[40] Walckenaer pointed out convincingly that there was no relationship at all between the actual writings of Fréret and those ascribed to him in the Revolutionary period.[41] The antireligious tracts and Revolutionary works were of a radically different nature. It was, in Walckenaer's view, an accident of bibliography and thanks to the scholarly sloppiness of the Baron de Sainte-Croix that anyone had taken this politically motivated crime against scholarship seriously. Walckenaer was able to find written admissions by Voltaire that the authorship of these works was other than Fréret. It is an interesting tale of fabrication, reputation, and refutation. For us, the real interest of the story lies in why Walckenaer felt it necessary to bring clarity to this obscure historical deception. Fréret's method and approach and indeed the questions Fréret deemed worthy of study were precisely those favored by Walckenaer. If we can describe Barbié du Bocage as a student and emulator of d'Anville, Walckenaer was following the footsteps of Fréret.[42] However, whereas Fréret's real publications in geography, chronology, and the history of France left a significant body of research, Walckenaer's writings, considered over his entire career, did not amount to a consistent and coherent research focus. While Walckenaer was himself in many senses part of the mainstream of research in historical geography, his dabblings in everything from natural history to sentimental fiction left a contribution that, as Walckenaer himself acknowledged, was scattered, relatively unsubstantive, and primarily generated by questions set by others.[43]

Alexander von Humboldt's Examen critique

The focus of Alexander von Humboldt's interests certainly lay in the natural world and it was in the realm of natural geography that he made his most innovative contributions to modern thought. However, he did conduct research into a branch of historical geography, the history of geographic thought, and a key moment in that history: the Columbian "discovery" of America. While there seem to be shafts of brilliance in everything that Humboldt wrote, in historical geography Humboldt was not the master and innovator he proved to be in natural geography.

There is a strikingly modern feature to Humboldt's study of Columbus: its strongly contextual nature. Humboldt was interested in much more than recounting the discovery and mapping of the Americas. He was interested in fully understanding that moment which had so changed the face of the modern world.[44] Columbus, in his view, did not stand alone but carried with him centuries of scholarship, understanding, myth, and misunderstanding, all of which had a profound effect on his decisions and thoughts. Humboldt fully explored that tradition, looking at the works read by Columbus and at the influences acting on contemporaries likely to have influenced Columbus. Humboldt was fascinated by the rootedness of thought. No matter how innovative, in his view a man always carried some of the prejudices of his time and of his civilization. These prejudices were often both his strength and his weakness. Humboldt sought to fully explore the complex interweave of science, myth, economy, religion, individual heroism, self-interest, and tradition within the history of discoveries.[45] In particular, he saw intellectual tradition as both a powerful guide and a limiting obstacle.[46] Thus, Humboldt sought in this work to reconstruct not just the conditions and nature of the Columbian ventures but the cosmography within which Columbus lived and acted. Humboldt was trying to write a truly contextual history: one capable of understanding a man such as Columbus.[47]

Humboldt's interest in Columbus was highly personal. In the first few pages of his five-volume *Examen critique de l'histoire de la géographie*, he framed this interest in more abstract intellectual terms: Humboldt hoped that his own exploration in the Americas would give him insight into Columbus's experiences and thinking. There was little distance in Humboldt's admiration for Columbus. He so strongly identified with Columbus that some of his descriptions of Columbus—extolling the innovative nature of the latter's thought and noting his indebtedness to both contemporaries and predecessors—are remarkable self-portraits.

> Columbus did not limit himself to the collection of isolated facts: he combined them, looking for the relationship between them. Sometimes he boldly proceeded to the level of the discovery of the general laws governing the physical world. This tendency to generalize the results of observation is all the more worthy of attention as before the fifteenth century, I would say even before Father Acosta, there is no other such attempt. In his reasoning about physical geography, of which I will provide a remarkable fragment, the great navigator, contrary to his wont, did not allow himself to be guided by scholastic philosophy. He found connections by virtue of his own theories and his own observations.[48]

Perhaps too they are unconscious declarations of how Humboldt wanted to be understood and remembered. In this next passage he finds two of his own principal concerns in Columbus: the development of new forms of expression to carry new modes of thought and the power of beautiful and elegant description.

Each new land discovered by Columbus seemed to him more beautiful than the one he had just described. He bewailed the fact that he could not *vary the forms of the language* so as to make the delicious impressions he had had as he hugged the coasts of Cuba and the little islands of the Bahamas reach the Queen's soul. *In these tableaux de la nature (and why not give this name to these descriptive pieces so full of charm and truth?)*, the old mariner sometimes demonstrated a stylistic talent that those initiated into the secrets of the Spanish language and who prefer the vigor of color to severe and composed correction, would know how to appreciate.[49]

In a sense, Humboldt's study of Columbus is most revealing of Humboldt himself. Although he was almost certainly unaware of this self-portraiture, Humboldt did regard this five-volume study as, if not self-indulgent, then a source of almost illicit personal pleasure.[50]

The combination of interest in graphic language, geographic discovery, and the transmission of ideas inevitably led Humboldt to the history of cartography. It is another of Humboldt's innovations to have engaged in some fairly sophisticated analysis of key maps in the discovery and dissemination of information on the Americas.[51] There was a growing interest in this line of enquiry among scholars such as D. Martin Fernandez de Navarette, Marie Armand Pascal d'Avezac-Macaya, Jomard, Walckenaer, and Santarem. However, Humboldt brought his skepticism to the study of the map and sought in these documents not only truth but misconception and prejudice.

Contextual breadth and personal engagement notwithstanding, Humboldt's *Examen critique* is one of the least readable of his works. In these five volumes Humboldt gave greatest rein to his tendency to digression. The volumes are peppered with mini essays on misconceptions in the history of mapping, on the depiction of particular peninsulas, on the etymology of particular place names, . . . Some of these asides are philosophical/historical gems on the nature of thought, on the peculiar coexistence of slavery and religion,[52] on the horror and contradictions of modern existence,[53] or on resistance to new ideas in science and society.[54] However, they are hidden in tomes of historical detail. On occasion he would begin a thought on one page and only complete it tens or even hundreds of pages later.[55] He would often provide an extraordinary amount of detail on a theme or subject, and only inform the reader of the relevance of all of this information after the fact.[56] The whole text wandered unpredictably. The reader, having learned of the context, life, science, old age, and decline of Columbus, would suddenly be fully immersed in a discussion of the circumstances of Columbus's birth.[57] There was always a logic to the flow. In this case, the fate of Columbus's legacy was affected by the conditions of his birth, but the reader would have to follow an apparently irrelevant description for some distance to seize that logic. In this work, Humboldt was so demanding of his readers, and so trying of their patience, that were they not centrally interested in the questions he was posing, they would be unlikely to progress beyond the first volume. It is no wonder that, of all of Humboldt's works, this one—the only one entirely devoted to a branch of what we now see as historical geography—is the least cited and least read of his works. It is likely that it was also so in his own day.

GLIMMERINGS OF INNOVATION: THE FORGOTTEN WORK AND IMAGINATION OF JEAN-ANTOINE LETRONNE

Historical geographers today are, for the most part, far more interested in national pasts and indigenous cultures than in the historical geography of antiquity. This is so much the case that to work in the geography of ancient Greece or Rome today would probably be to consign one's career to oblivion. Indeed, to a damned oblivion: to suggest any intellectual parentage from those civilizations to ours has become, in some circles, symptomatic of a profound racism. Further, "classics" is sometimes used in the corridors of North American universities as a shorthand suggesting irrelevancy, out-of-touchness, and lack of political purchase. Consequently, it is easy to dismiss the historical geography of the early nineteenth century on two counts: for the poor quality of much of the research and writing, and for its predominant preoccupation with antiquity. Both lines of reasoning, but especially the latter, are profoundly unhistorical. Whatever the prevailing attitude to classics today, there was an aura and passion surrounding its study in the early nineteenth century which gave the study of ancient geography a resonance and importance that would be the envy of most modern historical geographers. In ancient, classical, and medieval studies, the early nineteenth century was a period of exciting discoveries. Egypt was revealed through the Napoleonic expedition and especially the publication of the spectacular twenty-two-volume *Description de l'Égypte*. Within two decades of the expedition, efforts to decipher the hieroglyphs began to bear fruit and the monuments of ancient and classical Egypt, silent for centuries, began to speak to Europeans. Who had these people been? What was their relationship to the better-known civilizations of Greece and Rome? Was there an identifiable Egyptian heritage in European culture and thought?

What did the existence of ancient religions and ancient astronomy suggest about the uniqueness and authority of Christianity and the accuracy of biblical chronology? To say nothing of the craze for Egyptian-style representations, now known as Egyptomania.[58] But Europeans were rediscovering other civilizations in this period, including those of ancient Greece and Rome. Pompeii, first discovered in the sixteenth century and identified as Pompei in 1763, was being studied increasingly intensively. Ancient Greece was popularized in the work of Jean-Jacques Barthélemy and in writings of Johann Wolfgang von Goethe and Friedrich Schiller among many others.[59] It was also explored by Choiseul-Gouffier from 1776 to 1782, and immortalized in his popular and sumptuously illustrated *Voyage pittoresque de la Grèce*.[60] It was reexplored in the 1830s by an official French scientific expedition, modeled on the expedition to Egypt.[61] It even became the passion and final resting place of Byron, who linked the strivings of modern Greeks for independence from Turkish rule with the democratic ideals of ancient Greece. The classical world was important to the early nineteenth century in a more pragmatic way. Both the old European legal codes and the new European national legal codes were strongly influenced by Roman law. In a sense the new legal codes moved European law away from its Roman sources, yet the codes were in part constructed from just these. In parts of Germany, Roman law was the basis of the legal system into the late nineteenth century.[62] In the late eighteenth and early nineteenth centuries the history and literature of Greece were used as a source of "moral precepts," "exalted characters," and "exemplary attitudes."[63] They were a source of non-Christian saints and of symbols and myths for both Revolutionaries and Napoleon. In France, Greece and Rome also played an important role in political discourse in the pre-Revolutionary, Revolutionary, and Napoleonic periods and under the Restoration. This discourse had less to do with any scholarly study of Rome or Greece, than with contemplation on the nature of the state or on the nature of civil society under the guise of pseudo-academic reference to Athens or Sparta, or indeed China, Egypt, and the "primitive peoples" of the Americas . . . As Nicole Loraux and Pierre Vidal-Naquet have so clearly shown us, this discourse heavily laces the writings of Cornelius de Pauw, Pierre-Charles Levesque, the Marquis Marie Jean Antoine Nicolas de Caritat de Condorcet, Constantin Chasseboeuf de Volney, François Auguste Vicomte de Chateaubriand, Joseph de Maistre, and Benjamin Constant, among others.[64] Much of the prevalent interest in ancient civilizations was romantic and to our eyes both politically and historically naive. It nevertheless created an atmosphere of avid interest which meant that academic works which touched upon these civilizations and themes were read, and far more widely read than one might imagine. Thus, the nineteenth century, which witnessed the beginning of truly academic study of the ancient past, the creation of a professional cadre of classical scholars, philologists, and medievalists,[65] also saw an intensification of a more popular use of the ancient and medieval past to lend legitimacy to a political stance or argument. This, then, was the broader intellectual and social context of Letronne's works on Ptolemaic Egypt, Greece, Rome, and medieval Europe.

As already suggested above, the work of the now almost entirely forgotten Letronne stood radically apart from that of his colleagues. Although originally an historical geographer very much in the mold of d'Anville and Barbié du Bocage, Letronne was a scholar who grew with his times and his research and thus quickly outgrew the traditionalist conceptions of geography. Thoroughly engaged in the intellectual debates of his own time which revolved around a renewed fascination with the ancient origins of Western science, religion, and philosophical thought, each one of his articles was an adventure. Driven by intellectual curiosity and pleasure in an argument well made, Letronne abandoned the map, the narrow sense of geography as topological information, and a descriptive writing mode. As many of his contemporaries were beginning to ask questions about the nature of society and government, Letronne turned that curiosity on the ancient world. As some of his contemporaries sought to understand and map the infinitesimally small variations in the movements of the heavens, Letronne wondered precisely what the ancients had understood of all of that. Not satisfied to merely capture and reproduce monuments and inscriptions, to which activity the members of the expedition to Egypt had for the most part restricted themselves, Letronne sought to contextualize these remnants of an ancient civilization and exhaust them of any historical and geographic intelligence they might contain. In his attempts to understand societies and writings long past, he developed critical historical and linguistic analysis techniques that were in every sense modern and rigorous. Yet, while Letronne moved well beyond the bounds of geography's traditional interests, he never lost his interest in geography and saw, and described, most of his works as contributing to the advancement of the field. He frequently expressed his interest in a particular object or theme as related to larger questions in "history and geography." And history and geography were inseparable in Letronne's writing and thinking. Still, Letronne was from at least the 1820s on the margins of mainstream historical geography, cited superficially and rarely by most geographers. By the end of his career, he was scarcely regarded as a geographer at all, as is clear from the total silence regarding him from historians of geography. As we have seen in the case of proto–social scientific geography and natural geography, it was precisely on the margins, at some distance from the constraints imposed by the geographic tradition, that innovation was most likely to take place.

An Engaged Scholar

The dispassion and distaste for the quotidian that pervades Gosselin's work is entirely absent from that of Letronne. In Letronne we find a man who sought to play a role. In addition to being an active scholar with well over fifty publications to his name, Letronne was an academic administrator of first rank. He seems to have had attachments or even administrative posts in most of the major institutions of higher learning and research (with some focus on antiquity) of his period. He entered the Académie des inscriptions (as a government appointee) in 1816.[66] In 1819 he was named inspector at the military schools, perhaps by virtue of the high regard in which his "elementary" geography text was held.[67] From 1818 until his death he was an "author" for the *Journal des savants*, the established arbiter of scholarly standards in France. Named conservator of medals and antiquities at the Bibliothèque nationale, he soon became the Library's Director-General. While at the Bibliothèque nationale, he had its press develop and begin use of hieroglyphic fonts. He was among the first to use those fonts.[68] In 1831 he was named to the chair of "histoire et morale" at the Collège de France and held that chair until (in 1840) he was named to the chair of archaeology, recently made illustrious by Champollion and left empty by the latter's early death.[69] His courses there were deemed significant and successful enough to warrant the publication of both course summaries and class notes.[70] From 1838, he was a senior administrator at the Collège de France. By 1840 he was also part of an overseeing committee at the National Archives and director of the École de Chartes.[71] He gave new direction to this illustrious school for archivists and librarians and turned the institution into the nursery of a professional cadre—which cadre has helped to preserve and interpret national heritage since. Even at the Académie des inscriptions, his administrative mind found an occupation as he twice proposed the radical reform of that institution.[72]

Integral to his scholarly and more administrative accomplishments was a broad interdisciplinarity of interests. Somewhat like Humboldt, Letronne was prepared to train himself in the areas dictated by the particular research quest upon which he was engaged. Thus, in the course of his career he challenged Cuvier's assessment of the rate of growth of the Nile delta.[73]

He informed himself sufficiently about geology to understand geological formations in Egypt and the possible manner in which they were exploited for building materials and the potential obstacles to so doing.[74] He studied astronomy sufficiently to understand zodiac systems, their antiquity, their relationship to agriculture, and their numerical structures.[75] In trying to understand ancient monies, he did not stop short at the work of antiquities scholars like himself but read and integrated many of the ideas of prominent contemporary economists such as Adam Smith and Jean-Baptiste Say.[76] Finally, he watched the nascent field of archaeology with interest and worked hand in hand with archaeologists, travelers, and field artists whenever the opportunity offered itself.[77]

In general, Letronne appears to have been interested in the work and interests of his contemporaries and concerned to render his own work interesting to them. Thus, and in contrast to the work of most historical geographers, wherever possible Letronne linked his work on antiquities to current preoccupations. Proposals to cut the Suez Canal, which animated France and Britain from the time of the expedition to Egypt to the cutting of the canal by de Lesseps in mid-century, gave him the opportunity to write about the ancient canal, its course, its history of openings and closings, and the strategic anxieties it had occasioned in the past. In writing about the ancient calendrical system of Egypt, he speculated on its possible links to the Julian calendar,[78] thus, in a sense, bringing ancient Egypt into the everyday lives of calendar- and clock-bound Europeans. Capable of taking the lead and showing the way, Letronne was also prepared to participate in other peoples' projects. He wrote willingly for de Férussac's *Bulletin général et universel des annonces et des nouvelles scientifiques*. More significantly, much of his work was written in reaction to the work of others. Walckenaer's discovery and reproduction of Dicuil's *De mensura orbis terrae* (1807) stimulated his own work on the subject.[79] The *Description de l'Égypte* (1809–1822) and Hamilton's *Aegyptiaca* (1809) stimulated his interest in Egypt.[80] Finally, Champollion's efforts at decipherment stimulated an already growing interest in inscriptions from the 1820s until his death. He acknowledged these influences and clearly saw his work less as a personal achievement than as a contribution to a larger intellectual venture shared by much of his generation.

Respected by specialists, Letronne wrote to interest the layperson and, to that end, fully contextualized his studies, often to the point of recounting the adventure of the discovery of an inscription he was reproducing and interpreting. As we shall see, Letronne was deeply committed to the most rigorous and painstaking research, for example, into epigraphy. Yet despite his focus on the minutiae of words, spellings, and individual letters, for Letronne,

> . . . the science of words only attains its full value when it is used in the improvement of the science of things.[81]

And "things" derived their value from their association with "higher" civilization. Letronne was every inch an engaged scholar: broadly read, deeply committed to thorough and honest research, and open to a wide array of intellectual influences.

A Geographer by Training

Letronne's first intended career was that of an artist. When his father's death in Napoleon's forces put an end to that possibility, Jean-Antoine Letronne had already worked in Jacques Louis David's atelier for some six years. Jean-Antoine's second career was to be geography. After his father's death in 1801, he redirected his studies toward the more practical and remunerative goal of gaining entry into the École polytechnique. On the way, partly as a consequence of a course in geography that he took from Mentelle at his École central, Letronne turned to geography and especially the geography of the ancient world. Mentelle's influence on him must have been considerable: after taking Mentelle's course, Letronne joined his teacher as a collaborator on his "Dictionnaire de géographie" in 1806.[82] In 1812 he dedicated his first major study, on the geography of ancient Syracuse, to Mentelle. It was Mentelle who advised Letronne to not make the same mistake he had. He urged Letronne to devote his life to research that would earn him the respect of the scientific community rather than to education and the composition of popular geographies.[83] Letronne took Mentelle's advice. He studied Greek with Gail at the Collège de France and began to work on original Greek and Latin texts.

Letronne's early works were considered mainstream traditional geography by his contemporaries. In his teens and early twenties, he contributed to a number of general geographies and geographic dictionaries and wrote the antiquities section of Peuchet and Chanlaire's *Statistique des départements* (1807).[84] Unsurprisingly, given the map-focus of much of the geography of his era, the purpose of Letronne's first memoir, on the colony of Syracuse, was the production of a map of Syracuse in 414 b.c. This map, he argued, would elucidate Thucydides' account of the Athenian conquest of that Sicilian town. He had even caught enough of the general mood in geography to apologize for the inclusion of a textual discussion of the map.[85] It was the last time that Letronne would apologize for going beyond the map. In fact, even in this early study, Letronne's curiosity and rigor took him beyond Barbié du Bocage's. Letronne not only located the town and its fortification walls, he traced the evolution of the city, its ports, walls, and suburbs from the time of the siege (414 b.c.) to the time of Cicero (74 b.c.). Still, in this, his first major effort, he sought to avoid the charge of temerity and disrespect for scholars and scholarship in the well-established field of ancient history by claiming to bring nothing more than a geographer's sensitivity to the subject. In 1814 Letronne published a critical edition of Dicuil's recently rediscovered *De mensura orbis terrae*.[86] Also in 1814, he published the first edition of an elementary geography which was adopted as a basic geographic text at the military colleges. This book was enormously popular, going through six editions by 1820 and twenty-seven editions by 1857.[87] It was a traditional universal geography which had the great merit of concision and clarity. In the same period, Letronne published reviews of translations of ancient geographic texts[88] and even attempted a Gosselin-type analysis of ancient measures. This memoir won Letronne the Académie des inscriptions' annual prize for 1816. Finally, from 1815 to 1819, Letronne worked for the government as translator of the official French edition of Strabo's *Geography*.[89] In most respects, then, Letronne's early career was very typically that of an armchair geographer focused on the ancient world, and he saw himself as working in the tradition of Mentelle, Barbié du Bocage, Walckenaer, Malte-Brun, etc. By 1825, he regarded himself as something of an expert in ancient geography.

It is clear that as early as his first major scholarly work, his translation and commentary of Dicuil's cosmography in 1814, Letronne began to develop a more critical attitude to his geographical colleagues. Walckenaer's edition of Dicuil was clearly superficial, unrigorous, and relatively uninteresting compared to Letronne's.[90] Barbié du Bocage came into a great deal of criticism, well beyond geography, for his modern map of Greece. Nevertheless, as Letronne perhaps somewhat acidly commented in his eulogy to Barbié du Bocage,[91] the man seemed to have positively laid claim to the geography of ancient Greece. The work carried out for the *Description de l'Égypte*, particularly that by Jomard on the hieroglyphs and ancient measures, and by Fourier on the zodiacs, was increasingly criticized, not least by Letronne himself. Letronne rarely took his geographic colleagues to task and when he did he generally avoided naming names.[92] When Letronne mentioned geographers by name, it was with respect and with many compliments. Yet the simple act of redoing their work at an incomparably higher quality was the most damning of all possible criticism. His critical style is exemplified by his discussion of the *Description de l'Égypte* in his dedication and preface to his two-volume magnum opus on the Greek and Roman inscriptions of Egypt. One could imagine from his description of it that the *Description de l'Égypte* had been his guide and his inspiration.[93] Yet, Letronne's work on the Greek inscriptions made it clear that it was the sloppy copying of inscriptions by the members of that expedition and their inability to see what was utterly evident that had misled and retarded work on Egypt for at least two decades. In spite of their painstaking sketching, these geographers and engineers had missed the fact that hieroglyph writing had survived well into the Greek and Roman periods (as Greek writing very clearly underlay hieroglyphic writing on numerous inscriptions); that the temples which contained zodiacs were dedicated to and by Greeks (sometimes in visible lettering on the epistyle of the temple) and therefore could not possibly date back 10,000 years; that many of the temples most admired by the expedition members were Greek temples (which would have been clear had they bothered to read the Greek written on the temple walls); and that there was a clear stylistic periodization to all the temples which did not require a Champollion—nor mapping and mathematical manipulation—to decipher. Letronne admired less the work carried out by the French militarized scholars in Egypt than the solitary researches of Hamilton,[94] the earlier work by Richard Pococke,[95] and the later sketchings of travelers like Frédéric Cailliaud[96] and artists like Jean Nicolas Huyot and François Chrétien Gau.[97] Letronne did not engage in the same sort of war against Jomard as did Champollion, but there can have been no love lost between the men, and Letronne cannot have wanted to attach his name too firmly to the sort of geography practiced by Jomard.

New Directions

It was not the beautiful and monumental *Description de l'Égypte* that in the first instance drew Letronne's attention to Egypt and gave an altogether new direction to his work. It was, ironically, his attempts to fully understand the text of an Irish monk of the ninth century who probably never left Ireland. Dicuil's *De mensura orbis terrae* offered a classical scholar a number of puzzles. It contained information about Egypt, Northern Scotland, and Iceland from the middle of the eighth century. Further, it demonstrated knowledge of a wide array of Greek and Roman authors, yet was written in a Latin marred by numerous peculiar errors that made its author seem semiliterate. Letronne juggled the facts and analyzed the writing until he arrived at a plausible explanation for each apparent anomaly—which explanation is still accepted today. What particularly fascinated Letronne was Dicuil's description of the travels of the monk "Fidelis" to Egypt. Letronne used Fidelis's account of his travels to resolve two puzzles left mysterious by the expeditions to Egypt: the chronology of the opening and the closing of the freshwater canal from the Nile to Suez and the discrepancy between descriptions of the pyramids of Giza from the time of Herodotus and Diodorus through the Arab middle ages to modern times. He recognized that measurements of the height and even the base of the pyramids carried out prior to those taken on the Napoleonic expedition would be hopelessly inaccurate. But the top platform of Cheops, the largest of the pyramids, was easily measurable and had been measured repeatedly over the last 2,000 years. Using these measurements and the descriptions provided by ancient, medieval, Arab, and modern travelers, he was able to prove and date the loss of Cheops' marble facing.

Once his attention was focussed on Egypt, Letronne found himself with a plethora of problems to resolve. Egypt was covered with Greek inscriptions which certainly dated from the Greek dominance of that country. These inscriptions, being in Greek and therefore in a sense known, had attracted very little attention. While some were copied, few had been read or interpreted with any care. Apart from these Greek inscriptions, which had tended to be treated as the graffiti of Johnny-come-lately conquerors, it was assumed that the Ptolemies had left little mark on the land. How could this be, he wondered, particularly given the Greek propensity to build and impose their social and political institutions wherever they traveled. One problem quickly came to preoccupy Letronne: the puzzle of the zodiacs. Fourier and his colleagues had decided that, based on a comparison of how the sky should have looked approximately 5,000 years ago and the depiction offered by the zodiacs, the zodiacs, and therefore the temples in which they were housed, including the temples of Esne (Latopolis) and Tentyra, dated from approximately 5000 b.p.[98] Letronne looked at the evidence provided, not by Jomard or Fourier but by Jollois and Devilliers and subsequent travelers, and concluded that the zodiacs were a Greek product executed in the Roman period. Indeed, that zodiacs had had nothing to do with ancient Egypt but were an import into the Mediterranean perhaps from India. They seemed to have become popular during the Roman Empire, particularly at the beginning of the Christian era. The evidence he used was the existence of Greek zodiacs found in Greece and the Greek writing on both the zodiacs and on the exterior walls of the temples in Egypt housing the zodiacs. His resolution to this problem confirmed a suspicion which became the principal preoccupation of the rest of his scholarly career. This was that the Greeks had been so amazed by, and respectful of, Egyptian civilization that they—the conquerors—had adopted many aspects of Egyptian culture; from religion, to language, to architectural style.[99] Thus, much of what Europeans had most admired of "ancient Egypt" was the product, not of the pharaohs, but of an unusual and happy symbiosis between two great Mediterranean cultures. Structures constructed during the Greek period included those at Kalapsché, Dakkeh, Aadon in Nubia; many in Egypt, especially those at Philae and Elephantine; Typhonium of Karnak; and some parts of the buildings of Medyney-Abou and the buildings of Qournah. Contributions of the Roman period included exterior sculptures at the temple of Edfou, at the Typhonium of Karnak, and all those decorating the temples of Esné and Tentyra. In that vein, the Rosetta stone, which had attracted so much attention for the possibility it offered of finally deciphering the hieroglyphs, was interesting in and of itself. This was not to take anything away from the achievements of Champollion, for whom Letronne had both respect and friendship, but there was much to learn from the Greek on the stone. It revealed that rather than imposing their rule and language on Egypt, the Greeks, from the time of the Alexandrian invasion, had taken the trouble to translate each decree into the Egyptian hieroglyphs and to post them in every temple in the land. This cast the Greek conquest in a new light and also suggested that Egyptian culture (religion, language, art, bureaucracy, etc.) had probably survived not only the Persian invasion but well into the Roman period.[100] Letronne's resolution of the zodiac puzzle marked him in three significant ways: it left him with an abiding dislike of abstract and apparently rigorous (because mathematical) solutions to historical problems (further discussed below); it convinced him that epigraphy and paleography offered the best means of revealing both the geography and the history of Egypt; and it drew him from traditionally geographic approaches of mapping and describing the ancient world into the nature of Egyptian society in the ancient, Greek, and Roman periods. Thus did Letronne find his own path to the post-Enlightenment interest in the nature of society, so remarkably absent from the writings of most traditional geographers of this period.

Reading Egyptian Society

It was Letronne's discovery that the simplest inscriptions could, if properly translated and exhausted of all of their historical content, reveal a great deal about the religion, the administration, and the social structure prevailing during the periods of Greek and Roman domination. Letronne worked solidly on the Greek and Roman inscriptions of Egypt from approximately

1817 until his death in 1848. In 1842 he published the first two volumes of his *Recueil des inscriptions,* which was originally projected to be a four-volume work. This was a collection of all known Greek and Roman inscriptions in Egypt from the conquest of Alexander to the conquest of the Arabs. Although Letronne never traveled to Egypt and Nubia, many of these inscriptions had been discovered at his instigation. He reported that while both Hamilton and the engineers and geographers working on the *Description de l'Égypte* had collected 80 inscriptions between them, Letronne was presenting and fully interpreting 700 inscriptions.[101] Letronne's contribution was less the reproduction, correction, publication, and translation of these inscriptions—though that was a huge task in itself. His major contribution was his interpretation of these writings, which he classed according to their function: religious, administrative, or Christian. He died before the two volumes containing administrative and Christian inscriptions could be published. The religious inscriptions he did publish included those concerning the construction and dedication of Egyptian temples, sacerdotal acts, dedications or religious offerings, and acts of adoration or "souvenirs de visite" (graffiti by another name). Thus, there were among these inscriptions many which were suggestive of the administration of the Greek and Roman empires.

Letronne was particularly interested in the survival of Egyptian cults well into the Christian period and in the degree of religious tolerance and skepticism reigning in Greek and Roman Egypt. Fascinated by a temple to Isis in Upper Egypt covered with the graffiti of Greek religious pilgrims, he declared that Isis had won many Greek adherents who had practiced their religion freely and openly. From a small constellation of pilgrims' graffiti on the Hydreuma of Panium, Letronne learned that a group of Greek merchants had been travelling with two Jews. The Greeks had stopped at the temple to pay their respects to Pan and had allowed the Jews to inscribe their names too. They had even allowed the Jews to express their respect for the deity, without mentioning the name Pan—thus allowing them to worship their own God at a temple to Pan. This suggested a degree of tolerance rare in Letronne's own time.[102] Many of these prayers for protection or records of pilgrimage, thanks to both the purpose of these inscriptions and to ever-present human vanity, clearly identified their authors and all of the members of the family for whom protection was being sought. Thus, they provided a fund of information on people often otherwise completely unknown and sometimes—most excitingly—described elsewhere in a tomb, on a papyrus, or in an edict. The fact that many of the inscriptions on the temples to Isis dated from approximately sixty years after the edict of Theodosius suggested to Letronne that even after the beginning of the less tolerant Christian period, a greater degree of religious tolerance reigned in the southern reaches of Egypt. A certain diversity of religion and religious belief seemed, then, to have reigned in Roman Egypt prior to the Christian period.[103]

Letronne was equally interested in inscriptions which revealed details of Roman administration of Egypt, including tax assessment, tax collection, litigation, and also how that administration was experienced by the Egyptians. One inscription found in the Great Oasis clearly suggested that in the first century a.d. the Roman administration had found the amount of litigation over land assessment following the annual flood insupportable and had sought to reduce it. However, the measures planned did not include a cadastral survey of the *entire* realm out of respect for the ancient customs of the land, as was clear from this portion of the inscription:

> As for those who became alarmed when they heard talk of a measurement of the lands in Alexandrian territory, insofar as the ancient evaluation was maintained and never the chain of a land surveyor was carried on these lands, they should not address such supplications to us. These would be utterly useless, as no one would dare, nor would they have permission, to renew a territorial measurement; as you must always benefit from the advantages of that which has been done since ancient times.[104]

This inscription provided valuable information concerning the Roman intention to effect a cadastral survey in Egypt. It also suggested their preparedness to respect local arrangements concerning taxes where a custom was well entrenched. Beyond that, Letronne mined this inscription for information concerning the efficiency of the Roman administration. By comparing the date of the inscription and the date of the ascension of the Roman emperor, he was able to estimate the speed at which information passed from Rome to Alexandria in the first century a.d.: twenty-seven days!

Letronne was particularly interested in the religious and linguistic politics of the Roman Empire in Egypt. The inscription in the Nubian temple of Thalmis by the Nubian King Silco provided precious details on these aspects of Roman administration.[105] This inscription, written in Greek, announced the conquests of King Silco over his vanquished enemies. From this relatively uninteresting notice of the insignificant victories of an obscure king, Letronne was able to learn a great deal. From the style of the Greek he was able to date the inscription to the reign of Justinian. From the poor and stilted nature of the Greek he determined that Silco was not a native Greek speaker. From the sentence structure he established that Silco had learnt his Greek from the Scriptures and that he was certainly, therefore, a Christian and probably a recent one. Further, Silco was almost certainly a Christian king who now represented Greco-Roman culture in the borderlands of the empire. This told Letronne much about the extent of both the empire and its influence on language and religion in this period. Judging from the lack of information on Nubia and Abyssinia in Ptolemy and the works of other Greco-Roman geographers, few Greeks can have traveled to this area. Almost certainly the Greek language had been brought to the coastal regions by Greek merchants as early as the fourth and fifth centuries b.c. Where commerce had spread, so had the use of the Greek language. Later, in the Roman period, Christianity had found fertile ground among the Greek speakers and had spread relatively quickly in those regions, further stimulating the learning of the Greek language. It was an ingenious interpretation and one which fully suggested the extraordinary historical value of apparently uninteresting inscriptions.

Letronne managed an equally interesting reading of a Greco-Egyptian papyrus dating from 146 b.c.[106] The papyrus was an announcement of the escape of two slaves from Alexandria. This document, found in the Louvre, was presumed to have been preserved in a mummy's sarcophagus. It provided a variety of very valuable information on both Greco-Egyptian social structure and on the value of money in that period. The announcement described the slaves, what they were carrying, and the conditions under which particular rewards would be offered. From these relatively sparse details, Letronne was able—in Sherlock Holmes–like fashion—to reconstruct the scene of the crime and to create a profile of the slaves and their condition. The owner of the slaves was a well-known political figure. This allowed Letronne to date the proclamation according to when he was known, from other sources, to have been in Alexandria for negotiations. Letronne was also able to establish that one of the slaves had escaped from the baths, taking his master's clothes to cover his state (perhaps indicated by a tattoo on his chest, a necklace, bracelets or by lack of clothing), and his money belt, including ten pearls and some gold coins. One of the slaves, a tall good-looking man with a cleft in his chin, had tattoos on his hands. It was traditional to brand recaptured escapee slaves facially. Letronne surmised that perhaps this slave was one of his master's favorites and had been branded on his hands to minimize the severity of the required penalty. This time the escapee was assisted by a co-escapee, another slave who was probably an overseer. The tall favorite was a Syrian and Letronne was pretty sure he knew in which war he had been captured and enslaved. The reward offered suggested that the slaves were highly valued and that there was considerable risk that anyone discovering the slaves would have every reason to try to keep them. The reward differential between information leading to the slaves' discovery within or outside a temple also suggested that Egyptian temples, like Greek temples, were asylums for slaves, who could not be legally (or easily) apprehended from such locations. The rewards offered for the slaves also allowed Letronne to determine, for the first time, the relative value of three of the most common coins used in Egypt in that period. Letronne's interest in the relative and consumption value of coinage, in problems cause by coin shortages, and in what he called "public economy" was a marked departure from the preoccupations of many of his contemporaries, particularly Gosselin and Garnier, who seemed to believe that coinage had an absolute and invariable value.[107] In constructing this picture of slavery in Egypt, Letronne used the papyrus itself, a statue of a slave held in the Museum of Pie-Clementin [Pius-Clementine], coins found in Egypt and held at the Louvre, and accounts of similar fictional events in the writings of Xenophon, Artemidorus, Lucian, Petronius, Moschus, and Apuleius. Again, through a relatively simple document, long-known but never fully interpreted, Letronne was able to seize a few key facts about the social structure of Egypt under the Ptolemies.

A New Methodology and Rigor

Modern historians of epigraphy have described Letronne's *Recueil des inscriptions* as "a discourse on method, a program for the generations to come."[108] This work did embody a methodology and approach developed by Letronne over the thirty-plus years of analysis of Greek and Roman inscriptions. It was a methodology that was extremely modern in its essentials and that might have offered an excellent guide not only to students of epigraphy but to nascent historical geography, had geographers recognized its relevance. His methodology can best be summed up as follows: the juggling of multiple sources (with a special value placed on field sources) until, with careful interpretation, they align into a single coherent expression; the use of philological analysis to uncover subtleties of expression and unintended meanings in texts; a preparedness to hazard a rigorous guess and to be wrong; respect for colleagues and predecessors who may not have understood the whole picture but who certainly contributed to it; a savvy understanding of the nature and limitations of historical documents; and an intolerance for simple-minded systems that promise enlightenment without effort and rigor.

Letronne adapted the central element of his approach from d'Anville's well-known critical comparison method of constructing maps from multiple sources. When trying to resolve an historical or geographic problem,

Letronne sought multiple sources. Given that he generally had limited access to the inscriptions he was studying, he worked from multiple copies by different hands.[109] When he was trying to reconstruct a past event from written sources, generally to construct the context around an inscription, he used every possible source, from texts written by figures alive at the time, to papyri,[110] to coins and medallions, to travel accounts and archaeologists' reports, to sketches and paintings of the relevant locales,[111] to fictional accounts of similar events, to the writings of modern and not so modern historians. Wherever possible, he sought to link these sources. Certainly, his use of them was imaginative. In his earliest work, in trying to reconstruct the defensive fortifications built near Syracuse, he plumbed the ancient texts, not just for information on Syracuse, but for the contemporary wisdom on siege warfare at the time of Thucydides.[112] His work on coins was highly innovative. He was among the first to realize the extraordinary potential of these hardy little pieces of metal not only for the reconstruction of dynasties but for the study of resource distribution and scarcity, extent of commercial influence, and the identification of centers of commercial activity.[113] Similarly, and as was just beginning to become standard practice, when studying an ancient manuscript he tried to identify all extant copies and began his investigation by conducting a critical comparison of all copies and recensions. This certainly describes his methodology with Dicuil's *De mensura orbis terrae.*[114] It was his comparison of multiple manuscript versions which allowed Letronne to identify a large proportion of the apparent errors in Dicuil's writing as copyist's mistakes and even to compose a guide of likely copyist errors for manuscripts of the period.

Although, and perhaps because, Letronne never traveled to most of the regions he studied, the source he most valued was the informed traveler's account. From his earliest publication he sought to check ancient historians against surviving ruins. Sometimes, through his knowledge of the ancients or through philological insight, he could see what was on location more clearly than could travelers visiting the site.[115] This insight never gave Letronne a sense of superiority over field workers and explorers. On the contrary, he fully acknowledged their assistance and described their work as vital and even heroic. In this vein, he dedicated his study of the Egyptian zodiacs to Frédéric Cailliaud.[116] Letronne saw his work as part of a cycle of exploration in which he suggested the areas and themes of greatest interest to explorers, according to which they directed their travels and collected results, which results Letronne analyzed.[117] On the basis of his analysis, field workers and explorers would again return to the field. Indeed, his relationship with proto-archaeologists and explorers was so close that explorers would design their itineraries so as to be able to sketch inscriptions for him.

The use of multiple and varied sources was not mere pedantry or show-off erudition. Letronne believed that if one could find an explanation that reconciled *all* of the facts gathered from a large number and great diversity of sources, then that explanation was likely to be very close to the truth. In a sense then, Letronne was a great linguistic and historical puzzle solver. But, at the heart of his methodology lay less a method than a talent. Letronne had a remarkable instinct for interconnections, which may have been a by-product of his considerable reading, his linguistic acumen, and a remarkable memory. His work on the manuscript of an Irish monk led him to Egypt. His reconciliation of multiple and apparently conflicting evidence on the date of the closing of the Suez Canal taught him who this Dicuil was, when he lived, from whom he studied, and a great deal about not only the man's education and learning but about the depth of Irish civilization in that period. He presented this idea to his French audience in terms they could understand:

> It is true that Charlemagne's stimulation of scholarship did not reach Ireland. But this island did not need such stimulation. It was, in the eighth century, more enlightened, that is to say, less barbarous, than any other European country.[118]

The source of this instinct for interconnections may also have been an almost childlike curiosity. While studying the question of what ornamental stone had once covered the pyramid of Cheops, Letronne wondered where and how it had been mined. Indeed, how had it been transported.[119] Further, who had mined and transported it and under what conditions. He looked and found, in the exploration of Sir Gardner Wilkinson, markings on the walls of the quarries near the Red Sea that

> seem to indicate the number of stones cut by each worker. This would suggest that the men working there were condemned to a particular amount of work, according to their sentences.[120]

He suspected that a careful study of the stones cut from the various quarries of Egypt would help identify and distinguish building periods and that this would in turn provide key information on the life and death of major water and land transportation routes.[121] Thus did he begin to spin the whole historical geography of Greek and Roman Egypt, integrating into it the physical and social realms of contemporary existence.

Philological analysis was at the heart of Letronne's methodology. His identification of the influence of ancient texts upon Dicuil allowed him to reconstruct that author's library. It also allowed him to reconstruct the Latin of the Irish monks. His sensitive reading of King Silco's inscription allowed him to trace the Scriptures in the speech patterns of this approximately 1,300-year-dead recent Greek speaker, almost as today linguists might trace the neighborhood or region in which a person grew up by their accent and speech patterns.[122] His 1824 analysis of the Greek in the Rosetta stone, its punctuation, spelling, formalistic nature of the writing, and the system of dating used, taught him about Egyptian social structure, hierarchy, the education of the stone's author, and the manner and regularity with which information was shared with the conquered people of Egypt.[123] By all accounts, Letronne was a remarkable philologist who made a major contribution to correcting and annotating key Latin and Greek texts. As must already be clear, he did not stop at the production of a catalogue of inscriptions. It was his passion to interpret and elucidate these ancient and long-forgotten words.

With sources as old and varied as those used by Letronne, a scholar had to dare to be wrong to come to any significant conclusions. Letronne, in spite of his commitment to rigorous and painstaking scholarship was prepared to be wrong.[124] He frequently ended an article with a textual shrug and a description of the relative certainty he accorded each piece of his research. Beyond the egotism of the scholar, what counted was the contribution made to the larger scholarly venture.[125] This mattered far more than whether Walckenaer became offended because Letronne had dared to publish a critical annotated edition of Dicuil first. Sloppiness, scholarly pedantry, and amateurish dilettantism were all far greater enemies than error, as Letronne made clear in the reviews he wrote for the *Journal des savants* and for de Férussac's *Bulletin*. In these reviews, Letronne sternly pointed out egotism and name-dropping. Courier's translation of Herodotus was certainly erudite: he detailed, "point by point, everyone else's opinion," but always "without ever providing a satisfactory solution himself, the principal purpose of which seems to have been to prove that he has read a great deal." What this translator needed was not erudition but "a more delicate sense and a more practiced use of his own language."[126] Letronne was equally sharp in his encounter with Raoul Rochette, with whom he disagreed, among other things, over whether the temples of Egypt had once been painted. It was Letronne's correct contention that they had been. Letronne did not resign himself to the lack of rigor of many of his colleagues. He attacked this problem through his reviews, which some contemporaries found excessively critical.[127] He also attempted on at least two occasions in his career, in the late 1820s and in the mid-1830s, to reform the most prestigious and central institution of humanities scholarship in eighteenth- and nineteenth-century France: the Académie des inscriptions et belles-lettres.

This Academy, Letronne felt, was not realizing its potential. Instead of devoting itself properly to the historical sciences, it was following its puzzling and outmoded name into confusion and oblivion. Why not restructure it in the image of the Academy of Sciences or the Academy of Beaux Arts and even the recently reestablished Academy of Moral and Political Sciences, which had fixed subdivisions and a far more rigorous election procedure? Why not reduce its size and increase its rigor? A system allowing the judgment and balancing of scholarly merit could thereby be introduced and this would rid the Academy of Inscriptions of the nepotism and favoritism that had created great gaps in subject areas of major importance. What Letronne was asking for was the professionalization of the historical sciences and their reform from the very top. With the top reformed, candidates would flood in and the Academy might win some respect both from government and the public. Above all, he wanted the "amateurs," out of the Academy: "men who, while educated or spiritual, have left almost no mark on science."[128] What was needed was scholars who through their work and scholarly criticism would define the fields worthy of study. These would be specialists and researchers, not the learned pontificators of the past. Foreign chief archivists and librarians could serve as corresponding members. It was the work of such researchers and "professionals" in the seventeenth, eighteenth, and early nineteenth centuries that would define the starting structure of the new Académie des inscriptions, which name Letronne was prepared to keep if this apparent conservatism would allow the reform of the institution. This new Academy was to be structured around six major subdisciplines. At the heart and soul of it (occupying six chairs) were to be the historical sciences: chronology and geography. Following that and occupying five to six chairs were to be the history of philosophy, the history of religion, and legal history. Philology and archaeology (mostly focused on Greek and Roman culture) should also have six chairs each.[129] By virtue of its national importance, medieval studies and the critical history of France should occupy a larger (though unspecified) number of chairs. Finally, Oriental studies should have between six to eight chairs. In short, Letronne was calling for the same development of tough disciplinary structures in the historical sciences as were clearly well developed in the sciences, the new social sciences, and even in the fine arts. This disciplinarity would encompass geography but its logic was not dictated by geography or the geographic tradition. There is no clear statement in this plan of what Letronne thought of his geographical colleagues at the Academy. Many—but by no means all—of these geographers, both past and present, had been amateurs, singularly uninterested in "science" as Letronne defined it, and had left no mark at all on scholarship. If Letronne did not single them out in his plan, and if what he thought about them is not *entirely* clear, what one geographer thought of this plan *is* clear. Implementation of the plan would have required radical reform of the Academy. No doubt many members would have lost their seats. Walckenaer commented succinctly in his obituary of Letronne written for the Academy of Inscriptions:

> The Academy did not adopt this project. It was not even discussed—except in committee. The advantages were doubtful and its inconveniences real.[130]

Indeed, such inconveniences were precisely what Letronne was after.

If Letronne was critical of sloppy research—or intolerant of the complete absence of research from some of his colleagues at the Académie des inscriptions—he was generous to serious scholars. From his earliest publication, Letronne cited the research of predecessors and colleagues. He began many articles with a list of all previous work on the subject. Even when he was correcting or improving on their research, he paid them the courtesy of a reference and took a dim view of those who did not.[131] This was most uncharacteristic of an age in which the citation was considered optional and was typographically expensive. It was all part of his sense of a tradition of scholarship, of honor among professionals.

One of the flaws in the historical research of many of Letronne's contemporaries was what can only be described as a naive attitude to sources. Between Gosselin, who was prepared to believe, based on medieval maps and extant measures, that the ancients had had a better knowledge of Europe than modern Europeans, and Garnier, who thought that coins had an absolute and unchanging value, misconceptions about the role and nature of documents and artifacts abounded. By contrast, Letronne's approach was down-to-earth and commonsensical. He knew, almost instinctively, what information could be derived from Dicuil's *De mensura orbis terrae* and what could not. This sense is best exemplified by his work on the periplus of Scylax. This extended list of the place names and peoples of the Mediterranean and Black seas was generally regarded as an important geographical document which reflected the geographical knowledge of both its author and its time. There was considerable dispute as to the precise time of its composition. Some authors placed it just prior to Herodotus, others in the time of Polybius, and still others in the age of Alexander the Great's father, Phillip II of Macedonia. Letronne looked at the document and was stupefied by the assertions of his colleagues.

I don't think that any of them has seriously questioned whether the periplus has any compositional unity; whether it stems from a single source or multiple sources; if all parts of it date from the same or different periods. It is easy to see that all depends on the answers to these questions. But they [these questions] have occupied the attention of the Baron de Sainte-Croix and Mr. Gail less than of all the other scholars that have spoken of Scylax.[132]

This work, far from being a piece of geographical research or an important navigational aid, was obviously a compilation, composed for the education of children and reflecting no time or author in particular. What was the point of asking scholarly questions if one did not think about them before launching into detailed research? Indeed, what was the value of sources at all if one relied principally on silly preconceived ideas? The scholar should take the source as his guide and tailor his research to the types of questions it was capable of illuminating.

Letronne declared his opposition to "systems" as early as 1812 and in his work on the topography of Syracuse. In that work he described the "esprit de système" as something that has to be carefully avoided, by opening oneself to the critical opinions of predecessors.[133] Garnier's highly speculative work on the value of ancient monies led him to further refine his sense of what was dangerous about preconceived notions or systems and to suggest ways of avoiding them in research. There was, Letronne argued, a two-fold seduction to "systems." Most were based on a small amount of data yet purported to be able to explain far more than the data could support. Thus, they appeared to offer quick and easy solutions to problems that had no easy solution. The second seduction applied to systems that were already established, either in the academic community or in the scholar's mind. The familiarity of the preconceived notion was itself seductive. Letronne warned historical practitioners that they must resist seduction and be prepared to put their own opinions and views aside while looking at the evidence. Researchers must adopt the historical method:

> That method which consists of moving from the known to the unknown, letting oneself be directed by the facts, without allowing oneself the least hypothesis, is now the only admissible method for all historical questions where the solution depends on the concordance of positive data. In erudition, as in physics, in order to attain the truth, one must see only the facts, note them, collect them, and link them through a theory which is their common expression.[134]

Letronne was not here advocating pure empiricism or anything akin to positivism. The aim of historical research was theory, but theory had to be built from facts, their comparison and their concordance. This did not preclude investigation based on intuitive insight into a limited number of sources, so long as the limited basis of the conjecture was clear and its aim was to stimulate research that could prove or disprove the conjecture.[135] The resonance of these ideas with Champollion's and the similarity of expression used by Champollion and Letronne suggests that Champollion's own rigorous and painstaking research may well have hardened Letronne's convictions.[136] Prior to 1816 Letronne himself had been flirting with the kind of system he so roundly criticized in 1817. He gained entry into the Academy of Inscriptions with an essay on the extraordinary accuracy of the metric system employed by the Greeks and Romans in Egypt. His arguments in this essay were more numerical than historical. The prizewinning essay may have gained him entry into the Academy, but Letronne soon thought better of his speculations and never published it. Letronne was approached by an admirer, some thirty years later, who wanted him to revisit the famous essay. Letronne refused and, showing equal disdain for the flattery and the essay, handed the original manuscript to his visitor with the comment: "You want it? You can have it."[137] Letronne came to believe that it was a false faith in numbers and their apparent accuracy that was encouraging some historians to engage in ridiculous and irresponsible speculation. Those who,

> principally since Condorcet, [have sought] to apply the calculation of probabilities to questions of moral order, and above all to the degrees of certitude of historical facts. They have flattered themselves in thinking that they can calculate *the odds that* this event or that event occurred or did not occur. It is a misfortune that they do not see that this probability, resting as it does on an entirely arbitrary numerical basis, can only provide chimerical or illusory results. In no case could it replace that intimate and absolute conviction, which admits neither less nor more, produced by the examination of all of the diverse circumstances accompanying a real event.[138]

For Letronne, probabilistic arguments in history were an "abusive use of mathematical analysis."[139] Researchers who used this form of argument had fallen into the trap of anachronism and had replaced the hard work of contextualization with simulated certainty. Writing about the astronomy of the ancient Egyptians and their calendrical system, Letronne commented:

> We forget too often, in studying questions of this type, the role that religion played in Egyptian society. We only see the scientific side of their calendar and of their knowledge in general. We seek to find in them precision and a quest for exactitude, which are the characteristic traits of the modern mind. So, we see them taking measurements with the same rigor that we could apply with the help of our repeating circles, and our theodolites, armed with verniers and telescopes. It is but a small step to suppose that they had a

bureau of longitudes housed in each of their great colleges. These seductive dreams vanish with truly historical study of their calendar. Astronomy, which the ancient Egyptians did not distinguish from astrology, was more a matter of religion than of science for them. Thus, the movement of the solstice relative to the heliacal rising of Sirius which always returned on the same day, was for the Egyptians an indifferent event, without consequence, which they explained as an effect of the will of Osiris or by some other reasoning which either seems good or bad depending on the view one takes of them.[140]

This kind of distortion was not a minor concern to Letronne. What was at stake was the health and well-being of humanities research and particularly of historical research. Thus it was with a touch of acerbic humor and pride that he announced the success of the École de Chartes in training France's future archivists and historians in the following terms:

> Isn't it also admirable that this youth, now entirely free, having passed their exams, and abandoned to their own devices, has resisted, by force of the salutary stimulus they have received, all of the preoccupations of passing fads? They have not fallen into any of those literary, political, or religious aberrations which mire so many of the most distinguished of our young minds. No messieurs, not one of our young paleographers has entered into those conspiracies against common sense, which we witness daily. I have looked carefully. I have not found one of them who has even simply dreamed of inventing a philosophical system, nor even the teeniest new religion![141]

There was humor in the observation, but Letronne's comments did coincide with the development and popularization of the positivistic ideas of Auguste Comte and their practical realization in the pseudo-religiosity of Saint-Simon and his followers.

The final aspect of Letronne's method, and perhaps the most important, was the beauty of his work. He always allowed the reader a sense of discovery, a participation in the historical investigation, writing almost in the style of a detective novelist. The sources were often reproduced in full, their obscurities gradually peeled away and their significance slowly constructed. It was, the reader could see for him- or herself, from incidental details that a whole society was being reconstructed. There was much that was technical in Letronne's writing but the technical was set aside, thus allowing the reader to chose to read or to not read, and only what was essential to his purpose was presented.[142] Letronne respected his readers and had little inclination to waste their time. While he presented the technical basis of his assertions, from philological analyses, to footnotes, to the analyses of others, he also always provided sufficient contextualization of the problem at hand to allow the educated, but uninitiated, reader to fully enter into the argument being made. In addition, he often provided context to the historical investigation in which he was engaged. So it was that in describing the dedication to Osiris in the remains of a temple found during the cutting of the Alexandria Canal, Letronne succinctly recounted for the reader the circumstances of the cutting of the canal; previous attempts to cut the canal; Mehmet Ali's gift of the find to Sir Sidney Smith via M. Salt—in addition to providing the translation of the inscription and a full explanation of the people mentioned in the inscription and how they fit into the history, genealogy, and place names of Egypt.[143] Thus, while Gosselin gave the impression that no reader could possibly have the erudition to fully assess his arguments, Letronne opened his arguments to all and sought to engage them by fully contextualizing his work.

The Geographic Nature of Letronne's Work

It has already been emphasized that Letronne had something of a geographic education and a geographic apprenticeship and that geographers were a major influence in his early career. Arguably, although Letronne became very critical of mainstream geographical research, his work did remain highly geographic throughout his career—certainly by today's standards. In 1839 he was using ancient sources to resolve an issue of physical geography; namely, whether there was an ancient connection between the Red and Dead Seas.[144] From the late 1830s, he worked to reconstruct the Ptolemaic geography of the Red Sea area; not just its physical geography but settlement patterns, and the routes employed for travel and transport. In so doing, he cited the geographers who had worked on related questions, including d'Anville, Bruce, Browne, Mannert, de Rozière, Jomard, . . .[145] In 1841 he was using town names to date and delimit the imperial and commercial conquests of particular Ptolemaic kings.[146] He was aware that the reconstruction of past geographies could cast valuable and unanticipated light on political, economic, and even social phenomena. As he succinctly phrased it: "When a geographical point is well determined, it happens that all of the circumstances that might be tied to it fall in line with the established position."[147] In 1845 he sought to reconstruct the network of ancient roads emanating from Carthage in Tunisia by use of extant road markers, ancient histories, and inscriptions.[148] The similarity between this work and the work carried out by Ernest Carette on the modern routes used in Algeria in the 1840s is striking and suggests that Letronne's publication would have been regarded as geographic had geographers noticed it.[149] His publication on Asia Minor in 1845 was an attempt to reconstruct the geography of the region (although Letronne did not deny the value of this work to epigraphy, numismatics, and chronology). This attempt contrasts sharply with Barbié du Bocage's study of the geography of much the same region. Faced with a similar paucity of sources, instead of restricting himself to ancient sources, Letronne fully explored the writings of modern travelers and proto-archaeologists, thereby suggesting the means by which the geography of ancient Asia Minor could be further elucidated.[150] In a much more modern sense, even Letronne's work on the calendrical system of the ancient Egyptians, which occupied him from the late 1830s to 1848, was geographic. For Letronne, the calendrical system organized more than time: it embodied a complex time-space relationship which found expression in the urban/ceremonial center of the system, the geographical extent of its use, and the calendar's relationship to the agricultural system and natural geography of Egypt.[151] Beyond the geographic nature of his own research, Letronne also advocated the teaching of geography at the École de Chartes, and considered geography an essential part of any archivist's education.[152]

However, it would seem that once Letronne's preoccupation became less classically geographic, that is, once he began to use epigraphy, archaeology, numismatics, historical interpretation, and field exploration (albeit by others) to reconstruct not only past locations but the nature of past societies and their relationship to space, he moved out of the realm of contemporary geography. That is not to say that he was ostracized. Simply that geographers did not register his work as geographic. There is some evidence of this. While Letronne cited his geographic colleagues, they, for the most part,[153] ignored Letronne's work. Further, as Louis Robert, a geographer teaching at the Collège de France in the 1930s, commented, although Letronne's epigraphic work was highly geographic and while he developed a method, he left no students and had no followers.[154] Perhaps most telling is Baron Walckenaer's academic obituary read before the Academy of Inscriptions in 1850. In this account of Letronne's considerable achievements, instead of claiming Letronne for geography, Walckenaer commented that he had contributed to many fields without ever devoting himself to one in particular.[155] One is reminded of the interdisciplinarity in the realm of the natural sciences of that other great figure on another fringe of the nineteenth-century geographic tradition: Alexander von Humboldt.

CONCLUSION

Arguably no historical subject or theme is inherently more interesting than any other. It is the degree to which it resonates with modern concerns, or is believed to do so, that lends it interest. In the early nineteenth century, at the early stages of the rediscovery of indigenous cultures, it was the ancient civilizations, deemed ancestral to European culture, that were a source of fascination to many. Geographers contributed to the study of both ancient and medieval cultures through the study of old maps and travel accounts, through the production of critical editions of ancient geographies, and, with a peculiarly early-nineteenth-century preoccupation, through the study and analysis of ancient measurement systems. This geography, although sometimes inspired, was often narrowly map-based, untheoretical and descriptive, and pointlessly detailed and erudite. Considered scholarly and worthy of recognition in the Académie des inscriptions, the work of these geographers had limited appeal in both the larger academic world and for subsequent generations of geographers. In the midst of this largely uninspiring body of writing, the work and activities of Jean-Antoine Letronne stand out as something altogether different. Originally trained as a geographer, Letronne began his academic life as a geographer focussed on precisely those questions that preoccupied other contemporary geographers: the re-edition of ancient geographies; the mapping of important events in ancient history; the assessment of ancient measurement and coinage systems. Within a decade of the initiation of his academic career, Letronne had been captivated by the civilization of ancient Egypt and particularly the interaction of Egyptian, Greek, and Roman cultures along the banks of the Nile. His knowledge of Greek and Latin, his contacts with Champollion, his interest in the broader intellectual currents and interests of his day, soon drew him away from the narrower concerns of his geographic colleagues. In the course of an active intellectual and administrative life—very much on the margins of mainstream geography—Letronne proved himself an innovative and imaginative scholar.

Perhaps responding to the growing interest in society, Letronne developed an interest in the nature of ancient societies encompassing the economic life, administration, religious life, and even in some regards the everyday life of those living in the Greek and Roman empires. Using epigraphy as his access, he also explored the spatial reality of Greek and Roman rule including what could be revealed of the commercial and transportation systems. Most importantly, in the course of his life Letronne refined an approach to the study of the past that owed a great deal to Nicolas Fréret and Jean-Baptiste Bourguignon d'Anville. Their approach had been the critical comparison of multiple sources. Letronne took that technique further, and, going beyond traditional spatial sources of information, he interrogated coins, inscriptions, papyri, travel accounts (old and recent), ancient texts of a variety of sorts, drawings and maps, searching for a consonance of voice. Letronne had a mind for interconnections and was often able to seize the connection between apparently unrelated and obscure events. His approach entailed reading between the lines and, above all, looking critically at the way language, and particularly the Greek language, was used. Letronne showed that a single word could reveal whole dimensions of social reality if deeply interpreted.

Letronne's approach to the study of the past was highly disciplined and, as is clear from his attempted reform of the Académie des inscriptions, Letronne dearly wanted disciplined standards established in humanities research. Yet his approach was also interdisciplinary and by his second decade of active research he had strayed far from the preoccupations of mainstream geographers in the content, scope, approach, and detail of his research. Even his finely balanced attitude to theory, which rejected predetermined systems but embraced hypothesis and informed conjecture, was immeasurably distant from the thinking of Jean-Denis Barbié du Bocage and his colleagues. Letronne was in favor of discipline and disciplines based on demonstrated traditions of research. There is little doubt, from his abiding interest in geographic questions and his support of geography education at the École des Chartes, that Letronne saw geography as one of those demonstrated traditions. However, it seems clear that the geographers of his generation did not see Letronne as part of that tradition.

Conclusion

Whether I have done right by my Subject, must be left to the
judgment of the learned Reader. However, I cannot but hope, that
my attempting of it may be Encouragement for some able Pen to
perform it with more Success. —Jonathan Swift, *A Tritical Essay upon
the Faculties of the Mind* (1711)

This book has sought to track the activities and thoughts of geographers
through a period of uncertainty and change in a key moment in the history
of geography. It was not a discursive space or time shaped by one or even
a number of great thinkers. Nor was it one of startling growth or enduring
insight. Why, then, bother to study it? In 1989 I published an article which
compared geography as it was practiced by the French on the scientific
expedition to Egypt in roughly 1798 with that practiced around the mid-
1800s by the French on the scientific expedition to Algeria. The difference
was startling. By the mid–nineteenth century, geographers, if they had not
yet reinvented geography, if they had not yet transformed its methodologies
and curiosity, were in the process of doing so. However, no one seems
to know or care about the geography of early-nineteenth-century France.
Perfectly reasonable and educated French historians of geography informed
me that in this period geography did not exist and there were no geographers.
It seems to be that only periods considered heroic or foundational, periods
in which individuals can be identified as disciplinary touchstones, are those
that attract research. But the world goes on with or without heroes and
arguably, heroes—beyond the daily heroism of most lives—are invented
more than they are born.

My context was also important to the choice of this period. Working in
the social sciences and humanities at the end of the twentieth century, I am
aware that this time shares something with the early nineteenth century. If
discursive formations were ragged edged and fluid at the beginning of that
century, today, after a hundred years of high and strong disciplinary walls, the
walls are collapsing again, leaving many in academia in the later twentieth
century with a sense that disciplines are less relevant, that intellectual
networks are complex, and that the lines of power are being redrawn in
ways that are not necessarily predictable and that may threaten the status—

and perhaps the existence—of a number of established disciplines in the sciences, social sciences, and humanities. It is my sense that we may be able to learn more from a time of greater discursive fluidity than from the now, I would argue, relatively overstudied "discipline forming" period of the late nineteenth/early twentieth century. This is not to say that history must reflect back "our current beliefs and assumptions" to have any meaning for us.[1] Simply that it is long past time to look beyond the narrow disciplinarity that has made the late nineteenth and early twentieth centuries such a fixation for historians of geography, particularly in France.

So what was geography in the eighteenth century and what happened to it from roughly 1760 to the mid-nineteenth century? In France eighteenth-century geography was almost entirely consumed with the very real intellectual problem of how to describe the world. To that end, great strides were made in the course of the eighteenth century toward the development of a language of representation, both textual and graphic. In particular, topographic mapping acquired a definition, consistency, and regularity that was unprecedented. The concern with representational accuracy was expressed in terms of the increasing topological accuracy of maps, the growing variety and number of scales of depiction available to accommodate the diverse uses to which maps were increasingly put, and enhanced measurement, compilation, and map reproduction techniques. Geographers in this period published in a circumscribed number of genres. Apart from some travel accounts, which were often written by non-geographers, geographers' publications included universal geographies (structured regionally or alphabetically), compilation manuals, surveying and mapping manuals, maps made by critical comparison of written or drawn sources, and maps made with the assistance of field observation. The geography structured by this preoccupation with description was an established discursive formation with representation by the end of the eighteenth-century in both the Academy of Sciences and the Academy of Inscriptions. Its primary institutions in the eighteenth century were not universities but secondary schools; in particular, the Jesuit, Oratorian, and military colleges. Outside of those institutions geography, both literary and mathematical, was taught by apprenticeship and tutoring. Apart from its presence in these institutions and in the *Encyclopédie,* proof of the existence of the discursive formation is to be found in the shared community, standards, sense of professional turf, and concern with continuity evident in the correspondence and publications of geographers. Whether the geographers of the eighteenth century were fully aware of it or not, the debate which took place in France (and much of Europe) over the size and shape of the earth, between on the one side the Cassinis and on the other Newton, demonstrated the limits of representation. In the course of that debate geographers first discovered the limits of direct measurement and observation of the surface of the earth to its description. They discovered that understanding how the universe functioned might lead to a more complete and true description of the universe. It was to be a lesson learned and relearned again at various scales over the next 150 years.

By the last decade of the eighteenth century, geography's loss of direction and status was at crisis proportions. We are fortunate to be able to identify two incidents which strongly suggest the nature and severity of this loss of direction. The first was the manner in which the geography course proposed and delivered by two of France's foremost geographers at the École normale was received. Sharp criticism came from both the students in the course, including some of the other professors in that short-lived but elite institution, and from the *Décade philosophique*, the mouthpiece of those intellectuals and public figures who considered themselves the heirs of the Enlightenment. The consensus appears to have been that the geography taught at the École had nothing to offer modern science, was elementary and simplistic, and demonstrated little sense of its own aims relative to sister and far more distant sciences. The scientific career of Cassini IV and its demise in the years following the Revolution provides another window on geography's loss of direction and status. Once one of the most powerful and privileged of the large-scale geographers/astronomers of the eighteenth century, Cassini IV was cast out of science by the combined forces of the Revolution, a movement to democratize science, the bureaucratization of large-scale mapping, and the mathematization of astronomy and its movement toward explanatory physics. Cassini IV could do little in the face of forces of this magnitude than fulminate against injustice and the conspirators of the Revolution, whose principal aim was in his view the destruction of French society.

There were three principal "reactions" to this loss of direction and status. Malte-Brun represents one of the most conservative reactions. Perhaps it was his background and training in literary and historical studies that led him to believe that the solution to geography's problems could be found in the rejuvenation of one of the oldest traditions of geographic writing. He took his cue from Strabo and banished from his universal geography theory, explanation, and the most technical aspect of geography: cartography. He believed that by broad interdisciplinary reading, critical comparison of sources, and a poetic and beautiful form of expression, descriptive geography, or the description of the world, could once again be made inviting to the larger public. In the genre of the universal geography,

Malte-Brun set the standard for at least a century. Yet although his work was appreciated by subsequent generations of geographers, it suffered from weaknesses inherent in the genre and its very antiquity. In a period in which explanatory and theoretical science were moving to the fore, it was fairly exclusively descriptive. It coped with the increasing volume of research in natural history, geology, botany, and the nascent social sciences by summarizing the results of research by others and expanding in volume. And it marked off no turf for original research for geographers. Arguably, the genre, which demanded considerable time, gathering of voluminous sources, and reading for its execution, so occupied the attention of many geographers that they had little energy or inclination to consider some of the questions raised by military geographers and figures working on the margins of the discursive formation in social, natural, and historical geography. Perhaps most seriously, universal geographies as written by most geographers lacked a coherent and unifying argument, one that could make the universal geography more than a more or less ordered compilation of more or less digested information. The shortcomings of the genre are clearest and more egregious in the universal geographies of Guillaume Delisle, Edme Mentelle, and Abbé Jean-Joseph Expilly. It is clear from Le Père François's *La Science de la géographie* and Humboldt's *Cosmos* that with a great deal of thought and contemplation, the universal geography could rise above the orderly enumeration of features and characteristics of peoples and governments. However, for that, an aim larger than describing the observable earth, or cataloging all knowledge about the surface of the earth, was necessary.

A second highly conservative type of geography is represented by the thought and career of Edme-François Jomard. While Malte-Brun clung to the descriptive text as definitional of geography, Jomard considered the map and the mapping metaphor to be geography's heart and soul. The map, measurement, and the ordering of information that it implied played a role in all of his major publications and limited the degree to which he could correctly identify problems not subject to resolution by cartographic means and design solutions appropriate to their resolution. While Jomard's experience in Egypt and his abiding interest in that country was particular to him and his life experience, his commitment to mapping monuments, to mapping exploration, to the production and study of the maps of Western civilization, and to the science of mapping and categorizing peoples, which was to become ethnography, was typical of the kind of activities engaged in by geographers throughout the nineteenth century. These were activities that derived directly from the map as metaphor and the sense that the geographer's primary function was the mapping of phenomena. In a sense, then, there was as strong a following for Jomard's conception of geography as there was for Malte-Brun's tradition of literary geographic description.

Geography in the service of the state was an established tradition long before the Revolutionary and Napoleonic periods. However, for a variety of complex military, social, political, and economic reasons that relationship intensified during Napoleon's reign. The militarization of French society, in particular, had a major impact, less on the already established and relatively well-off scholars, than on the still-young men hoping to make a career in science. As a result, the careers of two military geographers, André de Férussac and Jean-Baptiste-Geneviève-Marcellin Bory de Saint-Vincent, are suggestive of the impact of the pressure of state needs and interests, essentially utility, on scholarly activity. The impact was not entirely negative. Both men were exposed to opportunities and influences that they would likely not have encountered had they not served as officers. The training they received and the nature of the military geography practiced by their colleagues the ingénieurs-géographes, as directed by Generals Pierre-Alexandre-Joseph Allent and Pascal Vallongue, encouraged them toward an analysis of the human and natural landscape not at all evident in the cabinet geographies of men like Edme Mentelle, Jean-Nicolas Buache de la Neuville, or even Conrad Malte-Brun. The sensitivity to landscapes and society that they developed in the course of their duties as intelligence officers influenced their thought and writings, even when these strayed far from what we today consider geography. In the long run, there was little long-term commitment on the part of the state (and its changing regimes) to geographic research even of the most evidently useful. If the cartographers working on the Carte de France could be "let go," how much more expendable must intelligence officers have seemed in times of relative peace. Still, de Férussac and Bory de Saint-Vincent were among the fortunate military personnel who retained a strong and sustaining relationship with the military which more than once helped them keep body and soul together. In the absence of sustained state funding to universities and a reasonable number of institutes of higher learning and study, direct state employment of geographers in the military bureaucracy and on military campaigns, maritime voyages, and colonial expeditions offered them some of the only means of funding expensive and exciting new exploration and research. The costs were of a more personal nature: health, freedom to pursue questions they, themselves, deemed of interest, and ideological coherence, were all likely to be compromised. State interest in geographic research did not die with Napoleon. Geographers would be called upon in every major war from the conquest of Algeria to the wars of the present day by France and by Western powers in general. The relationship between state power and geography is an old one which shows no sign of waning, although the particular roles that geographers have played have varied considerably.

While geographers were struggling to make their traditional approaches cope with a new situation, researchers on the margins of geography were beginning to explore new possibilities for the subject. In social science, as a result of his interest in the influences of external factors on "man," both natural and those emanating from government, Count Constantin-François de Chasseboeuf de Volney explored the relative impact of climate, soil, and government in Syria, Egypt, Corsica, and to a limited extent the United States. Although his work was incomplete and tentative, Volney began to develop an approach to the study of both nature and human society, and particularly the relations between them, that geographers might have found liberating had they paid sufficient attention to it. In particular, the importance Volney accorded prior to 1802 to being there and seeing for himself was a concern that did not trouble most early-nineteenth-century French cabinet geographers. Geographers had some opportunity to notice Volney's ideas as he taught alongside Mentelle and Buache de la Neuville at the École normale. Volney was one of the principal guides used by the officers on the Napoleonic expedition to Egypt and by the Dépôt de la guerre and thus the ingénieurs-géographes consulted Volney on a number of occasions, in particular on Egypt and its place names. Volney seems to have given up his geographic work as a result of Napoleon's hostility to the kind of social criticism engaged in by the Ideologues. The closure of the Class of Moral and Political Sciences of the Institut resulted in something of a hiatus in civilian studies of society and social problems. At that stage Volney appears to have lost interest in, or the drive to pursue, such questions. The sources of Chabrol de Volvic's interest in society and social problems were military and statist, as is clear from his experience and activities in Egypt, Napoléonville, and Montenotte. In the latter two regions he experimented with the circulation of people and goods and sought to plan space to better fit the area into the new national/imperial territory. Chabrol de Volvic's geography, then, was a practical geography, conceived and put into practice on the ground. His work was also intellectual and theoretically innovative. It was in the city of Paris and with the help of at least two geographers and the mathematician Jean-Baptiste Fourier that Chabrol de Volvic instituted and implemented one of the first urban statistical surveys in the Western world. This entailed far more than the collection of statistics: the *Recherches statistiques sur la ville de Paris et le département de la Seine* demonstrated sophisticated statistical analysis, superior display of quantitative information, and a type of thinking essential to planning the future use of heavily employed and sometimes contested public spaces. Writing in the 1820s and 1830s, the geographer Adrien Balbi seems to have absorbed nothing from the work and thought of Volney and little from Chabrol de Volvic's publications. This, in spite of the fact that he was aware of the work of both men. Balbi was limited in his regional and statistical works by a simplistic understanding of statistics as a body of hierarchically structured facts and in his ethnographic work by a tendency to merely collate information from often contradictory sources. He seems to have been reluctant or unable to engage in the thought necessary to develop and present the ideas and arguments he sketched out in his publications and to move beyond description and the accumulation of the views of authorities to make a theoretical argument based on observation and analysis. Nevertheless, his interest in statistics and his explorations in ethnography did point the way that a number of later nineteenth-century geographers were to take.

The early nineteenth century witnessed another great innovator, in the realm of natural geography, Alexander von Humboldt. Humboldt functioned principally in the realm of natural science, responding to theory and research being conducted in geognosy, geology, botany, zoology, geodesy, and astronomy. Consequently, he too must be considered marginal to mainstream French geography. Trained as a mining engineer in Göttingen, Humboldt brought to the study of the natural realm an approach and philosophy informed by Naturphilosophie. When compared to the publications of early-nineteenth-century cabinet geographers such as Balbi and even to those of the militarily trained natural geographers, Humboldt's writings stand out as something altogether different. There is in his corpus a consistency of purpose and a coherence that is perhaps the most salient characteristic of his work. Interested throughout his life in the natural world, he made some contribution to the systematic study of rocks, plants, animals, and climate. However, Humboldt's real interest lay in the physics of nature, or the interactions and interconnections between natural phenomena. The exploration of these led him beyond the obstacles which deterred most geographers: the problem of balancing empiricism with the elaboration of theory; the reconciliation of description and the exploration of explanation and cause; the observation of external form and at the same time internal functioning; the problem of studying movement, change, and distribution. . . . His work in these realms encouraged him to seek new forms of investigation and expression that could facilitate movement between discursive formations, including field observation and measurement, the use of mathematical expression, and thematic mapping. In addition to his commitment to empirical and theoretical scientific research, Humboldt had an aesthetic sensibility which together with his interest in nature and in human societies led him to focus on human-scale landscapes as much as on individual rocks, plants, etc. French geographers were aware of Humboldt's travels in South America. They cited him, sought his editorship and membership in their principal institutions, and some even corresponded with him. However, there is little sign that any French geographer prior to mid-century took Humboldt, his ideas, and his approach to natural science anywhere near as seriously as did Mary Somerville. The man, his explorations, and his aura seem to have held the attention of geographers, to the detriment of a consideration of his way of thinking and conducting research.

Historical geography was encumbered and limited at the beginning of the nineteenth century by the tradition in which it was written, and it suffered from a paucity of imagination. The mainstream figures, Pascal-François-Joseph Gosselin and Jean-Denis Barbié du Bocage, undertook work so narrow in conception and scope that it can have had but limited interest to non-geographers and geographers alike. Charles Athanase Walckenaer, on the contrary, seems to have shared with André de Férussac and Bory de Saint-Vincent an inability to concentrate his attention on a particular field or type of inquiry. When he did direct his attention to geography, rather than to fiction writing, the natural history of man, or the flora and fauna of Paris, he too tended to ask questions limited to the determination of the location of past activities, events, personages, or towns. Alexander von Humboldt's one work of historical geography was innovative thanks both to his highly contextual approach to the history of science and to his personal identification with the early explorers of the Americas. However, it was such a self-indulgent piece of erudition, in which he gave free reign to his tendency to contemplation and digression, that it too had little impact on either contemporaries or subsequent scholarship. Among the historical geographers, only one stands out as truly innovative, Jean-Antoine Letronne, who was such an original and independent thinker that, in spite of his early geographic training and clearly geographic publications, the geographers did not recognize him as within the fold. Letronne's originality was a consequence of both a fertile imagination, which opened him to the major preoccupations of his time and led him from question to question until he had revealed the history and nature of the Greek presence in Egypt, and a rigor and discipline of approach and method that lent his work the kind of coherence and integrity to be found in the writings of Alexander von Humboldt on the natural world. Letronne's curiosity and competence extended well beyond geography but remained firmly rooted in geographic questions, which seemed to be the glue holding his individual inquiries together. With the geographers substantially unaware of the meaning and relevance of his work to their own, there was little hope that the full power of Letronne's imagination would be recognized either by contemporaries or by immediate successors.

What then, was the discursive formation of geography at the end of the eighteenth century and how had that changed by mid-nineteenth century? What were the objects of geographic study and how did these change? At the beginning of the period, the earth was a thin surface with boundaries and limits, both human and physical, and it was the duty of geographers to locate, delineate, and describe those borders and limits and what they contained. By the mid-nineteenth century, the earth had acquired relief and depth which carried meaning far beyond the geographers' abilities to elucidate. The plant and animal world, formerly simply enclosed in borders that had little to do with their nature or evolution, had acquired spatial meaning linked to arguments about the history of the earth and of the movements and changes in life on the earth. In the eighteenth century, human society and behaviors were easily characterizable and could be fit within the boundaries and limits described by geographers. By the mid-nineteenth century, societies themselves were seen to be an object of study, and their complexity, diversity, and history were becoming a focus of attention. What were the operations and methodologies which geographers could engage in? Arguably, at the beginning of the period, geography could claim the map as its own invention, its primary tool, and the most important product of its efforts. At the end of the period, it had to share the map not only with commercial mapmakers but with a growing number of scientists and scholars whose interest in social and scientific processes encouraged them to create the new space of the thematic map. In the 1760s, geographic texts explaining the decisions made by cartographers were not rare and were regarded with esteem. By the mid-nineteenth century, such works had all but vanished except as technical manuals accompanying maps and were considered dry and dull. In the eighteenth century, universal geographies still seemed useful and scholarly. By the late nineteenth century, they had grown to mammoth proportions and had lost the coherence they might have had in earlier times. Even one of the most synthetic thinkers of the century, Alexander von Humboldt, could not make the genre work. What impact did all this have on geography's place within the hierarchy of sciences? There is evidence that geography lost status as well as direction in this period, that it failed to reinvent itself, that it stumbled along with the force of inertia.

Notes

INTRODUCTION

1. That is to say that I am less concerned than is Lesley Cormack to measure the presence and nature of geography in recognizably disciplinary institutions, in part because there were no major institutions in France with a long and significant history of geographic instruction, beyond those discussed in chapter 2 below. In addition, I agree with Foucault that a genealogical approach, which focuses on "the singularity of events" and which cultivates "the details and accidents that accompany every beginning," is a more exciting approach to history and one which better reflects the incoherence, dispersion, disorder, and the "derisive and ironic" in life and in science. That said, Lesley Cormack's *Charting an Empire* (1997) is a fine reconstruction of the geography taught, read, and written at Oxford and Cambridge from 1550 to 1620.

2. A good guide to the considerable literature in this vein prior to 1985 is Gary Dunbar's *The History of Modern Geography* (1985), esp. 211–352.

3. The history of ideas approach is exemplified by Clarence Glacken's *Traces on the Rhodian Shore* (1967). Other works in this vein are: Paul Claval, *Les Mythes fondateurs* (1980). For a deeply contextualized history of ideas approach see Berdoulay, *La Formation de l'école française* (1981).

4. Infamous for this is Richard Hartshorne's *The Nature of Geography* (1939) and his "The Concept of Geography as a Science" (1958). But there are, sadly, many other examples, including Margarita Bowen's *Empiricism and Geographical Thought* (1981).

5. "Tradition" is a concept reinvented by David Livingstone to allow discussion of the history of geography without immediately implying that there is an "essential nature of geography." Thus, he posits an organism that changes and evolves. Taking a history of ideas approach to the concept of "tradition" as does Livingstone—following a particular line of descent—de-emphasizes the exclusions, the contestations, the lulls, the gaps, and the collapses in the history of geography and understates the power of discursive formations to limit and form research and practice. I welcome Gillian Rose's contention that "tradition" implies exclusion. I think that discursive formations do also and I agree that those exclusions, whether gender-based or otherwise, should be a particular focus of attention. I believe that this book explores a number of exclusions of ideas, approaches, and practices from the mainstream of geography's tradition that were later—much later—to become part of the discursive formation of "geography" through other influences. See Livingstone, *The Geographical Tradition* (1992), and the reaction to it in the Institute of British Geographers by Felix Driver, David Matless (with whom I profoundly agree), Gillian Rose, and David Livingstone. Driver, "Geographical Traditions" (1995); Matless, "Effects of History" (1995); Driver, "Sub-merged iden-

tities" (1995); Rose, "Tradition and Paternity" (1995); Livingstone, "Geographical Traditions" (1995).

6. Foucault, "Politics and the Study of Discourse" (1991), 54.

7. Foucault, "Politics and the Study of Discourse" (1991), 56.

8. Foucault, *Les Mots et les choses* (1966); Foucault, "Governmentality" (1991).

9. De Dainville, *La Géographie des humanistes* (1940), 55–60; this is one of the sustained arguments in Livingstone, *The Geographical Tradition* (1992).

10. Mayhew, "Contextualizing Practice" (1994), 324.

11. See Latour, *Science in Action* (1987), 228–241.

12. A fact that Margarita Bowen deeply regrets. See Bowen, *Empiricism and Geographical Thought* (1981).

13. For Stafford, geography underwent a Foucaultian transformation in the early part of the nineteenth century from a focus on the external form of continents to their interior detail. Leaving aside the historical accuracy of this observation (that is, to what extent was geography not previously concerned with interior detail, in her sense of interior), it is hard to see in this apparent exterior-interior analysis a significant change in the nature of geography. Stafford, *Voyage into Substance* (1984).

14. Philippe Buache, "Essai de géographie physique" (1752).

15. Broc, *La Géographie des philosophes* (1974), 203.

16. Geography's position on the tree of knowledge was explicitly addressed by Kant, Cortambert, de Ferussac, Letronne, Etienne-Nicolas Calon, and Omalius d'Halloy, all of whom might be convincingly described as geographers. See Kant, *Critik der Urtheilskraft* (1790); Cortambert, *Mélanges géographiques* (1862); de Férussac, *Plan sommaire d'un traité de géographie* (1821); Letronne, *Projet de diviser* (1834); Calon, *Convention nationale* (1795); Omalius d'Halloy, "De la classification des connaissances" (1834). On Kant as a geographer see May, *Kant's Concept of Geography* (1970).

17. Keith Baker discusses the difference in approach of Gusdorf, *Introduction aux sciences humaines* (1960), and Gay, *The Enlightenment* (1966–69), on the one hand, who see nineteenth-century science as in continuity with Enlightenment thought and Foucault, who sees a profound break between the two. Baker seems to fall in with Foucault as he thinks the disagreement is derived from Gay and Gusdorf's lack of definition of the ideas characterizing the social sciences in the Enlightenment. See Baker, *Condorcet* (1975), x.

18. First developed in Foucault, *Les Mots et les choses* (1966), but qualified and elaborated upon in his *L'Archéologie du savoir* (1969). An excellent guide to the writings of Michel Foucault is Sheridan, *Michel Foucault* (1980). All future references to Foucault's work will be to the English-language editions.

19. Foucault, *The Order of Things* (1971), 50.

20. Foucault, *The Order of Things* (1971), xiv.

21. A theory-driven, rather than a narrative, history of thought is logically defended by Lemon, "The possibility of the History of Thought" (1995).

22. Collins, *Changing Ideals in Modern Architecture* (1965).

23. Both Fierro-Domenech and Lejeune are weak in this regard. See Fierro-Domenech, *La Société de géographie* (1983); Lejeune, "La Société de géographie" (1982); Lejeune, "Les Sociétés de géographie en France" (1986–87). Michael Hef-

fernan's work on the Société de géographie is much more balanced between content and context. See Heffernan, "The Science of Empire" (1994), and Heffernan, "The Spoils of War" (1995).

24. For a different but interesting interpretation of the historical and intellectual meaning of Buache's systems see Belyea, "Images of Power" (1992), 6–7.

25. Cassirer, *The Problem of Knowledge* (1950): 15

26. Linné, *Classes plantarum* (1738).

27. Adanson, *Familles des plantes* (1763).

28. Absolutely necessitating the use of theory. As William Boelhower has pointed out, "Theory . . . means making visible . . ." Boelhower, "Inventing America" (1988), 479.

29. See Rudwick, "The Emergence of a Visual Language" (1976); Rudwick, *The Great Devonian Controversy* (1985); Camerini, "Evolution, Biogeography, and Maps" (1993); Cambrosio, Jacobi, and Keating, "Ehrlich's 'Beautiful Pictures'" (1993).

30. Le Grand, "Is a Picture Worth a Thousand Experiments?" (1990).

31. This is, of course, something of a simplification. As Arthur Robinson explains, thematic mapping can be seen as emerging from a whole series of confluent events and innovations from new concepts to new technologies to broader social change. In science he focuses on the development of the law of universal gravitation, the refinement of quantitative description of observable facts, the development of mathematical forms of expression including differential calculus and probability, the development of universal standards of measurement, the rise of social statistics, and the growth in interest in the links between environment and disease. There were, in his view, equally important preconditions in industry, trade, and their supporting technologies including transportation. See Robinson, *Early Thematic Mapping* (1982), 26–43.

32. This perfection of language is best described in de Dainville, *Le Langage des géographes* (1964). The culmination of development was the 1802 reform of cartographic symbology. This is also well dealt with in Berthaut, *Les Ingénieurs-géographes* (1902), 243. The best original account of these reforms is to be found in *Mémorial du Dépôt de la guerre* (1831).

33. I say "his or her" but, unsurprisingly, women were absent from the early-nineteenth-century discursive formation of geography in France.

34. Foucault, *Archeology of Knowledge* (1972), 23.

CHAPTER ONE

1. Nicolas Desmarest, "Géographie physique," in *Encyclopédie ou dictionnaire raisonné* (1757), 7: 613.

2. As Newman has shown for historical thought and writing, explanation is implied in most modern forms of description. See Newman, *Explanation by Description* (1968).

3. Robert de Vaugondy, *Essai sur l'histoire de la géographie* (1755), viij.

4. Robert de Vaugondy, *Essai sur l'histoire de la géographie* (1755), 158–60.

5. The *Ratio atque institutio studiorum* stood as the pedagogical rule for the Jesuits from 1599 to 1832.

6. Briet, *Parallela geographiae* (1648–1649). Quoted from de Dainville, *La Géographie des humanistes* (1940), 187.

7. Labbé, *La Géographie royalle* (1646), 41. Quoted from de Dainville, *La Géographie des humanistes* (1940), 206.

8. De Dainville, *La Géographie des humanistes* (1940), 212.

9. François, *La Science de la géographie* (1652), 141; see also de Dainville, *La Géographie des humanistes* (1940), 295.

10. Coton, *Intérieure occupation d'une âme dévote* (1608), 298. Quoted from de Dainville, *La Géographie des humanistes* (1940), 100.

11. De Dainville, *La Géographie des humanistes* (1940), 321.

12. Desmarest, "La Géographie physique" (1757), 7: 614.

13. Desmarest, "La Géographie physique" (1757), 7: 614. This discussion of Desmarest's influential views on physical geography is based on two virtually identical sources, Desmarest's article in Diderot and Jean d'Alembert's *Encyclopédie*, and his contribution on the same subject in the much later *Encyclopédie méthodique* (Paris: Chez H. Agasse, 1795), see especially tome 1 of volume 75: 793–803.

14. Desmarest, "La Géographie physique" (1757), 7: 613–14.

15. Desmarest, "La Géographie physique" (1757), 7: 617.

16. See Hahn, *Anatomy of a Scientific Institution* (1971), 97; Pedley, *Bel et Utile* (1992); de Dainville, "Enseignement des 'géographes'" (1978), 414–54. A systematic search of the popular press in this period might well reward anyone interested in assessing the importance of this form of education. The following ad, for example, appeared in the *Feuille du Salut Public*, le 29 pluviôse an II: "Course in Geography: On 29 pluviôse, Citizen Mentelle will begin a course on mathematical, physical, political, ancient, and modern geography, in 18 sessions. These will take place from noon until two o'clock on uneven days of the decade. Applicants can sign up at Mentelle's (home and) geographic office, cours du Muséum (le louvre). Price 25 liv."

17. Diderot and Alembert, eds., *Supplement* (1780). Withers reproduces a portion of this graphic in "Geography in Its Time" (1993), 255–64.

18. Jean-Baptiste-René Robinet, *Supplément à l'Encyclopédie* (1777), 3: 204. See also Kathleen Hardesty, *The* Supplément *to the* Encyclopédie (1977), esp. 3.

19. *Encyclopédie méthodique* (1787, 1782–1832).

20. Hahn, *Anatomy of a Scientific Institution* (1971), 108–9.

21. De Dainville discusses this in his *Le Langage des géographes* (1964), 104. See also de Dainville, "De la Profondeur à l'altitude" (1958), 195–203.

22. De Dainville, *Langage des géographes* (1964), 98.

23. Baker has pointed out that it is a mistake to associate the Académie des sciences with professionalism: "The Academy offered no formal pattern of training for research; it neither conferred nor required educational certification as the prerequisite for entry into the profession; it provided no regular paid careers in which scientific research was organized as a full-time occupation." The Academy was a body of experts admitted on the basis of demonstrated expertise. Therefore, while membership in the Academy cannot be used to bolster an argument for the professional nature of geography in the eighteenth century, the membership

of geographers suggests the respect with which their activities were held among contemporaries. Baker, *Condorcet* (1975).

24. Jourdain, *Le Budget de l'instruction publique* (1857), 213–14.

25. Dussaud, *La Nouvelle Académie* (1946), 93ff.

26. Aucoc, *L'Institut de France* (1889), xii.

27. On Frérét, see Simon, *Nicolas Fréret* (1961); and Walckenaer, *Rapport fait à l'Académie* (1850), 70–216.

28. Culled from Barret-Kriegel, *Les Académies de l'histoire* (1988), 302–20. Most of these scholars are discussed further in either this chapter or chapter 8.

29. De Dainville also considers the Jesuit teaching in the area of philosophy to have been important to the shaping of geography but admits that this was an altogether different venture based neither on description and verbal repetition nor on observation and mathematical or graphic description but on argumentation. He further acknowledges that it was never called geography but philosophy. De Dainville, *La Géographie des humanistes* (1940), 459–60.

30. Hence the word trivial, derived from the Trivium, of which rhetoric formed a part.

31. Such as the material covered in Chales's *Cursus seu mundus mathematicus* (1674).

32. Such as Riccioli's *Geographiae et hydrographiae* (1661).

33. Artz, *The Development of Technical Education* (1966), 41–42.

34. Artz, *The Development of Technical Education* (1966), 43–46.

35. On the history of these corps, see Bret, "Le Dépôt générale de la Guerre" (1991), 113–57, and Berthaut, *Les Ingénieurs-géographes* (1902).

36. On the course of instruction undertaken in this school and in particular on the "concours" method of advancement and the role of mapmaking in this, see Yvon, "Les 'concours' de l'École" (1987).

37. Berthaut, *La Carte de France* (1898), 1: 5; and *Les Ingénieurs-géographes* (1902), 1: 22–27.

38. Berthaut, *Les Ingénieurs-géographes* (1902), 1: 137–38.

39. According to Pedley, *Bel et Utile* (1992), 60–61, Robert de Vaugondy was accused by Philippe Buache of modeling his *Essai* a little too closely on Bruzen de la Martinière's "Essai sur l'origine" (1722).

40. Excluded from discussion here is Jacques-Dominique Cassini, or Cassini IV, the youngest of the astronomer/geographer Cassinis. He assumes a central role in the argument in chapter 2.

41. Robert de Vaugondy, *Mémoire sur les différens accroissemens* (1760), 3.

42. Delisle, *Introduction à la géographie, . . . volume 1: Géographie* (1746). The publisher's "Eloge de Monsieur Delisle" describes both his teaching of royalty and his own apprenticeship with Cassini. On this also see Drapeyron, "L'éducation géographique" (1887), 241–56.

43. It is likely that the assistants he describes in Philippe Buache, "Observations sur l'étendue" (1741), 2: 449–52, were also his students. On his teaching of royalty see Drapeyron, "L'éducation géographique" (1887), 241–56.

44. César-François Cassini (Cassini de Thury) (III) described training these

"engineers" in *Description géométrique de la France*. . . . (1783), 17–18; at the end of the work Cassini reproduced his instructions to his engineers.

45. There are many possible ways of selecting the key eighteenth-century geographers. One method, choosing only those geographers mentioned by some of the most important men of letters, would harvest many of the same names. See Pedley, *Bel et Utile* (1992), 11 and note 1.

46. Bruzen de la Martinière, *Le Grand Dictionnaire* (1739), xj.

47. Robert de Vaugondy, *Mémoire sur les différens accroissemens* (1760), 27.

48. Bellin, *Description géographique de la Guyane* (1763), viij and 252.

49. Delisle, *Introduction à la géographie . . . volume 1: Géographie* (1746). Delisle is the author of this work, despite the views of the Michaud biographers (upon whom Numa Broc may also have relied). See Robert de Vaugondy, *Essai sur l'histoire de la géographie* (1755), 227–28.

50. Bellin, *Description géographique de la Guyane* (1763), ix.

51. For Robert de Vaugondy's many criticisms of d'Anville, see Robert de Vaugondy, *Essai sur l'histoire de la géographie* (1755), 121–24, 234–35; for his attempts to recognize d'Anville's contributions and to smooth his feathers, see 125. See also Marcel, "Correspondance de Michel Hennin" (1907), 446–47. On Buache's criticism of Robert de Vaugondy, see Pedley, *Bel et Utile* (1992), 76–78, 89–92.

52. Delisle, *Introduction à la géographie . . . volume 2: La Sphère* (1746), 9.

53. Delisle, "Lettre de M. Delisle au R. P.*" (1700), 243.

54. Bruzen de la Martinière, *Le Grand Dictionnaire* (1739), xj.

55. Delisle, "Seconde Lettre de M. De Lisle à M. Cassini" (1700), 224.

56. Robert de Vaugondy, *Essai sur l'histoire de la géographie* (1755), 155.

57. Robert de Vaugondy, *Essai sur l'histoire de la géographie* (1755), 230–31 slams copyists; Robert de Vaugondy, *Mémoire sur les différens accroissemens* (1760), 4, refuses to discuss poor-quality maps.

58. Bruzen de la Martinière, *Le Grand Dictionnaire* (1739), x.

59. Delisle's assurance is brought into relief by Nolin's expression of horror at this audacity. See Nolin, "Extrait d'une lettre de M. Nolin" (1700), 278.

60. Bellin, *Description géographique de la Guyane* (1763), last page of book.

61. Position held by Jean de la Grive. Pelletier, *La Carte de Cassini.* (1990), 86.

62. This is according to Lottin l'aîné, *Catalogue chronologique des libraires* (1789), 1: 180, 249, 267. The purview of the office included "Géographie, navigations et voyages." It may well not have existed prior to 1745 as the number of censors grew by more than 400 percent in the course of the eighteenth century (from 41 in 1727 to 178 in 1789) and their subject specialization was similarly refined. This is according to Herrman-Mascard, *La censure des livres* (1968). It is important to note that the position of royal censor did not carry a salary, although it was customary to award a censor a small annual salary after he had provided twenty years of service. On this, the administration of censorship in the eighteenth century, and the spirit guiding the censors, see Hanley, "The Policing of Thought" (1980), 265–95, and Goldgar, "The Absolutism of Taste" (1992), 87–110.

63. Works which were textual explanations of map compilation include d'Anville, *Notice de l'ancienne Gaule* (1760), iv, which was actually written prior to the

production of the map; Bellin's *Remarques sur la carte de l'Amérique* (1755); and Philippe Buache's *Explication et développement* (ca. 1754). The last two hundred pages of Robert de Vaugondy's *Essai sur l'histoire de la géographie* (1755) is a compilation memoir for his *Atlas universel* (1752). Another compilation memoir is Guillaume Delisle's *Nouvelles cartes des découvertes* (1753). Examples of guidebooks to cartographic data collection or representation are Dupain de Montesson's *L'Art de lever les plans* (1763), *La science de l'arpenteur* (1766), and *La science des ombres* (1750); and Bourcet's "Mémoires sur les reconnaissances militaires" (1875–1876).

64. De Dainville, *Le Langage des géographes* (1964), 55.

65. Pedley, *Bel et Utile* (1992), 61–64.

66. The distinction between "lever" and "faire" a map and its importance to eighteenth-century geographers was pointed out by de Dainville in *Le Langage des géographes* (1964), 48. On the methods employed by Cassini III and his surveyors, see Pelletier, *La Carte de Cassini* (1990), 100.

67. Pelletier, *La Carte de Cassini* (1990), 13.

68. Such as to be found on Bressani, "Novae Franciae Accurata Delineato" (1657); or, less strikingly, in Coronelli's "La Louisiana, parte settentrionale, scoperta sotto la protettionae di Luigi XIV, Ré di Franchia, etc. Descritta et dedicata dal P. Cosmografo Coronelli, All'Illustriss; et eccellentiss: S. Zaccaria Bernardi, sù dell'Ecc: S. Francesco." On Bressani's map, see Heidenreich and Dahl, "The French Mapping of North America" (1980), 2–11.

69. Nicolson, *Mountain Gloom and Mountain Glory* (1959); Broc, *Les Montagnes vues par les géographes* (1969); Roger, *Les sciences de la vie* (1963).

70. This latter definition was the norm in the seventeenth century and can still be seen on a map dating from as late as 1807: "Théâtre des opérations de l'armée du nord, et désert que le général Arnold traversa en marchant contre le Québec," which depicts the region liberally watered by the Chaudière, St. François, Kennebec, St. John, Allagash, and Penobscot rivers and suffering from anything but a shortage of water.

71. Such as in Villaret's "Carte géométrique du Haut-Dauphiné et du Comté de Nice," 9 sheets, 1: 86,000 (1749–1763). See Pelletier, *La Carte de Cassini* (1990), 86–87.

72. Bacler d'Albe, "Carte générale du Théâtre de la guerre en Italie" (1798).

73. Buache, "Cartes des nouvelles découvertes" (1752).

74. On, for example, his November 1755 "Canada, Louisiane et terre angloises."

75. These neologisms, for the most part identified by François de Dainville, are a small portion of the new geographic terms born in the late eighteenth and early nineteenth centuries.

76. Rizzi-Zannoni, "Carte de la Pologne" (1772).

77. Levallois, *Mesurer la terre* (1988), 23.

78. Calculations here are based on my reading of Smith, *From Plane to Spheroid* (1986), 78–83 and 92–94.

79. Subsequent nineteenth-century remeasurements and recalculations suggest that Maupertuis's underestimation of the degree was due to errors in astronomical observations, error due to the plumb line, error in the length of the meridian, and error from refraction. See Smith, *From Plane to Spheroid* (1986), 168–92.

80. *Journal des savants* (1697), 68–69; the review in the *Journal des savants* was of a book by Le Père Louis le Comte de la compagnie de Jesus, Mathématicien du Roy, *Nouveaux Mémoires sur l'état present de la Chine* (Paris: Chez Iean Anisson, 1696). The reviewer had criticized the Jesuit author for suggesting that latitude and longitude could be determined by two means: astronomically and by using old maps, rutters (route charts or lists), and accounts. On this see also de Dainville, *La Géographie des humanistes* (1940), 458–59.

81. Pedley, *Bel et Utile* (1992), 110–12.

82. Brunet, *L'Introduction des théories de Newton* ([1931] 1970), and Healy, "Mechanistic Science and the French Jesuits" (1956), 57, 65–67, 92, 106–9, 178–80.

83. D'Anville, *Proposition d'une mesure* (1735), xviij. Italics are mine.

84. Robert de Vaugondy, *Essai sur l'histoire de la géographie* (1755), 123.

85. Robert de Vaugondy, *Mémoire sur une question de géographie pratique* (1775), 37. The committee charged with ruling on the merit of Robert de Vaugondy's statement, of which d'Anville was a member, ruled in Robert de Vaugondy's favour. They did so, however, without admitting that there was anything misleading when geographers such as Rizzi-Zannoni claimed to be able (indeed to be obliged) to depict the flattening of the earth on their geographic maps. The ruling by de Lalande, Bailly and d'Anville read as follows: "These are the reflections against the use of this element expressed by M. Robert de Vaugondy who is jealous of public approbation and above all concerned with the perfection of his geographic maps. We cannot, however, blame geographers who propose to use it. We could not exclude the rigourous precision toward which the Academy tends ceaselessly in its work, but as, in this case, this precision is more metaphysical than practical, and as it may be destroyed by the inevitable error in observations, we think that in applauding the efforts of geographers who try to take into account the flattening of the earth, the Academy can still consider those maps which neglect this flattening to be good. We also think the Academy can give its approbation to M. de Vaugondy's reflections." This very tortuous form of expression comes from the original, not the translator (p. viij).

86. Varenius was truly exceptional in his critical approach to, for example, questions concerning the origin of mountains, the causes of the saltiness of the sea, and the composition of the interior of the earth. Indeed, he was exceptional enough to attract the attention and accolade of Sir Isaac Newton. Varenius, *A Complete System of General Geography: Explaining the Nature and Properties of the Earth* ("The Third edition, with large additions"). This addition was improved by Isaac Newton and Dr. Jurin (London, 1736). The first edition was published in 1664 by Les Elzevier—in Latin. The second edition (1681) was the first annotated by Newton. It is also in Latin.

87. These quasi-explanatory maps include: Abbé Louis de Courcillon de Dangeau's "Carte des Universités de France" (1697); an anonymous "Carte générale des Sévennes," ms at 1:54,000, which provided both a qualitative and a quantitative description of population; Jean Astruc's "Réprésentation des divers vents du Languedoc" (1737); Hensel, "Synopsis Universae Philologiae" (1741), which depicted the languages of the world; Scherer, "Representatio totius Africae" (1703, 1710, and 1730), which depicted the extent of the true faith shown by shade and

light; Guettard and Dupain-Triel's "Cartes minéralogiques de la France" (1780–1784); and Leroy, Tesier, and Buache's "Carte relative à l'orage du 13 juillet 1788" (1789). The most deliberately explanatory of these maps was Astruc's 1737 wind map. This does not exhaust the list of quasi-explanatory maps. For example, Didier Robert de Vaugondy produced three ethnographic maps in his school atlas, the *Nouvel Atlas portatif* (1762). These were world maps showing geographic distribution by religion, skin color, and facial type. Robert de Vaugondy's comments on the map do, however, reveal the extent to which explanation was off-limits: "It is an enigma difficult or perhaps impossible to resolve, if only to say that such has been the will of the Supreme Author, of which we only see effects, without being able (nor should we dare) to penetrate its depths." Robert de Vaugondy, *Institutions géographiques* (1766), 243.

CHAPTER TWO

1. This takes the form of frequent references to the parlous state of the field and the disrepute into which it has fallen. The sense of this decline is strongly expressed in the writings of Malte-Brun, as will be clear in chapter 3. But many other geographers expressed similar concerns. Indeed, explanations for the prevailing "disgust for this science" abound, and it was popularly attributed to the poor way in which geography had been taught. See in particular Barbié du Bocage, "Proposition d'une nouvelle méthode" (1795), 483–91. For echoes of this sense among British geographers of this period, see Bowen, *Empiricism and Geographical Thought* (1981). A strong sense of the intellectual flaccidity of geography in early-nineteenth-century France is also clear in Staum's discussion of the geography practiced in the Class of Moral and Political Studies of the National Institute. See Staum, "Human Geography in the French Institute" (1987), 332–40.

2. This is part of what Eric Brian refers to as a much more common and generalized exchange between administrators and academicians at the end of the Ancien Régime: "This was the down-beat in an exchange in which the latter [scholars] provided the former [administrators] with the instruments useful to government in exchange for the institutionalisation of some of their activities." Brian, *La mesure de l'état* (1994).

3. For a radically different, and I would argue erroneous, reading of the geography course at the École, see Sergio Moravia, "Philosophie et géographie" (1967), 949–53.

4. The information on the École normale comes principally from Fayet, *La Révolution française* (1960), 329–54; and Nordman, *Leçons d'histoire, de géographie* (1994); when it is not directly derived from the courses themselves.

5. Among the instructors were Lagrange, Monge, Haüy, Berthollet, Thouin, Volney, Sicard, Bernardin de Saint-Pierre, and Garat.

6. See in particular Nordman, "Buache de La Neuville" (1984), 105–10; Lagarde, "Le passage du Nord-Ouest" (1989), 19–43; Gaziello, *L'expédition de Lapérouse* (1984), especially 22 and 83; Drapeyron, "Projet d'établissement" (1887), 241–55; and Nordman, *Leçons d'histoire, de géographie* (1994), 135–41.

7. Pastoureau, "Histoire de la bibliothèque nationale" (1989), 62–69.

8. Malte-Brun, ed., *Annales des Voyages* (1810), 5: 209.

9. Bory de Saint-Vincent, *Essais sur les Isles Fortunées* (1803), 443.

10. Archives nationales, F17 1541, Paris 12 prairial an XII, Letter to the Minister of the Interior.

11. Nordman, *Leçons d'histoire, de géographie* (1994), 142.

12. *Séances des Écoles normales,* vol. 1 [1802].

13. Malte-Brun, *Précis de la géographie universelle* (1810–1829).

14. *Séances des Écoles normales,* 1: 65.

15. "Débats," *Séances des Écoles normales* (n.d.), 183.

16. *Séances des Écoles normales,* 1: 69.

17. *Séances des Écoles normales,* 1: 67.

18. "Débats," *Séances des Écoles normales,* 398.

19. *Séances des Écoles normales,* 2: 74.

20. *Séances des Écoles normales,* 2: 422.

21. "Débats," *Séances des Écoles normales,* 74.

22. "Débats," *Séances des Écoles normales,* 88.

23. *Séances des Écoles normales,* 1: 69.

24. *Séances des Écoles normales,* 1: 60–70.

25. *Séances des Écoles normales,* 1: 67.

26. *Séances des Écoles normales,* 2: 5, 279–308.

27. "Débats," *Séances des Écoles normales,* 165–75, 183–85, 288.

28. *La Décade philosophique, littéraire et politique* 30 (30 pluviôse, an III): 351–53.

29. *La Décade philosophique, littéraire et politique* 30 (30 pluviôse, an III): 352.

30. Nordman, *Leçons d'histoire, de géographie* (1994), 6.

31. A. Challe, "Lettres de Joseph Fourier" (1958), 105–34.

32. "Débats," *Séances des Écoles normales,* 439.

33. For a balanced approach to Cassini IV see Konvitz, *Cartography in France* (1987), 26–31. Gillespie considers both Cassini's commitment to science and his talent limited. Gillespie, *Science and Polity* (1980), 90–91, 95, and 125.

34. In this vein see Outram's excellent "The Ordeal of Vocation" (1983), 251–73.

35. The school, incidentally, which Delambre attended. See Cohen, "Jean-Joseph Delambre" (1971), 4: 14–18.

36. He did see his work as benefiting geography. He wrote about his efforts to reform the Observatory in 1784: "I also planned to establish a school of practical astronomy at the Observatory where mariners and those proposing to undertake distant voyages could come and train themselves in astronomical observations. Geography would have derived great benefit from this." Jacques-Dominique Cassini (IV), *Mémoires pour servir à l'histoire des sciences* (1810), 9.

37. According to the preface, "Éloge de Monsieur Delisle," in Guillaume Delisle's *Introduction à la géographie, . . . volume 1: Géographie* (1746), xxxj: "M. Delisle entered the Academy in 1700. A student of M. Cassini's, although he never engaged in observations, nor did he wish to. However, it was felt that the use that he knew how to make of observations could compensate for those that he did not make. As in the plan of the Academy there was no place for geography, they let him take one which, apparently, would return to an astronomer, in the absence of a geographer like him."

38. César-François Cassini de Thury (Cassini III), *Description géométrique de la France* (1783).

39. Arguably, though, there was less distance between Cartesian theoretical astronomy and geography in the seventeenth century.

40. Jacques-Dominique Cassini (IV), file D 5 33 entitled "Magnetisme, temperatures des caves, etc.: Projet d'un levé géométrique de la Carte de Toscane, Avant propos," Cassini IV Manuscripts, Observatoire de Paris.

41. Jacques-Dominique Cassini (IV), "Lettre à Mr. Sevatti," no date, attached to the "Projet d'un levé géométrique de la Carte de Toscane," Cassini IV Manuscripts, file D 5 33.

42. Jacques-Dominique Cassini (IV), "Projet d'une Carte Générale de La Toscane," Cassini IV Manuscripts, file D 5 33 (section on the primary triangulation).

43. Jacques-Dominique Cassini (IV), *Mémoires pour servir à l'histoire des sciences* (1810), 4–5.

44. There is recurring reference to this lineage expressed in exactly that manner in Cassini IV's writings.

45. Chapin, "The Academy of Sciences" (1967), 371–404.

46. Jacques-Dominique Cassini (IV), *Mémoires pour servir à l'histoire des sciences* (1810), 116–17.

47. Jacques-Dominique Cassini (IV), "A Madame de Cambry" (1842), no. 2, Lettres en vers et prose, 39–40.

48. Jacques-Dominique Cassini (IV), "Atlas céleste et catalogue" (ca. 1784).

49. Cassini de Thury, *Description géométrique de la France* (1783), 193.

50. Jacques-Dominique Cassini (IV), "Troisième Mémoire: Projet et Description d'un Nouvel Observatoire. Exposé des principes qui doivent diriger les architectes dans la construction et la distribution des édifices destinés aux observations," in *Mémoires pour servir à l'histoire des sciences* (1810), 63.

51. On the longstanding importance of Cassini's work for the Observatory, see Fayet, *La Révolution française* (1960), 139–41 and 409–10.

52. This is discussed at length in Gillespie, *Science and Polity* (1980), 118–23.

53. Jacques-Dominique Cassini (IV), *Extrait des observations astronomiques* (1786), 1.

54. Cassini III wrote concerning geographic engraving: "All art that does not assure artists recompense proportional to their talent will be forever neglected." Cassini de Thury, *Description géométrique de la France* (1783).

55. Jacques-Dominique Cassini (IV), "Objets économiques" (n.d.).

56. Devic, *Histoire de la vie* (1851), 263.

57. Jacques-Dominique Cassini (IV), *Collection de riens qui vaillent* (1842), 65–66.

58. Jacques-Dominique Cassini (IV), manuscript rough copy of a letter sent probably to Mr. Sevatti which asks for a clear commitment from the Grand Duke of Tuscany, filed with the "Projet d'un levé géométrique de la Carte de Toscane" in the Observatoire de Paris, Cassini IV Manuscripts, file D 5 33.

59. Jacques-Dominique Cassini (IV), *Lettre de Jean-Dominique Cassini à ses co-associés* (1840).

60. Jacques-Dominique Cassini (IV), *Mémoires pour servir à l'histoire des sciences* (1810), 135.

61. On the Corsica cadastral survey see Albitreccia, *Le plan terrier de la Corse* (1942); Huguenin, "French Cartography of Corsica" (1970), 123–37; Konvitz, *Cartography in France* (1987), 42 n. 29; Huguenin, "La cartographie ancienne de la Corse" (July 1962 and July 1963), 23: 85–98 and 26: 33–55, respectively; and Kain and Baigent, "The Cadastral Map" (1992), 221–24.

62. On Jacotin's career and activities in Egypt, see Godlewska, *The Napoleonic Survey of Egypt* (1988).

63. Jacotin, "Mémoire sur la carte de l'Égypte" (1822), *État moderne* 2: 1–118.

64. Jacotin's comments in this paragraph are from a report dated 20 June 1818 in the Archives de l'ancien Dépôt de la guerre, Service historique de l'armée de terre, Vincennes. Cited in Berthaut, *La Carte de France* (1898), 1: 64–65.

65. Berthaut, *La Carte de France* (1898), 1: 65.

66. The confiscation was also quite typical for that period and ran in the same direction as Napoleon's militarization of society. Napoleon, above all, could not see large-scale cartography as a secular civilian activity. Its military potential was simply too great. Consequently, this was not the only survey to undergo confiscation and militarization. Exactly the same thing happened to the survey of Egypt, which originally had been associated with the activities of the Commission of Arts and Sciences but which also was taken over by the Dépôt de la guerre, in spite of the vociferous and insistent complaints of the editor of the *Description de l'Égypte,* Edme-François Jomard.

67. Delambre, *Histoire de l'astronomie* (1827), 308. See also Devic, *Histoire de la vie* (1851), 378. According to Fayet, Lakanal was prepared to reintroduce Cassini to the Observatory in 1795. See Fayet, *La Révolution française* (1960), 151, 410.

68. Jacques-Dominique Cassini (IV), *Mémoires pour servir à l'histoire des sciences* (1810), 98–99.

69. Jacques-Dominique Cassini (IV), "Dialogue entre deux détenus" (1842), no. 3, Mélanges, 12.

70. Jacques-Dominique Cassini (IV), "Dialogue entre deux détenus" (1842), 15.

71. Jacques-Dominique Cassini (IV), "Dialogue entre deux détenus" (1842), 15–16.

72. Jacques-Dominique Cassini (IV), *Quelques idées sur l'instruction publique* (1802), 4–5.

73. Jacques-Dominique Cassini (IV), *Quelques idées sur l'instruction publique* (1802), 11–12.

74. Devic, *Histoire de la vie* (1851), 402.

75. Drapeyron, *Les deux Buaches* (1888). See also Gille, *Les Sources statistique* ([1964] 1980), 29; and Bourguet, *Déchiffrer la France* (1989), 23–30.

76. In fairness to Cassini and as Jules Simon points out, the functionary status of the Institut was far clearer and far more onerous than it had been for the Académie des sciences. Simon, *Une académie sous le Directoire* (1885), 101–13.

77. Jacques-Dominique Cassini (IV), "Sur le choix d'un État" (1842), 73.

78. Jacques-Dominique Cassini (IV), "Au secrétaire de la classe des sciences" (1842), no. 3, Mélanges, 77–80.

79. Jacques-Dominique Cassini (IV), *Manuel de l'étranger qui voyage en Italie* (1778), 21.

80. Jacques-Dominique Cassini (IV), "Mon Apologie" (1842), no. 3, Mélanges, 75.

81. Delambre, *Histoire de l'astronomie* (1827), 253–54; see also Cassini IV's response to Delambre's version of the history of eighteenth-century astronomy: Cassini IV, *Réflexions présentées aux éditeurs* (n.d.).

CHAPTER THREE

1. "In spite of the number of those who have undertaken this before us, if we in turn undertake to deal with the same subjects, we cannot be blamed, unless it is proven that our developments bring nothing new." Strabo [Strabon], *Géographie* (1969), 1: 85.

2. Frank Lestringant has written extensively on the cosmographical tradition in the Renaissance. He considers that toward the end of the Renaissance the genre of the cosmography was in decline. His argument is that with the great "discoveries" and the growth of knowledge they represented for European thought, cosmographies encountered three principal intellectual challenges to which they could not respond. These were (1) how to preserve a holism in the face of rapidly growing and highly diverse knowledge, (2) a growing distance between theoretical (cabinet) and practical (navigational) geographic knowledge, and (3) an inability to choose between and reconcile the authority of seeing for oneself and textual authority (generally from ancient texts) in a manner fitting to the scale and coherence of a cosmography. I agree that the first and third dilemmas were highly problematic for the logic of the cosmography or universal geography. However, it is important to note that geographers continued to write universal geographies or cosmographies and that these remained a genre closely attached to geography into the twentieth century. See especially Lestringant, "Le Déclin d'un savoir" (1993), 319–40.

3. Proisy d'Eppe, *Dictionnaire des girouettes* (1815), 256–73. This work was written to stigmatize the relatively powerful and prominent who since 1789 had shifted their political allegiances. Each name was followed by a series of their contradictory pronouncements. The book was enormously popular.

4. The *Journal de l'empire*, known as the *Journal des débats* from 1789 to 1805, was the most read newspaper in Paris of the late eighteenth and early nineteenth centuries. It was an anti-Revolutionary paper which attacked the spirit and writings of the Enlightenment. After 1805, in spite of its open support of Napoleon, it suffered increasing censorship. The paper was finally taken over by the state in 1811. See André Cabanis, "Journal des débats" (1987), 981–83; and Alfred, *Histoire politique* (1938).

5. *Journal de l'empire* (22 February 1807), 3.

6. Conrad Malte-Brun, review of *Campagne des armées françaises en Saxe* (1807), 3–4.

7. Conrad Malte-Brun, ed., *Annales des voyages* (1807–), 1: 122.

8. Conrad Malte-Brun, *Précis de la géographie universelle* (1810), 1: 5.

9. Six volumes were written before his death in 1826. Two volumes were written and published posthumously.

10. Drohojowska, "Malte-Brun. *Précis de la géographie universelle*" (1884), 18–19. Vivien de Saint Martin, in his table of contents, describes Malte-Brun as on a par with Ritter in his importance to descriptive geography. Louis Vivien de Saint-Martin, *Histoire de la géographie* (1873). Adrien Balbi played a particularly important role in establishing and building Malte-Brun's reputation. His works are full of laudatory references to Malte-Brun.

11. Broc, "Un Bicentenaire: Malte-Brun" (1975), 720.

12. See the article by Conrad Malte-Brun reviewing the new "Géographie de Strabon, traduite de Grec en français," *Journal de l'empire* (1 November 1807): 3–4; Malte-Brun, *Annales des voyages* (1807), 1: 120; Malte-Brun, *Précis de la géographie universelle* (1810), book 5; Malte-Brun, *Précis de la géographie universelle* (1829), 8: 155.

13. Archeologists have found Homer to be an excellent source of geographic and historical information but as Pierre Lévêque put it: "The Iliad and the Odyssey are far too composite in nature for the historian to undertake their study unwarily." See Pierre Lévêque, *L'Aventure grecque* (1964), 47.

14. "If political philosophy is essentially directed at those who govern and if geography also responds to the needs of government, that must be an advantage for geography. That advantage lies in the realm of action." Strabo, *Géographie* (1969), 1: 80.

15. Strabo, *Géographie* (1969), 1: 77–78.

16. Strabo, *Géographie* (1969), 1: 171.

17. "But this is even richer! He claims that he cannot see any practical value in the study of borders." Strabo, *Géographie* (1969), 1: 172.

18. Strabo, *Géographie* (1969), 2: 37–38.

19. Strabo, *Géographie* (1969), 1: 78.

20. Strabo draws a line between the cause-seeking Aristotelians and the clarity-seeking Stoics. Strabo, *Géographie* (1969), 2: 69.

21. Strabo, *Géographie* (1969), 1: 84.

22. Strabo, *Géographie* (1969), 1: 144.

23. Strabo, *Géographie* (1969), 2: 118.

24. Strabo, *Géographie* (1969), 1: 83

25. Thus, in his discussion of Hipparchus's contention that the ocean surrounding the inhabited world was not a single entity, Strabo wrote: *"For the concerns at hand,* all that we need to say is that, in terms of the uniformity of the regime, *it is better to believe* that the more there is a liquid element spread around the world, the more the stars will be solidly held by the vapors emanating from it." Strabo, *Géographie* (1969), 1: 72. Italics are mine.

26. Strabo, *Géographie* (1969), 2: 68.

27. Strabo, *Géographie* (1969), 1: 84 and 153.

28. Strabo, *Géographie* (1969), 1: 95.

29. Strabo, *Géographie* (1969), 1: 107–8.

30. Strabo, *Géographie* (1969), 1: 88.

31. Strabo, *Géographie* (1969), 1: 92.
32. Strabo, *Géographie* (1969), 2: 44, 50.
33. Strabo, *Géographie* (1969), 1: 65.
34. Strabo, *Géographie* (1969), 1: 63–65.
35. Strabo, *Géographie* (1969), 2: 91–92.
36. Strabo, *Géographie* (1969), 2: 92
37. Strabo, *Géographie* (1969), 1: 86–87, 141.
38. Strabo, *Géographie* (1969), 2: 53.
39. Varenius's *A Complete System of General Geography* (London, 1736) was probably one of the most respected geographies of the seventeenth and eighteenth centuries. It would be fair to say that this is almost certainly due to the respect and attention it received from Sir Isaac Newton. But why was Newton so impressed by this work? Was it because Varenius's picture of the solar system was Copernican rather than Ptolemaic? It is clear from the translator's comments in the third edition that the first and second edition of *A Complete System* was insufficiently Copernican for Newton: "In the *Astronomical Part* we have strengthened our Author's Arguments in Favour of the *Copernican* Hypothesis. . . . In the *Philosophical and Physical Part,* we have rejected the improbable Conjectures of the Antients, and the unwarrantable Suppositions of *Des Cartes,* which our Author seems to be fond of: Instead whereof, we have (with the learned Dr jurin) introduced the *Newtonian Philosophy* to solve the *Phaenomena,* as being much more eligible than the *Cartesian,* for the *Agreeable* and *Geometrical* Manner of its Conclusions." Thus, the appeal of the work for Newton must have resided less in its scientific quality (although, as I have already suggested, Varenius had a greater tendency to explanation than the norm), than in its ready accessibility or popularity and Varenius's primary preoccupation with physical phenomena. This would seem to be supported by another of the translator's comments: "The Reason why this great Man took so much Care in Correcting and Publishing our Author, was, because he thought him necessary to be read by his Audience, *the Young gentlemen of Cambridge,* while he was delivering Lectures upon the *same Subject* from the *Lucasian Chair.*" Young men in Cambridge would need to know something about the physical phenomena in the world. A corrected Varenius would provide an excellent introduction and guide.
40. Mentelle, *Elémens de géographie* (1783), vij–viij.
41. Delisle, *Introduction à la géographie . . . volume 1: Géographie* (1746), 1.
42. Mentelle, *Géographie comparée* (1778), 43.
43. Expilly, *La Polychrographie en six parties* (1756), 39.
44. Mentelle, *Géographie comparée* (1778), 34–35.
45. François, *La Science de la géographie* (1652), 1–3.
46. Delisle, *Introduction à la géographie* (1746), 1: xvj–xvij.
47. Delisle, *Introduction à la géographie* (1746), 1: vij.
48. Delisle, *Introduction à la géographie* (1746), 1: 11.
49. Expilly, *La Polychrographie* (1756), préface.
50. Mentelle, *Géographie comparée* (1778), 51.
51. Varenius, *A Complete System of General Geography* (1736), 10.
52. Delisle, *Introduction à la géographie* (1746), 1: 272.
53. Expilly, *La Polychrographie* (1756), préface.

54. Mentelle, *Cosmographie élémentaire* (1781), vij—viij.

55. Delisle, *Introduction à la géographie* (1746), 1: 7.

56. Delisle, *Introduction à la géographie* (1746), 1: 330.

57. Expilly, *La Polychrographie* (1756), 1.

58. Mentelle, *Géographie comparée* (1778), 38–40.

59. Mentelle, *Géographie comparée* (1778), 38–40.

60. Delisle, *Introduction à la géographie* (1746), 1: 22.

61. Mentelle, *Géographie comparée* (1778), 21–22.

62. Mentelle, *Géographie comparée* (1778), 43–44.

63. Malte-Brun, *Précis de la géographie universelle* (1810), 1: 2.

64. Malte-Brun, *Précis de la géographie universelle* (1810), 1: 525.

65. Malte-Brun, *Précis de la géographie universelle* (1817), 5: 415, for example.

66. Malte-Brun, *Précis de la géographie universelle* (1810), 1: 13.

67. Malte-Brun, *Précis de la géographie universelle* (1810), 2: 88.

68. Perhaps thanks to his heavy use of de Lalande's *Abrégé d'astronomie*, which, as Lalande disliked the Cassinis, would not have instilled great respect for them in de Malte-Brun.

69. Malte-Brun, *Précis de la géographie universelle* (1810), 1: 525.

70. This very interesting observation supports my discussion of the magnitude of the shift from a topographic to a thematic map in both my Introduction and in chapter 7. Malte-Brun, *Précis de la géographie universelle* (1810), 2: 134.

71. Malte-Brun, *Précis de la géographie universelle* (1810), 1: 8–10.

72. Malte-Brun, *Précis de la géographie universelle* (1810), 1: 6.

73. Malte-Brun, *Précis de la géographie universelle* (1810), 1: 523–24.

74. Malte-Brun, *Précis de la géographie universelle* (1810), 2: 474.

75. Malte-Brun, *Précis de la géographie universelle* (1810), 2: 495.

76. Malte-Brun, *Précis de la géographie universelle* (1810), 2: 159.

77. Strictly speaking, then, general principles could not precede facts.

78. Malte-Brun, *Précis de la géographie universelle* (1810), 2: 506.

79. Malte-Brun, *Précis de la géographie universelle* (1810), 2: 159.

80. Malte-Brun, *Précis de la géographie universelle* (1810), 2: 575.

81. Malte-Brun, *Précis de la géographie universelle* (1810), 2: 592.

82. Malte-Brun, *Précis de la géographie universelle* (1810), 2: 606.

83. "After having considered the general causes of winds, and those which modify the effects of these, let us now follow the traces of those which by their regularity and their generality are of greater interest to geography." Malte-Brun, *Précis de la géographie universelle* (1810), 2: 392.

84. Malte-Brun, *Précis de la géographie universelle* (1810), 2: 356.

85. Malte-Brun, *Précis de la géographie universelle* (1810), 2: 191.

86. Malte-Brun, *Précis de la géographie universelle* (1810), 2: 475.

87. Pinkerton and Malte-Brun were contemporaries and rivals while Mentelle's work was of an earlier generation and type.

88. "Avis de l'éditeur," in Balbi, *Abrégé de géographie* (1833).

89. "Plan de l'Abrégé," in Balbi, *Abrégé de géographie* (1833).

90. Particularly in its characterization of d'Anville as having no significant

predecessor, no model, and as creating modern scientific geography single-handedly. Vivien de Saint-Martin, *Histoire de la géographie* (1873), 426.

91. There is some confusion about whether Vivien de Saint-Martin was referring here to Malte-Brun or to Humboldt. This section of Vivien de Saint-Martin's conclusion is described in the extensive summary table of contents as a discussion of the contributions of Malte-Brun and Ritter. The text, however, refers to the contributions of Humboldt and Ritter. In each instance the name of Malte-Brun or of Humboldt, respectively, is mentioned only once. The emphasis in the text on descriptive rather than scientific geography and the description of the mystery person's contribution to overcoming an arid and fastidious nomenclature (precisely the phraseology used by Malte-Brun) would, however, suggest that it is to Malte-Brun that this text applies. Vivien de Saint-Martin, *Histoire de la géographie* (1873), 584.

92. Malte-Brun put a fine point on it by differentiating the work and life of a traveler-explorer and the work of the geographer on two separate occasions. On pages 13–14 of his *Société de géographie* (1823), he asked the Society what they would rather fund. Would it be, on the one hand, "one of those eternal shipwrecks at 'Cap Blanc,'" or, on the other, a scholarly picture of the diverse nature of the deserts which paints for you here, the Sahara, so similar to the basin of dried-out lakes, and there, the green savannas naked of large vegetation, and which brings all the phenomena of those vast portions of our world under general laws. Which of these two works should be published? But it is useless to discuss a question determined by our regulation. That fundamental law says that we will publish maps, and what would a map be without a scholarly analysis of the material of which it is composed?" Similarly, in his *Précis de la géographie universelle* (1810), 1: 525, he commented about the role and task of the geographer: "but how much we would prefer to be far from the tumult of the factions that divide the republic of letters, following, through a thousand perils, the glorious route drawn by the Columbuses and the Humboldts! How we envy you, you who, compass and telescope and even weapons in hand, go to accomplish the discovery of our world! It is for you that in the midst of her mysterious Alps, Central Asia hides its ancient treasures of knowledge, necessary to the completion of the history of our species. . . . For us, as iniquitous destiny has denied us a share of this work, we must look for consolation in the irksome task of describing those parts of the world that are already known." However irksome, the truly indispensable and intellectual role was that of the geographer. Referring again to Humboldt's travels in South America, Malte-Brun made the point that others might have carried out his research there just as profitably: "An illustrious member of our society visited the greater part [of South America] with the most religious care, with the rarest talent. But this scholarly traveler, didn't he himself tell us that within the cities of Mexico and Caracas there were very educated people very capable of describing their native country?" Malte-Brun, *Société de géographie* (1823), 9.

93. Green, *Chasing the Sun* (1996), 18–19; Meschonnic, *Des mots et des mondes* (1991), 147–50, 236–43; Kafker, "The Influence of the Encyclopédie" (1994), 389–99.

94. "Géographie voyages—Dialogue entre Messieurs S . . . et D. . . . ., sur le

Sieur Malte-Brun, auteur du *Précis de la géographie universelle,*" which took place 2 December 1812 in an assembly of some sort. Bibliothèque nationale, N. A. Fr. 22186.

95. Humboldt discussed the limitation of the approach typical of universal and regional geographies to earth description in his *Voyage aux régions équinoxiales* (1816), 21–23.

96. Humboldt, *Cosmos* (1849), 1: x.

97. See Hartshorne's *The Nature of Geography* (1939) and *Perspective on the Nature of Geography* (1959).

98. Knight, "Romanticism and the Sciences" (1990), 13–24; Morgan, "Schelling and the Origins" (1990), 25–37; Engelhardt, "Historical Consciousness" (1990), 55–68; Rehbock, "Transcendental Anatomy" (1990), 144–60; Nicholson, "Alexander von Humboldt" (1990), 169–85; Rupke, "Caves, Fossils" (1990), 241–59; Snelders, "Romanticism and Naturphilosophie" (1970), 193–215; Albury and Oldroyd, "From Renaissance Mineral Studies" (1977), 187–215; Ospovat, "Romanticism and German Geology" (1982), 105–17; Dagonet, "Valentin Haüy" (1972), 327–36; Lenoir, "The Göttingen School" (1981), 111–205; Baumgärtel, "Alexander von Humboldt" (1969), 19–35; Bruhns, *The Life of Alexander von Humboldt* (1873); Stevens, *The Humboldt Library* (1863).

99. Dettelbach, "Humboldtian Science" (1996), 287–304.

100. Mary Somerville took some inspiration from Humboldt, as is clear from the introduction and text of her *Physical Geography* (1867) and from the text of her *On the Connection of the Physical Sciences* (1846). Thomas Kuhn points out the significance of Mary Somerville's comments in her Preface to *On the Connection of the Physical Sciences* to the development of the theory of the conservation of energy. Kuhn, "Energy Conservation" (1977), 66–104. On Mary Somerville, see Patterson, "Mary Somerville" (1969), 309–39; Patterson, "Somerville" (1970), 521–25; Sanderson, "Mary Somerville" (1974), 410–20; and Livingstone, *The Geographical Tradition* (1992), 172–74.

101. De Dainville, *La Géographie des humanistes* (1940), 213, 218–19, 276–302.

102. François, *La Science de la géographie* (1652), 202–3.

103. François, *La Science de la géographie* (1652), 189.

104. François, *La Science de la géographie* (1652), 1–3, 5.

105. François, *La Science de la géographie* (1652), "Avis au lecteur."

106. François, *La Science de la géographie* (1652), 1–3.

107. François, *La Science de la géographie* (1652), 424–25.

108. François, *La Science de la géographie* (1652), 345–46.

109. François, *La Science de la géographie* (1652), 11.

110. This is a somewhat simplified version of his argument as François also recognized that heat emanated from the earth itself.

111. The last of these is perhaps less evident from this quotation, but it is clear from his discussion of this on pages 10–11 and in his section devoted to the importance of place and time on pages 87–89 that he included human communication and commerce within this category.

112. François, *La Science de la géographie* (1652), 24.

113. François, *La Science de la géographie* (1652), 85.

114. For an interesting explanation of the prevailing contemporary concern to enumerate, see Cormack, *Charting an Empire* (1997), 183 and 185: "[C]horographers, like lawyers in political philosophy, saw themselves empowered with certain privileges over nature by right of their taxonomic knowledge."

115. All previous quotations in this paragraph are in François, *La Science de la géographie* (1652), 1–3.

116. François, *La Science de la géographie* (1652), 1–3 and 22–23.

117. François, *La Science de la géographie* (1652), 142.

118. François, *La Science de la géographie* (1652), 142.

119. François, *La Science de la géographie* (1652), "Avis au lecteur," 1–3 and 31.

120. François, *La Science de la géographie* (1652), 231–32.

121. François, *La Science de la géographie* (1652), 33.

122. François, *La Science de la géographie* (1652), 35.

123. Nicholson, "Alexander von Humboldt" (1990), 170.

124. Humboldt saw this as in keeping with the aims of all of the experimental sciences. Humboldt, *Cosmos* (1849), 1: 29–30.

125. Humboldt, *Cosmos* (1849), 1: 360.

126. Humboldt, *Cosmos* (1849), 1: 36.

127. Humboldt, *Cosmos* (1849), 1: 36–37.

128. The first two volumes of the *Cosmos* can be considered a single essay as the pages are numbered consecutively and continuously.

129. Humboldt, *Cosmos* (1849), 1: 39.

130. On this, see especially Shaffer, "Romantic Philosophy" (1990), 38–54.

131. Humboldt, *Cosmos* (1849), 1: 42 and 2: 467.

132. Thus, he considered that the study of the character of rocks fell within the purview of geology while the study of the cosmos had greater affinity with geognosy (the science of the texture and succession of terrestrial strata) and physical geography (the science of geographical forms and outlines).

133. Humboldt, *Cosmos* (1849), 1: 28, 48–49.

134. Humboldt, *Cosmos* (1849), 2: 467.

135. Humboldt, *Cosmos* (1849), 2: 438.

136. Humboldt, *Cosmos* (1849), 1: 17.

137. Humboldt, *Cosmos* (1849), 1: 17.

138. Humboldt, *Cosmos* (1849), 1: 147.

139. Humboldt, *Cosmos* (1849), 1: 368.

140. Humboldt, *Cosmos* (1849), Preface.

141. Humboldt, *Cosmos* (1849), 1: 5.

142. Humboldt, *Cosmos* (1849), 1: 6.

143. Humboldt, *Cosmos* (1849), 1: 55

144. Humboldt, *Cosmos* (1849), 2: 435.

145. Humboldt, *Cosmos* (1849), 1: Preface.

146. Humboldt, *Cosmos* (1849), 1: 6.

147. Humboldt, *Cosmos* (1849), 1: 22.

148. Humboldt, *Cosmos* (1849), 1: 30.

149. Humboldt, *Cosmos* (1849), 2: 474.

150. This is not to suggest that Humboldt was unpolitical. For a discussion of

one of the political dimensions of Humboldt's life and work see Pratt, *Imperial Eyes* (1992).

CHAPTER FOUR

1. This was a trend that perhaps began with the printing press and has now culminated in our map-saturated modern society, in which maps of an extraordinary variety of types and qualities are to be found everywhere from restaurant placemats to hiking trails to logos to software packages. See Wood (with Fels), *The Power of Maps* (1992), 34–38.

2. For a full development of this argument see Harley, "The Map and the Development" (1987), 5–23.

3. *Bulletin de la Société de géographie de Paris* 1 (1822), 1–9.

4. La Renaudière, "Notice annuelle des travaux de la Société" (1827), 8: 320–21.

5. Bruguière, "Orographie de l'Europe" (1830), 3: 512–14.

6. Fierro-Domenech argues that the engineers formed a significant proportion of the Paris Geographical Society's membership in the 1820s, that they increased in actual number in the 1830s, but that sometime before the 1850s their membership declined radically because "the Society had nothing to offer them that appeared useful to their profession." *La Société de géographie* (1983), 23–24, 29, and 45.

7. La Renaudière, "Notice annuelle des travaux de la Société" (1827), 8: 326.

8. Fierro-Domenech, *La Société de géographie* (1983), and Lejeune, "Les Sociétés de géographie en France" (1986–1987).

9. Lejeune, "Les Sociétés de géographie" (1986–1987), 86, and Fierro-Domenech, *La Société de géographie* (1983), 11.

10. Lacroix, "Walckenaer" (n.d.), 44: 221–37.

11. "Cérémonie en l'honneur d'Edme-François Jomard" (1939), 29–46.

12. Élie de Beaumont, *Société de géographie* (1860).

13. Lacouture, *Champollion* (1988), 339.

14. Jomard, *Comparaison de plusieurs années d'observations* (1832).

15. Benoiston de Châteauneuf, *Considérations sur les enfants trouvés* (1824), and *Extraits des Recherches statistiques* (1824).

16. Jomard, *Du Nombre des délits criminels* (1827).

17. Jomard, *Comparaison de plusieurs années* (1832), 8. He used the inadequacy of modern statistics to argue for better and more complete state collection of statistics: "Why don't the ministries of commerce, the navy, and finance publish analogous documents annually and, further, why don't we take advantage of the organization of the national guard, which can make the gathering together of these statistics easier than in any other country in the world."

18. Dupin, *Tableau comparé de l'instruction populaire* (1828).

19. Jomard, *Comparaison de plusieurs années* (1832), 37.

20. Fourier, "Premier mémoire sur les monumens astronomiques" (1818), Antiquités, Mémoires, 2: 71–86. Jomard, whose basic education included a healthy measure of mathematics and astronomy, also carried out a similar analysis in his "Essai d'explication d'un tableau astronomique" (1809), Antiquités, Mémoires, 1: 255–61.

21. He unwittingly revealed the dangerous circularity of his reasoning in his memoir when he argued that the view, and proof, that the Egyptians were very proficient in geometry "is essential to the explanation of the results contained in this memoir." Jomard, "Mémoire sur le système métrique" (1809), Antiquités, Mémoires, 1: 699.

22. The idea of "truth," and the belief that he was bringing scientific clarity to the befuddled humanities, was dear to Jomard. Speaking about the possibility of determining the state of the "sciences" of the "ancients," he commented that "geometry, more than any other branch of knowledge, offers the means of achieving truth. In effect, the theorems of geometry do not allow vague interpretations to take hold." Jomard, "Mémoire sur le système métrique" (1809), 699.

23. This is clear from his quotation of Gosselin, who attributed the apparent inaccuracy of Greek measures to the modern inability to interpret them properly: "The itinerary measures used by the ancients were more correct than we thought. . . . It is often difficult . . . to decide if the errors apparently in the itineraries are the responsibility of the ancients or are due to the limited nature of our current knowledge." Jomard, "Mémoire sur le système métrique" (1809), 495. It is even clearer in Jomard's conclusion concerning the real purpose of the pyramids: they were nothing less than an enormous measurement standard, something like the national meter, the national pound, etc. Jomard, "Mémoire sur le système métrique" (1809), 531.

24. Jomard put it in the clearest of terms: "These results confirm . . . that the great geographic distance measures used by the ancient Greek writers rested on the value of the Egyptian degree." Jomard, "Mémoire sur le système métrique" (1809), 502.

25. I say "relatively" critical as Jomard was capable of baldly uncritical declarations such as: "The comparison of the numerous distances provided by the authors [meaning ancient authors], with the map which we measured geometrically in Egypt, will immediately give the value of the great itinerary measures such as the 'schoene,' the stade, the mile, etc." Jomard, "Mémoire sur le système métrique" (1809), 498.

26. Letronne, *Notice sur la traduction d'Hérodote* (1823).

27. Jomard, *Procédé pour prendre des empreintes* (1846).

28. See Jomard, *Notice historique sur la vie et les voyages de René Caillé* (1830).

29. Jomard, *Notice sur le second voyage de M. F. Cailliaud* (1823); and Cailliaud and Jomard, *Voyage à Méroe* (1826–1827), 4 vols., and 2 vols. of plates.

30. For an indication of the scope of this activity see Fierro-Domenech, *Inventaire des manuscrits* (1984).

31. For the argument that Jomard had geo-ethnographic interests as early as the expedition to Egypt, see Dias, "Une Science nouvelle: la géo-ethnographie de Jomard" (1998).

32. This is clear from the definition of "ethnographie" adopted by the new Société d'ethnographie in 1859. Société d'ethnographie, Paris, *Société d'ethnographie fondée en 1859* (1868), 1: 1.

33. Jomard, *Classification méthodique* (1862), 3–4.

34. Jomard, *Classification méthodique* (1862), 3.

35. Jomard, *Classification méthodique* (1862), 3–4.

36. Jomard, *Classification méthodique* (1862), 8 and 15.

37. On the significance and power of the museum as metaphor in the nineteenth century, see Georgel, "The Museum as Metaphor" (1994).

38. Jomard, *Classification méthodique* (1862), 10.

39. Jomard, *Les Monuments de la géographie* (n.d).

40. Jomard, *Introduction à l'Atlas* (1879), 6–7.

41. Jomard listed the following predecessors: "Gottschling, Hauber, L. Hubner, Mayer, L. Schlicht, Fréret, Formaleoni, Zanetti, Andrés, Cladera, Mannert, Sprengel, D. Vincent, de Murr, Tiraboschi, Zurla, Morelli, Heeren, Malte-Brun, Hoffmann, Lelewel, Baldelli, A. Pezzana, Pinkerton, Playfair, Loewenberg, etc."

42. Jomard, *Introduction à l'Atlas* (1879), 4.

43. Jomard, *Introduction à l'Atlas* (1879), 49. For more on Jomard's activities as conservator of France's national map collection, see Pelletier, "Jomard et le Département" (1979), 18–27.

44. Jomard, *Introduction à l'Atlas* (1879), 8–9.

45. If there was a conflict of interest, it would have been between his own personal interests as a map collector and/or purchaser and the interests of the Bibliothèque nationale (royale) map collection, of which he was chief.

46. For an analysis of the quality of the atlas on its own terms, as a facsimile, see Godlewska, "Jomard: The Geographical Imagination" (1995),120–22.

47. Jomard, *Sur la publication des Monuments* (1847), 12.

48. Santarem, *Essai sur l'histoire de la cosmographie* (1848–1852).

49. Santarem, *Essai sur l'histoire de la cosmographie* (1848), 2: 130.

50. Jomard, *Sur la publication des Monuments* (1847), 3–5.

51. Santarem, *Essai sur l'histoire de la cosmographie* (1848), 1: xxxviii.

52. Santarem, *Examen des assertions* (1847), 26.

53. Santarem, *Examen des assertions* (1847), 29.

54. The antagonism directed at Champollion by Jomard is frequently described in Lacouture, *Champollion* (1988), though not particularly well explained. It seems, as Lacouture describes it, little more than the otherwise inexplicable intolerance of the brilliance of a younger man.

55. Jomard, *Note sur la nouvelle direction* (1859).

56. Indeed, Champollion himself admitted this in 1817 or 1818. See Lacouture, *Champollion* (1988), 230.

CHAPTER FIVE

1. Harley, "Silences and Secrecy" (1988), 57–76; and "Meaning and Ambiguity in Tudor Cartography" (1983), 22–45.

2. Helgerson, "Nation or Estate?" (1993), 68–74; Rundstrom, "The Role of Ethics" (1993), 21–28; Orlove, "The Ethnography of Maps" (1993), 29–46; Edney, "The Patronage of Science" (1993), 61–67; Helgerson, "The Land Speaks" (1986), 51–85; Harley, *Maps and the Columbian Encounter* (1990).

3. Alison Blunt, Gillian Rose, et al. have explored the complex relationship between geography, imperialism, and gender in Blunt and Rose, *Writing Women and Space* (1994). For more general treatments of the subject, see Bell, Butlin,

and Heffernan, *Geography and Imperialism* (1995); and Godlewska and Smith, *Geography and Empire* (1994).

4. Cormack, "The Fashioning of an Empire" (1994), 15–30.

5. Jacob, *Géographie et ethnographie* (1991).

6. Drapeyron, *Les Deux Buaches* (1888). See also Gille, *Les Sources statistiques* ([1964] 1980), 29; and Bourguet, *Déchiffrer la France* (1989), 23–30.

7. Broc, *La Géographie des philosophes* (1974), 481–82.

8. Eyries, "Mentelle, François-Simon" (n.d.), 27: 661–63.

9. Konvitz, *Cartography in France* (1987), 1–31.

10. Broc, *La Géographie des philosophes* (1974).

11. We know, for example that both Malte-Brun and Walckenaer were offered such posts and that Malte-Brun declined his due to its inadequate remuneration. Møller, *La Critique dramatique et littéraire* (1971).

12. As is clear in Bourcet, "Mémoires sur les reconnaissances" (1875–1876; first published in 1827, but written and widely circulated before 1780); and Allent, "Essai sur les reconnaissances militaires" (1829), 1: 387–522. See also Quimby, *The Background to Napoleonic Warfare* (1957).

13. The works which have explored some of these developments in terms of their impact on cartography include Harley, Petchenik, and Towner, *Mapping the American Revolutionary War* (1978); and McNeill, *The Pursuit of Power* (1982). Although the subject is far from exhausted.

14. This sense of capturing an historic past and of thus claiming it for France was very clear in the writings of the scholars involved on the expedition to Egypt.

15. On this and on the role of population growth and changing social structure in military innovation see McNeill, *The Pursuit of Power* (1982), especially chapters 5 and 6.

16. Lynn, *The Bayonets of the Republic* (1984).

17. See École polytechnique, Paris, *Livre du centenaire* (1894–1897); Fourcy, *Histoire de l'École polytechnique* ([1828] 1987); Hayek, *The Counter Revolution of Science* (1952).

18. On the impact of this ideologization of war see Clausewitz, *On War* (1984), 592–93.

19. Best, *War and Society* (1982), 92–93.

20. Malte-Brun, "Aperçu des agrandissements" (1808), 204–52.

21. Corvisier, *Dictionnaire d'art et d'histoire militaires* (1988), 721.

22. There is a substantial literature on this peculiarly Napoleonic brand of pillage including Mackay Quynn, "The Art Confiscations of the Napoleonic Wars" (1945), 437–60; and Boyer, "Les responsabilités de Napoléon" (1964), 241–62. On its effects in Italy, see Fugier, *Napoléon et l'Italie* (1947), 39–40.

23. This was almost certainly in part the result of a number of government circulars in 1799 which tried to mobilize functionaries and of a variety of organizations behind this collective effort, including, for example, engineers, tax officials, road inspectors, the agricultural societies, and the professors and librarians of the écoles centrales. Gille, *Les Sources statistiques de l'histoire de la France* (1980), 119–20. The birth of the *Annales des statistiques* provides a sense of this increasingly general interest.

24. This is perhaps best captured in Abel Gance's film *Napoleon,* in the scene in which Napoleon, preparing for his first Italian campaign, lies on the floor in a room filled with maps cascading from the desk to the floor. Napoleon himself is both lying on maps and partially covered by them. In the background sits a globe, traditional symbol of hegemonic power.

25. I have not explored the middle-class origins and views of many of the military geographers that I have read from this period but I strongly suspect that most would fit Adas's characterization of them as "individuals who through hard work had risen above their modest family origins [and who] placed a high premium on improvement." Adas, *Machines as the Measure of Men* (1989), 184.

26. The main source for Napoleon's views on descriptive geography is Lefranc, *Histoire du Collège de France* (1893).

27. Cameralism was a body of literature emanating from the German states in the eighteenth century focused on the sources of state wealth. The term cameralist is also used to describe a cadre of state bureaucrats concerned with the wealth and management of the state.

28. On the service of civilian geographers, see Godlewska, "Napoleon's Geographers: Imperialists and Soldiers of Modernity" (1994), 39–49.

29. Presumably either the War of the Second Coalition or the Campaign against Austria.

30. Berthier, "Rapport du Ministre de la guerre" (1802–1803), 1: 213.

31. For more on the mapping carried out in Egypt, see Godlewska, *The Napoleonic Survey of Egypt.* (1988).

32. Bret, "Le Dépôt générale de la guerre" (1991), 113–57.

33. See especially "Programme de l'instruction à donner aux élèves de l'école des ingénieurs-géographes à établir au Dépôt général de la guerre," sent by Sanson to Mathieu Dumas 12 frimaire an 11. Service historique de l'armée de terre, MR 1978 Doc 5, and Dépôt général de la guerre, 2xa 3.

34. General Sanson was a scion of the famous Sanson family of mapmakers.

35. Vallongue, "Coup d'oeil sur les systèmes" (1831), 2: 153.

36. Duboy de Laverne, "Réflexions sur un ouvrage" (1829), 1: 189; and Parigot, "La Bataille de Leuthen" (1829), 1: 190–211.

37. Duboy de Laverne, "Réflexions sur un ouvrage" (1829), 1: 181–89.

38. Berthier, "Rapport du Ministre de la guerre"(1802–1803), 1: 213; Vallongue, "Avertissement" (1829), 1: 231.

39. Lettre du général Sanson au général Mathieu Dumas, Conseiller d'Etat, le 25 prairial an 11 (14 juin 1803). Manuscrits des archives de la guerre, Service historique de l'armée de terre, Vincennes, MR 1978. On the opposition to the reorganization of the Dépôt de la guerre, Sanson wrote: "With this corps will die all hope of achieving the Cadastre of France." On the Mechain and Delambre measurements and on the cadastre, see also Konvitz, *Cartography in France* (1987), 50–51, 59. On the involvement of the engineers in the map of France see Pelletier, *La Carte de Cassini* (1990), 139–56 and 195–209. For a full sense of the activities of these engineers, nothing replaces Berthaut, *Les Ingénieurs-géographes* (1902).

40. During the Napoleonic period, the geographical engineers wrote the most remarkable of these memoirs. Among them were Jacotin's memoir on the map

of Egypt and Soulavie's on l'Atlas napoléon. Jacotin, "Mémoire sur la carte de l'Égypte" (1822), *État moderne*, 2: 1–118. On this memoir, see Godlewska, *The Napoleonic Survey of Egypt* (1988). Soulavie, "Mémoire sur l'emploi des matériaux dans la construction de l'Atlas napoléon," January 1813, Manuscrits des archives de la guerre, MR 1629, Service historique de l'armée de terre, Vincennes. The sense of the limitation of maps and the military importance of these cartographic memoirs was well expressed by Vallongue, "Avertissement" (1829), 1: 233.

41. In this way the genre is similar to the departmental surveys carried out by Napoleon's prefects. Bourguet, *Déchiffrer la France* (1989). In his 1827 reissue of his "Essai sur les reconnaissances militaires," Allent compared the statistical work of the engineers with that of the prefects. Allent, "Essai sur les reconnaissances militaires" (1829), 1: 512, note 1.

42. This innovation, substantially neglected by historians of geography, is linked to the attempt to create a permanent institutional structure for the geographical engineers. It derives, in particular, from the attempted establishment of what Calon (the head of the Dépôt de la guerre between 1793 and 1797) called the "The Geographical Engineering Institute." Calon had an encyclopedic conception of human, physical, and applied geography. This was the spirit of the geography to be taught in the proposed school of geography at the Dépôt de la guerre in 1802. See Calon, *Observations à la Convention nationale* (1795); and Lettre du général Pascal Vallongue au Conseiller d'état le général Dumas, 17 décembre 1802, and attached to that letter there is a "Programme de l'instruction que je propose pour l'École des Géographes à établir au Dépôt de la Guerre." The program was composed of five parts: Mathematics, Physics, The Art of Drawing; Accessory Knowledge Considered in its Generality. Those sections directly related to the topographic memoirs were Physics and Accessory Knowledge. These were retained in the syllabus adopted by the School for geographical engineers at the Dépôt de la guerre in 1809. Manuscrits des archives de la guerre, MR 1978, Service historique de l'armée de terre, Vincennes.

43. For example, Samson des Hallois, Discours sur la situation de final [Finale] pour servir d'explication du plan . . . 24 juin 1706." Manuscrits des archives de la guerre, MR 1400, Service historique de l'armée de terre, Vincennes. According to Colonel Berthaut, in 1782 the Chevalier de Bouligney specifically ordered that any memoirs produced in conjunction with large-scale mapping of the Alsace region focus entirely on military questions and in particular on obstacles of possible military importance. Berthaut, *Les Ingénieurs-géographes* (1902), 1: 81–82.

44. Essentially, the exploration of the relations between the individual, nature, and society with as its principal aim the improvement of society.

45. The clearest example of this is to be found in Martinel. Mémoire historico-statistique de la Commune de Mondovi fesant partie du Champ de bataille de Mondovi. 3 thermidor an XIII (22 July 1805). Manuscrits des archives de la guerre, MR 1364 Doc 24, Service historique de l'armée de terre, Vincennes, 38, 49.

46. Chevalier Allent advised the engineers to avoid useless details: "The determination of what is useful among the political, fiscal, judicial, and administrative

details of a particular country (on its population, the classes it is composed of, and the domestic, social, and political customs that distinguish it . . .) is up to the engineer to judge and that is an integral part of his occupation." Allent, "Essai sur les reconnaissances militaires" (1829), 1: 512–13, note 1.

47. "Considérations sur la Forêt-Noire et sur le Tyrol" (1829), 2: 193.

48. There are two very clear examples of this: in the 1802 Commission's attempt to bring uniformity into the language of cartography; and in Vallongue's concern with "the language of geography." See "Procès-verbal des conférences de la commission chargée . . . à la perfection de la topographie . . ."; "Essai sur les échelles graphiques"; "Explication des teintes et des signes conventionnels"; "Notice sur les caractères et les hauteurs des écritures pour les plans et cartes topographiques et géographiques;" and "Coup d'oeil sur les systèmes"—all in *Mémorial du Dépôt de la guerre* (1831), 2: 1–39; 47–98; 99–102; 103–24; 155–56.

49. Soulavie (ingénieur-géographe), "Notice sur la topographie considérée" (1829), 1: 266–312.

50. Barbié du Bocage, "Notice historique et analytique" (1829), 1: 4–10.

51. Vallongue, "Coup d'oeil sur les systèmes" (1831), 2: 143.

52. Vallongue, "Coup d'oeil sur les systèmes" (1831), 2: 150.

53. Vallongue, "Coup d'oeil sur les systèmes" (1831), 2: 142 and 148.

54. Allent, "Essai sur les reconnaissances militaires" (1829), 1: 470 and 488; Allent, "Sur les échelles de pente" (1829), 2: 98.

55. Thus during his stay in Bavaria, Soulavie acquired "approximately 200 pieces on the topography of that country," and the geographical engineers Lecesne and Epailly were sent to northern Europe to gather all the topographic documents they could find on those countries. Berthaut, *Les Ingénieurs-géographes* (1902), 1: 240 and 2: 84–87.

56. See, for example, Joseph-Charles-Marie Bentabole, Mémoire militaire et statistique de la Commune de Cosseria fesant partie du champ de bataille de Cosseria . . . 24 octobre 1804. Manuscrits des archives de la guerre, MR 1364, Service historique de l'armée de terre, Vincennes, 2–5.

57. Joseph-Charles-Marie Bentabole, Mémoire statistique, historique et militaire du canton de Finale pour le champ de bataille de Scherer fait à Finale le 15 mai 1807. Manuscrits des archives de la guerre, MR 1383, Service historique de l'armée de terre, Vincennes, 40.

58. This concern with analysis for the improvement of agriculture is clear in Martinel, Mémoire historico-statistique de la commune de Mondovi fesant partie du champ de bataille de Mondovi, 22 juillet 1805. Manuscrits des archives de la guerre, MR 1364, Service historique de l'armée de terre, Vincennes, document 24.

59. Allent, "Essai sur les reconnaissances militaires" (1829), 507–8.

60. Allent, "Essai sur les reconnaissances militaires" (1829), 510–11. Respect for and knowledge of local customs and laws was fundamental. The French authorities were little inclined to rescue those geographers who found themselves in difficulty as a result of their ignorance of these. Lettre de Muriel à Schouany Paris 2 décembre 1805. Manuscrits des archives de la guerre, MR 1366, Service historique de l'armée de terre, Vincennes.

61. Instruction des élèves ingénieurs-géographes, no date. Manuscrits des archives de la guerre, MR 1978, Service historique de l'armée de terre, Vincennes,·

62. Soulavie, "Notice sur la topographie" (1829), 1: 266–67.

63. Vallongue, "Avertissement" (1829), 1: 231.

64. Vallongue, "Procès-verbal des conférences" (1831), 2: 3.

65. "Considérations sur la Forêt-Noire" (1831), 2: 193.

66. Each of their topographic memoirs included an essay on local history, which explored the history of those aspects of the countryside that attracted the attention of the geographical engineers, and an essay on local military history. On the inadequacy of armchair history, see Vallongue, "Avertissement" (1829), 1: 113; and Lagardiolle, "Notes sur les principaux historiens" (1829), 1: 160–61.

67. On the military career of this remarkable man, see Horward, "Jean-Jacques Pelet" (1989): 1–22.

68. See chapter 6 and also Godlewska, "Map, Text and Image" (1995), 5–28.

69. His military record, held at the Service historique de l'armée de terre, Vincennes, describes the injury as to his shoulder, but his biographers all describe it as a bullet which went through the chest.

70. Malte-Brun, *Annales des voyages,* 57th book.

71. For further biographical information on André-Étienne-Just-Pascal-Joseph-François d'Audebard baron de Férussac, see the personal dossier at the Service historique de l'armée de terre, Vincennes, in the alphabetical register of personal dossiers 1791–1847 under "Daudebard" and the following biographical notices: Rabbe Alphonse, Vieilh de Boisjolin, and Sainte-Preuve, "Férussac, André-Étienne-Just-Pascal-Joseph-François d'Audebard, baron de," in *Biographie universelle et portative des contemporains* (1834), 2: 1678–82; Quérard, *La France littéraire* (1828), 401–2; Weiss, "Férussac" (n.d.), 14: 31–32; Louandre and Bourquelot, *La littérature française contemporaine* (1848), 3: 490; and Dupin, "Nécrologie de M. A.-É.-J.-P.-J.-F. d'Audebard, baron de Férussac" (1836): 491.

72. Such was the case with "Notice sur une inscription et sur une medaille phéniciennes, trouvées en Andalousie," Rabbe, Vieilh de Boisjolin, and Sainte-Preuve, *Biographie universelle* (1834), 2: 1679.

73. Férussac, *Extraits du journal de mes campagnes en Espagne* (1812).

74. According to the best of de Férussac's biographers, Vieilh de Boisjolin, Sainte-Preuve, and Rabbe, this was apparently the fate of a year's worth of research into the history and geography of Silesia. This biography was written within de Férussac's lifetime probably with de Férussac's assistance and cooperation, as a copy of the volume within which the biography was published was among the books in de Férussac's library at his death. See Férussac, *Catalogue des livres composant la bibliothèque du feu M. le baron de Férussac* (1836).

75. Férussac, *Histoire naturelle générale et particulière des mollusques* (1820–1851), 2: vj–vij.

76. It is very difficult and perhaps fruitless to separate out the work of father and son: while Jean-Louis d'Audebard de Férussac (1745–1815) was alive, father and son worked as a team.

77. Férussac, *Histoire naturelle générale et particulière des mollusques* (1820–1851), 2: préface.

78. On this see Férussac, "Rapport fait à l'Institut de France" (1814); and Férussac, *Notice analytique sur les travaux* (1824).

79. Férussac, *Notice analytique sur les travaux* (1824), 10ff.

80. This entire discussion takes place on pages 67–75 of Férussac, *Mémoires géologiques sur les terreins* (1814).

81. All of the quotations from this paragraph are taken from Férussac, *Notice sur les travaux* (1924), 14–15. De Férussac, himself, copied them from his own papers both published and unpublished.

82. Férussac, "Distribution géographique des productions aquatiques" (1825), 7: 269–70; and Férussac, *Notice sur les travaux* (1924), 15.

83. The dates and details of his applications are to be found in his *Notice analytique sur les travaux* (1824). At the Muséum d'histoire naturelle, B.970 (4).

84. Férussac, *Note supplémentaire à la notice* (1825).

85. Letter in which de Férussac acknowledges the Institut's refusal for the fourth time. [Muséum d'histoire naturelle, B.970 (4)] Dated only Paris, ce 17 février.

86. See especially the comments of d'Orbigny in Férussac and d'Orbigny, *Histoire naturelle générale et particulière* (1835–1848).

87. Férussac, *Histoire naturelle générale et particulière des mollusques* (1820–1851), 1: preface.

88. *Mémorial du Dépôt de la guerre* (1831).

89. Férussac, *Extraits du journal* (1812), viij-ix.

90. Férussac, *Extraits du journal* (1812), xi.

91. Férussac, *De la Nécessité de fixer et d'adopter un corps de doctrine* (1819); and Férussac, *Plan sommaire d'un traité de géographie* (1821), which includes an intriguing "Tableau générateur et analytique des sciences qui ont pour objet l'univers et les êtres en général, le globe et les sociétés humaines en particulier."

92. Férussac, *Histoire naturelle générale et particulière des mollusques* (1820–1851), 2: preface.

93. Férussac, "Mollusques et Conchifères" (1825), 7: 261.

94. Férussac, *De la Nécessité d'une correspondance régulière* (n.d.), 5.

95. Férussac, *Bulletin des sciences mathématiques, astronomiques, physiques et chimiques* (1824–1831).

96. Férussac, *Bulletin des sciences naturelles et de géologie* (1824–1831).

97. Férussac, *Bulletin des sciences médicales* (1824–1831).

98. Férussac, *Bulletin des sciences agricoles et économiques* (1824–1831).

99. Férussac, *Bulletin des sciences technologiques* (1824–1831).

100. Férussac, *Bulletin des sciences géographiques, etc. économie politique, voyages* (1824–1831).

101. Férussac, *Bulletin des sciences historiques, antiquités, philologie* (1824–1831).

102. Férussac, *Bulletin des sciences militaires* (1824–1831).

103. Férussac, *De la Nécessité d'une correspondance régulière* (n.d.), 42.

104. On this, see Taton, *Les Mathématiques dans le Bulletin de Férussac* (1947); and Weiss "Férussac" (n.d.) 14: 31–32, which, unfortunately, is full of errors. One of the most interesting comments on the impact of the *Bulletin* on Paris academics comes

from Adrien Balbi, who was in Paris and working on his linguistic ethnographic atlas at the time (see chapter 6). Adrien Balbi, "Introduction," in *Atlas ethnographique du globe* (1826), cxxiii.

105. On the early history of book collection and bibliography, during the sixteenth, seventeenth and eighteenth centuries in France, see Barret-Kriegel, *Les Académies de l'histoire* (1988), 265–92; and Estivals, *Le Dépôt légal sous l'ancien régime* (1961).

106. Férussac, *De la Nécessité d'une correspondance régulière* (n.d.), 17.

107. Férussac, *De la Nécessité d'une correspondance régulière* (n.d.), 24.

108. Férussac, *De la Nécessité d'une correspondance régulière* (n.d.), 22.

109. De Férussac's argument is remarkably in keeping with Ben-David's in Ben-David, "The Rise and Decline of France" (1970), 160–79.

110. For the sizeable contribution made by Champollion to the seventh section of the *Bulletin*, see Lacouture, *Champollion, une vie de lumières* (1988), 464.

111. Taton, *Les mathématiques dans le Bulletin* (1847), 5.

112. One of de Férussac's early biographers recounts a story that is worth repeating: "M. Le baron Alexandre de Humboldt found the de Férussac Bulletin on the borders of China, in a little Russian post consisting of a few men. This fact, which we heard the illustrious story teller himself recount, gives an idea of the merited popularity and immense influence achieved by this enterprise." Philippe Le Bas, "L'Univers. Histoire et description de tous les peuples" (1842), 7: 792.

113. In 1835 he was officially described by the Corps royale d'état-major as "without means." Rapport particulier d'Audebard de Férussac, 1835. Dossier Personnel, Service historique de l'armée de terre, Vincennes.

114. By 1830 he was a lieutenant-colonel and he received the Légion d'honneur in 1831.

115. Adrien Balbi devoted several pages of his *Abrégé de géographie rédigé sur un nouveau plan* (1833), iii—xi, to a discussion of de Férussac's *De la Nécessité de fixer et d'adopter un corps de doctrine*. While acknowledging that de Férussac had expressed "reasonable complaints" about the quality of some research in geography, he basically dismissed his plan out of hand as too ambitious. Further, he rejected the important role given by de Férussac to statistics.

116. See Férussac, "Mollusques et Conchifères" (1825), 7: 254–70; and Férussac, *Bulletin général et universel* (1822–1823), prospectus and Tableau des sections du Bulletin.

117. Among the Napoleonic military/engineering geographers to achieve a significant reputation in these latter three fields were Boblaye and Coraboeuf for geodesy; Jacotin, Lapie, and Bacler d'Albe for large-scale mapping; and Coquebert de Montbret and Chabrol de Volvic for administration.

118. Bory de Saint-Vincent, *Essais sur les Isles Fortunées* (1803) and Bory de Saint-Vincent, *Voyage dans les quatre principales îles* (1804).

119. Bory de Saint-Vincent, *Relation du voyage de la Commission scientifique de Morée* (1836).

120. Bory de Saint-Vincent, *Guide du voyageur en Espagne* (1823), 649.

121. Bory de Saint-Vincent, *Justification de la conduite* (1815), 14.

122. Bory de Saint-Vincent, *Description du plateau de St.-Pierre de Maestricht* (1819).

123. When Bory de Saint Vincent was being hounded from country to country by regimes responsive to the pressures exerted by Louis XVIII's government, Humboldt was able to provide him with asylum for a time and a network of colleagues and friends in Berlin. In friendship and gratitude, Bory de Saint Vincent dedicated his first volume of the *Annales générales des sciences physiques* to Humboldt.

124. Drouin, "Bory de Saint-Vincent et la géographie botanique" (1998), 139–57.

125. Although he did compose an elementary treatise on reptiles, *Traité élémentaire d'erpétologie* (1842), and, as already mentioned, he worked collaboratively on floras during the expeditions to Morea and Algeria.

126. Berthaut, *Les Ingénieurs-géographes* (1902), 2: 189.

127. Bory de Saint-Vincent, *Nouvelle carte d'Espagne et de Portugal* (1824); and Bory de Saint-Vincent, *Notice sur la nouvelle carte d'Espagne* (1824).

128. Bory de Saint-Vincent, *Voyage dans les quatre principales îles* (1804).

129. Bory de Saint-Vincent, *Expédition scientifique de Morée. Section des sciences physiques* (1832–1836); Bory de Saint-Vincent, *Relation du voyage de la Commission scientifique de Morée* (1836—).

130. Bory de Saint-Vincent, *Note sur la Commission scientifique de l'Algérie* (n.d.); Bory de Saint-Vincent, "Rapport concernant la géographie et la topographie" (1838); Bory de Saint-Vincent, "Flore de l'Algérie, Cryptogamie" (1846–1867); and Bory de Saint-Vincent, *Sur l'anthropologie de l'Afrique* (n.d. [1845]).

131. Bory de Saint-Vincent, *Essais sur les isles fortunées et l'antique Atlantide* (1803).

132. Bory de Saint-Vincent, *Guide du voyageur en Espagne* (1823); Bory de Saint-Vincent, *Résumé géographique de la péninsule ibérique* (1826).

133. The prospectus for this atlas appeared as a title page-cum-flier: *Atlas géographique, statistique et progressif des départments de la France et de ses colonies. Sous la direction de M. Pierre Tardieu, accompagné d'une notice historique sur la France, par M. Bory de Saint-Vincent, membre de l'Institut*, in Bory de Saint-Vincent, *La France par tableaux géographiques et statistiques* (1844). The atlas finally appeared as Perrot, *Atlas géographique, statistique et progressif* ([1845]).

134. Although Jean-Marc Drouin has argued convincingly that his interest in plant geography was as acute as Humboldt's but that he sided with de Candolle in a debate over the nature of plant geography. See Drouin, *Réinventer la nature* (1991), 60ff.

135. Bory de Saint-Vincent, "Notice biographique sur Malte-Brun" (1827).

136. This reference is unclear. He may instead have been referring to Francisco Pons (1768–1855) who also counted geography among his more historical and literary interests.

137. I have been unable to positively identify this Spanish geographer.

138. Geology is relatively absent from Bory de Saint-Vincent's *Dictionnaire classique d'histoire naturelle* but receives some attention, for example in his *Description du plateau de St.-Pierre de Maestricht*.

139. Bory de Saint-Vincent, *Guide du voyageur en Espagne* (1823), 226.

140. In his *Dictionnaire classique d'histoire naturelle,* Bory de Saint-Vincent looks forward to a time when it may be possible to determine mammalian regions (300), links the concept of the human race to such a regionalism (300), and, together with Guillemin, hails de Candolle's concept of the botanical region (272, 289–90). See also Bory de Saint-Vincent, *Guide du voyageur en Espagne* (1823), 201.

141. Bory de Saint-Vincent, *Guide du voyageur en Espagne* (1823), 202.

142. Bory de Saint-Vincent, *Guide du voyageur en Espagne* (1823), 204–5.

143. Bory de Saint-Vincent, *Guide du voyageur en Espagne* (1823), 209.

144. Bory de Saint-Vincent, "Notice sur la Nouvelle carte d'Espagne" (1823), 19.

145. Bory de Saint-Vincent, *Guide du voyageur en Espagne* (1823), 282–83.

146. Bory de Saint-Vincent, *Dictionnaire classique d'histoire naturelle* (1825), 7: 240

147. See Bory de Saint-Vincent, "Rapport concernant la géographie et la topographie" (1838), 52–53.

148. It is important to note that this was written very soon after the Napoleonic expedition to Egypt. Bory de Saint-Vincent, *Essais sur les isles fortunées* (1803), 47–48.

149. Bory de Saint-Vincent, *L'Homme (homo), essai zoologique sur le genre humain* (1827), 67.

150. See Bory de Saint-Vincent, *Guide du voyageur en Espagne* (1823), 234 and 642.

151. Bory de Saint-Vincent, *Essais sur les isles fortunées* (1803), 460.

152. Bory de Saint-Vincent, *Essais sur les isles fortunées* (1803), 240–41.

153. See Bory de Saint-Vincent, "Rapport concernant la géographie et la topographie" (1838), 10; and, for another example, his *Voyage dans les quatre principales îles* (1804), 23.

154. Bory de Saint-Vincent, *Guide du voyageur en Espagne* (1823), 67–68.

155. Bory de Saint-Vincent, *Description du plateau de St.-Pierre de Maestricht* (1819), 49–50.

156. Bory de Saint-Vincent, *Voyage dans les quatre principales îles* (1804), 87.

157. See for example Bory de Saint-Vincent, *Relation du voyage* (1836), 108–10; Bory de Saint-Vincent, *Essais sur les isles fortunées* (1803), 105; and Bory de Saint-Vincent, *Voyage dans les quatre principales îles* (1804), 24.

158. According to his biographer, Bory de Saint-Vincent spent so much time in prison to force his creditor to pay his living expenses. For the scholar of limited means working on a number of manuscripts, and with a daughter prepared to provide food, company, and correspondence, prison was an ideal situation.

159. Role, *Un Destin hors série* (1973), 111–12.

160. Role, *Un Destin hors série* (1973), 119–20.

CHAPTER SIX

1. See Staum, "Human Geography in the French Institute" (1987), 332–40.

2. Gaulmier, *L'Idéologue Volney* (1951), 347–51.

3. On the reciprocal influence of Volney and the other ideologues, see especially and most recently Barthélémy Jobert's excellent introduction to Volney's "Leçons

d'histoire," in Nordman, *Leçons d'histoire, de géographie, d'économie politique* (1994), 43–46.

4. Among the publications which discuss Volney's geographical writings in any depth are Gaulmier, *L'Idéologue Volney* (1951); Moravia, "Philosophie et géographie" (1967), 937–1011; Albitreccia, "Ajaccio, étude de géographie" (1938), 361–72; and Vallaux, "Deux précurseurs" (1938), 83–93. Works which discuss them more in passing include: Broc, *La Géographie des philosophes* (1974); Chinard, *Volney et l'Amérique* (1923); Kühn, *Volney und Savary* (1938); Godlewska, "Napoleonic Geography" (1990), 1: 281–302; Deneys, "Le Récit d'histoire" (1989), 43–71; Deneys, "Géographie, histoire et langue" (1989), 73–90; Moravia, "La méthode de Volney" (1989), 19–31; and Désirat and Hordé, "Volney, l'étude des langues" (1984), 133–41. This last work claims Volney as a precursor of anthropology.

5. Volney, *Voyage en Syrie* (1787).

6. Volney, *Tableau du climat* (1803).

7. Volney, *Questions de statistique* (1795).

8. Volney, "Etat physique de la Corse" (1821).

9. In keeping with the views of Cabanis, climate was broadly defined not as an effect of latitude alone but as "the combination of physical circumstances connected with each locale." Cabanis, *Rapports du physique* (1802), 564.

10. Volney, *Tableau du climat* (1803), préface; Volney, "État physique de la Corse" (1821), 317–18.

11. Volney, "Questions de statistique" (1813), 389. This work is part of a genre which would reward analysis by geographers. The genre of the research guide for travelers, statesmen, and spies was already highly evolved. Volney himself refers to a predecessor, Berchtold, *Essai pour diriger et étendre les recherches* (1797). Berchtold in turn refers to a predecessor, Lettsom, *The Naturalist's and Traveller's Companion* (1774). Volney's work would also be worth comparing to that of de Gérando, "Considérations sur les diverses méthodes" (n.d.), avertissement; Quesnay, *Questions intéressantes* (1762); David [Michaelis pseud.], *Recueil des questions proposées* (1763); and Thomas Jefferson's *Notes on the State of Virginia* (1784–1785). That, however, far from exhausts the genre.

12. Claval, *Les mythes fondateurs* (1980), 85.

13. Plongeron sees Volney and Cabanis as the two ideologues with a strong almost semi-Romantic interest in nature. Plongeron, "Nature, métaphysique et histoire" (1973), 398–403.

14. Volney, *Voyage en Syrie* (1787), iv—v.

15. Volney, *Voyage en Syrie* (1787), 348–54.

16. Volney, *Voyage en Syrie* (1787), 19–31.

17. Volney, *Voyage en Syrie* (1787), 52–66; 297–324.

18. *Séances des écoles normales* (1802), 2: 76.

19. Mentelle, *La Géographie enseignée* (1795), Éclaircissements préliminaires.

20. Which, as Sergio Moravia has pointed out, was a concern and a methodology that stretched back to Descartes and Locke and found a new importance among the ideologues. Moravia, "Philosophie et géographie" (1967), 990–91.

21. Volney, *Voyage en Syrie* (1787), 457.

22. Volney, *Voyage en Syrie* (1787), 104–7 note; 141–42.

23. Indeed, this is nowhere clearer than in Volney, *Les Ruines* (1791), in which he proposed a solution to social corruption the world over: the annihilation of religion.

24. Volney, *Voyage en Syrie* (1787), 182–83, 151; and Volney, *Tableau du climat* (1803), 182.

25. Volney, *Tableau du climat* (1803), 2, 420.

26. Volney, *Voyage en Syrie* (1787), 418–19.

27. An unhealthy society for Volney was one in which the comfort and security of all men no longer took precedence over the wealth and power of a few. The key indicators of such a society were a destructive and thoughtless use of nature, a neglect of family and of family values, and a dominant military caste.

28. Volney, *Voyage en Syrie* (1787), 418–19.

29. Volney, *Voyage en Syrie* (1787), 241.

30. On the value to Volney of the testimony of common folk rather than highly placed officials, see Besnard, *Souvenirs* (1979 [1880]), 1: 336.

31. Volney, "Réponse de Volney" (1821), 2: 16.

32. Volney, *Voyage en Syrie* (1787), 424. His own multiple and critical use of sources is clear in each of his publications and is frequently described. See for example, Volney, *Tableau du Climat* (1803), 1, 209.

33. Volney, *Tableau du climat* (1803), 656.

34. See Bouteiller, "La société des observateurs" (1956), 448–65; and Stocking, "French Anthropology in 1800" (1955), 134–50. Witness also the passion of this era for exploratory voyages and the attentions of de Gérando, Volney, and others to the writing of instructions for prospective travelers. This story is well told by both Broc, in *La Géographie des philosophes* (1974); and by Moravia, in both "Philosophie et géographie" (1967), and *Il Pensiero degli idéologues* (1974).

35. I borrow this expression from Lefranc, "Le Voyageur Volney" (1897), 1: 87.

36. In *Les Ruines* the young man who has contemplated the ruins of the great Asian civilizations laments the fate of all empires, which is to fall to ruins. He is taken by an all-knowing and all-seeing spirit far enough above the earth to contemplate the hemisphere from Europe to East Asia. The importance of this image, which is metaphorical, is highlighted by Volney's inclusion in *Les Ruines* of an engraving of the globe as seen by the young man and the spirit. Volney, *Les Ruines* (1791). See the two-page illustration across pages 24 and 25. The globe, drawn and engraved by Delettre, is centered on the Middle East. It does not depict the Americas at all and France is peculiarly misshapen. A legend composed of fifteen numbers with accompanying place names adds to the small number of place names on the globe itself.

37. This is not to imply that Volney was a sophisticated twentieth-century or even late-nineteenth-century social analyst. His sense of the complexity of social structure and evolution was limited, as was his understanding of the very different functions of particular social phenomena, such as religions, in different societies and circumstances. Nor, as Anne and Henry Deneys point out, had Volney resolved the problem of the very different nature of analysis necessary for the physical and political realms. See in particular Henry Deneys, "Le Récit d'histoire" (1989), 50, par. 2; and Anne Deneys, "Géographie, histoire et langue" (1989), 73–90. But then, their dichotomous nature was not yet apparent, or rather had not yet been created

by nineteenth-century modes of scientific thought. Although, arguably, Volney's inability in his *Tableau du climat* to match his analysis of the soil and climate of the United States with an associated political analysis suggests that the problem was already beginning to arise.

38. This is not to imply that the only motive for his travels was research. It seems clear that Volney was probably acting as a spy for the French government both in Egypt and in Corsica and it is probable that most of his trips were funded by that means. See Sibenaler, *Il se faisait appeler Volney* (1992); Laurens, *Les Origines intellectuelles* (1987), chap. 11 (171–85).

39. Volney, *Les Ruines* (1791), 35.

40. Volney, *Voyage en Syrie* (1787), 348–54.

41. Volney, *Voyage en Syrie* (1787), 348.

42. Volney, *Les Ruines* (1791), 112.

43. Volney, *Voyage en Syrie* (1787), 378; and Volney, *Tableau du climat* (1803), 463–65.

44. Pauw, *Recherches philosophiques* (1768–1769).

45. Rousseau, *Émile* (1966).

46. Volney, *Tableau du climat* (1803), 476.

47. Volney, *Tableau du climat* (1803), 426.

48. Anne Deneys presents a subtle and most interesting analysis of Volney's failure to complete this work. While I agree that Volney was limited by his political and social context and his concern with the march of "civilization," I do not see Volney's interests as so tied to the holistic and descriptive preoccupations of the eighteenth century as to render them without relation to the modern science of geography. Indeed, as I have argued here, much in his methods and interests suggests the contrary. Deneys, "Géographie, histoire et langue" (1989), 73–90.

49. This is nowhere more clearly expressed than in chapter 14 of *Les Ruines*.

50. Volney, *Voyage en Syrie* (1787), 2: 327–28.

51. In fact, he did not complete his medical degree but had three years of medical training. Gaulmier, *Un Grand témoin* (1959), 23. On the influence of Condillac and of Enlightenment ideologue thought in general—from Voltaire to d'Holbach to Helvétius—on Volney, see Barthélémy Jobert's introduction to Volney's "Leçons d'histoire," in Daniel Nordman, *Leçons d'histoire, de géographie, d'économie politique* (1994), 38–46.

52. Volney, *Tableau du climat* (1803), 145.

53. Volney, *Tableau du climat* (1803), 248; 152–53.

54. Volney, *Tableau du climat* (1803), 238.

55. Volney, *Voyage en Syrie* (1787), 310–12.

56. Volney, *Voyage en Syrie* (1787); see in particular section 1 of chapter 17.

57. Volney, *Tableau du climat* (1803), 334–52.

58. Volney, *Voyage en Syrie* (1787); see in particular section 3 of chapter 17.

59. Volney, *Voyage en Syrie* (1787), 17–18.

60. Volney, *Tableau du climat* (1803), 26.

61. Volney, *Tableau du climat* (1803), 276.

62. *Séances des écoles normales* (1802), 76–77.

63. Volney, *Tableau du climat* (1803) 1, 3–4, 37, 58, 74, 97 note 1; and Volney, *Voyage en Syrie* (1787), viii, 19–31, 34, etc.

64. Gaulmier, *L'Idéologue Volney* (1951), 255 note 5, and 429–30.

65. Godlewska, *The Napoleonic Survey of Egypt* (1988), 124.

66. Gaulmier, *L'Idéologue Volney* (1951), 237, 535 note 3.

67. Picavet suggests that after the dissolution of the Third Class of the Institut, it was virtually impossible to publish in realms in any way touching upon political subjects. That would substantially eliminate any proto–social sciences. Picavet, *Les Idéologues* (1891), 80, 138. On the centrality of the Third Class of the Institut to ideologue philosophy and their social project, see Gusdorf, "La Conscience révolutionnaire" (1978), 305–10. On the nature of the research carried out in the Third Class, and in particular on the geography practiced there, see Staum, "Human Geography in the French Institute" (1987), 332–40.

68. Volney, *Tableau du climat* (1803), préface.

69. With occasional exceptions such as his anti-monarchist *Histoire de Samuel* (1819).

70. He acknowledges the assistance of Jomard for his work in Egypt; the heavy use of the statistics and fieldwork of the military geographers for his work in Italy; and there are suggestions that his publication on the département of the Seine was as much the work of Baron Charles Athanase Walckenaer as it was Chabrol de Volvic's. For this interpretation, see Lacroix, "Walckenaer" (n.d.), 221–37.

71. There is some scholarly disagreement over the nature of property rights under the Egyptian Islamic system and under the French regime. Abd Al-Rahim considers that the French moved Egypt further toward private land ownership through the confiscation of some Mameluke lands and their redistribution among peasantry capable of paying the land taxes. Abd Al-Rahim, "Land Tenure in Egypt" (1984). Cuno considers that while Egyptian peasants did not enjoy "private property" rights prior to the nineteenth century, their usufruct rights were so extensive that they may be considered tantamount to private property rights. He further argues the continuity of development between the traditional Islamic land law of the seventeenth and eighteenth centuries and the developments of the early nineteenth century. Cuno, "Egypt's Wealthy Peasantry" (1984). While reviewing many of the same facts, Richard Debs sees the early nineteenth century and the French invasion and subsequent French influence through the rule of Muhammad Ali as a rift with Islamic development and the beginning of a strong "Western" and "modern" influence. Debs, *The Law of Property in Egypt* (1963), 48.

72. I would not be comfortable describing all of these individuals as geographers. Chabrol de Volvic was more inclined to describe himself, and anyone who "studies the many factors which influence climate . . . the action of that climate on animate beings, the new men that he finds himself surrounded by . . . ," as a philosopher. Chabrol de Volvic, "Essai sur les moeurs" (1822), 361.

73. Chabrol de Volvic, "Essai sur les moeurs" (1822), 377.

74. Chabrol de Volvic, "Essai sur les moeurs" (1822), 363.

75. The degree to which this was colonization and expressed as such is clear in the words used in a series of reports from government agents sent to the Vendée which deal with the problem of "civilizing the natives." See especially "Réponse du Cn.

Giraud aux questions qui lui ont été faites sur le Morbihan," Archives nationales, AF IV 1053. Cited in Morachiello and Teyssot, "State Town: Colonization" (1979), 33 and footnote 40.

76. Morachiello and Teyssot, "State Town: Colonization" (1979), 31.

77. On this, see Godlewska, "Map, Text and Image" (1995), 5–28.

78. Chabrol de Volvic, *Statistique . . . de Montenotte* (1824), v.

79. See Chabrol de Volvic, "Mémoire sur la Spezzia," Mémoires, MR 1400, Service historique de l'armée de terre, Vincennes.

80. Boudard, "La Réunion de Gênes" (1968), 158.

81. On statistics collection as a part of regular administrative accounting, see Gille, *Les Sources statistiques* (1964 [1980]), 23–42, 116–22, 131–36, 149–50; and Bourguet, *Déchiffrer la France* (1988).

82. Chabrol de Volvic's achievements in Liguria are generously described in Boudard, "Un Préfet napoléonien" (1956), 119–30.

83. Chabrol de Volvic, *Statistique . . . de Montenotte* (1824), 1: i.

84. Chabrol de Volvic, *Statistique . . . de Montenotte* (1824), 1: ij.

85. Chabrol de Volvic, *Statistique . . . de Montenotte* (1824), 2: 466–67.

86. Chabrol de Volvic, *Statistique . . . de Montenotte* (1824), 2: 275.

87. For further details on this mission see Boudard, "La Mission du préfet" (1969), 181–88.

88. Papayanis, *Horse-drawn Cabs* (1996), 31–53, 81, 86, and 95.

89. See Coleman, *Death is a Social Disease* (1982), 142–44.

90. On Walckenaer's administrative career, see Tulard, *Paris et son administration* (1976), 414–15.

91. Chabrol de Volvic, *Recherches statistiques sur . . . Paris* (1821), introduction.

92. Chabrol de Volvic, *Recherches statistiques sur . . . Paris* (1821), vjj—vjjj.

93. Lepetit, *Chemins de terre* (1984). For a different view of the role of maps in urban development in France, see Konvitz, *The Urban Millennium* (1985), 78–95.

94. Chabrol de Volvic, *Recherches statistiques sur . . . Paris* (1821), 107.

95. Chabrol de Volvic, *Recherches statistiques sur . . . Paris* (1821), ix.

96. This sense was most clearly expressed by Adrien Balbi in his negative reaction to the importance André de Férussac gave statistics in his proposed geography text for future general staff officers. Barbié du Bocage's characterization of geography as vestibule of the sciences captures some of the same sense: "Linked to all the sciences, geography serves, as it were, as introduction to each of them and prepares the way to their fruitful study. It is a vestibule from which more than a hundred doors lead to all of the branches of human knowledge." *Bulletin de la société de géographie* 1 (1822): 9. In a similar vein S. Berthelot wrote about the geography of the Society's founders in these terms: "to make known countries in terms of their position, their climate, and their resources, to stimulate new discoveries, to encourage voyages, to direct, encourage, and recompense the zeal of voyageurs, and *to gather together as general information all their individual observations,* that was our founders' aim." Berthelot, "Rapport sur les travaux" (1843), 333–34.

97. See Bernard Lepetit, "Missions scientifiques et expéditions militaires" (1998), 102–3.

98. For a full discussion of all of these problems, see Tulard, *Paris et son*

administration (1976), 487 (not fully understanding crowding), 466 (recognition of mistakes), 452–53 (building craze), 442–44 (borrowing to build), 387 (differential bread tax proposal).

99. Tulard, *Paris et son administration* (1976), 408, 513–14.

100. Adrien Balbi describes Malte-Brun as such in the Discours préliminaire of his *Essai statistique sur le royaume de Portugal* (1822), xx.

101. La Renaudière, "Brun ou Brunn (Malte-Conrad)" (n.d.), 8.

102. See in particular Malte-Brun's letter to Balbi, discussed in greater length below, predicting the considerable problems Balbi would encounter in composing his *Atlas ethnographique du globe*. See Balbi, *Introduction à l'Atlas ethnographique* (1826), 2–13.

103. See Godlewska, "L'Influence d'un homme" (1991), 62–79.

104. Balbi, *A Statistical Essay on the Libraries* (1986 [1835]); and Balbi, *Essai statistique sur le royaume de Portugal* (1822).

105. Balbi, *Introduction à l'Atlas ethnographique du globe*, Partie historique et littéraire, vol. 1, Discours préliminaire et introduction (1826); and Balbi, *Atlas ethnographique du globe* (1826).

106. Balbi, *A Statistical Essay on the Libraries* (1986), 32; and Balbi, *Essai statistique sur le royaume de Portugal* (1822), 1: 218–19.

107. Balbi, *Essai statistique sur le royaume de Portugal* (1822), 1: xviii, 144 and 401.

108. Balbi, *Essai statistique sur le royaume de Portugal* (1822), 2: 21.

109. Balbi, *Essai statistique sur le royaume de Portugal* (1822), 1: 143.

110. Balbi, *Essai statistique sur le royaume de Portugal* (1822), 1: 235–39.

111. Balbi, *Essai statistique sur le royaume de Portugal* (1822), 1: 241.

112. Balbi, *Essai statistique sur le royaume de Portugal* (1822), 1: 155.

113. Balbi, *Essai statistique sur le royaume de Portugal* (1822), 1: 300.

114. Balbi, *Essai statistique sur le royaume de Portugal* (1822), 1: 148.

115. Balbi, *Essai statistique sur le royaume de Portugal* (1822), 1: 148, 149.

116. Balbi, *A Statistical Essay on the Libraries* (1986), 31.

117. Balbi, *Introduction à l'Atlas ethnographique du globe* (1826), 61.

118. This is not to imply that Balbi was free of racism or cultural intolerance. Many of Balbi's language characterizations were heavily value-laden, he was uncritical about colonialism and the Portuguese involvement in the slave trade, and he was quite capable of describing the North Africans as a degraded race. See Balbi, *Essai statistique sur le royaume de Portugal* (1822), 2: 24–25, 27, 29–30; and Balbi, *Introduction à l'Atlas ethnographique du globe* (1826), cii and cxxx.

119. Balbi, *Introduction à l'Atlas ethnographique du globe* (1826), xlvii, cxxi, cxlii–cxliii.

120. Balbi, *Introduction à l'Atlas ethnographique du globe* (1826), xxi.

121. Balbi, *Introduction à l'Atlas ethnographique du globe* (1826), cxxvii.

122. Balbi describes the idea as dating from at least 1817, when he introduced into his universal geography "a table of the principal known languages subdivided in five sections corresponding to the five great divisions of the globe." He argued that it was the first time such a table had appeared in any *Abrégé de géographie*. Balbi, *Introduction à l'Atlas ethnographique du globe* (1826), cxvi.

123. Balbi, *Introduction à l'Atlas ethnographique du globe* xvi–xix.

124. Balbi, *Introduction à l'Atlas ethnographique du globe* (1826), xxi, lxxiv–lxxv.

125. Balbi did not use the word "probability" frequently. This author found only one use. In discussing the amount paid annually for book purchase by the Imperial Library in Vienna, Balbi commented, "However enormous this last sum may appear, there are authentic facts one can draw from to demonstrate, if not its mathematical exactness, at least its probability." The phrase he wanted was "approximate magnitude." Balbi, *A Statistical Essay on the Libraries* (1986), 13.

126. Balbi, *A Statistical Essay on the Libraries* (1986), 58.

127. There are many examples of such tables in Balbi's works. The table where this most caught my attention is his "Tableau comparatif du nombre des étudians qui ont fréquenté les écoles de plusieurs états de l'Europe," in Balbi, *Essai statistique sur le royaume de Portugal* (1822), 2: 150.

128. This is well expressed in his ethnographic atlas which, by virtue of its attempt to classify, Balbi deemed: " . . . linguistics, elevated to the rank of science . . ." Balbi, *Introduction à l'Atlas ethnographique du globe* (1826), xlii–xliv.

129. Balbi, *Essai statistique sur le royaume de Portugal* (1822), 1: 103–6, 190; Balbi, *A Statistical Essay on the Libraries* (1986), 62–63, 65.

130. Balbi, *Essai statistique sur le royaume de Portugal* (1822), 1: 85–89.

131. Balbi, *Essai statistique sur le royaume de Portugal* (1822), 2: 153.

132. Balbi, *Essai statistique sur le royaume de Portugal* (1822), 1: 84.

133. This tabular structure was common to a number of works produced in this period, including Jarry de Mancy, *L'Atlas historique* (1831–1835); and Denis, *Tableau historique analytique* (1830). For the link to Balbi's work, see Balbi, *Introduction à l'Atlas ethnographique du globe* (1826), cxli. These works were both trying to cope with information overload through the creation of a semi-graphic synthesis and were taking their cue from a partial understanding of the nature of botanical classification.

134. Remusat, Review of Balbi (1827), 282–91.

135. Remusat, Review of Balbi (1827), 284–85.

136. Remusat, Review of Balbi (1827), 286.

137. Balbi, *Essai statistique sur le royaume de Portugal* (1822), 142 and viii.

138. See Balbi's appendix: "Essay of a Statistical Table of the World . . . ," in Balbi, *A Statistical Essay on the Libraries* (1986), 138–46.

139. Balbi, *Essai statistique sur le royaume de Portugal* (1822), 2: 20.

140. Balbi, *Introduction à l'Atlas ethnographique du globe* (1822), 1: 41–42.

141. Linguistic geography became an established and respected subdiscipline toward the end of the nineteenth century as a result, largely, of the work of Gilliéron and Edmont. It is striking that Balbi is not mentioned as a predecessor in the works of either of these scholars. Nor is he even mentioned in the early histories of linguistic geography by Terracher, *Histoire des langues* (1929), and Dauzat, *La Géographie linguistique* (1922).

142. Contrast especially his many references to the Church, religion, and the Church hierarchy in Portugal in the sections in volume one on "administration and government structure," with his discussion of the Church in volume 2 in the section devoted to "ecclesiastical geography."

143. Balbi, *Essai statistique sur le royaume de Portugal* (1822), 1: contrast pages 143 and 400.

144. Balbi, *Introduction à l'Atlas ethnographique du globe* (1826), contrast xix and xx with lxxv.

145. Balbi, *Introduction à l'Atlas ethnographique du globe* (1826), lxxxi, xcviii, cxxiv.

146. Balbi, *Introduction à l'Atlas ethnographique du globe* (1826), xviii–xix; lxxxvi.

147. Balbi, *Introduction à l'Atlas ethnographique du globe* (1826), xx versus xcv; cxv and 44.

148. Balbi was very deeply committed to geography: all of his references to the established academic geographers were laudatory, sometimes overwhelmingly so; he tended to regard the academic geographers working in Paris as authorities on par with the greatest minds of the century; he classed geography as one of the most useful and important sciences; and he lamented the disrespect into which geography and statistical geography had fallen as a result of the unrigorous work of the popularizers who called themselves geographers. See in particular Balbi, *Introduction à l'Atlas ethnographique du globe* (1826), xlviii–xliv; and Balbi, *A Statistical Essay on the Libraries* (1986), 27, 62, 140, 147.

149. See Balbi's appendix "Essay of a Statistical Table of the World . . . ," in Balbi, *A Statistical Essay on the Libraries* (1986), 132.

150. The three former comprise sections of his *Essai statistique sur le royaume de Portugal* (1822). The latter is to be found in his *A Statistical Essay on the Libraries* (1986), 109.

151. Balbi, *Essai statistique sur le royaume de Portugal* (1822), 1: 214.

152. Balbi, *A Statistical Essay on the Libraries* (1986), 81.

153. Balbi, *Essai statistique sur le royaume de Portugal* (1822), 2: 232, 234.

154. Balbi, *Introduction à l'Atlas ethnographique du globe* (1826), cxviii–cxx. Given his sense of the importance of Wilhelm von Humboldt's work, it is interesting that when Balbi acknowledged the assistance of ninety-two scholars, Wilhelm von Humboldt is listed as having provided only a vocabulary (list of twenty-six words), not advice or criticism. See ibid., cxxi.

CHAPTER SEVEN

1. De Candolle, "Géographie botanique" (1820), 422. What de Candolle wanted, and missed, was a complete description of all of the physical features of the earth. Such description, of course, would require understanding sufficient to identify salient features. What he was asking for, then, was something along the lines of the 1:80,000 geological map of France for the whole world. This sort of work was only beginning in de Candolle's time and ultimately rested on the significant theoretical and empirical work carried out in geology in the early nineteenth century. Of interest too is de Candolle's characterization of this sort of description or mapping as "that part of general physics that is really part of geography."

2. See for example Omalius d'Halloy, "De la classification" (1834); Omalius d'Halloy, "Note additionnelle" (1838); Cortambert, *Mélanges géographiques* (1862); and Férussac, "Tableau générateur et analytique" (1821). Humboldt in his *Cosmos*

(39–40) listed the following as important attempts to classify the sciences: Le Père Gregoire Reisch, *Margarita Philosophica* 1486, 1504 . . . 1535 [he noted that only the 1513 edition had important geographic content]; Sir Françis Bacon, *Of the Advancement and Proficiency of Learning or the Partitions of the Sciences* (London: T. Williams, 1674); Denis Diderot and Jean d'Alembert, "Essai d'une distribution généalogique des sciences et des arts principaux," in Diderot and d'Alembert, *Supplement* (1780); André-Marie Ampère, *Carmen mnemonicum. Classification des connaissances humaines, ou Tableaux synoptiques des sciences et des arts* (n.l., n.d.); Reverend William Whewell, *The Philosophy of the Inductive Sciences Founded upon Their History* (London: J. W. Parker, 1840); Roswell Park, *Pantology, or a Systematic Survey of Human Knowledge* (Philadelphia: A. MacCay, 1847).

3. See for example Mackinder, *On the Scope* (1951 [1887]).

4. Bochner, *Eclosion and Synthesis* (1969); Cunningham and Jardine, "Introduction: The Age of Reflexion" (1990), 1–9; Outram, *Georges Cuvier* (1984), see especially p. 4.

5. Humboldt, *Essai sur le géographie des plantes* (1805), 13–14.

6. It is interesting here to compare the geographer Pinkerton's exclusion of travel from the purview of geography in Pinkerton, *Modern Geography* (1807), xxxii. Humboldt's irritated slap at Pinkerton in his *Essai sur la géographie des plantes* (1805), x–xi, suggests just how ridiculous he found such attitudes: "Working on location, I should not have expected to be reproached with bitterness* [footnote: *Géographie moderne de Pinkerton, translated by Walckenaer; vol. 6, pp. 174–77] for having found the course of rivers and the direction of mountain [chains] very different from the map by La Cruz; but it is the lot of travelers to displease when they observe facts that are contrary to received opinion."

7. De Candolle and von Humboldt, "Géographie Botanique" (1820), 422.

8. On the term "geognosy" and the complex contexts and history of its meanings, see Greene, *Geology* (1982), 38–39.

9. Humboldt takes the term and the mode of study from his teacher Abraham Gottlob Werner who shared with Humboldt a mining training. Humboldt, *A geognostical essay* (1823), 66–67, 80–82.

10. Humboldt, *A Geognostical Essay* (1823), 2.

11. Girault-Soulavie, *Histoire naturelle de la France méridionale* (1780), 1: 15–16.

12. De Candolle, "Géographie Botanique" (1820), 361.

13. De Candolle, "Géographie Botanique" (1820), 423.

14. Omalius d'Halloy, *De la classification* (1834), 5.

15. In using these last five terms, in addition to the more commonly used "géognosie," de Férussac was trying, in the first instance, to distinguish, yet combine, the results of sciences focussed on the inanimate earth and those focussed on living beings. The second part of each term was designed to facilitate differentiation between sciences of formation, descriptive sciences, or those focussed on "the manner of existing," and sciences which sought laws of existence. See "Tableau générateur et analytique" (1821).

16. Humboldt, *Cosmos.* (1849), 1: 28.

17. Humboldt, *Cosmos,* (1849), 1: 41

18. Humboldt's sense that his research was innovative, indeed that he was

inventing a new field, is clear from his description of the aims of his American expedition. See Humboldt, *Voyage aux régions équinoxiales* (1816), 1: 3

19. Humboldt, *A Geognostical Essay* (1823), vi.

20. Humboldt, *A Geognostical Essay* (1823), 20 and 23.

21. This is clear throughout his work but is expressed most directly in Humboldt, *Voyage aux régions équinoxiales* (1816), 1: 3–5.

22. The best expression of this intent is to be found in his concern to link his famous thematic diagram showing plant incidence as related to altitude entitled "Géographie des plantes équinoxiales. Tableau physique des Andes et Pays voisins. Dressé d'après des observations et des Mesures prises sur les lieux depuis le 10e degré de latitude boréale jusqu'au 10e de latitude australe en 1799, 1800, 1801, 1802 et 1803" with his topographic block diagrams depicting landscape form. See Hanno Beck and Wilhelm Bonacker's facsimile reproduction of Humboldt, *Atlas géographique et physique* (1969), lxxii.

23. Humboldt, *Voyage aux régions équinoxiales* (1816), 1: 99.

24. Smith, *European Vision* (1988), 4, 18, 27–29, 203–12.

25. Michael Dettelbach, in his superb essay in *Visions of Empire*, sees this commitment to the unity of the cosmos as part of a larger Humboldtian project investigating global physics. He argues that Humboldt was deeply influenced by both Laplacian physics, which sought to understand the world through the multiplication of measurement-based observations, and the German aesthetic response (in particular the aesthetic writings of Friedrich Schiller) to the Napoleonic imposition of modern centralized (anti-local) state control in the German principalities. See, Dettelbach, "Global Physics" (1996), 258–92.

26. "The physical sciences are held together by the same bonds that unite all the phenomena of nature." Humboldt, *Voyage aux régions équinoxiales*, (1816), 1: 5

27. As did many nineteenth-century geographers. See Rudwick, *The Great Devonian Controversy* (1985), 20; and Humboldt, *A Geognostical Essay* (1823), 1, 5, 34, 85.

28. One of his clearest statements on the nature of theory, its origins in empirical research, and its great distance from speculation came from a gentle but unequivocal refutation of Buachian theory. See Hanno Beck and Wilhelm Bonacker's facsimile reproduction of Humboldt, *Atlas géographique et physique* (1969), xlix–l.

29. Humboldt, *A Geognostical Essay* (1823), 78–80.

30. Humboldt, *A Geognostical Essay* (1823), 18.

31. Humboldt, "Des lignes isothermes" (1817 [1813]), 469.

32. Humboldt, "Des lignes isothermes" (1817 [1813]), 469.

33. Humboldt, "Des lignes isothermes" (1817 [1813]), 470–71.

34. Humboldt, *A Geognostical Essay* (1823), vi–vii.

35. Humboldt, *Voyage aux régions équinoxiales* (1816), 1: 6.

36. Humboldt, "Géographie des plantes équinoxiales" (1805 [1973]).

37. De Candolle and Humboldt, "Géographie botanique" (1820), 383–84, 413.

38. Minguet, *Alexandre de Humboldt* (1969), 76–77.

39. Humboldt, *The Fluctuations of Gold* (1900), 30. This work was first published as "Ueber die Schwankungen der Goldproduktion mit Rücksicht auf staatswirth-

schaftliche Probleme," in *Deutsche Vierteljahres Schrift* 1 (1838) IV, 1–40. I am indebted to Dr. Ingo Schwartz for this reference.

40. Humboldt and Bonpland, *Recueil d'observations* (1811), 298–304.

41. Humboldt, *The Fluctuations of Gold* (1900), 43.

42. Godlewska, "Map, Text and Image" (1995), 5–28.

43. Beck and Bonacker, *Atlas géographique et physique* (1969), plate 6 and discussion.

44. Humboldt, "Des lignes isothermes" (1813), 462.

45. Humboldt, "Des lignes isothermes" (1813), 462–63, 542, 602, and 496; Beck and Bonacker, *Atlas géographique et physique* (1969), lxxx.

46. Beck and Bonacker, *Atlas géographique et physique* (1969), lxxii.

47. Humboldt, *A Geognostical Essay* (1823), 58.

48. Humboldt, *Political Essay* (1811), 46.

49. Humboldt, "Sur la respiration des crocodiles" (1811), 253–59, 49–92.

50. Humboldt, *Voyage aux régions équinoxiales* (1816), 7.

51. Humboldt, *A Geognostical Essay* (1823), 76.

52. In his essay "Sur deux nouvelles espèces de crotales [rattlesnakes]," speaking of the different incidence of snakes between different parts of the world, he demonstrated at once the rigor and subtlety of comparisons and the critical role of sensitivity to the scale of evidence available to support such comparisons. Humboldt, "Sur deux nouvelles espèces de crotales" (1811), 1–8.

53. Humboldt, *Volcans des Cordillères* (1854), 5.

54. Beck and Bonacker, *Atlas géographique et physique* (1969), 6.

55. Beck and Bonacker, *Atlas géographique et physique* (1969), lxxviii.

56. Claval, *Régions, nations* (1968); and Chevalier, "L'Abbé Soulavie" (1986), 81–100.

57. De Candolle, "Géographie Botanique" (1820).

58. Omalius d'Halloy, *Division de la terre* (1839); and Omalius d'Halloy, *Observations sur un essai* (1823).

59. Humboldt, *Voyage aux régions équinoxiales* (1816). Humboldt periodically expresses his disgust with travel accounts, as on pp. 6–7. On pp. 48–51 Humboldt discusses the contradiction between the scientific study of peoples and their customs and the focus and approach of the travel narrative.

60. This is clear from his failure to complete his *Voyage aux régions équinoxiales* and from the choppy nature of its prose. Indeed, throughout this work, Humboldt seems unable to choose between a systematic focus on the one hand and narrative on the other, making abrupt transitions from one approach to the other. Systematic analysis was clearly his preference: he tells the reader that he reluctantly chose to write a narrative due to the delay and difficulty in publishing much of the systematic work.

61. Humboldt and Bonpland, *Recueil d'observations de zoologie* (1811); Humboldt, "Mémoire sur l'os hyoïde" (1811), 1–13; Cuvier, "Recherches anatomiques" (1811), 93–126.

62. Humboldt, *Sur les singes* (1811), 306.

63. De Candolle, "Géographie botanique" (1820), 365–66, 369, 371–72.

64. Humboldt, *A Geognostical Essay* (1823), 54.

65. Humboldt, *A Geognostical Essay* (1823), 81.

66. Humboldt, *A Geognostical Essay* (1823), 83–84.

67. For an excellent discussion of these proto-thematic maps, which Gilles Palsky describes as "cartes spéciales" and "cartes singulières," see Palsky, "Aux origines de la cartographie thématique" (1996), 129–45; and Palsky, *Des chiffres et des cartes* (1996).

68. Arthur Robinson and Helen Wallis argue that Humboldt's use of the isotherm and his presentation of a paper on isolines to the French Academy of Sciences in 1817 was "the catalyst for many similar uses of the isarithm in thematic cartography." Robinson and Wallis, "Humboldt's Map of Isothermal Lines" (1967), 122.

69. Wood (with Fels), *The Power of Maps* (1992), 25.

70. Beck and Bonacker, *Atlas géographique et physique* (1969), lxxxii.

71. Beck and Bonacker, *Atlas géographique et physique* (1969), lv.

72. Beck and Bonacker, *Atlas géographique et physique* (1969), lxvii.

73. Humboldt, "Des lignes isothermes" (1813), 462.

74. Humboldt, "Des lignes isothermes" (1813), 463.

75. Humboldt, "Des lignes isothermes" (1813), 510–11.

76. In time the space to be found in some thematic maps would become increasingly removed from geographic space, as in the case of genetic maps in which space is chemically rather than topologically defined.

77. Robinson and Wallis, "Humboldt's Map of Isothermal Lines" (1967), 122.

78. "Bandes isothermes, et distribution de la chaleur sur le globe," in Humboldt, "Des lignes isothermes" (1813), 602.

79. "Most natural phenomena have two distinct parts: one of which can be submitted to rigorous calculation; the other can only be attained by induction and analogy." Humboldt, "Des lignes isothermes" (1813), 545.

80. In his comments on this graphic he remarked: "This map shows the ebb and flow of metallic wealth. Here we can see the general west to east movement, which is in opposition to the flow of the oceans, the atmosphere, and the civilization of our species!" Beck and Bonacker, *Atlas géographique et physique* (1969), lxxxiii.

81. Beck and Bonacker, *Atlas géographique et physique* (1969), lxvi.

82. This graphic was adapted from one developed by August Crome. For this link I am grateful to Sybilla Nikolow, "Rendering the Strength of the State Visible: August Crome's Statistical Maps Around 1800," conference paper presented at New Perspectives on Alexander von Humboldt. An International Symposium, Georg-August-Universität Göttingen, 29–31 May 1997.

83. Beck and Bonacker, *Atlas géographique et physique* (1969), lxxxiii-lxxxiv.

84. Humboldt, *A Geognostical Essay* (1823), 476–77.

85. Beck and Bonacker, *Atlas géographique et physique* (1969), plate 21.

86. In his description of those diagrams in which he exaggerated the height of the formations to give a clearer sense of their appearance, he suggested why he ultimately abandoned his graphic pasigraphy. See Beck and Bonacker, *Atlas géographique et physique* (1969), plates 12–15 and p. lxx.

87. Humboldt, *A Geognostical Essay* (1823), 466–67.

88. Humboldt, *A Geognostical Essay* (1823), vi.

89. Beck and Bonacker, *Atlas géographique et physique* (1969), lxxii.
90. Humboldt, *Essai sur la géographie des plantes* (1805), v–vi.
91. Humboldt, *Essai sur la géographie des plantes* (1805), 19–23.
92. Humboldt, *Essai sur la géographie des plantes* (1805), 27–30.
93. Humboldt, *Essai sur la géographie des plantes* (1805), 30.
94. Foucault, "Fantasia of the Library" (1977), 87–109.
95. Humboldt, *Essai sur la géographie des plantes* (1805), 35.

CHAPTER EIGHT

1. This was the case with the map of Greece which Barbié du Bocage produced for the War Department in 1808. The War Department had recognized the poor quality of its information and, given the recent travels of Choiseul-Gouffier and others to Morea, had asked Barbié du Bocage to compile a map from these sources. See Berthaut, *Les Ingénieurs-géographes* (1902), 1: 277–78. On the poor quality of the map and others made from it by Barbié du Bocage, see Bory de Saint-Vincent, *Relation* (1836), 48; Walckenaer, *Géographie ancienne* (1839), lix; and in the manuscripts of the Société de géographie, Doc 3126 Colis 19, letter from Lapie to the President of the Commission (not personally named), Paris le 15 septembre 1826.

2. For a small selection of these, consult my bibliography. The Département des cartes et plans at the Bibliothèque nationale in Paris has a large proportion of Lapie's maps.

3. Indeed, it is precisely Lapie's lack of "erudition" which, in the eyes of geographic contemporaries and successors, limited his professional stature. See in particular the words of his biographer Alfred Maury, "Lapie" (1834), 23: 228–29.

4. Butlin, *Historical Geography* (1993), 12.

5. Barbié du Bocage, "Proposition d'une nouvelle méthode" (1795), 483–91. Note especially the last page, on which he refers to himself as having learnt geography "alone and without master."

6. Gosselin, "Géographie ancienne" (1810).

7. Walckenaer, *Dicuili Liber de mensura* (1807); *Recherches sur la géographie ancienne* (1822); "Notice bibliographique" (1825); *Géographie ancienne historique et comparée des Gaules* (1839).

8. Humboldt, *Examen critique* (1836), 5 volumes.

9. Jomard, *Discours prononcés* (1826), 2.

10. Barthélemy, *Voyage du jeune Anacharsis* (1788).

11. Barthélemy, *Voyage du jeune Anacharsis* (1788), vii.

12. Barbié du Bocage, *Recueil de cartes géographiques* (1788).

13. Barbié du Bocage, "Carte des marches et de l'Empire d'Alexandre-Le-Grand" (1804).

14. On the drama surrounding Fréret's authorship of antireligious tracts, see Walckenaer, *Rapport fait à l'Académie des inscriptions* (n.l., n.d. [but sometime after 1842]). For Sainte-Croix's role, see p. 45; a careful and muted expression of the reservation accorded Sainte-Croix's research is to be found in Letronne, "Observations historiques et géographiques sur le périple" (1825), 4.

15. Sainte-Croix, *Examen critique des anciens* (1775). This book won the Académie des inscriptions' prize for 1770. The subject was set by the Académie.

16. Sainte-Croix, *Examen critique des anciens* (1775), 101–6.

17. Barbié du Bocage, "Précis de Géographie ancienne" (1828), 609–734.

18. Nor, it would appear, did the political world have much interest in him. Jules Simon commented in 1885: "Even the Terror did not for an instant distract him from his habits. . . . And something remarkable happened to him: he was neither proscribed, nor arrested, nor accused." He was merely requisitioned by the War Department in 1794, along with most of the other geographers not already working for the state. It was after both he and his manuscripts had been deemed of little value to the war effort that he was assigned the editorship of Strabo's geography. Simon, *Une Académie sous le directoire* (1885), 404; on the activities of these geographers at the Dépôt de la guerre, see Berthaut, *Les Ingénieurs-géographes* (1902), 138, 152–53.

19. See Humboldt, *Examen critique* (1836), 1: 140, 144–45; Letronne, *Recueil des inscriptions* (1842), 1: 174–75, 188; Malte-Brun, *Précis de la géographie universelle* (1810), 1: 17 (but also many other references in this and his other works); Bory de St. Vincent, *Géographie* (1834), 5; Santarem, *Essai sur l'histoire de la cosmographie* (1848–1852), xv–xvi; Reference to the basis of Carette's work by Ravoisier, *Beaux-Arts Architecture et Sculpture* (1846–53), 2–5; Jarry de Mancy, *Atlas historique et chronologique* (1831); and Jomard, *Introduction à l'atlas* (1879).

20. Staum, "The Class of Moral and Political Sciences" (1980), 394.

21. In conjunction with the memoirs discussed here, Gosselin produced over seventy-five maps which either reproduced the world as known to particular Greek thinkers (Eratosthenes; Hipparchus; Polybius; Strabo; Marinus of Tyre; Ptolemy; Herodotus; Pytheas; Hecatus; Philemon; Timaeus; "Xenophonte Lampsaceno"; Pliny; Avienus; Tacitus; Megasthènes; "Déimaque"; "Patrocles"; Artemidorus; Diodorus; Agrippa; Onesicritus; Pomponius Mela; Solin; "Aethicus"; "Paul Orise"; Martianus Capella; the anonymous cosmographer of Ravenna; Isidora of Seville), or provided cartographic information for those reading about those periods. For the maps he had produced up to 1814, see Gosselin, *Atlas ou recueil des cartes* (1814).

22. Gosselin, *De l'Évaluation et de l'emploi des mesures* (1813). Read before the Institut de France 29 July 1804.

23. Gosselin, *Recherches sur le principe* (1817), 1: 501.

24. Letronne, "Mémoire sur cette question" (1822).

25. Letronne, *Notice sur la traduction d'Hérodote* (1823), 12. That this was very much Letronne's style is suggested by a description of Letronne in Walckenaer's notice on him, written for the Institut soon after Letronne's death. In that notice, Walckenaer says that one of the reasons he was appreciated in the academic world was for "those indirect criticisms, made with malicious modesty . . ." Walckenaer, "Notice historique sur la vie" (1850), 60.

26. Balbi, *Abrégé de géographie* (1833), xx.

27. Walckenaer, *Géographie ancienne historique et comparée des Gaules* (1839), xvii–xx. Walckenaer criticized not the numerical method employed by Gosselin, or the arguments that the ancients had extraordinarily accurate topographic information, but that they acquired this information from eastern civilizations. "That conjecture, which no ancient text supports, is not at all necessary to take into account the facts which M. Gosselin has so well discerned." xviij–xix.

28. Simon, *Une Académie sous le directoire* (1885), 393.

29. Malte-Brun, *Précis de la géographie universelle* (1810), 1: 525.

30. Humboldt, *Examen critique* (1836), 1: xxiii–xxiv.

31. Walckenaer, *Faune parisienne* (1802).

32. See in particular Walckenaer, *L'Ile de Wight* (1813), a romantic novel very much in the style of Bernardin de Saint-Pierre's *Paul et Virginie* (1823) with its morally uplifting tale and attention to landscape and natural beauty. Walckenaer's social and political conservatism is nowhere clearer than in this work. See also Walckenaer, *Essai sur l'histoire de l'espèce humaine* (1798). Like Bory de Saint-Vincent, he felt the need to engage in the current debate about the nature and distinctiveness of the human species. Rather than arguing, as did Bory de Saint-Vincent, for polygenesis, Walckenaer argued the distinctiveness of the human species as a whole by virtue of art, law, religion, industry, and human intelligence. It was a very general essay, too abstract and contradictory to make a real contribution to either scientific debate or social criticism. This is perhaps clearest in his footnoting style. Thus, his contention that peoples living in large open territories would ultimately be controlled by despotic government was supported by the following footnote: "Read the history of China, that of the Assyrians, the Babylonians, etc., and all of the great empires of Asia" (p. 195).

33. Most of this work was also geographic and is therefore discussed under that rubric.

34. Pinkerton, *Abrégé de géographie moderne* (1827).

35. Walckenaer, *Introduction à l'analyse géographique des itinéraires* (1839); Walckenaer, *Recherches sur la géographie ancienne* (1822).

36. Walckenaer, *Dicuili Liber de mensura* (1807).

37. Walckenaer, "Notice bibliographique" (1825).

38. Walckenaer, *Rapport fait à l'Académie des inscriptions* (n.l., n.d.).

39. Walckenaer, *Paroles prononcées à la Société de géographie* (1847).

40. A subject, apparently, still of some controversy. See Gopnik, "The First Frenchman" (1996), 44–53.

41. Walckenaer, *Rapport fait à l'Académie des inscriptions* (n.l., n.d.), 32.

42. Fréret's twenty-nine geographic publications included: studies of ancient measures and distances; analyses of ancient maps; descriptions and historical explanation of geographic inscriptions found in France; accounts of ancient routes; some descriptions of dramatic changes to physical geography in the past; studies on the locations of ancient peoples; studies identifying and locating particular ancient towns, rivers . . . ; studies on the meaning of particular words in use in Gaulish sources; reviews of works of key geographers; extracts of voyages, etc.

43. Walckenaer wrote in the introduction to his more substantive work on the geography of ancient Gaul in 1839 (by which time he was sixty-eight years old): "Among those studies, perhaps too varied, to which I devoted myself, there is none that usurped more of my leisure moments than geography. I dare say that I have always followed, with studious constancy, the great progress that that science has made in our day. I have tried to contribute to this progress myself and by seconding the work of others. Still, I have not yet published anything on one of the branches

of geographic science that has been the principal focus of my effort." Walckenaer, *Introduction à l'analyse gèographique des itinéraires* (1839), i.

44. Humboldt, *Examen critique* (1836), 1: préface.

45. Humboldt, *Examen critique* (1836). This is the overarching purpose of the entire book. It is most succinctly expressed in volume 3 on pages 12–14 and 248–49.

46. Humboldt, *Examen critique* (1836), 4: 108.

47. Humboldt, *Examen critique* (1836), 3: 352.

48. Humboldt, *Examen critique* (1836), 3: 25.

49. Humboldt, *Examen critique* (1836), 3: 228–31.

50. Humboldt, *Examen critique* (1836), 1: x.

51. Humboldt, *Examen critique* (1836). See in particular 1: 326–27; 3: 175ff. and 214ff.

52. Humboldt, *Examen critique* (1836), 3: 262–319 and especially 307–9.

53. Humboldt, *Examen critique* (1836), 3: 317–18 "Such is the complexity of human destiny that those same cruelties that bloodied the conquest of the two Americas have reappeared in those times we consider characterized by prodigious progress in enlightenment, by a general softening in customs. Yet a single man, barely in the middle of his career can have seen the Terror in France, the inhuman expedition to Saint-Domingue, political reactions and continental civil wars in America . . . Passions have coincided with an irresistible effort in the same circumstances whether in the nineteenth or the sixteenth century."

54. Humboldt, *Examen critique* (1836), 1: 254.

55. Humboldt, *Examen critique* (1836), vol. 1: 73–75, 97, and 272ff. (historical and contemporary influences on Columbus); vol. 4: 28–30, 187, 273ff. (on whether Amerigo Vespucci was trying to steal Columbus's thunder).

56. Humboldt, *Examen critique* (1836), 2: 50–56.

57. Humboldt, *Examen critique* (1836), 3: 352–98.

58. Humbert, Pantazzi, and Ziegler, *Egyptomania* (1994).

59. Barthélemy, *Voyage du jeune Anacharsis* (1788); Goethe, *Iphigenie auf Tauris* (1901 [1787]). On the importance of the Greek ideal in the work of Goethe and Schiller, see Gelzer, "Die Bedeutung der Klassischen Vorbilder" (1979).

60. Choiseul-Gouffier, *Voyage pittoresque de la Grèce* (1782–1822). Another similar work of the period is by Melling, *Voyage pittoresque de Constantinople* (1819).

61. France, Ministère de l'éducation nationale, *Expédition scientifique de Morée* (1831–1838).

62. Coing, "Roman Law and the National Legal Systems" (1979), 29–37.

63. Leigh, "Jean-Jacques Rousseau" (1979), 157 [155–68].

64. Loraux and Vidal-Naquet, "La Formation de l'Athènes bourgoise" (1979), 169–222.

65. Sandys, *A History of Classical Scholarship* (1967)

66. Barthélemy-Saint-Hilaire, "Letronne" [1859], 365–73.

67. Mentioned in de Férussac, *Bulletin général et universel des annonces* (1822), sixth section, third livraison.

68. He used the typographic hieroglyphs in Letronne, *Inscription grecque accompagné des noms hiéroglyphiques* (1843).

69. Robert, *L'Épigraphie grecque* [1939].

70. Letronne, *Recueil des inscriptions grecques et latines* (1842) 1: xxi.

71. See Barthélemy-Saint-Hilaire, "Letronne" [1859], 365–73; and Letronne, *Discours prononcé à la séance d'inauguration de l'école* (1847).

72. Walckenaer, "Notice historique sur la vie" (1850).

73. Letronne, *L'Isthme de Suez* (n.d.), 5–6.

74. Letronne, *Recueil des inscriptions grecques et latines* (1842) 1: 147.

75. Letronne, *Recherches pour servir à l'histoire de l'Égypte* (1823), xvj; and Letronne, *Nouvelles recherches sur le calendrier* (1863), 2.

76. Letronne, *Considérations générales sur l'évaluation des monnaies grecques et romaines* (1817), 8–11 and chapter 5.

77. See for example Letronne, *Extrait d'une notice écrite par M. Letronne* (1863).

78. Letronne, *Nouvelles recherches sur le calendrier* (1863), 2.

79. Letronne, *Recherches géographiques et critiques sur le livre* (1814), v.

80. Letronne, *Recueil des inscriptions grecques et latines* (1842), 1: iv–vii.

81. Letronne, *Recueil des inscriptions grecques et latines* (1842), 1: xli.

82. See Barthélemy-Saint-Hilaire, "Letronne" [1859] and Walckenaer, "Notice historique sur la vie" (1850).

83. Interestingly, the French philologist Ernest Renan wrote several decades later about the dangers of subordinating philology to education in the following terms: "Suppose that philology's only value was derived from its value in education. That would deprive it of its dignity. That would reduce it to pedagogy. That would be the worst of humiliations. The value of knowledge must be found in itself, not in the use one can make of it for the education of children. . . . What a strange vicious circle! If things are only good if they can be professed, and if only those who teach them, study them, what is the point of teaching them?" Cited in Seznec, "Renan et la philologie classique" (1979), 356 [349–62].

84. Walckenaer, "Notice historique sur la vie" (1850), 71ff.

85. Letronne, *Essai critique sur la topographie de Syracuse* (1812), 5.

86. Letronne, *Recherches géographiques et critiques* (1814).

87. I have consulted the second edition: Letronne, *Cours élémentaire de géographie* (1820). On the popularity of this work, see Barthélemy-Saint-Hilaire, "Letronne" [1859].

88. See the *Journal des savants* for the 1820s.

89. Strabo [Strabon], *Géographie* (1805–1819).

90. Walckenaer, *Dicuili Liber de mensura* (1807).

91. Jomard, *Discours prononcés aux funérailles* (1826), 4.

92. Such as when he censured "geographers" for their failure to incorporate any of the results of the Wilkinson expedition to Egypt into the maps and geographical treatises. Letronne, *Recueil des inscriptions grecques et latines* (1842) 1: 136.

93. Letronne, *Recueil des inscriptions grecques et latines* (1842), 1: dédication.

94. Hamilton [pseud.], Mathew Carey, *Aegyptiaca* (1810).

95. Pococke, *Aegypti ac nobilissimi* (1746).

96. Cailliaud, *Voyage à l'oasis de Thèbes* (1821).

97. Gau, *Antiquités de la Nubie* (1822–1827).

98. Fourier, "Premier Mémoire sur les monumens astronomiques" Vol., Anti-

quités, Mémoires, 2: 71–87 (1818); Jollois and Devilliers, "Description des monumens astronomiques" Vol., Antiquités, Descriptions, 1: 1–16 (1809).

99. Letronne, *Observations critiques et archéologiques* (1824).

100. Letronne, *Recueil des inscriptions grecques et latines* (1842), 1: vi–vii and
xi–xiii.

101. Letronne, *Recueil des inscriptions grecques et latines* (1842), 1: xxx. On
the respect with which Letronne's work is still regarded today, see J. Pouilloux,
"L'Épigraphie grecque"(1990), 319–24.

102. Letronne, *Recueil des inscriptions grecques et latines* (1842), 2: 254–55.

103. Letronne, *Recueil des inscriptions grecques et latines* (1842), 2: 205–10.

104. Letronne, *Deux inscriptions grecques gravées sur le Pylône* (n.d.), 28–29.

105. Letronne, *Matériaux pour l'histoire du Christianisme en Égypte* (1832), see
especially 40–55.

106. Letronne, *Récompense promise à qui découvra* (1833).

107. Letronne, *Considérations générales sur l'évaluation des monnaies* (1817), 14.

108. Pouilloux, "L'Épigraphie grecque" (1990), 320–21.

109. See especially Letronne, *Matériaux pour l'histoire du Christianisme* (1832);
and Letronne, "Explication d'une inscription grecque en vers" (1825).

110. For Letronne's sense of the importance of linking inscriptions and papyri,
see Letronne, *Recueil des inscriptions grecques et latines* (1842), 1: xxxiii.

111. Letronne, *Recherches géographiques et critiques* (1814), 86–90; see also
Letronne, *Examen du texte de Clément d'Alexandrie* (1828).

112. Letronne, *Essai critique sur la topographie de Syracuse* (1812), 78.

113. Picard, "Numismatique et épigraphie" (1990), 252–53.

114. Letronne, *Recherches géographiques et critiques* (1814), 31.

115. Letronne, *Mémoire sur le tombeau d'Osymandyas* (1822), 1–24.

116. Letronne, *Observations critiques et archéologiques* (1824), 6 and dédication.

117. On this, see Letronne, *Recueil des inscriptions grecques et latines* (1842),
1: iii–iv, xiii–xvi, xxix, xxxi–xxxii, xxxii–xxxiii; Letronne, *Recherches pour servir à
l'histoire de l'Égypte* (1823); Letronne, *Matériaux pour l'histoire du Christianisme*
(1832); Letronne, *Sur la séparation primitive des bassins* (1839).

118. Letronne, *Recherches géographiques et critiques* (1814), 35.

119. Letronne, *Recherches géographiques et critiques* (1814), 110.

120. Letronne, *Recueil des inscriptions grecques et latines* (1842), 1: 147.

121. See especially Letronne, *Recueil des inscriptions grecques et latines* (1842), 1:
138 and 174–99.

122. Letronne, *Matériaux pour l'histoire du Christianisme* (1832), 9ff.

123. Letronne, *Inscription grecque de Rosette* (1840).

124. Louis Robert is worth quoting on this: "Even after a century, the work of
Letronne is of primary importance. An aura of youth emanates from it. Not only is
his method that which we employ in our studies, but many of the interpretations he
developed, his conjectures and the corrections that he proposed have been confirmed
by subsequent work. It has been rightly said of him: 'he claimed the right to make
mistakes, but he seldom used that right.' One could add that even his errors were
instructive." Robert, *L'Épigraphie grecque* [1939], 8.

125. For clear statements on this, see Letronne, *Recherches géographiques et critiques* (1814), i, vi, 31.

126. Letronne, *Notice sur la traduction d'Hérodote* (1823), 3.

127. See Robert, *L'Épigraphie grecque* [1939], 12; and Letronne, *Matériaux pour l'histoire du Christianisme* (1832), 26.

128. Letronne, *Projet de diviser en sections* (1834), 2–3.

129. Letronne had a profound prejudice against what were then believed to be nonliterate cultures. He wrote about the exclusion of the cultures of Africa and North America from his plan in the following terms: "All that which, in this part of the world, is not Asiatic or European, belongs to people who, having neither literature nor history, are beyond the Academy's field of research. This also can be said of the Americas, which did not enter the realm of history until the fifteenth century. The languages of the Americas, being linked to no literature, and given the small number of its monuments with unknown dates and which are unattached to any history, could certainly be of interest to a philosopher but could never be the object of truly historical or literary research." Letronne, *Projet de diviser* (1834), 26–27.

130. Walckenaer, "Notice historique sur la vie" (1850), 101.

131. Letronne, "Observations historiques et géographiques sur le périple" (1825), 20.

132. Letronne, "Observations historiques et géographiques sur le périple" (1825), 5.

133. Letronne, *Essai critique sur la topographie de Syracuse* (1812), 92.

134. Letronne, *Considérations générales sur l'évaluation des monnaies* (1817), viii.

135. Letronne, *Sur la séparation primitive des bassins* (1839), 3–4.

136. Lacouture, *Champollion* (1988), 230.

137. Letronne, *Recherches critiques, historiques et géographiques sur les fragments d'Héron d'Alexandrie* (1851), v.

138. Letronne, *Recueil des inscriptions grecques et latines* (1842), 1: xliii.

139. Letronne, *Recueil des inscriptions grecques et latines* (1842), 1: xliii.

140. Letronne, *Nouvelles recherches sur le calendrier* (1863), 120.

141. Letronne, *Discours prononcé à la séance d'inauguration de l'école* [1847], 11.

142. Letronne, "Premier mémoire: Inscription grecque" (1832), 9.

143. Letronne, *Recueil des inscriptions grecques et latines* (1842), 1: 1.

144. Letronne, *Sur la séparation primitive des bassins* (1839).

145. Letronne, *Recueil des inscriptions grecques et latines* (1842), 1: 137 and 174–88.

146. Letronne, *L'Isthme de Suez* [1841].

147. Letronne, *Recueil des inscriptions grecques et latines* (1842), 1: 187.

148. Letronne, *Observations historiques et géographiques sur l'inscription d'une borne militaire* [1844–1845].

149. See Carette, "Recherches sur la géographie et le commerce de l'Algérie" (1844) and Carette, "Étude des routes suivies par les Arabes" (1844).

150. Letronne, *Sur quelques points de la géographie ancienne* (1845).

151. Letronne, *Nouvelles recherches sur le calendrier des anciens Éyptiens* (1863), 144.

152. Letronne, *Discours prononcé* ([1847])

153. Alexander von Humboldt cited Letronne's 1831 article on Atlas and seems to have had great respect for Letronne's scholarship and views. Indeed, he referred to Letronne as "mon savant et illustre ami." See Letronne, "Essai sur les idées cosmographiques" [1831], section 7; and Humboldt, *Examen critique* (1836), 1: 30, 37, 174–75, 179, 193; 2: 90; 3: 118–29. The Baron Walckenaer also cited Letronne, but only to disagree with him; see Walckenaer, *Géographie ancienne* (1839), 30.

154. Robert, *L'Épigraphie grecque* [1939], 6, 15 and 228.

155. Walckenaer, "Notice historique sur la vie" (1850), 61.

CONCLUSION

1. Skinner, in Mayhew, "Contextualizing Practice" (1994), 326.

References

Adanson, Michel. 1763. *Familles des plantes.* Paris: Chez Vincent.

Adas, Michael. 1989. *Machines as the Measure of Men: Science, Technology, and Ideologies of Western Dominance.* Ithaca and London: Cornell University Press.

Albitreccia, Antoine. 1938. Ajaccio. Étude de géographie humaine. *Annales de géographie* 47 (15 juillet): 361–72.

———. 1942. *Le Plan terrier de la Corse au 18e siècle: étude d'un document géographique.* Paris: Presses universitaires de France.

Albury, W. R., and D. R. Oldroyd. 1977. From Renaissance Mineral Studies to Historical Geology, in the Light of Michel Foucault's *The Order of Things. The British Journal for the History of Science* 10 (3, 36): 187–215.

Alembert, Jean Le Rond d'. 1753. *Essai sur la société des gens de lettres et de grands, sur la réputation, sur les Mécènes, et sur les récompenses littéraires.* Paris.

Alfred. 1938. *Histoire politique, anecdotique et littéraire du 'Journal des débats.'* Paris.

Allent, Général Pierre-Alexandre-Joseph. 1800 (an IX). *Mémoire sur la réunion de l'artillerie et du génie, adressé au premier consul de la République française.* Paris: Duprat.

———. 1829. Essai sur les reconnaissances militaires. In *Mémorial du Dépôt de la guerre imprimé par ordre du Ministre 1802–03.* Paris: Reprint. Ch. Picquet.

———. 1829. Sur les échelles de pente, et les autres lignes caractéristiques de la surface du sol. In *Mémorial du Dépôt de la guerre imprimé par ordre du Ministre 1802–03.* Paris: Reprint. Ch. Picquet.

———. N.d. *Chambre des Pairs . . . Séance du . . . 2 mars 1818. Discours prononcé par M. le chevalier Allent, . . . pour la défense du projet de loi relatif au recrutement de l'armée . . .* Vol. 2, *Chambre des Pairs. Impressions diverses. Session de 1817.* Paris: Imprimerie de P. Didot l'aîné.

Al-Rahim, A. Abd. 1984. Land Tenure in Egypt and Its Social Effects on Egyptian Society: 1798–1813. In *Land Tenure and Social Transformation in the Middle East,* edited by T. Khalidi. Beirut: American University of Beirut.

Amarigilo, Jack, Stephen Resnick, and Richard D. Wolff. 1990. Division and Difference in the Discipline of Economics. *Critical Inquiry* 17 (1):108–37.

Amat, Roman d'. 1959. Chabrol, Gilbert-Joseph-Gaspard de. In *Dictionnaire de biographie française,* edited by M. Prevost and R. d'Amat. Paris: Librairie Letouzey.

Ampère, André-Marie. 1834–1843. *Essai sur la philosophie des sciences, ou Exposition analytique d'une classification naturelle de toutes les connaissances humaines . . .* 2 vols. Paris: Bachelier.

André. Statistique de Verone et ses environs, par André, Cicille et Turgot, ingénieurs-géographes. Service historique de l'armée de terre, Vincennes. Manuscript, 1380.

———. 1801. Mémoires topographiques, statistiques et militaires concernant la région située entre l'Adda et l'Adige, par les ingénieurs-géographes, Martinel, Duvivier, Laignelot, André, Cabos, Cicille, Lerouge, Pasquier, Santerre, et Velasco. Service historique de l'armée de terre, Vincennes. Manuscript, 1377 (27 cahiers, 1 carton).

———. 1802. Mémoires et notes sur la feuille 116 de la carte d'Italie, par André, Cabos, Chaboud, Cicille, Duvivier, Laignelot, Lerouge, Prato, Santerre, Velasco, ingénieurs-géographes, 15–18 janvier 1802. Service historique de l'armée de terre, Vincennes. Manuscript, 1379 (4 cahiers de 10, 13, 15 et 21 pages).

———. 1807. Cahiers topographiques, par arrondissement, des départements du Léman et du Mont-Blanc, par André et Bertre, ingénieurs-géographes (18 cahiers). Service historique de l'armée de terre, Vincennes. Manuscript, 1365.

Annales générales des sciences physiques. 1819–1821. Edited by J.-B.-G.-M. Bory de Saint-Vincent, Drapier, and V. Mons. 8 vols. Bruxelles: Weissenbruch.

Anville, Jean-Baptiste Bourguignon d'. 1735. *Proposition d'une mesure de la terre dont il résulte une diminution considérable dans sa circonférence sur les parallèles . . .* Paris: Chaubert.

———. 1738. *Réponse de M. d'Anville, . . . au Mémoire envoyé à l'Académie royale des sciences, par M. Simonin . . . contre la mesure conjecturale des degrez de l'Équateur, en conséquence de l'étendue de la mer du Sud.* N.p.

———. 1743. *Mémoire instructif pour dresser sur les lieux des cartes particulières et topographiques d'un canton de pays renfermant dix ou douze paroisses . . .* N.p.

———. 1760. *Notice de l'Ancienne Gaule, tirée des monumens Romains.* Paris: Desaint et Saillant.

———. 1766. *Mémoires sur l'Égypte ancienne et moderne, suivis d'une description du golfe Arabique ou de la mer Rouge . . .* Paris: Imprimerie royale.

———. 1768. *Géographie ancienne abrégée . . .* 3 vols. Paris: Chez Merlin.

———. 1777. *Considérations générales sur l'étude et les connoissances que demande la composition des ouvrages de géographie . . .* Paris: Imprimerie de Lambert.

———. 1838. Description de la Gaule, à l'époque ou les Francs s'y sont établis. In *Collection des meilleurs dissertations, notices et traités particuliers relatifs à l'histoire de France, composée en grande partie de pièces rares, ou qui n'ont jamais été publiées séparément pour servir à compléter toutes les collections de mémoires sur cette matière,* edited by J.-M.-C. Leber. Paris: J.-G. Dantu.

Après de Mannevillette, Jean-Baptiste-Nicolas-Denis d'. 1765. *Mémoire sur la navigation de France aux Indes . . .* Paris: Imprimerie royale.

Arbellot, Guy. 1973. La Grande Mutation des routes de France au milieu de XVIIIe siècle. *Annales, Economies, Société, Civilisations* 28: 765–90.

Arbellot, Guy, Bernard Lepetit, and Jacques Bertrand. 1987. Routes et communications. In *Atlas de la Révolution française,* edited by S. Bonin and C. Langlois. Paris: Éditions de l'École des hautes études en sciences sociales.

Artz, Frederick B. 1966. *The Development of Technical Education in France, 1500–1850.* Cambridge, MA, and London: MIT Press.

Astruc, Jean. 1737. Représentation des divers vents du Languedoc et tentative d'en expliquer le mécanisme par la disposition du relief.

Aucoc, Léon. 1889. *L'Institut de France. Lois, statuts et règlements concernant les anciennes académies et l'Institut, de 1635 à 1889. Tableau des Fondations*. Paris: Imprimerie nationale.

Audouin, Jean Victor, Isiard Bourdon, Alexandre Brongniart, Augustine Pyramus de Candolle, and Jean-Baptiste-Geneviève-Marcellin Bory de Saint-Vincent, eds. 1822–1831. *Dictionnaire classique d'histoire naturelle, par Messieurs Jean Victor Audouin, Isiard Bourdon, Alexandre Brongniart, Augustine Pyramus de Candolle . . . et Bory de Saint-Vincent.* Paris: Rey et Gravier and Beaudouin frères.

Aulard, François Victor Alphonse. 1911. *Napoléon 1er et le monopole universitaire; origines et fonctionnement de l'Université impériale.* Paris: Librairie Armand Colin.

Auroux, Sylvain. 1984. Linguistique et anthropologie en France (1600–1900). In *Histoires de l'anthropologie (XVI—XIX siècles): Colloque sur la pratique de l'anthropologie aujourd'hui, 19–21 novembre 1981, Sèvres,* edited by B. Rupp-Eisenreich. Paris: Klincksieck.

Bacler d'Albe, Louis-Albert-Ghislain Baron de. Rapport de M. d'Albe à l'Empereur, 20 février 1811 (2 pages in manuscript). Service historique de l'armée de terre, Vincennes. Manuscript, 1221.

———. 1798 (an VI). Carte générale du théâtre de la guerre en Italie et dans les Alpes depuis le passage du Var le 29 7bre 1792 V.S. jusqu'à l'entrée des Français à Rome le 22 pluviôse an 6 . . . Milan: Chez l'auteur.

———. 1802. Carte du théâtre de la guerre en Italie lors des premières campagnes de Bonaparte.

Baker, Keith Michael. 1975. *Condorcet: From Natural Philosophy to Social Mathematics.* Chicago and London: University of Chicago Press.

Balbi, Adrien. 1822. *Essai statistique sur le royaume de Portugal et d'Algarve, comparé aux autres états de l'Europe, et suivi d'un coup d'oeil sur l'état actuel des sciences, des lettres et des beaux-arts parmi les Portugais des deux hémisphères.* Paris: Chez Rey et Gravier, Libraires.

———. 1826. *Atlas ethnographique du globe, ou Classification des peuples anciens et modernes d'après leurs langues, précédé d'un discours sur l'utilité et l'importance de l'étude des langues appliquée à plusieurs branches des connaissances humaines; d'un aperçu sur les moyens graphiques employés par les différens peuples de la terre; d'un coup-d'oeil sur l'histoire de la langue slave, et sur la marche progressive de la civilisation et de la littérature en Russie, avec environ sept cents vocabulaires des principaux idiomes connus, et suivi du tableau physique, moral et politique des cinq parties du monde . . .* Paris: Rey et Gravier.

———. 1826. *Introduction à l'Atlas ethnographique du globe (contenant un discours sur l'utilité et l'importance de l'étude des langues appliquée à plusieurs branches des connaissances humaines; un aperçu sur les moyens graphiques employés par les différens peuples de la terre; des observations sur la classification des idiomes décrits dans l'atlas; un coup-d'oeil sur l'histoire de la langue slave et sur la marche progressive de la civilisation et de la littérature en Russie . . .), ou Classification des peuples anciens et modernes d'après leur langue.* Paris: Chez Rey et Gravier.

————. 1833. *Abrégé de géographie rédigé sur un nouveau plan . . .* Paris: Chez Jules Renouard.

————. 1986 [1835]. *A Statistical Essay on the Libraries of Vienna and the World [Original title: Essai statistique sur les bibliothèques de Vienne précédé de la statistique de la Bibliothèque impériale comparé aux plus grands établissemens de ce genre anciens et modernes . . .].* Translated by Larry Barr (eds. Larry and Janet L. Barr). Jefferson, North Carolina, and London: McFarland and Co., Inc.

Barbié du Bocage, Jean-Denis. 1788. *Recueil de cartes géographiques, plans, vues et médailles de l'ancienne Grèce, relatifs au voyage du jeune Anacharsis, précédé d'une analyse critique des cartes.* Paris: Chez de Bure.

————. 1795. *Notice sur une géographie en Grec vulgaire . . . Imprimée à Vienne en Autriche en 1790.*

————. 1795. Proposition d'une nouvelle méthode pour enseigner la géographie, par J. D. Barbié, chargé de la partie géographique à la Bibliothèque nationale. *Magazin encyclopédique, ou Journal des sciences, des lettres et des arts* 2: 483–91.

————. 1803. Carte semi-topographique de la Morée, dressée et gravée au dépôt général de la guerre . . . 1807; une grande feuille.

————. 1804. Carte des marches et de l'Empire d'Alexandre-Le-Grand, dressée pour l'ouvrage de Mr. de Ste. Croix, intitulé *Examen des historiens d'Alexandre,* par J. D. Barbié du Bocage, Géographe des relations-extérieures. An XIIIème.

————. 1809. *Note sur la géographie de Pinkerton (Turquie d'Asie).*

————. 1811. Précis de la géographie ancienne. In *Abrégé de géographie moderne, rédigé sur un nouveau plan . . . ,* edited by J. Pinkerton and C. A. Walckenaer. Paris.

————. 1819. *Plans itinéraires du Bosphore de Thrace et de la mer Marmara, avec le canal des Dardanelles, indiquant les points de vue des tableaux du 'Voyage pittoresque de Constantinople.' Analyse des plans, et mémoire sur la topographie de Constantinople, du Bosphore et de ses environs.* Paris: Chez les éditeurs.

————. 1822. [Presidential Address]. *Bulletin de la société de géographie* 1: 9–10.

————. 1828. Précis de Géographie ancienne. In *Abrégé de la géographie moderne ou description historique, politique, civile et naturelle des empires, royaumes, états et leurs colonies, avec celle des mers et des îles de toutes parties du monde,* edited by J. Pinkerton, C. A. Walckenaer, and a. J.-B.-B. Eyriés. Paris: Dentu, Imprimeur-Libraire.

————. 1829. Notice historique et analytique sur la construction des cartes géographiques. In *Mémorial du Dépôt de la guerre imprimé par ordre du Ministre 1802–03.* Paris: Ch. Picquet.

Barret-Kriegel, Blandine. 1988. *Les Académies de l'histoire . . .* Paris: Presses universitaires de France.

Barthélemy, Jean-Jacques. 1788. *Voyage du jeune Anacharsis en Grèce, dans le milieu du quatrième siècle avant l'ère vulgaire.* Paris: Chez de Bure.

Barthélemy-Saint-Hilaire, Jules. [1859]. Letronne, Antoine-Jean. In *Biographie universelle (Michaud) ancienne et moderne, ou histoire, par ordre alphabétique, de la vie publique et privée de tous les hommes qui se sont fait remarquer par leurs écrits, leurs actions, leurs talents, leurs vertus ou leurs crimes,* edited by C. Desplaces. Paris: Chez Madame C. Desplaces; and Leipzig: Librairie de F. A. Brockhaus.

Bauer, Henry. 1990. A Dialectical Discussion on the Nature of Disciplines and Disciplinarity: The Antithesis. *Social Epistemology* 4 (2): 215–27.

Baumgärtel, Hans. 1969. Alexander von Humboldt: Remarks on the Meaning of Hypothesis in His Geological Researches. In *Toward a History of Geology (Proceedings of the New Hampshire Inter-Disciplinary Conference on the History of Geology, September 7–12, 1967)*, edited by C. J. Schneer. Cambridge, MA, and London: MIT Press.

Beche, Sir Henry Thomas de la, ed. 1836. *A Selection of Geological Memoirs Contained in the Annales des Mines written by Brongniart, Humboldt, von Buch, and others.* London: William Phillips.

Becher, Tony. 1989. *Academic Tribes and Territories: Intellectual Enquiry and the Cultures of Disciplines.* Milton Keynes, England: Open University Press.

Bell, Morag, Robin A. Butlin, and Michael J. Heffernan, eds. 1995. *Geography and Imperialism 1820–1940.* Manchester: Manchester University Press.

Bellanger, Claude, Jacques Godechot, Pierre Guiral, and Fernand Terrou. 1969–1976. *Histoire générale de la presse française.* Paris: Presses universitaires de France.

Bellin, Jacques-Nicolas. 1755. *Remarques sur la carte de l'Amérique septentrionale comprise entre le 28e et le 72e degré de latitude avec une description géographique de ces parties, . . .* Paris: Imprimerie de Didot.

———. 1763. *Description géographique de la Guyane, contenant les possessions et les établissemens des françois, des espagnols, des portugais, des hollandois dans ces vastes pays . . .* Paris: Imprimerie de Didot.

———. N.d. *L'Enfant géographe, ou Nouvelle méthode d'enseigner la géographie . . .* Paris: Bellin.

Ben-David, Joseph. 1970. The Rise and Decline of France as a Scientific Centre. *Minerva* 8 (2): 160–79.

———. 1971. *The Scientist's Role in Society: A Comparative Study.* Englewood Cliffs, NJ: Prentice-Hall.

Benoiston de Châteauneuf, Louis-François. 1824. *Considérations sur les enfants trouvés dans les principaux états de l'Europe, . . . mémoire lu à l'Académie royale des sciences, dans la séance du 11 août 1823.* Paris: L'auteur.

———. 1824. *Extraits des Recherches statistiques sur la ville de Paris et le département de la Seine, recueil de tableaux dressés et réunis d'après les ordres de M. le comte de Chabrol . . . [Extrait du Bulletin universel des sciences et de l'industrie, 1824].* Paris: au bureau de "Bulletin."

Bérard, François, Denis Feissel, Pierre Petitmengin, and Michel Sève. 1986. *Guide de l'Epigraphiste. Bibliographie choisie des épigraphes antiques et médiévales.* Paris: Presses de l'École normale supérieure.

Berchtold, Leopold von. 1797. *Essai pour diriger et étendre les recherches des voyageurs qui se proposent l'utilité de leur patrie . . .* 2 vols. Paris: Du Pont.

Berdoulay, Vincent. 1981. *La Formation de l'école française de géographie 1870–1914.* Paris: Bibliothèque nationale.

Bergeron, Louis. 1979. Les Miroirs de la ville: un débat sur le discours des anciens géographes. *Urbi* 2 (décembre).

Berjaud, Léo. 1950. *Boutin, agent secret de Napoléon 1er et précurseur de l'Algérie française.* Paris: F. Chambriand.

Bernand, André. 1992. *La Prose sur pierre dans l'Égypte hellénistique et romaine.* 2 vols. Paris: CNRS.

Bernardin de Saint-Pierre, Jacques-Henri. 1823. *Paul et Virginie.* Paris: L. Janet.

Bertaud, Jean-Paul, Daniel Reichel, and Jacques Bertrand. 1989. L'Armée et la guerre. In *Atlas de la Révolution française,* edited by S. Bonin and C. Langlois. Paris: Éditions de l'École des hautes études en sciences sociales.

Berthaut, Colonel Henri-Marie-Auguste. 1898–1899. *La Carte de France, 1750–1898, étude historique . . .* 2 vols. Paris: Imprimerie du Service géographique de l'armée.

———. 1902. *Les Ingénieurs-géographes militaires, 1624–1831, étude historique . . .* 2 vols. Paris: Imprimerie du Service géographique de l'armée.

———. 1912. *Connaissance de terrain et lecture des cartes.* Edited by Service géographique de l'armée. Paris: Imprimerie du Service géographique.

Berthelot, Sabin. 1839. Rapport sur les travaux de la Société de géographie de Paris et sur le progrès de la science pendant l'année 1839. *Bulletin de la Société de géographie de Paris* 12: 306–7.

———. 1843. Rapport sur les travaux de la Société de géographie de Paris et sur le progrès de la science pendant l'année 1843 . . . *Bulletin de la Société de géographie de Paris* 20: 333–34.

Berthier, Alexandre. 1802–1803. Rapport du Ministre de la guerre au Consuls de la République sur les travaux du Dépôt général de la guerre pendant le cours de l'an X (1802). In *Mémorial du Dépôt de la guerre,* edited by France. Ministère de la guerre. Dépôt de la guerre. Paris: Ch. Picquet.

Bertin, Jacques. 1968. *Sémiologie graphique. Les diagrammes, les réseaux, les cartes. Avec la collaboration de Marc Barbut.* Paris: La Haye, Mouton.

Besnard, François-Yves. 1979. *Souvenirs d'un nonagénaire.* Vol. 1: 336. Paris: H. Champion, 1880. Reprint. Marseilles: Lafitte Reprints.

Best, Geoffrey. 1982. *War and Society in Revolutionary Europe, 1770–1870.* New York: St. Martin's Press.

Bevis, Richard. 1999. *The Road to Egdon Heath: The Aesthetics of the Great in Nature.* Montreal and Kingston: McGill/Queen's University Press.

Biographie universelle et portative des contemporains. 1834. Férussac, André-Étienne-Just-Pascal-Joseph-François d'Audebard, baron de. In *Biographie universelle et portative des contemporains, ou Dictionnaire historique des hommes vivants, et des hommes morts depuis 1788 jusqu'à nos jours . . . ,* edited by A. Rabbe, C.-A. Vielih de Boisjolin, and F.-G. B. de Sainte-Preuve. Paris: Chez F. G. Levrault.

Blainville, Henri-Marie Ducrotay de. 1845. *Histoire des sciences de l'organisation et de leurs progrès, comme base de la philosophie, . . . rédigée d'après ses notes et ses leçons faites à la Sorbonne de 1839 à 1841, avec les développements nécessaires et plusieurs additions . . .* 3 vols. Paris et Lyon: Périsse frères.

Blair, John. 1779. On the Rise and Progress of Geography. In *The Chronology and History of the World, from the Creation to the Year of Christ 1779 . . . ,* edited by J. Blair. London.

Blanchard, Anne. 1979. *Les Ingénieurs du "roy" de Louis XIV à Louis XVI: Étude du corps des fortifications.* Montpellier: Université Paul-Valéry.

Blunt, Alison, and Gillian Rose, eds. 1994. *Writing Women and Space: Colonial and Postcolonial Geographies.* New York and London: The Guilford Press.

Bochner, Salomon. 1969. *Eclosion and Synthesis: Perspectives on the History of Knowledge.* New York and Amsterdam: W. A. Benjamin, Inc.

Boelhower, William. 1988. Inventing America: A Model of Cartographic Semiosis. *Word and Image* 4, 2: 475–97.

Bois-Aymé, M. du. 1809. Mémoire sur la ville de Qoçeyr et ses environs et sur les peuples nomades qui habitent cette partie de l'ancienne Troglodytique. In *Description de l'Égypte,* edited by Edme-François Jomard, État Moderne, 1: 193–202. Paris: Imprimerie impériale.

———. 1809. Mémoire sur les anciennes branches du Nil et ses embouchures dans la mer. In *Description de l'Égypte,* edited by Edme-François Jomard, Antiquités, Mémoires, 1: 277–90. Paris: Imprimerie impériale.

———. 1809. Mémoire sur les tribus arabes des déserts de l'Égypte. In *Description de l'Égypte,* edited by Edme-François Jomard, État Moderne, 1: 577–607. Paris: Imprimerie impériale.

———. 1813. Notice sur le séjour des Hébreux en Égypte et sur leur fuite dans le désert. In *Description de l'Égypte,* edited by Edme-François Jomard, Antiquités, Mémoires, 1: 291–324. Paris: Imprimerie impériale.

Booker, Peter Jeffery. 1961–1962. Gaspard Monge (1746–1818) and His Effect on Engineering Drawing and Technical Education. *Transactions of the Newcomen Society* 34: 15–36.

Bory de Saint-Vincent, Jean-Baptiste-Geneviève-Marcellin. 1803 (an XI). *Essais sur les Isles Fortunées et l'antique Atlantide, ou, Précis de l'histoire générale de l'archipel des Canaries . . .* Paris: Baudouin.

———. 1804 (an XIII). *Voyage dans les quatre principales îles des mers d'Afrique, fait par ordre du Gouvernement, pendant les années neuf et dix de la République (1801 et 1802), avec l'histoire de la traversée du capitaine Baudin jusqu'au Port-Louis de l'île Maurice . . .* 3 vols. Paris: F. Buisson.

———. 1815. *Justification de la conduite et des opinions de M. Bory de Saint-Vincent . . . compris dans l'ordonnance du 24 juillet 1815.* Paris: Chez les marchands de nouveautés.

———. 1819. *Description du plateau de St.-Pierre de Maestricht . . .* Reprint from the *Annales générales des sciences physiques,* vol. 1. Bruxelles: Weissenbruch, Imprimerie du roi et de la ville.

———. 1823. *Guide du voyageur en Espagne . . .* 2 vols. Paris: Louis Janet.

———. 1823. Notice sur la nouvelle carte d'Espagne. In *Histoire d'Espagne depuis la plus ancienne époque jusqu'à la fin de l'année 1809, . . . traduite de l'anglais et continuée jusqu'à la Restauration de 1814, ouvrage revu et corrigé par le Comte Mathieu Dumas, auteur du Précis des événements militaires,* edited by J. Bigland. Paris: Firmin Didot père et fils.

———. 1824. *Notice sur la nouvelle carte d'Espagne jointe à cet ouvrage.*

———. 1824. *Nouvelle Carte d'Espagne et de Portugal, dressée et dessinée par Bory de Saint-Vincent.* Paris: Giraldon-Bovinet.

———. 1825. Géographie, sous les rapports de l'histoire naturelle. In *Dictionnaire classique d'histoire naturelle . . .* , edited by J.-B.-G.-M. Bory de Saint-Vincent. Paris: Rey et Gravier et Beaudouin frères.

———. 1826. *Résumé géographique de la péninsule ibérique, contenant les royaumes de Portugal et d'Espagne . . .* Paris: A. Dupont et Roret.

———. 1827. *L'Homme (homo), essai zoologique sur le genre humain; 2e édition, enrichie d'une carte nouvelle pour l'intelligence de la distribution des espèces d'hommes à la surface du globe terrestre . . .* 2nd ed. 2 vols. Paris: Rey et Gravier.

———. 1827. Notice biographique sur Malte-Brun. *Revue encyclopédique* (décembre).

———. 1832–1836. Section des sciences physiques . . . In *Expédition scientifique de Morée*, edited by France. Ministère de l'éducation nationale. Paris: F. G. Levrault.

———. 1835. *Compendio d'herpetologia o historia naturale dei rettili.* Milano: Presso A. F. Stella e Figli.

———. 1836. *Atlas.* Edited by France. Minist[e]re de l'éducation nationale, *Expédi n.d.tion scientifique de la Morée (1831–1838).* Paris and Strasbourg.

———. 1836. *Relation du voyage de la Commission scientifique de Morée dans le Péloponnèse, les Cyclades et l'Attique . . .* Edited by France. Ministère de l'éducation nationale. 3 vols. Vol. 1, *Expédition scientifique de Morée (1831–1838). Section des sciences physiques.* Paris: F.-G. Levrault.

———. 1838. *Géographie.* Edited by F. M. d. l. é. nationale. 3 vols. Vol. 2, part 1, *Expédition scientifique de Morée (1831–1838). Section des sciences physiques.* Paris: F.-G. Levrault.

———. 1838. Rapport concernant la géographie et la topographie. *Rapports de la commission chargée de rédiger des instructions pour l'exploration scientifique de l'Algérie (Extrait des Comptes rendues des séances de l'Académie des sciences séance du 23 juillet 1838), 49–53.*

———. [1838]. Note sur la Commission exploratrice et scientifique de l'Algérie, présentée à S. Exc. le ministre de la Guerre. Paris: Impr. de Cosson, n.d.

———. 1842. *Traité élémentaire d'erpétologie; ou, d'histoire naturelle des reptiles.* Paris: Mairet et Fournier.

———. 1844 (Prospectus). *La France par tableaux géographiques et statistiques, gravés au burin et coloriés . . . avec une introduction historique servant à coordonner les différentes parties de l'ouvrage, publié sous la direction de M. le baron Bory de Saint-Vincent . . .* Paris: A. Boulland.

———. 1845. Introduction historique servant à coordonner les différentes parties de l'ouvrage . . . In *Atlas géographique, statistique et progressif des départements de la France et de ses colonies . . .* , edited by A.-M. Perrot, Achin, and J. B. P. Tardieu. Paris: Dépôt de la guerre.

———. 1846–1867. Flore de l'Algérie, Cryptogamie. In *Botanique,* in *Exploration scientifique de l'Algérie pendant les années 1840, 1841, 1842 . . . Sciences naturelles,* edited by France. Commission scientifique de l'Algérie. Paris: Imprimerie impériale.

———. *Dictionnaire classique d'histoire naturelle, rédigé par une société de natural-*

istes, avec une nouvelle distribution des corps naturels en cinq règnes, . . . Paris: Imprimerie de Rignoux. N.d.

———. [1845]. *Sur l'anthropologie de l'Afrique.* N.l. N.d.

Botting, Douglas. 1973. *Humboldt and the Cosmos.* London: Michael Joseph.

Bouchu, Baron. 1816. *Prospectus de l'École royale polytechnique.* Paris: Madame V. Courcier.

Boudard, René. 1956. Un Préfet napoléonien en Ligurie: le Comte de Chabrol-Volvic. *Revue de l'Institut Napoléon,* 119–30.

———. 1968. La Réunion de Gênes à l'empire français en juin 1805. *Revue de l'Institut Napoléon.*

———.1969. La Mission du préfet Chabrol, "Géolier" de Pie VII, à Savone, (1809–1812). *Revue de l'Institut Napoléon* 112 (juillet): 181–88.

Boudet. 1818. Notice historique de l'art de la verrerie né en Égypte. In *Description de l'Égypte,* edited by Edme-François Jomard, Antiquités, Mémoires, 2: 17–38. Paris: Imprimerie impériale.

Bourcet, Lieutenant-général Pierre-Joseph de. 1875–1876. Mémoires sur les reconnaissances militaires, attribués au général Bourcet. *Journal de la librairie militaire,* 1re–2e années.

Bourguet, Marie-Noëlle. 1989. *Déchiffrer la France: La statistique départementale à l'époque napoléonienne.* Paris: Éditions des archives contemporaines.

———, Bernard Lepetit, Daniel Nordman, Maroula Sinarellis. 1998. *Invention scientifique de la Méditerranée. Égypte, Morée, Algérie.* Paris: Éditions de l'École des hautes études en sciences sociales.

Bouteiller, Mlle M. 1956. La Société des observateurs de l'homme, ancêtre de la société d'anthropologie de Paris. *Bulletins et mémoires de la société d'anthropologie de Paris* 10 (7): 448–65.

Bowen, Margarita. 1981. *Empiricism and Geographical Thought: From Francis Bacon to Alexander von Humboldt.* Cambridge: Cambridge University Press.

Boyer, Ferdinand. 1964. Les Responsabilités de Napoléon dans le transfert à Paris des oeuvres d'art de l'étranger. *Revue d'histoire moderne et contemporaine* 11 (octobre–décembre): 241–62.

Brandenburg, David J. 1950. Agriculture in the *Encyclopédie:* An Essay in French Intellectual History. *Agricultural History* 24: 96–108.

Bret, Patrice. 1991. Le Dépôt général de la guerre et la formation scientifique des ingénieurs-géographes militaires en France (1789–1830). *Annals of Science* 48: 113–57.

Brian, Éric. 1994. *La Mesure de l'état. Administrateurs et géomètres au XVIIIe siècle.* Paris: Albin Michel.

Briet, Le Père Philippe. 1648–1649. *Parallela geographiae veteris et novae.* Paris: S. Cramoisy and G. Cramoisy.

Broc, Numa. 1969. *Les Montagnes vues par les géographes et les naturalistes de langue française au XVIIIe siècle. Contribution à l'histoire de la géographie.* Paris: Bibliothèque nationale.

———. 1974. *La Géographie des philosophes, géographes et voyageurs français au XVIIIe siècle.* Paris: Éditions Ophrys.

————. 1975. Un Bicentenaire: Malte-Brun (1775–1975). *Annales de géographie* 84: 714–20.

————. 1978. Nationalisme, colonialisme et géographie: Marcel Dubois (1856–1916). *Annales de géographie* 87: 326–33.

Brockway, Lucile H. 1979. *Science and Colonial Expansion: The Role of the British Royal Botanical Gardens.* London and New York: Academic Press.

Bruguière, Louis. 1830. Orographie de l'Europe. In *Recueil de voyages et de mémoires.* Paris: Société de géographie.

Bruhns, K. 1873. *The Life of Alexander von Humboldt.* London.

Brunet, Pierre. 1970. *L'Introduction des théories de Newton en France au XVIIIe siècle avant 1738.* Paris: A. Blanchard, 1931. Reprint. Geneva: Slatkine Reprints.

Bruzen de la Martinière, Antoine-Augustin. 1722. Essai sur l'origine et les progrès de la géographie. In *Mémoires historiques et critiques.* Amsterdam: J.-F. Bernard.

————. 1739–1741. *Le Grand dictionnaire géographique, historique et critique . . .* 6 vols. Paris: P.-G. Lemercier (et Boudet).

Buache de la Neuville, Jean-Nicolas. 1775. *Mémoire sur les pays de l'Asie et de l'Amérique situés au nord de la mer du Sud, accompagné d'une carte de comparaison des plans de MM. Engel et de Vaugondy, avec le plan des cartes modernes, . . .* Paris: L'auteur.

————. 1791. Extract from a Memoir Concerning the Existence and Situation of the Solomon Islands, Presented to the Royal Academy of Sciences, January 9, 1781. In *Discoveries of the French in 1768 and 1769 to the South-East of New Guinea with the Subsequent Visits to the Same Lands by English Navigators, Who Gave Them New Names,* edited by Charles-Pierre d'Eveux Claret de Fleurieu. London: J. Stockdale.

Buache, Philippe. 1741. Observations sur l'étendue et la hauteur de l'inondation du mois de décembre 1740. *Mémoires de l'Académie royale des sciences* 2: 449–52.

————. 1741. Plan du cours de la Seine dans la traversée de Paris . . . In *Mémoires de l'Académie royale des sciences.*

————. 1752. Cartes des nouvelles découvertes entre la partie oriental de l'Asie et l'occidentale de l'Amérique. Avec des vues sur la Grande Terre reconnue par les Russes en 1741 et sur la mer de l'Ouest et autres communications de Mers. Présentée à l'Académie des sciences le 9 août 1752. Approuvée 6 septembre 1752.

————. 1752. Essai de géographie physique. [Includes two maps: Planisphère physique où l'on voit du Pôle Septentrional ce que l'on connoit de terres et de mers . . . and Carte physique et profile du canal de la Manche et d'un partie de la mer du Nord.] *Mémoires de l'Académie royale des sciences* 2: 619–35.

————. 1752–1753. *Considérations géographiques et physiques sur les nouvelles découvertes au nord de la Grande mer, appelée vulgairement la mer du Sud, avec des cartes qui y sont relatives . . .* Paris: Desaint et Saillant.

————. [1754 ca.] *Explication et développement de la carte générale de l'histoire sainte . . . en février 1754.* N.p.: n.d.

Buffon, Georges-Louis Leclerc, Comte de. 1811. Histoire naturelle de l'homme. In *Oeuvres complètes de Buffon.* Paris: J.-F. Bastien.

Buisseret, David. 1965. Les Ingénieurs du roi au temps de Henri IV. *Bulletin de la section de géographie, 1964:*13–81.

Buisson, Ferdinand. 1882–1893. *Dictionnaire de pédagogie et d'instruction primaire* . . . 5 vols. Paris: Hachette.

Bulick, Stephen. 1982. *Structure and Subject Interaction: Toward a Sociology of Knowledge in the Social Sciences.* New York: Marcel Dekker, Inc.

Bureaux de Pusy, Jean-Xavier. 1790. *Considérations sur le corps royal du génie, présentées au Comité militaire, par un membre de ce comité* . . . Paris: Imprimerie nationale.

———. N.d. *De la Réunion des mineurs du corps royal du génie et de celle du génie à l'artillerie.* N.p.

———. N.d. Projet de décret sur la conservation et le classement des places de guerre et postes militaires . . . présenté à l'Assemblée nationale au nom de son Comité militaire. In *Rapport sur la conservation et le classement des places de guerre* . . . Paris: Imprimerie nationale.

Burton, June K. 1979. *Napoleon and Clio: Historical Writing, Teaching, and Thinking during the First Empire.* Durham, NC: Carolina Academic Press.

Bus, Charles du. 1926. Les Collections d'Anville à la Bibliothèque nationale. *Bulletin de géographie historique et descriptive:* 93–145.

Büsching, Anton Friedrich. 1768–1779. *Géographie universelle, traduite de l'allemand de Mr. Büsching sur sa cinquième édition, avec des augmentations et corrections, qui ne se trouvent pas dans l'original.* 14 vols. Strasbourg: Chez Bauer & Compagnie.

Butlin, Robin. 1993. *Historical Geography through the Gates of Space and Time.* London: Edward Arnold.

Cabanis, André. 1987. Journal des débats. In *Dictionnaire Napoléon,* edited by J. Tulard. Paris: Fayard.

Cabanis, Pierre-Jean-Georges. 1802 (an X). *Rapports du physique et du moral de l'homme* . . . 2 vols. Paris: Crapart, Caille et Ravier.

Cailliaud, Frédéric. 1821. *Voyage à l'oasis de Thèbes et dans les déserts situés à l'occident de la Thébaïde* . . . Paris: Imprimerie impériale.

Cailliaud, Frédéric, and Edme-François Jomard. 1826–1827. *Voyage à Méroe, au Fleuve Blanc* . . . *à Syouah* . . . *fait dans les années 1819, 1820, 1821 et 1822.* 4 vols. and 2 vols. of plates. Paris: Imprimerie royale.

Calon, Étienne-Nicolas de. 1795 (an III). *Convention nationale. Observations à la Convention nationale, sur le projet d'établissement d'une école centrale des travaux publics, . . . 7 vendémiaire, an III.* Paris: Imprimerie nationale.

Cambrosio, Alberto, Daniel Jacobi, and Peter Keating. 1993. Ehrlich's "Beautiful Pictures" and the Controversial Beginnings of Immunological Imagery. *Isis* 84, no. 4 (December): 662–99.

Camerini, Jane R. 1993. Evolution, Biogeography, and Maps: An Early History of Wallace's Line. *Isis* 84, no. 4 (December): 700–727.

Carette, Antoine Ernest Hippolyte. 1844. Étude des routes suivies par les Arabes dans la partie méridionale de l'Algérie et de la Régence de Tunis pour servir à l'établissement du réseau géographique de ces contrées accompagnée d'une carte itineraire par E. Carette capitaine du génie, Membre et secrétaire de la Commission scientifique de l'Algérie. In vol. 1 of the *Sciences historiques et géographiques* of *Exploration scientifique de l'Algérie pendant les années 1840, 1841, 1842* publiée par ordre du gouvernement et avec le concours d'une Commission Académique. Paris: Imprimerie Imperiale.

————. 1844. Recherches sur la géographie et le commerce de l'Algerie méridionale par E. Carette capitaine du Génie, Membre et secrétaire de la Commission scientifique d'Algérie. Suivi d'une Notice Géographique sur une partie de l'Afrique Septentrionale par E. Renou Membre de la Commission Scientifique et accompagnées de trois cartes. In vol. 2 of the *Sciences historiques et géographiques* of *Exploration scientifique de l'Algérie pendant les années 1840, 1841, 1842* publiée par ordre du gouvernement et avec le concours d'une Commission Académique. Paris: Imprimerie Imperiale.

Cassini de Thury, César-François. 1783. *Description géométrique de la France . . .* Paris: Imprimerie J. Ch. Desaint.

————. N.d. *Description d'un instrument pour prendre hauteur et pour trouver l'heure vraie sans aucun calcul . . .* Paris: Imprimerie de A. Boudet.

Cassini, Jacques-Dominique. [1784]. Atlas céleste et catalogue des étoiles fixées. Pour le 1er janvier 1784. Dressé dans un ordre nouveau, et suivant une méthode bien plus simple et plus commode pour reconnoitre les constellations, les plus petites étoiles qui les composent et déterminer dans le ciel telle position que l'on jugera à propos. Observatoire de Paris. Manuscript, D 1 29.

————. [1803]. Lettre au Premier Consul, frimaire an 11. Observatoire de Paris. Manuscript, 1051 bis.

————. N.d. Magnétisme, températures des caves, etc.: Projet d'un levé géométrique de la Carte de Toscane. Observatoire de Paris. Manuscript, D 5 33.

————. N.d. "Objets économiques," in "Projet d'un levé géométrique de la Carte de Toscane" at the Observatoire de Paris: Cassini IV D 5 33 File entitled Magnétisme, températures des caves, etc.

————. 1770. *Voyage fait par ordre du roi en 1768, pour éprouver les montres marines inventées par M. Le Roy, . . . Avec le mémoire sur la meilleure manière de mesurer le tems en mer, qui a remporté le prix double en au jugement de l'Académie royale des sciences . . . par M. Le Roy l'aîné . . .* Paris: C.-A. Jombert.

————. 1772. *Voyage en Californie pour l'observation du passage de Vénus sur le disque du Soleil, le 3 juin 1769, contenant les observations de ce phénomène et la description historique de la route de l'auteur à travers le Mexique, par feu M. Chappe d'Auteroche, de l'Académie royale des sciences. Rédigé et publié par M. de Cassini fils, . . .* Paris: Charles-Antoine Jombert.

————. 1778. *Manuel de l'étranger qui voyage en Italie.* Paris: Veuve Duchesne.

————. 1786–1791. *Extrait des observations astronomiques et physiques, faites par ordre de Sa Majesté, à l'Observatoire royal, en l'année 1785. Sous le Ministère de M. le Baron de Breteuil.* 5 vols. Paris: Imprimerie royale.

————. 1802 (an X). *Quelques idées sur l'instruction publique communiquées au Conseil général du département de l'Oise par son vice-président J.-Dom. Cassini, Membre de l'Institut national, prairial an X.* N.p.: Imprimerie de Langlois.

————. 1810. *Mémoires pour servir à l'histoire des sciences et à celle de l'Observatoire royale de Paris, suivis de la vie de J.-D. Cassini écrite par lui-même, et des éloges de plusieurs académiciens morts pendant la Révolution,* . . . Paris: Bleuet.

————. 1818. *Réclamation en faveur de la Compagnie des associés pour la confection de la carte générale de la France, dite carte de Cassini.* Paris: Imprimerie de Migneret.

————. 1830. *Les Veillées du village, ou Dialogues sur divers sujets pour l'instruction et l'amusement des habitans de la campagnes.* Lille: L. Lefort.

————. 1840. *Lettre de Jean-Dominique Cassini à ses co-associés, membres de la Compagnie formée pour la confection de la carte générale de la France.* Paris: Imprimerie de Langlois.

————. 1842. À Madame de Cambry, en lui envoyant ma Carte réduite de la lune. In *Collection de riens qui vaillent, par un centenaire oisif.* Clermont-Oise: A. Carbon.

————. 1842. Au secrétaire de la classe des sciences de l'Institut 1er mai 1815. In *Collection de riens qui vaillent, par un centenaire oisif.* Clermont-Oise: A. Carbon.

————. 1842. *Collection de riens qui vaillent, par un centenaire oisif.* Clermont-Oise: A. Carbon.

————. 1842. Dialogue entre deux détenus sur la croyance en Dieu. In *Collection de riens qui vaillent, par un centenaire oisif.* Clermont-Oise: A. Carbon.

————. 1842. Mon apologie à un de mes confrères de l'Institut. In *Collection de riens qui vaillent, par un centenaire oisif.* Clermont-Oise: A. Carbon.

————. 1842. Sur le choix d'un état. In *Collection de riens qui vaillent, par un centenaire oisif.* Clermont-Oise: A. Carbon.

————. N.d. *Réflexions présentées aux éditeurs des futures éditions de "L'Histoire de l'astronomie au XVIIIe siècle."* Paris: Imprimerie de Béthune.

Cassini, Jacques-Dominique, Pierre-François-André Méchain, and Adrien-Marie Le Gendre. N.d. *Exposé des opérations faites en France, en 1787, pour la jonction des observatoires de Paris et de Greenwich,* . . . *Description et usage d'un nouvel instrument propre à donner la mesure des angles, à la précision d'une seconde.* Paris: Imprimerie de l'Institution des sourds-muets.

Cassini, Jacques-Dominique, Nicholas-Antoine Nouet, Jean Perny de Villeneuve, and Alexandre Ruelle. 1786 (–1791). *Extrait des observations astronomiques et physiques, faites par ordre de Sa Majesté à l'Observatoire royal en l'année 1787, 1788, 1789, 1790, 1791 sous le Ministère de M. le Baron de Breteuil.* Paris: Imprimerie royale.

Cassirer, Ernst. 1950. *The Problem of Knowledge* (New Haven: Yale University Press).

Cérémonie en l'honneur d'Edme-François Jomard, à Versailles. 1939. *Revue des études napoléoniennes* 44: 29–46.

Chabrol de Volvic, Cte. Gilbert-Joseph-Gaspard. Mémoire sur la Spezzia. Service historique de l'armée de terre, Vincennes. Manuscript, Mémoires, MR 1400.

————. 1822. Essai sur les moeurs des habitans modernes de l'Égypte. In *Description de l'Égypte,* edited by Edme-François Jomard, État Moderne, vol. 2, pt. 2: 361–526. Paris: Imprimerie impériale.

————. 1824. *Statistique des provinces de Savone, d'Oneille, d'Acqui, et de partie de la province de Mondovi, formant l'ancien département de Montenotte . . .* 2 vols. Paris: Imprimerie de J. Didot aîné.

————, and Edme-François Jomard. 1809. Description d'Ombos et des environs. In *Description de l'Égypte,* edited by Edme-François Jomard, Antiquités, Descriptions, 1: 1–26. Paris: Imprimerie impériale.

Chales, Le Père Claude François Milliet de. 1674. *Cursus seu mundus mathematicus.* Lugduni: Ex Officina Anissonina.

Challe, A. 1958. Lettres de Joseph Fourier. *Bulletin de la société des sciences historiques . . . de l'Yonne* 12: 105–34.

Chandler, David G., ed. 1979. *Dictionary of the Napoleonic Wars.* London and New York: Macmillain Publishing Co.

Chapin, Seymour L. 1967. The Academy of Sciences during the Eighteenth Century: An Astronomical Appraisal. *French Historical Studies* 5 (4): 371–404.

Charléty, Sébastin. 1896. *Histoire du Saint-Simonisme (1825–1864) . . .* Paris: Hachette.

Charmasson, T., A.-M. Lelorrain, and Y. Ripa. 1987. *L'Enseignement technique de la Révolution à nos jours 1789–1926.* Paris: Institut national de recherche pédagogique.

Chartier, Roger. 1978. Les Deux Frances: Histoire d'une géographie. *Cahiers d'histoire* 23: 393–415.

Chevalier, Michel. 1986. L'Abbé Soulavie, précurseur ardéchois de la géographie moderne (1752–1813). *Revue du Vivarais* 40, no. 2 (avril—juin): 81–100.

Chinard, Gilbert. 1923. *Volney et l'Amérique d'après des documents inédits et sa correspondance avec Jefferson.* Baltimore, Maryland: The Johns Hopkins Press; and Paris: Les Presses universitaires de France.

Choiseul-Gouffier, Marie-Gabriel-Florent-Auguste, comte de. 1782–1822. *Voyage pittoresque de la Grèce.* Paris: J.-J. Blaise.

Clausewitz, Carl von. 1984. *On War.* Edited by M. Howard and Peter Paret. Princeton, NJ: Princeton University Press.

Claval, Paul. 1968. *Régions, nations, grands espaces, géographie générale, géographie des ensembles territoriaux . . .* Paris: M.-T. Génin.

————. 1980. *Les Mythes fondateurs des sciences sociales.* Paris: Presses universitaires de France.

Cohen, I. Bernard. 1971. Jean-Joseph Delambre. In *Dictionary of Scientific Biography.* New York: Scribner.

Coing, H. 1979. Roman Law and the National Legal Systems. In *Classical Influences on Western Thought A.D. 1650–1870,* edited by R. R. Bolgar. Cambridge: Cambridge University Press.

Coirault, Gaston. N.d. *Les écoles centrales dans la centre-ouest, c'est-à dire dans le ressort de l'Académie de Poitiers . . . An IV à an XII . . .* Tours: Imprimerie Arrault et Cie.

Coleman, William. 1982. *Death is a Social Disease: Public Health and Political Economy in Early Industrial France.* Madison: University of Wisconsin Press.

Colin, Epidariste. N.d. *Extrait des Procès-verbaux de la Société d'émulation de l'Ile-de-France*. N.l.

Collins, Peter. 1965. *Changing Ideals in Modern Architecture, 1750–1950*. Montreal: McGill University Press.

Compère, Marie-Madeleine. 1989. La Question des disciplines scolaires dans les écoles centrales. Le Cas des langues anciennes. *Histoire de l'éducation* 42 (mai): 139–81.

Connelly, Owen. 1965. *Napoleon's Satellite Kingdoms*. New York: The Free Press.

Considérations sur la Forêt-Noire et sur le Tyrol. 1829. In *Mémorial du Dépôt de la guerre, imprimé par ordre du Ministre 1802–03*. Paris: Ch. Picquet.

Corbin, Alain. 1982. *Le Miasme et la jonquille. L'Odorat et l'imaginaire social XVIIIe–XIXe siècles*. Paris: Éditions Aubier Montaigne.

Cordier, Louis. 1818. Description des ruines de Sân (Tanis des anciens). In *Description de l'Égypte*, edited by Edme-François Jomard, Antiquités, Descriptions, 2: 1–18. Paris: Imprimerie impériale.

Cormack, Lesley Barbara. 1988. Non sufficit Orbem: Geography as an Interactive Science at Oxford and Cambridge 1580–1620. PhD dissertation, Department of Geography, University of Toronto, Toronto.

———. 1994. The Fashioning of an Empire: Geography and the State in Elizabethan England. In *Geography and Empire*, edited by A. Godlewska and N. Smith. Oxford, UK, and Cambridge, MA: Blackwell.

———. 1997. *Charting an Empire. Geography at the Universities, 1580–1620*. Chicago: University of Chicago Press.

Cortambert, Eugène. 1846. *Cours de géographie, comprenant le description physique et politique et la géographie historique des diverses contrées du globe, . . .* Paris: L. Hachette.

———. 1862. *Mélanges géographiques: Place de la géographie dans la classification des connaissances humaines*. Paris: Imprimerie de L. Martinet.

Corvisier, André. 1976. *Armies and Societies in Europe, 1494–1789*. Translated by Abigail T. Siddall. Bloomington and London: Indiana University Press.

———. 1988. *Dictionnaire d'art et d'histoire militaires*. Paris: Presses universitaires de France.

Costaz, Louis. 1809. Environs d'Erment. Note sur les restes de l'ancienne ville de Tuphium. In *Description de l'Égypte*, edited by Edme-François Jomard, Antiquités, Descriptions, 1: 17–18. Paris: Imprimerie impériale.

———. 1809. Grottes d'Elethyia. Mémoire sur l'agriculture, sur plusieurs art et sur plusieurs usages civils et religieux des anciens Égyptiens. In *Description de l'Égypte*, edited by Edme-François Jomard, Antiquités, Mémoires, 1: 49–78. Paris: Imprimerie impériale.

———. 1809. Mémoire sur la Nubie et les Barabras. In *Description de l'Égypte*, edited by Edme-François Jomard, État Moderne, 1: 399–408. Paris: Imprimerie impériale.

———. 1834. *Mémoire sur la construction des tables statistiques et sur la mesure des valeurs, . . .* Paris: Imprimerie de Moquet.

Coton, Le Père Pierre. 1608. *Intérieure occupation d'une âme dévote*. N.p.

Cournot, Antoine-Augustin. 1851. *Essai sur le fondements de nos connaissances et sur les caractères de la critique philosophique,* . . . Paris: L. Hachette.

Coutelle, J. M. J. 1812. Observations sur la topographie de la presqu'île de Sinaï, les moeurs, les usages, l'industrie, le commerce et la population des habitans. In *Description de l'Égypte,* edited by Edme-François Jomard, État Moderne, 2: 276–304. Paris: Imprimerie impériale.

———. 1818. Observations sur les pyramides de Gyzeh et sur les monumens et les constructions qui les environnent. In *Description de l'Égypte,* edited by Edme-François Jomard, Antiquités, Mémoires, 2: 39–56. Paris: Imprimerie impériale.

Crosland, Maurice. 1972. 'Nature' and Measurement in Eighteenth-Century France. *Studies on Voltaire and the Eighteenth Century* 87: 277–309.

———. 1976. La Science et le pouvoir: de Bonaparte à Napoléon III. *La Recherche* 71 (octobre): 842 ff.

Crouzet, E. 1904. *Travaux des topographes du génie militaire en France au XIXe siècle.* Paris: Librairie militaire Berger-Levrault et Cie.

Cunningham, Andrew, and Nicholas Jardine. 1990. Introduction: The Age of Reflexion. In *Romanticism and the Sciences,* edited by A. Cunningham and N. Jardine. Cambridge: Cambridge University Press.

Cuno, Kenneth M. 1984. Egypt's Wealthy Peasantry, 1740–1820: A Study of the Region of al-Mansura. In *Land Tenure and Social Transformation in the Middle East,* edited by T. Khalidi. Beirut: American University of Beirut.

Cuvier, Georges Léopold. 1811. Recherches anatomiques sur les reptiles regardés encore comme douteux par les naturalistes; faites à l'occasion de l'Axolotl, rapporté par M. De Humboldt du Mexique par M. Cuvier. In *Recueil d'observations de zoologie et d'anatomie comparée, faites dans l'océan Atlantique, dans l'intérieur du nouveau continent et dans la mer du Sud pendant les années 1799, 1800, 1801, 1802 et 1803* (Paris: F. Schoell and G. Dufour), 93–126.

———. 1819. Éloge historique de Nicolas Desmaret (*sic*), lu le 16 mars 1818. In *Recueil des éloges historiques lus dans les séances publiques de l'Institut royal de France.* Strasbourg-Paris.

———. 1989 (1828). Rapport historique sur les progrès des sciences naturelles depuis 1798 et sur leur état actuel. In *Rapports à l'Empereur sur le progrès des sciences, des lettres et des arts depuis 1789.* Paris: Belin.

Dacier, Bon-Joseph. 1810. *Rapport historique sur les progrès de la littérature ancienne, depuis 1789 et sur l'état actuel, présenté à sa Majesté l'Empereur et Roi, et son Conseil d'état, le 20 février 1808, par la classe d'histoire et de littérature ancienne de l'Institut, rédigé par M.Dacier.* Paris: Imprimerie impériale.

———. 1989 (1808). Histoire et littérature ancienne. In *Rapports à l'Empereur sur le progrès des sciences, des lettres et des arts depuis 1789.* Paris: Belin.

Daclin, Charles. 1838. *Table générale, analytique et raisonnée des matières contenues dans les trente-six premières années du Bulletin de la Société d'encouragement pour l'industrie nationale, comprenant les noms des auteurs mentionnés dans l'ouvrage . . .* Paris: Madame Huzard.

Dagonet, F. 1972. Valentin Haüy, Etienne Geoffroy Saint-Hilaire, Augustin Pyramus De Candolle: Une conception d'ensemble mais aussi un ensemble de conceptions. *Revue d'histoire des sciences* 25: 327–36.

Dainville, François de. 1940. *La Géographie des humanistes. Les Jésuites et l'éducation de la société française.* Paris: Beauchesne et ses fils.

————. 1958. De la Profondeur à l'altitude: des origines marines de l'expression cartographique du relief terrestre par côtes et courbes de niveau. In *Le Navire et l'économie maritime du Moyen Âge au dix-huitième siècle.* Paris: École pratique des hautes études.

————. 1964. *Le Langage des géographes. Termes, signes, couleurs des cartes anciennes, 1500–1800.* Paris: Éditions A. et J. Picard et Cie.

————. 1965–1966. Cartographie historique et histoire de l'éducation. *Annuaire des hautes études, 4ième section:* 351–57.

————. 1978. Enseignement des 'géographes' et des 'géomètres.' In *L'Éducation des jésuites (XVIe—XVIIIe siècles).* Paris: Les Éditions de Minuit.

Dangeau, Abbé Louis de Courcillon de. 1697. Carte des Universités de France.

Danjon, A. 1946. Le Verrier, créateur de la météorologie. *La météorologie,* ser. 4 (1): 863–82.

d'Anville. *See under* Anville.

Daudin, Henri. 1983. *De Linné à Jussieu; Méthodes de classification et idée de série en botanique et en zoologie (1740–1790).* Paris: Alcan, 1926–1927. Reprint. Paris: Éditions des archives contemporaines, Presses universitaires de la France.

Daunou, Pierre-Claude-François. 1842–1845. *Cours d'études historiques . . .* 3 vols. Vol. 2. Paris: F. Didot frères.

————. 1982. Loi du 3 brumaire an IV et Rapport sur l'instruction publique. In *Une Éducation pour la démocratie. Textes et projets de l'époque révolutionnaire,* edited by B. Baczko. Paris: Garnier.

Dauzat, Albert. 1922. *La Géographie linguistique.* Paris: E. Flammarion.

Davico, Rosalba. 1968. Démographie et économie; ville et campagne en Piémont à l'époque française. *Annales de démographie historique* 4: 139–64.

————. 1981. *'Peuple' et notables (1750–1816): Essais sur L'Ancien régime et la Révolution en Piémont.* Paris: Bibliothèque nationale.

David, Florence N. 1962. *Games, Gods and Gambling. The Origins and History of Probability and Statistical Ideas from the Earliest Times to the Newtonian Era.* New York: Hafner Publishing Company.

David, Johann [Michaelis pseud.]. 1763. *Recueil des questions proposées à une Société des savans qui par ordre de Sa Majesté Danoise font le voyage de l'Arabie.* Frankfurt.

Davies, Gordon L. 1969. *The Earth in Decay: A History of British Geomorphology 1578–1878.* London: MacDonald.

Davis, John L. 1984. Weather Forecasting and the Development of Meteorological Theory at the Paris Observatory, 1853–1878. *Annals of Science* 41 (4): 359–82.

De Candolle, Augustin Pyramus. 1813 (but actually published in 1817). Mémoire sur la Géographie des plantes de France, considérée dans ses rapports avec la hauteur absolue. In *Mémoires de physique et de chimie, de la Société d'Arcueil.* Paris: J. Klostermann fils.

————, and Friedrich Wilhelm Heinrich Alexandre Humboldt, Baron von. 1820. Géographie botanique. In *Dictionnaire des sciences naturelles,* edited by Plusieurs professeurs du Jardin du Roi. Paris; Strasbourg: Le Normant et Levrault.

Deacon, Margaret. 1971. *Scientists and the Sea, 1650–1900: A Study in Marine Science.* London and New York: Academic Press.

La Décade philosophique, littéraire et politique. 1794–1804. 42 vols. Paris.

———. 1795 (an III). Instruction publique: École normale. *La Décade philosophique, littéraire et politique* 30 (30 pluviôse): 351–53.

Debs, Richard Abraham. 1963. The Law of Property in Egypt: Islamic Law and Civil Code. PhD dissertation, Princeton University.

Delambre, Jean-Baptiste-Joseph. 1827. *Histoire de l'astronomie au dix-huitième siècle, . . .* Paris: Bachelier.

———. 1989 [1802?]. Rapport historique sur les progrès des sciences mathématiques depuis 1789 et sur leur état actuel. In *Rapports à l'Empereur sur le progrès des sciences, des lettres et des arts depuis 1789.* Paris: Belin.

Delaporte. 1812. Abrégé chronologique de l'histoire des Mamelouks d'Égypte, depuis leur origine jusqu'à la conquête des Français. In *Description de l'Égypte,* edited by Edme-François Jomard, État Moderne, 2: 121–89. Paris: Imprimerie impériale.

Delaunay, Pierre. 1970. Un projet de division géométrique du territoire français à la fin du 18e siècle. *Bulletin de la librairie ancienne et moderne,* no. 121 (janvier): 2–6.

Deleuze, Gilles. 1975. Ecrivain non: Un nouveau cartographe. *Critique* 332: 1205–27.

Delille, G. 1975. Cadastre napoléonien et structures économiques et sociales dans le Royaume de Naples. *Annuario dell'Istituto storico italiano per l'età moderna et contemporanea,* 23–24.

Delisle, Guillaume. 1700. Lettre de M. Delisle à M. Cassini sur l'embouchure de la rivière du Mississippi. *Journal des savants,* 211–18.

———. 1700. Lettre de M. Delisle au R. P.* sur la longitude de Paris. *Journal des savants,* 243.

———. 1700. Réponse de M. Delisle à la plainte de M. Nolin. *Journal des savants,* 293.

———. 1700. Réponse de M. Delisle à la seconde lettre de M. Nolin. *Journal des savants,* 351–53.

———. 1700. Seconde Lettre de M. De Lisle à M. Cassini pour justifier quelques endroits de ses globes et cartes. *Journal des savants,* 219–26.

———. 1700. Troisième lettre du Sr. Delisle à M. Cassini sur la question que l'on peut faire si le Japon est une Isle. *Journal des savants,* 236–42.

———. 1705. *Remarques sur le théâtre historique, pour l'an 400 de l'ère chrétienne, . . .* Paris.

———. 1746. *Introduction à la géographie, avec un traité de la sphère,* vol. 1, *Géographie.* 2 vols. Paris: Étienne-François Savoye.

———. 1746. *Introduction à la géographie, avec un traité de la sphère,* vol. 2, *La Sphère.* 2 vols. Paris: Etienne-François Savoye.

———. 1746. Review of Guillaume Delisle's *Introduction à la géographie avec un traité de la sphère* (1746). In *Journal des sçavans.*

———. 1753. *Nouvelles Cartes des découvertes de l'amiral de Fonte et autres navigateurs . . . dans les mers septentrionales, avec leur explication, qui comprend l'histoire des voyages . . . avec la description des pays, l'histoire et les moeurs des habitants . . .* Paris.

Denaix, M. A. 1827. *Essais de géographie méthodique et comparative, accompagnés de tableaux historiques faisant connaître la succession des différens états du monde depuis les temps les plus reculés jusqu'à nos jours, et suivi d'une théorie du terrain appliquée aux reconnaissances militaires. Dédiés à S.E. Le Comte Guilleminot.* Paris.

Deneys, Anne. 1989. Géographie, histoire et langue dans le *Tableau du climat et du sol des États-Unis. Corpus. Revue de philosophie* 11–12:73–90.

Deneys, Henry. 1989. Le Récit d'histoire selon Volney. *Corpus. Revue de philosophie* 11–12:43–71.

Denis, Ferdinand-Jean. 1830. *Tableau historique analytique et critique des sciences occultes, où l'on examine l'origine, le développement, l'influence et le caractère de la divination, de l'astrologie, des oracles.* Paris.

Depping, Guillaume. 1881. Le Mouvement géographique et les sociétés de géographie en France. In *Journal officiel. Lois et décrets,* edited by the French government.

Desaive, Jean-Paul, Jean-Pierre Goubert, Emmanuel Le Roy Ladurie, Jean Meyer, Otto Muller, and Jean-Pierre Peter, eds. 1972. *Médecins, climat, et épidémies à la fin du XVIIIe siècle.* Paris and The Hague: Mouton.

Désirat, Claude, and Tristan Hordé. 1984. Volney, l'étude des langues dans l'observation de l'homme. In *Histoires de l'anthropologie (XVI—XIX siècles): Colloque sur la pratique de l'anthropologie aujourd'hui, 19–21 novembre 1981, Sèvres,* edited by B. Rupp-Eisenreich. Paris: Klincksieck.

Desmarest, Nicolas. 1753. *Dissertation sur l'ancienne jonction de l'Angleterre à la France . . . avec des plans et des cartes topographiques.* Amiens: Chez la Vve. Godart.

———. 1756. *Conjectures physico-mécaniques sur la propagation des secousses dans les tremblements de terre, et sur la disposition des lieux qui ont ressenti les effets . . .* N.p.

———. 1757. Géographie physique. In *Encyclopédie, ou dictionnaire raisonné des sciences, des arts et des métiers, par une société de gens de lettres,* edited by D. Diderot and J. d'Alembert. Paris: Briasson, David l'aîné, Le Breton, and Durand.

———. 1795. Géographie-physique. In *Encyclopédie méthodique.* Paris: Chez H. Agasse.

———. 1798–1828. Dictionnaire de la géographie physique. Vols. 2–4. In *Encyclopédie méthodique.* Paris: Chez H. Agasse.

Destombes, Marcel. 1977. De la Chronique à l'histoire: le globe terrestre monumental de Bergevin (1784–1795). *Archives internationales d'histoire des sciences* 27 (100): 113–34.

Dettelbach, Michael. 1996. Humboldtian Science. In *Cultures of Natural History,* edited by N. Jardine, J. A. Secord, and E. C. Spary. Cambridge: Cambridge University Press.

———. 1996. Global Physics and Aesthetic Empire: Humboldt's Physical Portrait of the Tropics. In David Philip Miller and Peter Hanns Reill, eds., *Visions of*

Empire: Voyages, Botany, and Representations of Nature (Cambridge: Cambridge University Press), 258–92.

Devic, Jean-François-Schlister. 1851. *Histoire de la vie et des travaux scientifiques et littéraires de J.-D. Cassini IV, ancien directeur de l'observatoire.* Clermont: A. Daix.

Dias, Nélia. 1998. Une Science nouvelle: la géo-ethnographie de Jomard. In Marie-Noëlle Bourguet, Bernard Lepetit, Daniel Nordman, Maroula Sinarellis, *Invention scientifique de la Méditerranée. Égypte, Morée, Algérie*, 159–83. Paris: Éditions de l'École des hautes études en sciences sociales.

Diderot, Denis, and Jean d'Alembert, eds. 1780. *Supplément.* Vol. 2, *Encyclopédie, ou dictionnaire raisonné des sciences, des arts, et des métiers.* Paris.

Dijksterhuis, J. 1961. *The Mechanisation of the World Picture.* Translated by C. Dikshoorn. Oxford: Clarendon Press.

Dogan, Mattei, and Robert Pahre. 1990. *Creative Marginality: Innovation at the Intersections of Social Sciences.* Boulder: Westview Press.

Dolomieu, Déodat Gratet de. 1783. *Voyage aux îles de Lipari fait en 1781, ou Notices sur les îles Æoliennes pour servir à l'histoire des volcans, suivi d'un mémoire sur une espèce de volcan d'air, et d'un autre sur la température du climat de Malthe et sur la différence de la chaleur réelle et de la chaleur sensible . . .* Paris: rue et hôtel Serpente.

Dondin-Payre, Monique. 1988. *Un Siècle d'épigraphie classique: Aspects de l'oeuvre des savants français dans les pays du bassin Méditerranéen. Exposition organisée pour le Centenaire de l'Année épigraphique, 21–26 octobre 1988.* Paris: Institut de France.

Drapeyron, Ludovic. 1887. Projet d'établissement en Afrique, 1790. *Revue de géographie* 20: 241–55.

———. 1888. *Les Deux Buaches et l'éducation géographique de trois rois de France Louis XVI, Louis XVIII, Charles X, avec documents inédits.* Paris: Institut géographique de Paris.

Driver, Felix. 1995. Geographical Traditions: Rethinking the History of Geography. *Transactions of the Institute of British Geographers*, n.s., 20, 403–4.

———. 1995. Sub-merged Identities: Familiar and Unfamiliar Histories. *Transactions of the Institute of British Geographers*, n.s., 20, 410–13.

Drohojowska, Antoinette Joséphine Anne Symon de Latreiche, Ctesse. 1884. Malte-Brun. *Précis de la géographie universelle.* In *Les Savants modernes et leurs oeuvres.* Lille et Paris: Librairie de J. Lefort.

Drouin, Jean-Marc. 1998. Bory de Saint-Vincent et la géographie botanique. In Marie-Noëlle Bourguet, Bernard Lepetit, Daniel Nordman, Maroula Sinarellis, *Invention scientifique de la Méditerranée. Égypte, Morée, Algérie*, 139–57. Paris: Éditions de l'École des hautes études en sciences sociales.

———. 1991. *Réinventer la nature. L'Écologie et son histoire.* Paris: Desclée de Brouwer.

Du Halde, Jean-Baptiste. 1736. *Description géographique, historique, chronologique, politique et physique de l'Empire de la Chine et de la Tartarie chinoise, enrichie des cartes générales et particulières de ces pays, de la carte générale & des cartes particulières du Thibet, & de la Corée; ornée d'un grand nombre de figures & de vignettes gravées en taille-douce. Avec un Avertissement préliminaire, ou l'on rend compte des principales améliorations qui ont été faites dans cette nouvelle édition.* Vol. 1. La Haye: Chez Henri Scheurleer.

Duboy de Laverne, Philippe-Daniel. 1829. *Réflexions sur un ouvrage traduit de l'Allemand intitulé Esprit du système de la guerre moderne.* In *Mémorial du Dépôt de la guerre imprimé par ordre du Ministre 1802–03.* Paris: Ch. Picquet.

Dufour, Léon. 1902. Lettres d'un botaniste pendant la guerre d'Espagne: Lettres de Léon Dufour à Bouchet (avril 1808 à octobre 1813) communiquées par M. le Professeur L.-G. Pélissier. *Revue des études napoléoniennes* 3: 157–68.

Dulaure, Jacques-Antoine. 1788–1789. *Description des principaux lieux de France . . .* 6 vols. Paris: Lejay.

Dumont d'Urville, Contre-amiral Jules-Sébastien-César. 1830. *Voyage de la corvette l'Astrolabe exécuté par ordre du Roi pendant les années 1826–1827–1828–1829, sous le commandement de M. J. Dumont d'Urville, . . . Histoire du voyage.* 5 vols. Vol. 1. Paris: J. Tastu.

Dunbar, Gary S. 1985. *A History of Modern Geography. An Annotated Bibliography of Selected Works.* New York and London: Garland Publishing, Inc.

Dupain de Montesson, Louis Charles. 1750. *La science des ombres par rapport au dessein* (sic). Paris: C.-A. Jombert.

———. 1757. *Les Amusements militaires, ouvrage également agréable et instructif, servant d'introduction aux sciences qui forment les guerriers, . . .* Paris: Guillaume Desprez.

———. 1763. *L'Art de lever les plans de tout ce qui a rapport à la guerre et à l'architecture civile et champêtre.* Paris: C.-A. Jombert.

———. 1783. *Vocabulaire de guerre, ou Recueil des principaux termes de guerre, de marine, d'artillerie, de fortification, d'attaque & de défense des places, et de géographie.* Paris: Couturier fils.

———. 1813. *La Science de l'arpenteur dans toute son étendue . . .* Paris: Goeury.

Dupain-Triel, Jean-Louis. 1786. *La France connue sous ses plus utiles rapports, ou Nouveau Dictionnaire universel de la France, dressé d'après la carte en 180 feuilles de Cassini . . .* Paris: Imprimerie de L. Cellot.

———, and C. F. Abancourt. 1791. *Recherches géographiques sur les hauteurs des plaines du royaume, sur les mers et leurs côtes presque pour tout le globe et sur les diverses espèces de montagne. Ouvrage accompagné de cartes et de figures explicatives, à l'usage de l'instruction publique de la jeunesse.* Paris: Imprimerie de J.-B. Hérault.

Dupin, Baron Charles. 1828. *Tableau comparé de l'instruction populaire avec l'industrie des départemens, d'après l'exposition de 1827, présenté dans la seconde séance du cours de géométrie et de mécanique appliquées aux arts, professé pour les ouvriers, . . . le 23 décembre 1827.* Paris: J. Tastu.

———. 1836. Nécrologie de M. A.-É.-J.-P.-J.-F. d'Audebard, Baron de Férussac. *Moniteur universelle* 79 (samedi 19 mars 1836):491.

———. N.d. (date stamped 1863 by the BN). *Rapport sur la proposition d'accorder à la section de géographie et navigation le même nombre de membres qu'à toutes les autres sections . . .* Paris: Mallet-Bachelier.

———(Rapporteur). N.d. *Rapport verbal fait à l'Académie des sciences, le lundi 29 août 1825, au nom d'une commission composée de MM. le Baron Coquebert de*

Montbret, Desfontaines, Ampère et le Baron Ch. Dupin sur Le Bulletin universel des sciences et de l'industrie . . . Paris.

Dupont de Nemours, Pierre-Samuel. 1768. *De l'Origine et des progrès d'une science nouvelle . . .* London and Paris: Desaint.

Durand, Yves. 1984. *Vivre au pays au XVIIIe siècle. Essai sur la notion de pays dans l'ouest de la France.* Paris: Presses universitaires de France.

Durieux, Joseph. 1930. Deux officiers de l'empereur: Boutin et Vaissière. *Revue des études napoléoniennes* 30–31: 290–94.

Durkheim, Émile. 1938. *L'Évolution pédagogique en France.* 2 vols. Paris: Presses universitaires de France.

Dussaud, René. 1946. *La Nouvelle Académie des inscriptions et belles-lettres (1795–1914).* Paris: Librairie Orientaliste Paul Geuthner.

Dussieux, Louis-Étienne. 1882. *Les Grands Faits de l'histoire de la géographie, recueil de documents . . .* 3 vols. Paris: V. Lecoffre.

Edney, Matthew H. 1993. The Patronage of Science and the Creation of Imperial Space: The British Mapping of India, 1799–1843. In *Cartographica Monograph, 44: Introducing Cultural and Social Cartography,* edited by R. A. Rundstrom. Toronto: University of Toronto Press.

———. 1994. Mathematical Geography and the Social Ideology of British Cartography, 1780–1820. *Imago Mundi* 46: 101–16.

———. 1997. *Mapping an Empire: The Geographical Construction of British India, 1765–1843.* Chicago and London: University of Chicago Press.

Élie de Beaumont, Léonce. 1860. *Société de géographie. Assemblée générale du 16 décembre 1859. Discours d'ouverture.* Paris: Imprimerie de L. Martinet.

Encyclopédie méthodique, ou par ordre de matières, par une société de gens de lettres, de savans et d'artistes. 1787, 1782–1832. 192 vols. Paris: Liège.

Engelhardt, Dietrich von. 1990. Historical Consciousness in the German Romantic *Naturforschung.* In *Romanticism and the Sciences,* edited by A. Cunningham and N. Jardine. Cambridge: Cambridge University Press.

Estivals, Robert. 1961. *Le Dépôt légal sous l'ancien régime de 1537 à 1791.* Paris: Librairie Marcel Rivière et Cie.

Expilly, Abbé Jean-Joseph. 1756. *La Polychrographie en six parties.* Paris: Joseph Payen.

———. 1757. *Le Géographe manuel, contenant la description de tous les pays du monde, leurs qualités, leur climat, le caractère de leurs habitans, leurs villes capitales, avec leurs distances de Paris, et des routes qui y mènent tant par terre que par mer, les changes et les monnois des principales places de l'Europe en correspondance avec Paris, la manière de tenir les écritures de chaque nation de l'Europe, etc . . . .* Paris: Bauche.

Eymery, Alexis. 1815. Malte-Brun. In *Dictionnaire des girouettes, ou Nos Contemporains peint d'après eux-mêmes . . . par une société de girouettes,* edited by A. Eymery. Paris: Alexis Eymery.

Eyries, Jean-Baptiste-Benoît. N.d. Mentelle, François-Simon. In *Biographie universelle (Michaud) ancienne et moderne, ou histoire, par ordre alphabétique, de la vie publique et privée de tous les hommes qui se sont fait remarquer par leurs écrits, leurs actions, leurs talents, leurs vertus ou leurs crimes,* edited by C. Desplaces. Paris: Chez Madame C. Desplaces; and Leipzig: Librairie de F. A. Brockhaus.

Fabre, Jean-Antoine. 1797. *Essai sur la théorie des torrens et des rivières, contenant les moyens les plus simples d'en empêcher les ravages, d'en rétrécir le lit et d'en faciliter la navigation, le hallage et la flottaison; accompagné d'une discussion sur la navigation intérieure de la France, et terminé par la projet de rendre Paris port maritime . . . par le citoyen Fabre.* Paris: Bidault.

Faille, René, and Nelly Lacroq. 1979. *Les Ingénieurs-géographes Claude François et Claude-Félix Masse.* La Rochelle: Rupella.

Falcucci, Clément. 1939 (1932). *L'Humanisme dans l'enseignement secondaire en France au XIXe siècle.* Toulouse: Edouard Privat.

Fanon, Frantz. 1961. *Les Damnés de la terre.* Paris: F. Maspero.

Fauché, Jean-Baptiste, Adolphe Brongniart, Louis-Anastase Chaubard, and Jean-Baptiste-Geneviève-Marcellin Bory de Saint-Vincent. 1832. *Botanique.* Edited by France. Ministère de l'éducation nationale. Vol. 3, section 2, part 1, *Expédition scientifique de Morée (1831–1838). Section des sciences physiques.* Paris: F.-G. Levrault.

Fayet, Joseph. 1960. *La Révolution française et la science, 1789–1795.* Paris: Librairie Marcel Rivière.

Férussac, André-Etienne-Just-Paschal-Joseph-François d'Audebard, Baron de. Note sur une nouvelle espèce de ver terrestre du Brésil. [Includes plate 116, figures 2 and 3, by de Férussac]. In *Annales générales des sciences physiques,* edited by Bory de Saint-Vincent.

———. 1812. *Extraits du journal de mes campagnes en Espagne, contenant un coup d'oeil général sur l'Andalousie, une dissertation sur Cadix et sur son île, une relation historique du siège de Saragosse, . . .* Paris: F. Buisson.

———. 1814. *Mémoires géologiques sur les terreins formés sous l'eau douce par les débris fossiles des mollusques vivant sur la terre ou dans l'eau non-salée.* Paris: Chez Poulet.

———. 1814. Rapport fait à l'Institut de France, Classe des Sciences physiques et mathématiques, le 10 février 1813, sur un Mémoire de M. Daudebart de Férussac, intitulé: Considérations générales sur les fossiles des terreins d'eau douce. In *Mémoires géologiques sur les terreins formés sous l'eau douce par les débris fossiles des mollusques vivant sur la terre ou dans l'eau non-salée.* Paris: Chez Poulet.

———. 1819. *De la Nécessité de fixer et d'adopter un corps de doctrine pour la géographie et la statistique, avec un essai systématique sur cet objet et des programmes pour des cours sur ces deux sciences, dans leur application à l'art de la guerre.* Paris: Chez Magimel, Anselin et Pochard, Libraires pour l'art militaire.

———. 1820–1851. (Oeuvre posthume . . . Continué, mis en ordre et publié . . . A.-E.-J.-P.-J.-F. d'A. et G.-P. Deshayes.) *Histoire naturelle générale et particulière des mollusques terrestres et fluviatiles.* 4 vols. Paris: J.-B. Baillière.

———. 1821. Tableau générateur et analytique des sciences qui ont pour objet l'univers et les êtres en général, le globe et les sociétés humaines en particulier, in Férussac, *Plan sommaire d'un traité de géographie et de statistique à l'usage des officiers des état-majors de l'armée, précédé d'un essai sur la doctrine, le but et la marche de ces sciences,* . . . [Second title page: *De la Géographie et de la statistique, considérées dans leurs rapports avec les sciences qui les avoisinent de plus près, ou Essai sur la doctrine le but et la marche de ces sciences.*] Paris: Anselin et Pochard.

———. 1821. *Plan sommaire d'un traité de géographie et de statistique à l'usage des officiers des état-majors de l'armée, précédé d'un essai sur la doctrine, le but et la marche de ces sciences,* . . . [Second title page: *De la Géographie et de la statistique, considérées dans leurs rapports avec les sciences qui les avoisinent de plus près, ou Essai sur la doctrine le but et la marche de ces sciences.*] Paris: Anselin et Pochard.

———. 1822–1823. *Bulletin général et universel des annonces et des nouvelles scientifiques, dédié aux savants et tous les pays et à la librairie nationale et étrangère . . .* 2 vols. Paris: J. Didot.

———. 1823. *Coup d'oeil sur l'Andalousie, précédé d'un Journal historique du siège de Saragosse . . .* Paris: Ponthieu.

———. 1823. *Notice sur Cadix et sur son île.* Paris: Ponthieu.

———. 1824. *Notice analytique sur les travaux de M. de Férussac, un extrait des rapports fait à l'académie par ses commissaires sur ceux de ces travaux qui lui ont été présentés.* Paris: Imprimerie de Fain.

———. 1824–1831. *Bulletin des sciences agricoles et économiques. Quatrième section du 'Bulletin universel des sciences et de l'industrie,' publié sous la direction de M. Le Bon de Férussac, . . .* 8 vols. Paris: au bureau du Bulletin, 3, rue de l'Abbaye.

———. 1824–1831. *Bulletin des sciences géographiques, etc. économie politique, voyages. Sixième section du 'Bulletin universel des sciences et de l'industrie.' publié sous la direction de M. Le Bon de Férussac, . . .* 12 vols. Paris: au bureau du Bulletin, 3, rue de l'Abbaye.

———. 1824–1831. *Bulletin des sciences historiques, antiquités, philologie. Septième section du 'Bulletin universel des sciences et de l'industrie,' publié sous la direction de M. Le Bon de Férussac, . . .* 8 vols. Paris: au bureau du Bulletin, 3, rue de l'Abbaye.

———. 1824–1831. *Bulletin des sciences mathématiques, astronomiques, physiques et chimiques. Première section du 'Bulletin universel des sciences et de l'industrie,' publié sous la direction de M. le B.on de Férussac, . . .* 8 vols. Paris: au bureau du Bulletin, 3, rue de l'Abbaye.

———. 1824–1831. *Bulletin des sciences médicales. Troisième section du 'Bulletin universel des sciences et de l'industrie,' publié sous la direction de M. Le Bon de Férussac, . . .* 11 vols. Paris: au bureau du Bulletin, 3, rue de l'Abbaye.

———. 1824–1831. *Bulletin des sciences naturelles et de géologie. Deuxième section du 'Bulletin universel des sciences et de l'industrie,' publié sous la direction de M. Le Bon de Férussac, . . .* 11 vols. Paris: au bureau du Bulletin, 3, rue de l'Abbaye.

———. 1824–1831. *Bulletin des sciences technologiques. Cinquième section du 'Bulletin universel des sciences et de l'industrie,' publié sous la direction de M. Le Bon de Férussac, . . .* 8 vols. Paris: au bureau du Bulletin, 3, rue de l'Abbaye.

———. 1825. Distribution géographique des productions aquatiques—Mollusques et Conchifères. In Bory de Saint-Vincent, "Géographie, sous les rapports de l'histoire naturelle." In *Dictionnaire classique d'histoire naturelle,* edited by Bory de Saint-Vincent et al. Paris: Rey et Gravier et Baudouin frères.

———. 1825. *Note supplémentaire à la notice des travaux de M. de Férussac.* Paris: Imprimerie de Fain.

———. 1833. *Mémoire sur la colonisation de la Régence d'Alger, principes qui doivent servir de règles pour cette colonisation, système de défense à adopter pour garantir la colonie, . . .* Paris: Delaunay.

———. 1834. *De l'État actuel de la France et de la nécessité de s'occuper de son avenir, . . .* Paris: Chez Paulin.

———. 1836. *Catalogue des livres composant la bibliothèque du feu M. le baron de Férussac, Ancien Directeur du Bulletin universel des sciences et de l'industrie, colonel d'état-major, membre d'un grand nombre de sociétés savantes, nationales et étrangères. Dont la vente se fera le lundi 16 mai 1836, et jours suivans, à six heures de relevée, en son domicile, rue de l'Université, n. 25; par le ministère de M. Ducrocq, Commissaire-priseur, rue des Bons-Enfans, n. 28.* Paris: Chez K.-B. Baillière.

———. N.d. [After 1829]. *De la Nécessité d'une correspondance régulière et sans cesse active entre tous les amis des sciences et de l'industrie; des progrès successifs de l'esprit humain, envisagés dans leurs rapports avec le besoin, et des moyens successivement inventés pour y satisfaire: discours prononcé à la séance annuelle de la Société créée pour la propagation des connaissances scientifiques et industrielles, 1er mai 1829 . . .* Paris: Imprimerie de A. Firmin-Didot.

———, and Alcide d'Orbigny. 1835–1848. *Histoire naturelle générale et particulière des Céphalopodes acétabulifères vivants et fossiles.* Paris: Chez J.-B. Baillière.

———, and Baron Charles Dupin. [1831]. *Rapport fait au nom de la commission préparatoire (M. Sapey (prés), Comte Hector d'Aulnay, Baron de Férussac, Cunin de Gridaine, Baron Oberkampf, Paixhans, Comte de Rambuteau, Baron Charles Dupin rapporteur). Séance du 21 février 1831. Chambre des Députés session 1830, no. 249.*

Fierro-Domenech, Alfredo. 1983. *La Société de géographie: 1821–1946.* Paris: Librairie H. Champion.

———. 1984. *Inventaire des manuscrits de la Société de géographie.* Edited by France. Bibliotheque nationale. Departement des cartes et plans. Paris: Bibliothèque nationale.

Finger, Matthias. 1989. L'Approche biographique face aux sciences sociales: Le Problème du sujet dans la recherche sociale. *Cahiers Vilfredo Pareto. Revue européenne des sciences sociales* 27 (83): 217–46.

Foucault, Michel. 1966. *Les Mots et les choses; une archéologie des sciences humaines.* Paris: Gallimard.

———. 1969. *L'Archéologie du savoir.* Paris: Gallimard.

———. 1971. *The Order of Things. An Archeology of the Human Sciences (A translation of 'Les Mots et les choses').* New York: Pantheon Books.

———. 1972. *Archaeology of Knowledge.* London: Tavistock.

———. 1977. Fantasia of the Library. In *Language, Counter-Memory, Practice. Selected Essays and Interviews,* edited by D. F. Bouchard. Ithaca, New York: Cornell University Press.

———. 1986. Nietzsche, Genealogy, History. In *The Foucault Reader,* edited by Paul Rabinow. Harmondsworth: Penguin, 76–100.

———. 1991. Governmentality. In *The Foucault Effect: Studies in Governmentality with Two Lectures and an Interview by Foucault,* edited by Graham Burchell, Colin Gordon, and Peter Miller. London: Harvester Wheatsheaf.

Fouguères, M. 1943. Les Plans cadastraux de l'ancien régime. *Mélanges d'histoire sociale* 3: 54–69.

Fourcy, Ambroise. 1987. *Histoire de l'École polytechnique.* Paris: L'auteur, 1828. Reprint. Paris: Eugène Belin.

Fourier, Jean-Baptiste-Josephe. 1809. Recherches sur les sciences et le gouvernement de l'Égypte. In *Description de l'Égypte,* edited by Edme-François Jomard, Antiquités, Mémoires, 1: 803–24. Paris: Imprimerie impériale.

———. 1818. Premier Mémoire sur les monumens astronomiques de l'Égypte. In *Description de l'Égypte,* edited by Edme-François Jomard, Antiquités, Mémoires, 2: 71–86. Paris: Imprimerie impériale.

———. N.d. Préface historique. In *Description de l'Égypte,* edited by Edme-François Jomard. Paris: Imprimerie impériale.

Fox, Robert. 1992. *The Culture of Science in France, 1700–1900.* Aldershot, Great Britain: Variorum.

———. 1973. Scientific Enterprise and Patronage of Research in France 1800–1870. *Minerva* 11 (4): 442–73.

France, Commission des sciences et des arts d'Égypte. (General Editor: Edme-François Jomard). 1809–1822. *Description de l'Égypte, ou, Recueil des observations et des recherches qui ont été faites en Égypte pendant l'expédition de l'armée française, publié par les ordres de Sa Majesté l'empereur Napoléon le Grand.* 1st ed. Paris: Imprimerie impériale.

France, Commission scientifique de l'Algérie. 1844–1881. *Exploration scientifique de l'Algérie pendant les années 1840, 1841, 1842; publié par ordre du gouvernement et avec le concours d'une commission académique . . .* 40 vols. Paris: Imprimerie royale.

France, Ministère de l'éducation nationale. 1831–1838. *Expédition scientifique de Morée.* Paris: F.-G. Levrault.

France, Ministère de la guerre. Comité technique du génie, ed. 1803–1892. *Mémorial de l'officier du génie, ou Recueil de mémoires, expériences, observations, et procédés généraux propres à perfectionner la fortification et les constructions militaires.* Vol. 1. Paris: Goujon fils.

France, Ministère de la guerre. Dépôt de la guerre. 1829 (1802/3–1926). *Mémorial du Dépôt général de la guerre.* Vol. 1. Paris: Ch. Picquet.

France, Ministère de la guerre. État-major de l'armée. Section historique. 1901. *Liste chronologique des tableaux formant la collection du Ministère de la guerre (peintures, aquarelles, dessins) représentant les batailles, combats et sièges livrés par l'armée française, 1628–1887.* Paris: Imprimerie nationale.

François, Le Père Jean. 1652. *La Science de la géographie divisée en trois parties, qui expliquent les divisions, les universalitez, & les paticularitez du globe terrestre.* Rennes: Jean Hardy.

Fugier, André. 1947. *Napoléon et l'Italie.* Paris: J.-B. Janin.

Fuller, Steve. 1988. Disciplinary Boundaries. A Conceptual Map of the Field. In *Social Epistemology.* Bloomington: Indiana University Press.

———. 1991. Disciplinary Boundaries and the Rhetoric of the Social Sciences. *Poetics Today* 12 (2): 301–25.

Gaffarel, Paul. 1883. *L'Algérie, histoire, conquête et colonisation* . . . Paris: Firmin-Didot.

Gallois, Lucien. 1908. *Régions naturelles et noms de pays. Étude sur la région parisienne* . . . Paris: A. Colin.

Garnier, Jacques. 1987. État-Major. In *Dictionnaire Napoléon*, edited by J. Tulard. Paris: Fayard.

Gau, François Chrétien. 1822–1827. *Antiquités de la Nubie, ou Monumens inédits des bords du Nil situés entre la première et la seconde cataracte dessinés et mesurés, en 1819, par F. C. Gau* . . . *Ouvrage faisant suite au grand ouvrage de la commission d'Égypte.* Stuttgart: J. G. Cotta.

Gaulmier, Jean. 1951. *L'Idéologue Volney, 1757–1820. Contribution à l'histoire de l'orientalisme en France.* Beyrouth: Imprimerie catholique.

———. 1959. *Un Grand Témoin de la Révolution et de l'empire.* Paris: Hachette.

Gay, Peter. 1966–1969. *The Enlightenment: An Interpretation.* 2 vols. New York: Knopf.

Gaziello, Catherine. 1984. *L'expédition de Lapérouse, 1785–1788, réplique française aux voyages de Cook.* Paris: C. T. H. S. Documentation français.

Geison, Gerald L. 1984. *Professions and the French State, 1700–1900.* Philadelphia: University of Philadelphia Press.

Gelzer, T. 1979. Die Bedeutung der Klassischen Vorbilder beim alten Goethe. In *Classical Influences on Western Thought A.D. 1650–1870*, edited by R. R. Bolgar. Cambridge: Cambridge University Press.

Geoffroy Saint-Hillaire, Étienne, Isadore Geoffroy Saint-Hillaire, Chanoine Florent Deshayes, Georges Bibron, and Jean-Baptiste-Geneviève-Marcellin Bory de Saint-Vincent. 1832. *Zoologie. Animaux vertébrés, mollusques et polypiers* . . . Edited by France. Ministère de l'éducation nationale. Vol. 3, section 1, *Section des sciences physiques. Expédition scientifique de Morée (1831–1838).* Paris: F.-G. Levrault.

Géographie Voyages—Dialogue entre Messieurs S . . . et D. . . . ., sur le Sieur Malte-Brun, auteur du *Précis de la géographie universelle*, 2 décembre 1812. Bibliothèque nationale, Département des manuscrits, N. A. Fr. 22186.

Georgel, Chantal. 1994. The Museum as Metaphor in Nineteenth-Century France. In *Museum Culture. Histories, Discourses, Spectacles*, edited by Daniel J. Sherman and Irit Rogoff. Minneapolis: University of Minnesota Press, 113–22.

Gérando, Joseph-Marie, Baron de. 1838. *Notice nécrologique sur M. le Chevalier Allent* . . . Paris: Imprimerie de Vve. Agasse.

———. N.d. Considérations sur les diverses méthodes à suivre dans l'observation des peuples sauvages [28 fructidor an VIII]. In *Société des observateurs de l'homme. Avertissement.* Paris.

Gerbod, Paul. 1965. *La Condition universitaire en France au XIXème siècle, étude d'un groupe socio-professionnel, professeurs et administrateurs de l'enseignement secondaire public de 1842 à 1880* . . . (PhD dissertation. Lettres. Paris). Paris: Brive, Imprimerie Chastrusse et Cie.

Gersmehl, Phillip, William Kammratsh, and Herbert Gross. 1980. *Physical Geography*. Philadelphia: Saunders College.

Gieryn, Thomas F. 1983. Boundary Work and the Demarcation of Science from Non-Science: Strains and Interests in the Professional Ideologies of Scientists. *American Sociological Review* 48 (6): 781–95.

Gille, Bertrand. 1980 (1964). *Les Sources statistiques de l'histoire de la France, des enquêtes du XVIIe siècle à 1870*. Genève: Librairie Droz.

Gillespie, Charles Coulston. 1980. *Science and Polity in France at the End of the Old Regime*. Princeton, NJ: Princeton University Press.

Ginet, N. 1783. *Nouveau Manuel de l'arpenteur, ou Supplément à celui imprimé chez M. Cellot en 1770* . . . Paris: Chez Lamy.

Gingerich, Owen. 1982. The Historical Tension between Astronomical Theory and Observation. In *Revealing the Universe. Prediction and Proof in Astronomy*, edited by J. Cornell and A. P. Lightman. Cambridge, MA: MIT Press.

Girard, P. S. 1809. Mémoire sur le nilomètre de l'île d'Éléphantine et les mesures égyptiennes. In *Description de l'Égypte*, edited by Edme-François Jomard, Antiquités, Mémoires, 1: 1–48. Paris: Imprimerie impériale.

———. 1809. Mémoire sur les mesures agraires des anciens Égyptiens. In *Description de l'Égypte*, edited by Edme-François Jomard, Antiquités, Mémoires, 1: 325–56. Paris: Imprimerie impériale.

———. 1812. Mémoire sur l'agriculture, l'industrie et le commerce de l'Égypte. In *Description de l'Égypte*, edited by Edme-François Jomard, État Moderne, 2: 491–714. Paris: Imprimerie impériale.

Girault-Soulavie, Abbé Jean-Louis. 1780–1784. *Histoire naturelle de la France méridionale, ou Recherches sur la minéralogie du Vivarais, du Viennois, du Valentinois, . . . etc. sur la physique de la mer Méditerranée, sur les météores, les arbres, les animaux, l'homme et la femme de ces contrées* . . . 8 vols. Paris: J.-F. Quillau (and Nismes: C. Belle, Vol. 2 only).

Glacken, Clarence J. 1967. *Traces on the Rhodian Shore. Nature and Culture in Western Thought from Ancient Times to the End of the Eighteenth Century*. Berkeley: University of California Press.

Glazer, Nathan. 1974. The Schools of the Minor Professions. *Minerva* 12 (3): 346–64.

Glick, Thomas F. 1986. History and Philosophy of Geography. *Progress in Human Geography* 10 (2): 267–77.

Goby, Jean-Edouard. 1949. *Les Doctrines saint-simoniennes et les ingénieurs. Extrait de la Revue des conférences françaises en Orient, mars 1949*, 3–15. Paris.

———. 1951. Un compagnon de Bonaparte en Égypte: Dubois-Aymé. *Cahiers d'histoire égyptienne* (série 3, mars): 221–54.

Godechot, Jacques. 1951. *Les Institutions de la France sous la Révolution et l'Empire*. Paris: Presses Universitaires de France.

Godlewska, Anne. 1988. *The Napoleonic Survey of Egypt. A Masterpiece of Cartographic Compilation and Early Nineteenth-Century Fieldwork*. Vol. 25, *Cartographica Monograph*, nos. 38–39. Toronto: University of Toronto Press.

———. 1989. Traditions, Crisis, and New Paradigms in the Rise of the Modern French Discipline of Geography 1760–1850. *Annals of the Association of American Geographers* 79, 2 (June): 192–213.

———. 1990. Napoleonic Geography and Geography under Napoleon. In *Nineteenth Consortium on Revolutionary Europe.* Athens, GA: University of Georgia, Department of History, 1: 281–302.

———. 1991. L'Influence d'un homme sur la géographie française: Conrad Malte-Brun (1775–1826). *Annales de géographie* 558 (mai): 62–79.

———. 1994. Napoleon's Geographers: Imperialists and Soldiers of Modernity. In *Geography and Empire,* edited by A. Godlewska and N. Smith. Oxford, UK, and Cambridge, MA: Blackwell, 31–53.

———. 1995. Jomard: The Geographical Imagination and the First Great Facsimile Atlases. In Joan Winearls, ed., *Editing Early and Historical Atlases.* Toronto: University of Toronto Press, 109–35.

———. 1995. Map, Text and Image. The Mentality of Enlightened Conquerors: A New Look at the 'Description de l'Égypte.' *Transactions of the Institute of British Geographers,* n.s., 20 (1):5–28.

———, and Neil Smith, eds. 1994. *Geography and Empire.* Oxford, UK, and Cambridge, MA: Blackwell.

Goethe, Johann Wolfgang von. 1901 (1787). *Iphigenie auf Tauris.* Boston: D.C. Heath and Co.

Goldgar, Anne. 1992. The Absolutism of Taste: Journalists as Censors in 18th-Century Paris. In *Censorship and the Control of Print in England and France 1600–1910,* edited by R. Myers and M. Harris. Winchester: St. Paul's Bibliographies.

Gopnik, Adam. 1996. The First Frenchman. Why is all Paris battling over a fifth-century reformed barbarian? *The New Yorker* (October 7):44–53.

Gosselin, Pascal-François-Joseph. 1790. *Géographie des Grecs analysée, ou les Systèmes d'Eratosthènes, de Strabon, et de Ptolémée comparés entre eux et avec nos connoissances modernes . . .* Paris: Debure l'aîné.

———. 1798 (an VI)—1813. *Recherches sur la géographie systématique et positive des anciens pour servir de base à l'histoire de la géographie ancienne . . .* 2 vols. Paris: Imprimerie de la République.

———. 1810. Géographie ancienne. In *Rapport historique sur les progrès de la littérature ancienne, depuis 1789 et sur l'état actuel, présenté à sa Majesté l'Empereur et Roi, et son Conseil d'état, le 20 février 1808, par la classe d'histoire et de littérature ancienne de l'Institut,* edited by B.-J. Dacier. Paris: Imprimerie impériale.

———. 1813. *De l'Évaluation et de l'emploi des mesures itinéraires grecques et romaines. Observations générales sur la manière, de considérer et d'évaluer les anciens stades itinéraires—Éclaircissemens sur les différentes roses des vents des anciens.* Paris: Imprimerie impériale.

———. 1814. *Atlas ou recueil des cartes géographiques.* Paris: Imprimerie royale.

———. 1817. Recherches sur le principe, les bases et l'évaluation des différens systèmes métriques linéaires de l'antiquité. In *Géographie de Strabon,* edited by P.-F.-J. Gosselin. Paris: De l'imprimerie impériale.

Graham, Loren, Wolf Lepenies, and Peter Weingart, eds. 1983. *Functions and Uses of Disciplinary Histories.* Dordrecht: D. Reidel.

Green, Jonathan. 1996. *Chasing the Sun. Dictionary-Makers and the Dictionaries They Made.* London: Jonathan Cape.

Greene, Mott T. 1982. *Geology in the Nineteenth Century. Changing Views of a Changing World.* Ithaca and London: Cornell University Press.

Gregory, K. J. 1985. *The Nature of Physical Geography.* London: Edward Arnold.

Guettard, Jean-Étienne, and Jean-Louis Dupain-Triel. 1780–1784. *Cartes minéralogiques de la France.*

Guibert-Sledzewski, Elisabeth. 1985. Les idéologues, une approche de l'homme un et indivisible: le cas de Volney. *La Pensée* 246 (juillet—août): 102–12.

Guigne fils, M. de. 1810. Réflexions sur les anciennes observations astronomiques des chinois, et sur l'état de leur empire dans les temps les plus reculés. *Annales des voyages* 8:145–89.

Guilhem de Clermont-Lodève, Guillaume-Emmanuel-Joseph, Baron de Sainte-Croix. 1775. *Examen critique des anciens historiens d'Alexandre le Grand.* Paris: Chez Dessain Junior.

Gusdorf, Georges. 1960. *Introduction aux sciences humaines, essai critique sur leurs origines et leur développement . . .* Paris: Les Belles lettres.

———. 1966. De l'histoire des sciences à l'histoire de la pensée. In *Les Sciences humaines et la pensée occidentale.* Paris: Payot.

———. 1978. La Conscience révolutionnaire des idéologues. In *Les Sciences humaines et la pensée occidentale.* Paris: Payot.

Hahn, Roger. 1971. *The Anatomy of a Scientific Institution: The Paris Academy of Sciences 1666–1803.* Berkeley and Los Angeles: University of California Press.

Hamilton, William Richard. 1809. *Remarks on Several Parts of Turkey. Part 1. Aegyptiaca.* London: T. Payne.

Hanley, William. 1980. The Policing of Thought: Censorship in Eighteenth-Century France. *Studies on Voltaire and the Eighteenth Century* 183:265–95.

Harbison, Robert. 1977. The Minds's Miniatures: Maps. In *Eccentric Spaces.* New York: Alfred A. Knopf.

Hardesty, Kathleen. 1977. *The Supplément to the Encyclopédie.* The Hague: Martinus Nijhoff.

Hare, F. Kenneth. 1973. Energy Based Climatology and Its Frontier with Ecology. In *Directions in Geography,* edited by R. J. Chorley. London: Methuen and Co. Ltd.

Harley, John Brian. 1983. Meaning and Ambiguity in Tudor Cartography. In *English Map-Making 1500–1650,* edited by S. Tyacke. London: The British Library.

———. 1987. The Map and the Development of the History of Cartography. In *The History of Cartography,* edited by J. B. Harley and D. Woodward. Chicago: University of Chicago Press.

———. 1988. Silences and Secrecy: The Hidden Agenda of Cartography in Early Modern Europe. *Imago Mundi* 40:57–76.

———. 1990. *Maps and the Columbian Encounter: An Interpretive Guide to the Travelling Exhibition.* Milwaukee: Golda Meir Library, University of Wisconsin.

———, Barbara Bartz Petchenik, and Lawrence W. Towner. 1978. *Mapping the*

American Revolutionary War. Chicago and London: University of Chicago Press.

Hartshorne, Richard. 1939. *The Nature of Geography: A Critical Survey of Current Thought in the Light of the Past.* Lancaster, PA: The Association of American Geographers.

———. 1958. The Concept of Geography as a Science of Space, from Kant and Humboldt to Hettner. *Annals of the Association of American Geographers* 48 (2): 97–108.

———. 1959. *Perspective on the Nature of Geography.* Chicago: Published for the Association of American Geographers by Rand McNally.

Haüy, Abbé Réne-Just. 1783. *Essai d'une théorie sur la structure des crystaux appliquée à plusieurs genres de substances crystallisées . . .* Paris: Gogué et Née de la Rochelle.

Hayek, Friedrich August. 1952. *The Counter Revolution of Science: Studies on the Abuse of Reason.* Glencoe, IL: The Free Press.

Healy, George Robert. 1956. Mechanistic Science and the French Jesuits. A Study of the Responses of the *Journal de Trévoux* (1701–1762) to Descartes and Newton. PhD dissertation, University of Minnesota.

Heffernan, Michael J. 1994. The Science of Empire: The French Geographical Movement and the Forms of French Imperialism, 1870–1920. In *Geography and Empire,* edited by A. Godlewska and N. Smith. Oxford, UK, and Cambridge, MA: Blackwell.

———. 1995. The Spoils of War: The Société de Géographie de Paris and the French Empire, 1914–1919. In *Geography and Imperialism 1820–1940,* edited by M. Bell, R. A. Butlin, and M. J. Heffernan. Manchester: Manchester University Press.

Heidenreich, Conrad E., and Edward H. Dahl. 1980. The French Mapping of North America in the Seventeenth Century. *The Map Collector,* December, 2–11.

Helgerson, Richard. 1986. The Land Speaks: Cartography, Chorography, and Subversion in Renaissance England. *Representations* 16: 51–85.

———. 1993. Nation or Estate? Ideological Conflict in the Early Modern Mapping of England. In *Cartographica Monograph, 44: Introducing Cultural and Social Cartography,* edited by R. A. Rundstrom. Toronto: University of Toronto Press.

Hensel, Gottfried. 1741. *Synopsis Universae Philologiae, in qua Miranda Unitas et Harmonia Linguarum Totius Orbis Terrarum Occulta, e Literarum, Syllabarum, Vocumque Natura et Recessibus Eruitur . . . cum mappis geographico-polyglottis.* Nuremberg: apud Heredes Homannianos.

Heriot, Angus. 1961. *Les Français en Italie 1796–1799.* Translated by Jacques Broussen. Paris: Gallimard.

Herrman-Mascard, Nicole. 1968. *La Censure des livres à Paris à la fin de l'Ancien Régime (1750–1789).* Paris: Presses universitaires de France.

Holdich, Col. Sir Thomas. 1899. The Use of Practical Geography Illustrated by Recent Frontier Operations. *Geographical Journal* 13 (5):465–80.

Horward, Donald D. 1989. Jean-Jacques Pelet: Warrior of the Sword and Pen. *Journal of Military History* 53 (1):1–22.

Huguenin, Marcel. 1958. La Cartographie des Alpes françaises avant Cassini. *Bulletin d'information de l'Association des ingénieurs-géographes,* no. 10 (mars), 89–106.

———. 1962 and 1963. La Cartographie ancienne de la Corse. *Bulletin d'information de l'association des ingénieurs-géographes* 23 and 26:85–98 and 33–55.

———. 1970. French Cartography of Corsica. *Imago Mundi* 24:123–37.

Humbert, Jean-Marcel, Michael Pantazzi, and Christiane Ziegler. 1994. *Egyptomania. Egypt in Western Art 1730–1930.* Paris, Ottawa, and Vienna: Réunion des Musées Nationaux.

Humboldt, Friedrich Wilhelm Heinrich Alexandre, Baron von. 1805 (an XIII). *Essai sur la géographie des plantes, accompagné d'un tableau physique des régions équinoxiales fondé sur des mesures exécutées, depuis le dixième degré de latitude boréale jusqu'au dixième degré de latitude australe, pendant les années 1799, 1800, 1801, 1802 et 1803 par Al. de Humboldt et A. Bonpland.* Paris: Chez Levrault et Schoell.

———. 1805 (1973). Géographie des plantes équinoxiales. Tableau physique des Andes et Pays voisins. Dressé d'après des observations et des Mesures prises sur les lieux depuis le 10e degré de latitude boréale jusqu'au 10e de latitude australe en 1799, 1800, 1801, 1802 et 1803. In Alexander von Humboldt, *Voyage de Humboldt et Bonpland. Voyage aux régions équinoxiales du nouveau continent.* Cinquième partie, *Essai sur la géographie des plantes* (Paris, 1805). Facsimilé intégral de l'édition Paris 1805–1834 (Amsterdam: Theatrum orbis terrarum Ltd, 1973).

———. 1808. *Tableaux de la nature, ou Considérations sur les déserts, sur la physionomie les végétaux et sur les cataractes de l'Orénoque,* . . . Paris: F. Schoell.

———. 1811. Mémoire sur l'os Hyoïde et le Larynx des oiseaux, des singes et du crocodile. In *Recueil d'observations de zoologie et d'anatomie comparée, faites dans l'océan Atlantique, dans l'intérieur du nouveau continent et dans la mer du Sud pendant les années 1799, 1800, 1801, 1802 et 1803.* Paris: F. Schoell and G. Dufour.

———. 1811. Sur deux nouvelles espèces de crotales. In Alexander von Humboldt and Aimé Bonpland, *Recueil d'observations de zoologie et d'anatomie comparée, faites dans l'océan Atlantique, dans l'intérieur du nouveau continent et dans la mer du Sud pendant les années 1799, 1800, 1801, 1802 et 1803.* Paris: F. Schoell and G. Dufour.

———. 1811. Sur la respiration des crocodiles par A de Humboldt and Observations sur l'anguille électrique . . . du Nouveau continent. In Alexander von Humboldt and Aimé Bonpland, *Recueil d'observations de zoologie et d'anatomie comparée, faites dans l'océan Atlantique, dans l'intérieur du nouveau continent et dans la mer du Sud pendant les années 1799, 1800, 1801, 1802 et 1803.* Paris: F. Schoell and G. Dufour.

———. 1811. *Sur les singes qui habitent les rives de l'Orénoque, du cassiquaire et du rio Negro; par A. De Humboldt,* in Alexander von Humboldt and Aimé Bonpland, *Recueil d'observations de zoologie et d'anatomie comparée, faites dans l'océan Atlantique, dans l'intérieur du nouveau continent et dans la mer du Sud pendant les années 1799, 1800, 1801, 1802 et 1803.* Paris: F. Schoell and G. Dufour.

————. 1814. *Atlas géographique et physique des régions équinoxiales du nouveau continent, fondé sur des observations astronomiques, des mesures trigonométriques et des nivellemens barométriques.* Paris: Chez F. Schoell.

————. 1816. *Voyage aux régions équinoxiales du Nouveau Continent fait en 1799, 1800, 1801, 1802, 1803 et 1804, par Al. de Humboldt et A. Bonpland . . . avec un atlas géographique et physique.* Paris: F. Schoell (N. Maze, J. Smith et Gide fils).

————. 1817 (1813). Des lignes isothermes et de la distribution de la chaleur sur le globe. In *Mémoires de physique et de chimie de la Société d'Arcueil.* Paris: J. Klostermann fils.

————. 1823. *A Geognostical Essay on the Superposition of Rocks in Both Hemispheres.* London: Longman, Hurst, Rees, Orme, Brown, and Green, Paternoster-Row.

————. 1836–1839. *Examen critique de la géographie du Nouveau Continent, et des progrès de l'astronomie nautique au 15me et 16me siècles.* Paris: Gide.

————. 1843. *Asie centrale. Recherches sur les chaînes de montagnes et la climatologie comparée . . .* Paris: Gide.

————. 1849–1852. *Cosmos. Sketch of a Physical Description of the Universe.* Translated by E.C. Otte. Vols. 1 (1849), 2 (1849), and 4 (1852). London: Henry G. Bohn.

————. 1854. *Volcans des Cordillères de Quito et du Mexique . . .* Paris: Gide et J. Baudry.

————. 1864. *Cosmos. Essai d'une description physique du monde. Tr de M. H. Faye . . . et de M. Ch. Galuski . . .* Translated by Harvé Auguste Étienne Albans Faye and Louis Charles Galuski. Vol. 3. Paris: Théodore Morgand.

————. 1876. *Cosmos: Sketch of a Physical Description of the Universe . . .* Translated by E. C. Otté and W. S. Dallas. Vol. 5. London: George Bell and sons.

————. 1900. *The Fluctuations of Gold.* Translated by William Maude. New York: The Cambridge Encyclopedia Co.

————. 1966 (1811). *Political Essay on the Kingdom of New Spain . . .* New York: AMS Press, Inc.

————. 1969. *Atlas géographique et physique du Royaume de la Nouvelle-Espagne.* Facsimile reproduction by Hanno Beck and Wilhelm Bonacker. Stuttgart: Brockhaus.

————. 1973 (1805). Essai sur la géographie des plantes. In *Voyage de Humboldt et Bonpland. Voyage aux régions équinoxiales du nouveau continent. Cinquième partie. [1973 Facsimilé intégral de l'édition Paris 1805–1834].* Amsterdam: Theatrum orbis terrarum Ltd.

————, and Ainé Bonpland. 1811. *Recueil d'observations de zoologie et d'anatomie comparée, faites dans l'océan Atlantique, dans l'intérieur du nouveau continent et dans la mer du Sud pendant les années 1799, 1800, 1801, 1802 et 1803 . . .* Paris: F. Schoell.

Hunt, Lynn. 1994. The Virtues of Disciplinarity. *Eighteenth-Century Studies* 28 (1):1–8.

Huot, Jean-Jacques-Nicolas. 1828. Dictionnaire de la géographie physique. In *Encyclopédie méthodique.* Paris: Chez H. Agasse.

———. 1837. *Atlas complet du Précis de la géographie universelle de Malte-Brun. Dressé conformément au texte de cet ouvrage et entièrement revu et corrigé par M. J.-J.-N Huot.* Paris: Aimé André, Vve. Le Normant.

Idrisi, Abou ʿAbd Allah Mohammad ibn Mohammad ibn Idris, al-. 1836–1840. Géographie d'Edrisi, traduite de l'arabe en français, d'après deux manuscrits de la Bibliothèque du Roi et accompagnée de notes . . . par P. Amédée Jaubert, . . . In *Recueil de voyages et de mémoires publiés par la Société de géographie.* Paris: Imprimerie royale.

Jacob, Christian. 1991. *Géographie et ethnographie en Grèce ancienne.* Paris: Armand Colin.

———. 1992. *L'Empire des cartes. Approche théorique de la cartographie à travers l'histoire.* Paris: Albin Michel.

Jacotin, Pierre. 1818. Carte topographique de l'Égypte (50 sheets). In *Description de l'Égypte,* edited by Edme-François Jomard. Paris: Imprimerie impériale.

———. 1822. Mémoire sur la carte de l'Égypte. In *Description de l'Égypte,* edited by Edme-François Jomard, État Moderne, vol. 2, pt. 2: 1–118. Paris: Imprimerie impériale.

Jameson, Fredric. 1984. Postmodernism, or the Culture Logic of Late Capitalism. *New Left Review* 146 (July–August):53–92.

Jarry de Mancy, Adrien. 1831–1835. *Atlas historique et chronologique des littératures anciennes et modernes, des sciences et des beaux-arts, d'après la méthode et sur le plan de l'atlas de A. Lesage (comte de las Casas), et propre à former le complément de cet ouvrage . . . (avec la collaboration de F. Denis et E. Héreau).* Paris: Chez Jules Renouard.

Jefferson, Thomas. 1784–1785. *Notes on the State of Virginia; Written in the Year 1781, Somewhat Corrected and Enlarged in the Winter of 1782, for the Use of a Foreigner of Distinction, in Answer to Certain Quiries Proposed by Him . . . 1782.* Paris.

Jollois, Jean Baptiste Prosper, and René Edouard Devilliers. 1809. Description des monumens astronomiques découverts en Égypte. In *Description de l'Égypte,* edited by Edme-François Jomard, Antiquités, Descriptions, 1: 1–16. Paris: Imprimerie impériale.

———. 1809. Description d'Esné et de ses environs. In *Description de l'Égypte,* edited by Edme-François Jomard, Antiquités, Descriptions, 1: 1–26. Paris: Imprimerie impériale.

———. 1809. Description générale de Thebes. In *Description de l'Égypte,* edited by Edme-François Jomard, Antiquités, Descriptions, 1: 1–448. Paris: Imprimerie impériale.

———. 1809. Recherches sur les bas-reliefs astronomiques des Égyptiens. In *Description de l'Égypte,* edited by Edme-François Jomard, Antiquités, Mémoires, 1: 427–94. Paris: Imprimerie impériale.

———. 1818. Description des antiquités de Denderah. In *Description de l'Égypte,* edited by Edme-François Jomard, Antiquités, Descriptions, 2: 1–62. Paris: Imprimerie impériale.

Jomard, Edme-François. *Discours sur M. le baron Walckenaer au nom de la Société de géographie.* Paris.

———. 1809–1822. *Description de l'Égypte.* See France, Commission des sciences et des arts d'Égypte.

———. 1809. Description de l'île d'Éléphantine. In *Description de l'Égypte,* edited by Edme-François Jomard, Antiquités, Descriptions, 1: 1–20. Paris: Imprimerie impériale.

———. 1809. Description de Syène et des cataractes. In *Description de l'Égypte,* edited by Edme-François Jomard, Antiquités, Descriptions, 1: 1–28. Paris: Imprimerie impériale.

———. 1809. Essai d'explication d'un tableau astronomique peint au plafond du premier tombeau des rois de Thèbes à l'ouest de la vallée, suivi de recherches sur le symbole des équinoxes. In *Description de l'Égypte,* edited by Edme-François Jomard, Antiquités, Mémoires, 1: 255–64. Paris: Imprimerie impériale.

———. 1809. Exposition du système métrique des anciens Égyptiens, contenant des recherches sur leurs connaissances géométriques, géographiques et astronomiques, et sur les mesures des autres peuples de l'antiquité. In *Description de l'Égypte,* edited by Edme-François Jomard. Paris: Imprimerie impériale.

———. 1809. Mémoire sur le lac de Moeris comparé au lac du Fayoum. In *Description de l'Égypte,* edited by Edme-François Jomard, Antiquités, Mémoires, 1: 79–114. Paris: Imprimerie impériale.

———. 1809. Observations sur les Arabes de l'Égypte moyenne. In *Description de l'Égypte,* edited by Edme-François Jomard, État Moderne, 1: 545–76. Paris: Imprimerie impériale.

———. 1818. Description générale de Memphis et des pyramides, accompagnée de remarques géographiques et historiques. In *Description de l'Égypte,* edited by Edme-François Jomard, Antiquités, Descriptions, 2: 1–96. Paris: Imprimerie impériale.

———. 1818. Mémoire sur la population comparée de l'Égypte ancienne et moderne. In *Description de l'Égypte,* edited by Edme-François Jomard, Antiquités, Descriptions, 2: 87–142. Paris: Imprimerie impériale.

———. 1818. Mémoire sur les inscriptions anciennes recueillies en Égypte. In *Description de l'Égypte,* edited by Edme-François Jomard, Antiquités, Mémoires, 2: 1–16. Paris: Imprimerie impériale.

———. 1818. "Remarques sur les signes numériques des anciens Égyptiens," fragments d'un ouvrage ayant pour titre: Observations et recherches nouvelles sur les hiéroglyphes, accompagnées d'un tableau méthodique des signes. In *Description de l'Égypte,* edited by Edme-François Jomard, Antiquités, Mémoires, 2: 57–70. Paris: Imprimerie impériale.

———. 1819. *Notice sur les signes numériques des anciens Égyptiens, précédée du plan d'un ouvrage ayant pour titre: Observations et recherches nouvelles sur les hiéroglyphes, accompagnées d'un tableau méthodique des signes . . .* Paris: Imprimerie de Baudouin frères.

———. 1822. Description abrégée de la ville et de la citadelle du Kaire, accompagnée de l'explication des plans de cette ville et de ces environs, et des renseignemens, sur sa distribution, ses monumens, sa population, son commerce et son industrie. In *Description de l'Égypte,* edited by Edme-François Jomard, État Moderne, vol. 2, pt. 2: 579–778. Paris: Imprimerie impériale.

———. 1823. *Notice sur le second voyage de M. F. Cailliaud.* Paris: Goetschy.

———. 1827. *Du Nombre des délits criminels comparé à l'état de l'instruction primaire; par un membre de la Société formée, à Paris, pour l'amélioration de l'enseignement élémentaire.* Paris: Chez L. Colas.

———. 1830. *Recueil d'observations et de mémoires sur l'Égypte ancienne et moderne, ou Description historique et pittoresque de plusieurs des principaux monumens de cette contrée . . .* 6 vols. Paris: C.-L.-F. Pancoucke.

———. 1830. *Remarques géographiques.* Paris.

———. 1831. Article in which the status of the élèves égyptiens in Paris is reported [No title]. *Société d'encouragement pour l'industrie nationale* 30:412–13.

———. 1832. *Comparaison de plusieurs années d'observations faites sur la population française à divers âges sous le rapport du degré d'instruction. Lu à l'Académie des sciences le 27 août 1832 . . .* Paris: Imprimerie de Decourchant.

———. 1832. *Géographie de la France . . .* [Published within the same volume (vol. 12) are Barbier du Bocage, *Traité de géographie générale;* and V. Parisot, *Géographie de l'Europe*]. Bibliothèque populaire, ou l'Instruction mise à la portée de toutes les classes et de toutes les intelligences. Paris: 30, rue et place St-André-des-Arts.

———. 1834. Note sur l'aptitude des Arabes pour les sciences naturelles; par M. Jomard, chargé de la direction de l'école égyptienne. *Société d'encouragement pour l'industrie nationale* 33:320–23.

———. 1835. *Nouveaux Tableaux de lecture, assujettis aux systèmes et aux procédés de l'enseignement mutuel . . .* Paris: L. Colas.

———. 1836. *Coup-d'oeil impartial sur l'état présent de l'Égypte comparé à sa situation antérieure.* Paris: Imprimerie de Béthune et Plon.

———. 1839. *Notice historique sur la vie et les voyages de René Caillé.* Paris: Delaunay.

———. 1839. *Rapport fait à l'Académie royale des inscriptions et belles lettres . . . au sujet d'un pied romain.* Paris: Imprimerie royale.

———. 1842. *Discours sur Wilhelm, la vie et les travaux de G.-L.-B. Wilhelm, prononcé à l'assemblée général de la Société pour l'instruction élémentaire, le 5 juin 1842, par M. Jomard, . . . avec un appendice et le chant funèbre composé par M. Charles-Malo . . .* Paris: Perrotin.

———. 1844. *Notice sur la vie et les ouvrages de Claude-Louis Berthollet.* Annecy.

———. 1846. *Instructions sur les fouilles que l'on pourrait entreprendre dans la Cyrénaïque, afin d'y découvrir des monuments anciens, et sur les recherches archéologiques auxquelles on pourrait se livrer dans ce pays.* N.l.

———. 1846. *Procédé pour prendre des empreintes. [Extrait du Bulletin de la Société de géographie.]* Paris: Imprimerie de Bourgogne et Martinet.

———. 1847. *Sur la publication des Monuments de la géographie. [Extrait du Bulletin de la Société de géographie, 22 juillet 1847.]* Paris: Imprimerie de L. Martinet.

———. 1857. *Louis Nötinger, ingénieur des forages de l'isthme de Suez. Notice biographique.* Paris: Imprimerie de H. Plon.

———. 1859–1861. *Compte-rendu des séances de la société d'ethnographie américaine et orientale.* Vol. 1 8 (1859), and vol. 2 (1860–1861). Paris.

———. 1860. *Emplacement de l'ancienne ville de Péluse. A Monsieur le rédacteur en chef de l'Isthme de Suez.* Paris: Imprimerie de N. Chaix.

———. 1868. *Société d'ethnographie fondée en 1859. Exposé général. Actes constitutifs et statuts—liste des membres—catalogue des publications—récompenses décernées par la société—prix mis au concours—séances de la société—bibliothèque et collections—adresses des fonctionnaires—expositions et conférences—conditions à remplir pour faire membre de la société. Décembre 1868.* Paris: Bureau de la société d'ethnographie 47, Quai des Augustins.

———. 1879. *Introduction à l'Atlas des Monuments de la géographie . . . publiée par les soins et avec des remarques de M. E. Cortambert . . .* Paris: Arthus Bertrand.

———. [1822]. *Note sur un manuscrit égyptien sur papyrus, renfermant des plans de monumens avec des mesures écrites en chiffres hiéroglyphes. [Extrait de La Revue encyclopédique (47eme cahier).]* N.p.

———. [1836]. *Observations sur les chemins de fer de la Belgique et sur le projet de chemin de fer de Paris à Bruxelles, . . .* Paris: Imprimerie de Mme Huzard.

———. [1847]. *Extrait d'un mémoire sur l'uniformité à introduire dans les notations géographiques. [Extrait du Bulletin de la Société de géographie, avril 1847.]* Paris: Imprimerie de L. Martinet.

———. [1847]. *Instructions pour le voyage de M. Prax dans le Sahara septentrional. [Extrait du Bulletin de la Société de géographie, mars 1847.]* Paris: L. Martinet.

———. [1849]. *Instructions rédigées par une commission de la Société de géographie pour le voyage de M. Panet, du Sénégal en Algérie, sur la demande du Ministre de la Marine. [Extrait du Bulletin de Société de géographie cahiers de septembre et octobre 1849.]* Paris: Imprimerie de L. Martinet.

———. [1859]. *Note sur la nouvelle direction à donner à la recherche des sources du Nil. [Extrait d'une lettre adressée à la Commission centrale de la Société de géographie.]* Paris: Imprimerie de L. Martinet.

———. [1862]. *Classification méthodique des produits de l'industrie extra-européenne ou objets provenant des voyages lointains, suivie du plan de la classification d'une collection ethnographique complète, fragment lu à la société d'ethnographie, le 12 avril 1862 . . . [Extrait en partie du la Revue oriental américaine.]* Paris: Challamel aîné.

———. N.d. *Discours sur la vie et les travaux de Louis-Benjamin Francoeur, prononcé à l'assemblée générale de la Société pour l'instruction élémentaire, le 15 juin 1851.* Paris: Imprimerie de Schneider.

———. N.d. *Institut royale de France. Académie royale des sciences. Discours de M. Girard, . . . [de Bon. Cuvier et de E.-F. Jomard] prononcé au funérailles de M. le baron Fourier, le 18 mai 1830.* Paris: Imprimerie de A. F. Didot.

———. N.d. *Les Monuments de la géographie, ou recueil des anciennes cartes européennes ou orientales.* Paris.

———. N.d. *Notice sur les nouvelles découvertes faites en Égypte et sur l'influence qu'elles peuvent avoir sur l'étude des antiquités historiques [Extrait de la Revue encyclopédique, mai 1814.]* Paris: Imprimerie de Baudouin frères.

———. 1826. *Discours prononcés aux funérailles de M. J.-D. Barbié du Bocage, . . . suivi d'une notice sur sa vie et ses ouvrages.* Paris.

————. 1834. Extrait d'un plan d'une société de géographie projetée à Paris en 1785. *Bulletin de la Société de Géographie de Paris*, 2nd series, 1:411.

————. (Mathieu of Paris). N.d. A hand-drawn copy of the Matthieu of Paris map of the pilgrimage to Jerusalem. Manuscript map at the Bibliothèque nationale, Cartes et plans. This copy was made from the thirteenth-century original at the British Museum upon Jomard's request and sold to Jomard by a Mr. Wright. Copyist unspecified. Bibliothèque nationale, Département des cartes et plans.

Jordanova, Ludmilla J., and Roy S. Porter, eds. 1979. *Images of the Earth: Essays in the History of the Environmental Sciences*. Chalfont St Giles: British Society for the History of Science.

Jourdain, Charles-Marie-Gabriel Bréchillet. 1857. *Le Budget de l'instruction publique et des établissements scientifiques et littéraires depuis la fondation de l'Université impériale jusqu'à nos jours . . .* Paris: L. Hachette.

————. 1862–1866. *Histoire de l'Université de Paris au XVIIe et au XVIIIe siècle, . . .* Paris: L. Hachette.

Journal des savants. 1697. Review of *Nouveaux Mémoires sur l'état présent de la Chine*, by Le Père Louis le Comte de la compagnie de Jésus, Mathématicien du Roy, (Paris: Chez Iean Anisson, 1696). In *Journal des savants* (janvier), 68–69.

Journal militaire. 1818. Ordonnance du Roi portant formation d'un Corps royal d'état-major, et d'une école d'application pour le service de l'État-major général de l'armée (Bulletins des lois, 7e séries, no. 212). Paris le 6 mai. *Journal militaire* (first semester), 357–64.

————. 1818. Règlement approuvé par le Roi en son conseil le 23 septembre 1818, sur la répartition du service, l'École d'application, la Solde et l'uniforme du corps royal . . . en exécution des dispositions de l'Ordonnance du 6 mai 1818 . . . *Journal militaire* (second semester): 263–69.

Jourquin, Jacques. 1987. Bacler d'Albe (Louis-Albert-Ghislain, baron). 1761–1824. In *Dictionnaire Napoléon*, edited by J. Tulard. Paris: Fayard.

Julia, Dominique. 1976. *La Réforme de l'enseignement au siècle des Lumières*. Paris: Micro-éditions Hachette, no. 19.

————, Huguette Bertrand, Serge Bonin, and Alexandra Laclau. 1987. L'Enseignement 1760–1815. In *Atlas de la Révolution française*, edited by Serge Banin and Claude Langlois. Paris: Éditions de l'École des hautes études en sciences sociales.

Kafker, Frank A. 1994. The Influence of the Encyclopédie on the Eighteenth-Century Encyclopedic Tradition. *Notable Encyclopedias of the Eighteenth Century: Eleven Successors of the Encyclopédie*. Oxford: Voltaire Foundation, 315: 389–99.

Kain, Roger, and Elizabeth Baigent. 1992. *The Cadastral Map in the Service of the State: A History of Property Mapping*. Chicago: University of Chicago Press.

Kant, Immanuel. 1790. *Critik der Urtheilskraft*. N.l.

Kennedy, Emmet. 1977. Destutt de Tracy and the Unity of the Sciences. *Studies on Voltaire and the Eighteenth Century* 171: 223–39.

Kennett, Austin. 1925. *Bedouin Justice: Laws and Customs among the Egyptian Bedouin*. Cambridge: Cambridge University Press.

Kersaint, Georges. 1966. *Antoine François de Fourcroy (1755–1809), sa vie et son oeuvre . . .* Paris: Éditions du Muséum.

Khan, Mirza Abou Taleb. 1809. Preuves de la liberté des femmes en Orient, ou leur sort comparé à celui des Anglaises. *Annales des voyages* 9:27–45.

Kirchner, Walther. 1950. Mind, Mountain, and History. *Journal of the History of Ideas* 11:412–47.

Kitchin, Joanna. 1966. *Un Journal philosophique: La Décade (1794–1807).* Abbeville: Imprimerie F. Paillart.

Klein, Julie Thompson. 1990. The Disciplinary Paradox. In *Interdisciplinarity: History, Theory, Practice.* Detroit: Wayne State University Press.

Knight, David. 1990. Romanticism and the Sciences. In *Romanticism and the Sciences,* edited by A. Cunningham and N. Jardine. Cambridge: Cambridge University Press.

Knight, Isabel F. 1968. *The Geometric Spirit: The Abbé de Condillac and the French Enlightenment.* New Haven: Yale University Press.

Konvitz, Josef W. 1985. *The Urban Millennium: The City-Building Process from the Early Middle Ages to the Present.* Carbondale: Southern Illinois University.

———. 1987. *Cartography in France, 1660–1848: Science, Engineering, and Statecraft.* Chicago and London: University of Chicago Press.

Koyré, Alexandre. 1968. *Metaphysics and Measurement. Essays in Scientific Revolution.* London: Chapman and Hall.

Kühn, Herbert. 1938. *Volney und Savary als Wegbereiter des Romantischen Orienterlebnisses in Frankreich.* Leipzig.

Kuhn, Thomas S. 1970. *The Structure of Scientific Revolutions.* 2nd ed. Chicago and London: University of Chicago Press.

———. 1977. Energy Conservation as an Example of Simultaneous Discovery. In *The Essential Tension: Selected Studies in Scientific Tradition and Change.* Chicago and London: University of Chicago Press.

———. 1977. *The Essential Tension: Selected Studies in Scientific Tradition and Change.* Chicago and London: University of Chicago Press.

———. 1977. Second Thoughts on Paradigms. In *The Essential Tension: Selected Studies in Scientific Tradition and Change.* Chicago and London: University of Chicago Press.

La Barre, Antoine Joseph Le Febvre de. 1666. *Description de la France équinoxiale, ci-devant appelée Guyanne, et par les Espagnols, El Dorado. Nouvellement remise sous l'obéïssance du roy, par le sieur Le Febvre de La Barre, . . . avec la carte d'icelvy . . . et un discours très-utile . . . pour ceux qui voudront établir des colonies . . .* Paris: Jean Ribov.

La Pérouse, Jean-François de Galaup, Cte de. 1797 (an V). *Voyage de La Pérouse autour du monde, . . .* Paris: Imprimerie de la République.

La Renaudière, Philippe-François Lason. 1827. Notice annuelle des travaux de la Société. *Bulletin de la Société de géographie de Paris* 8:320–21.

———. N.d. Brun ou Brunn (Malte-Conrad). In *Biographie universelle (Michaud) ancienne et moderne, ou histoire, par ordre alphabétique, de la vie publique et privée de tous les hommes qui se sont fait remarquer par leurs écrits, leurs actions, leurs talents, leurs vertus ou leurs crimes,* edited by C. Desplaces. Paris: Chez Madame C. Desplaces; and Leipzig: Librairie de F. A. Brockhaus.

La Roncière, Charles Germain Marie de. 1924. *L'Ancienne Académie de Marine. (Extrait de L'Académie de Marine 1: 1922.)* Paris: Société d'éditions géographiques, maritimes et coloniales.

Labbé, Le Père Philippe. 1646. *La Géographie royalle présentée au . . . roy . . . Louis XIV.* Paris: M. Hénault.

Laborde, Cte. Alexandre-Louis-Joseph de. 1827–1830. *Itinéraire descriptif de l'Espagne, 3e édition revue, corrigée et considérablement augmentée, par M. le Cte Al. de Laborde, précédée d'une notice sur la configuration de l'Espagne et son climat, par M. de Humboldt, d'un aperçu sur la géographie physique, par M. le colonel Bory de Saint-Vincent, et d'un abrégé historique de la monarchie espagnole et des invasions de la péninsule jusqu'à nos jours . . .* 3rd ed. 6 vols. Paris: Chez Firmin Didot père et fils (et frères).

Lacouture, Jean. 1988. *Champollion, une vie de lumières.* Paris: Bernard Grasset.

Lacroix, Alfred. 1932. . . . *Figures de savants.* Vol. 1. Paris: Gauthiers-Villars et Cte.

Lacroix, Abbé Louis-Antoine Nicolle de. 1780. *Géographie moderne, précédée d'un petit Traité de la sphère et du globe et terminée par une géographie sacrée et un géographie ecclésiastique . . .* Nouvelle édition. 2 vols. Paris: Delalain l'aîné.

Lacroix, Paul. N.d. Walckenaer, Charles-Anathase. In *Biographie universelle (Michaud) ancienne et moderne, ou histoire, par ordre alphabétique, de la vie publique et privée de tous les hommes qui se sont fait remarquer par leurs écrits, leurs actions, leurs talents, leurs vertus ou leurs crimes,* edited by C. Desplaces. Paris: Chez Madame C. Desplaces; and Leipzig: Librairie de F. A. Brockhaus.

Lacroix, Silvestre-François. 1811. *Introduction à la géographie mathématique et critique et à la géographie physique.* Paris: J.-G. Dentu.

Lagarde, Lucie. 1989. Le Passage du Nord-Ouest et la mer de l'Ouest dans la cartographie française du XVIIIe siècle, contribution à l'étude de l'oeuvre des Delisle et Buache. *Imago Mundi* 41:19–43.

Lagardiolle. 1829. Notes sur les principaux historiens anciens et modernes considérés militairement. In *Mémorial du Dépôt de la guerre imprimé par ordre du Ministre 1802–03.* Paris: Ch. Picquet.

Laissus, Joseph. 1970. Un Astronome français en Espagne: Pierre-François-André Méchain. In *Compte-rendus, Sciences,* edited by 94th Congrès national des Sociétés savantes, 1969. Paris: Bibliothèque nationale.

Laissus, Yves. 1964. Le Jardin du Roi. In *Enseignement et diffusion des sciences en France au XVIIIe siècle,* edited by R. Taton. Paris: Hermann.

Lallemand, Paul-Joseph. 1888. *Histoire de l'éducation dans l'ancien Oratoire de France.* Paris: E. Thorin.

Lancret, Michel-Ange. 1809. Description de l'île de Philae. In *Description de l'Égypte,* edited by Edme-François Jomard, Antiquités, Descriptions, 1: 1–60. Paris: Imprimerie impériale.

———. 1809. Mémoire sur le système d'imposition territoriale et sur l'administration des provinces de l'Égypte dans les dernières années du gouvernement des Mamelouks. In *Description de l'Égypte,* edited by Edme-François Jomard, État Moderne, 1: 233–60. Paris: Imprimerie impériale.

————. 1809. Notice sur la branche Canopique. In *Description de l'Égypte*, edited by Edme-François Jomard, Antiquités, Mémoires, 1: 251–54. Paris: Imprimerie impériale.

————, and Du Bois-Aymé. 1818. Description d'Héliopolis. In *Description de l'Égypte*, edited by Edme-François Jomard, Antiquités, Descriptions, 2: 1–18. Paris: Imprimerie impériale.

Langins, Janis. 1980. Sur la première organisation de l'École polytechnique, texte de l'arrêté du 6 frimaire an III. *Revue de l'histoire des sciences* 33:289–313.

Lapage, Geoffrey. 1961. *Art and the Scientist*. Bristol: John Wright and Sons.

Lapie, Pierre. Manuscript materials concerning Pierre Lapie. Service historique de l'armée de terre, Vincennes. Manuscript, Série 1, Dossier No. 53350.

————. Manuscript materials concerning Pierre Lapie. Bibliothèque nationale, Département des manuscrits: Titres originaux de Dom Villevieille: Manuscrits français: 26280 Lapie (1336).

————. Manuscript materials concerning Pierre Lapie. Bibliothèque nationale, Département des cartes et plans, papers of the Société de géographie: Doc 1320 Ms. in-8 46; Doc 1456 Colis 1 bis; Doc 3102 Colis 19; Doc 3126 Colis 19, letter from Lapie to the President of the Commission (not personally named), Paris le 15 septembre 1826.; Doc 3320 Colis 19 bis; Doc 1628 Colis 3 bis; Doc 1919 Colis 5 bis.

————. 1802 (an X). Carte du voyage au Sénégal de Labarthe. In-8o. Paris.

————. 1803 (an XI). *Notes géographiques pour servir d'index à la carte de Syrie, relative à l'histoire de l'expédition de Bonaparte en Orient*. Paris: Chez Lapie et Piquet.

————. 1805 (an XIII). Atlas. In *Histoire des guerres des Gaulois et des François en Italie*, edited by G. B. A. Jubé de la Perrelle. Paris.

————. 1812. Atlas complet du Précis de la géographie universelle . . . In *Précis de la géographie universelle*, edited by C. Malte-Brun. Paris: F. Buisson.

————. 1812. Atlas supplémentaire du *Précis de la géographie universelle* de M. Malte-Brun, dressé conformément au texte de cet ouvrage . . . In *Précis de la géographie universelle*, edited by C. Malte-Brun. Paris: F. Buisson.

————. 1812. Carte de la Russie d'Europe avec l'Autriche, la Suède, le Danemark et la Norvège, la Prusse, le grand duché de Varsovie, les Provinces illyriennes, et une partie de la Confédération du Rhin et de la Turquie d'Europe . . . Paris: Thardieu.

Latour, Bruno. 1987. *Science in Action: How to Follow Scientists and Engineers through Society*. Cambridge, MA: Harvard University Press.

Laudan, Rachel. 1987. *From Mineralogy to Geology: The Foundations of a Science, 1650–1830*. Chicago and London: University of Chicago Press.

Laudy, Lucien. 1923. Napoléon et l'université de Bruxelles. *Revue des études napoléoniennes* 20–21:166–67.

Laurens, Henry. 1987. *Les Origines intellectuelles de l'expédition d'Égypte: l'Orientalisme Islamisant en France, 1698–1798*. Istanbul-Paris: Éditions Isis.

Laurens, Henry, Charles C. Gillespie, Jean-Claude Golvin, and Claude Traunecker. 1989. *L'Expédition d'Égypte, 1798–1801*. Paris: Armand Colin.

Lavallée, Joseph, Jean-Baptiste-Joseph Breton de la Martinière, Louis Brion de la Tour, and Louis (père) Brion de la Tour. 1772–1802. *Voyage dans les départements de la France.* 13 vols. Paris.

Lavedan, Pierre. 1952. *Histoire de l'urbanisme. Époque contemporaine.* Vol. 3. Paris: H. Lawrens.

Lavoisier, Antoine-Laurent. 1791. *Résultats extraits d'un ouvrage intitulé: "De la Richesse territoriale du royaume de France . . ." remis au comité de l'imposition, . . .* Paris: Imprimerie nationale.

Lazarsfeld, Paul Felix. 1970. *Philosophie des sciences sociales.* Paris: Gallimard.

Le Bas, Philippe. 1835. *Monuments d'antiquité figurée, recueillis en Grèce par la Commission de Morée, et expliqués par Ph. Lebas. 1er cahier. Bas-reliefs du temple de Phigalie.* Paris.

————. 1842. L'Univers. Histoire et description de tous les peuples. In *Dictionnaire encyclopédique de la France.* Paris: F. Didot.

Le Grand, Homer E. 1990. Is a Picture Worth a Thousand Experiments? In *Experimental Inquiries: Historical, Philosophical and Social Studies of Experimentation in Science,* edited by H. E. L. Grand. Dordrecht and Boston: Kluwer Academic Publishers.

Le Père, J. M. 1809. Mémoire sur la communication de la mer des Indes à la Méditerranée par la mer Rouge et l'isthme de Soueys. In *Description de l'Égypte,* edited by Edme-François Jomard, État Moderne, 1: 21–186. Paris: Imprimerie impériale.

Le Puillon de Boblaye, Émile, and Théodore Virlet. 1833. *Géologie et minéralogie.* Edited by France. Ministère de l'éducation nationale. Vol. 2, part 2, *Section des sciences physiques. Expédition scientifique de Morée (1831–1838).* Paris: F.-G. Levrault.

Le Puillon de Boblaye, Théodore. 1858. *Esquisse historique sur les Écoles d'artillerie pour servir à l'histoire de l'École d'application de l'artillerie et du génie.* Metz: Rousseau-Pallez.

Leclant, J. 1990. Allocution d'ouverture: séance de l'Académie des Inscriptions et Belles-Lettres. In *Actes du colloque international du centenaire de l'année épigraphique. Paris: 19–21 octobre 1988. Un siècle d'épigraphie classique: aspects de l'oeuvre des savants français dans les pays du bassin méditerranéen de 1888 à nos jours.* Paris: Presses universitaires de France.

Leclerc, Gérard. 1972. *Anthropologie et colonialisme. Essai sur l'histoire de l'africanisme.* Paris: Fayard.

Leclerc, Sébastien. 1669. *Pratique de la géométrie sur le papier et sur le terrain, avec un nouvel ordre et une méthode particulière.* Paris: Thomas Jolly.

Lecuyer, Bernard-Pierre. 1963. *La Recherche sociale empirique en France sous l'ancien régime.* Paris: École pratique des hautes études, VIe section, Séminaire de la sociologie empirique en France.

Lefranc, Abel-Jules-Maurice. 1893. *Histoire du Collège de France, depuis ses origines jusqu'à la fin du premier Empire.* Paris: Hachette.

————. 1897. Le Voyageur Volney et la critique de l'histoire. *L'Enseignement philosophique* 38 (1).

Leigh, R. A. 1979. Jean-Jacques Rousseau and the Myth of Antiquity in the

Eighteenth Century. In *Classical Influences on Western Thought AD. 1650–1870,* edited by R. R. Bolgar. Cambridge: Cambridge University Press.

Lejeune, Dominique. 1982. La Société de géographie de Paris: un aspect de l'histoire sociale française. *Revue de l'histoire moderne et contemporaine* 29.

———. 1986–1987. Les Sociétés de géographie en France, dans le mouvement social et intellectuel du XIXème siècle. PhD dissertation, History. Université de Paris X–Nanterre, Paris.

Lemay, Edna Hindie. 1984. Le Monde extra-Européen dans la formation de deux révolutionnaires. In *Histoires de l'anthropologie (XVI-XIX siècles): Colloque sur la pratique de l'anthropologie aujourd'hui, 19–21 novembre 1981, Sèvres,* edited by B. Rupp-Eisenreich. Paris: Klincksieck.

Lemon, Michael C. 1995. The Possibility of the History of Thought. In *The Discipline of History and the History of Thought.* London and New York: Routledge.

Lenglet-Dufresnoy, Abbé Nicolas. 1768. *Méthode pour étudier la géographie . . . 4e édition, revue, corrigée et augmentée.* Edited by J.-L. Barbeau de la Bruyère. 4th ed. 10 vols. Paris.

———. 1778. *Tablettes chronologiques de l'histoire universelle . . . Nouvelle édition . . .* Edited by J.-L. Barbeau de la Bruyère. New ed. 2 vols. Paris.

Lenoir, Timothy. 1981. The Göttingen School and the Development of Transcendental *Naturphilosophie* in the Romantic Era. *Studies in the History of Biology* (5):111–205.

———. 1993. The Discipline of Nature and the Nature of Disciplines. In *Knowledges: Historical and Critical Studies in Disciplinarity,* edited by E. Messer-Davidow, D. R. Shumway, and D. J. Sylvan. Charlottesville and London: University Press of Virginia.

Léon, Antoine. 1968. *La Révolution français et l'éducation technique.* Paris: Société des études robespierristes.

Lepetit, Bernard. 1984. *Chemins de terre et voies d'eau: Réseaux de transports et organisation de l'espace en France, 1740–1840.* Paris: École des hautes études en sciences sociales.

———. 1998. Missions scientifiques et expéditions militaires, Remarques sur leurs modalités d'articulation. In Marie-Noëlle Bourguet, Bernard Lepetit, Daniel Nordman, Maroula Sinarellis, *Invention scientifique de la Méditerranée. Égypte, Morée, Algérie,* 97–116. Paris: Éditions de l'École des hautes études en sciences sociales.

Lequin, Yves, ed. 1984. *Un Peuple et son pays.* 3 vols. Vol. 1, *Histoire des Français XIX–XX siècles.* Paris: Armand Colin.

Leroy, Tesier, and Buache. 1789. Carte relative à l'orage du 13 juillet 1788.

Lestringant, Frank. 1993. Le Déclin d'un savoir. La Crise de la cosmographie à la fin de la Renaissance. In *Écrire le monde à la Renaissance. Quinze études sur Rabelais, Postel, Bodin et la littérature géographique.* Caen: Paradigme.

Letronne, Jean-Antoine. 1812. *Essai critique sur la topographie de Syracuse au commencement du cinquième siècle avant l'ère vulgaire.* Paris: Pélicier.

———. 1814. *Recherches géographiques et critiques sur le livre: De Mensura orbis terrae composé en Irlande, au commencement du IXe siècle, par Dicuil.* Paris: G. Mathiot.

————. 1817. *Considérations générales sur l'évaluation des monnaies grecques et romaines et sur la valeur de l'or et de l'argent avant la découverte de l'Amérique [en réponse au Mémoire de M. le comte Garnier] lues à l'Académie, dans les séances des 30 mai, 13 et 27 juin et 11 juillet 1817.* Paris: Imprimerie de Firmin-Didot.

————. 1820. *Cours élémentaire de géographie ancienne et moderne, rédigé sur un nouveau plan, approuvé par la commission de l'Instruction publique et adopté pour les Écoles royales militaires. Seconde édition, revue et corrigée, avec une carte.* 2nd ed. Paris: Aumont Vve Nyon jeune.

————. 1822. *Deux inscriptions grecques gravées sur le pylône d'un temple égyptien dans la Grande Oasis, et contenant des décrets rendus par le préfet de l'Égypte, sous les règnes de Claude et de Galba, découvertes par M. Cailliaud en juillet 1818, restituées et réduites par M. Letronne.* Paris: Imprimerie royale.

————. 1822. Mémoire sur cette question: Les Anciens ont-ils exécuté une mesure de la terre postérieurement à l'établissement de l'École d'Alexandrie? *Mémoires de l'Institut royal de France, Académie des inscriptions et belles-lettres* 6:261–323.

————. 1822. *Mémoire sur le tombeau d'Osymandyas décrit par Diodore de Sicile. Remarques sur plusieurs inscriptions grecques du colosse de Memnon et sur celle du Nilomètre d'Elephantine.* Paris: Imprimerie royale.

————. 1823. *Notice sur la traduction d'Hérodote de M. A.-F. Milot et sur le prospectus d'une nouvelle traduction d'Hérodote, de M. P.-L. Courier . . .* Paris: Imprimerie de Firmin Didot.

————. 1823. *Recherches pour servir à l'histoire de l'Égypte pendant la domination des Grecs et des Romains, tirées des inscriptions grecques et latines relatives à la chronologie, à l'état des arts, aux usages civils et religieux de ce pays.* Paris: Chez Boulland-Tardieu.

————. 1824. *Observations critiques et archéologiques sur l'objet des représentations zodiacales qui nous restent de l'antiquité, à l'occasion d'un zodiaque égyptien peint dans une caisse de momie qui porte une inscription grecque du temps de Trajan . . .* Paris: Auguste Boulland.

————. 1825. Observations historiques et géographiques sur le périple attribué à Scylax. *Journal des savants*, 4.

————. 1828. *Examen du texte de Clément d'Alexandrie, relatif aux divers modes d'écriture chez les Egyptiens.* N.l.

————. 1832. *Matériaux pour l'histoire du Christianisme en Égypte, en Nubie et en Abyssine, contenus dans trois mémoires académiques sur des inscriptions grecques des Ve et VIe siècles.* Paris: Imprimerie royale.

————. 1832. Premier mémoire: Inscription Grecque déposée dans le temple de Talmis en Nubie par le Roi Nubien Silco, considérée dans ses rapports avec l'introduction du Christianisme et la propagation de la Langue grecque parmi les peuples de la Nubie et de l'Abyssinie. In *Matériaux pour l'histoire du Christianisme en Égypte, en Nubie et en Abyssine, contenus dans trois mémoires académiques sur des inscriptions grecques des Ve et VIe siècles,* edited by J.-A. Letronne. Paris: Imprimerie royale.

————. 1833. *Récompense promise à qui découvrira ou ramènera deux esclaves échappés d'Alexandrie le XVI épiphi de l'an XXV d'évergète II (10 juin de l'an 146 avant notre ère), Annonce contenue dans un Papyrus Grec traduit et expliqué par M. Letronne.* Paris: Imprimerie royale.

———. 1834. *Projet de diviser en sections l'Académie des inscriptions et belles-lettres, présenté a cette Académie en 1829* . . . Paris: Firmin-Didot et frères.

———. 1839. *Sur la séparation primitive des bassins de la mer Morte et de la mer Rouge, et sur la différence de niveau entre la mer Rouge et la Méditerranée* . . . Paris: Gide.

———. 1840. *Inscription grecque de Rosette. Texte et traduction littérale, accompagnée d'un commentaire critique, historique et archéologique.* Paris: Firmin Didot Frères.

———. 1842. *Recueil des inscriptions grecques et latines de l'Égypte étudiées dans leur rapport avec l'histoire politique l'administration intérieure, les institutions civiles et religieuses de ce pays depuis la conquête d'Alexandre jusqu'à celle des Arabes.* Paris: Imprimé par autorisation du Roi à l'Imprimerie royale (Chez Firmin Didot Frères, Libraires de l'Institut de France).

———. 1843. *Inscription grecque accompagné des noms hiéroglyphiques de Marc–Auréle et de Lucius Vérus trouvée à Philes en Égypte expliquée.* Paris: Imprimerie royale.

———. 1844–1845. *Observations historiques et géographiques sur l'inscription d'une borne militaire qui existe à Tunis et sur la voie romaine de Carthage à Theveste (Tebesa)* . . . *Extrait de la "Revue archéologique."* Paris: Imprimerie de Crapelet.

———. 1845. *Sur quelques points de la géographie ancienne de l'Asie Mineure* . . . N.l.

———. 1846. *Procédé pour prendre des empreintes. [Extrait du Bulletin de la Société de géographie.]* Paris: Imprimerie de Bourgogne et Martinet.

———. 1851. *Recherches critiques, historiques et géographiques sur les fragments d'Héron d'Alexandrie ou du système métrique considéré dans ses bases, dans ses rapports avec les mesures itinéraires des Grecs et des Romains et dans les modifications qu'il a subies depuis le règne des pharaons jusqu'à l'invasion des Arabes. Ouvrage posthume de M. Letronne couronné en 1816 par l'Académie des inscriptions et belles-lettres revu et mis en rapport avec les principales découvertes faites depuis par A.J.H. Vincent.* Paris: Imprimerie nationale.

———. 1863. *Extrait d'une notice écrite par M. Letronne, Membre de l'Institut de France sur la restauration faite par le Cte. de Schlick de quarante-deux maisons découvertes dans les ruines de Pompeï et d'Herculanum.* N.l.

———. 1863. *Nouvelles Recherches sur le calendrier des anciens Egyptiens, sa nature, son histoire et son origine.* Paris: Imprimerie impériale.

———. [1825]. *Explication d'une inscription grecque en vers, découverte dans l'île de Philae par M. Hamilton* . . . *Extraite de la suite des "Recherches pour servir à l'histoire de l'Égypte, pendant la domination des Grecs et des Romains."* Paris: Imprimerie de Fain.

———. [1826]. *Lettre à M. Joseph Passalacqua, sur un papyrus grec et sur quelques fragmens de plusieurs papyrus appartenant à sa collection d'antiquités égyptiennes* . . . *[28 mars 1826]. (Extrait du "Catalogue raisonné de la collection de M. J. Passalacqua".)* Paris: Imprimerie de C.-J. Trouvé.

———. [1831]. *Essai sur les idées cosmographiques qui se rattachent au nom d'Atlas, considérées dans leur rapport avec les représentations antiques de ce personnage fabuleux* . . . Paris: Imprimerie de Fermin-Didot.

———. [1841]. *L'Isthme de Suez, le canal de jonction des deux mers sous les Grecs, les*

Romains et les Arabes. (Extrait de la "Revue des deux mondes," 15 juillet 1841.) Paris: Imprimerie de H. Fournier.

———. [1847]. *Discours prononcé . . . à la séance d'inauguration de l'École, présidée par S.E. M. le ministre de l'Instruction publique, le 5 mai 1847.* Paris: Imprimerie de P. Dupont.

———. N.d. *Examen critique des Prolégomènes de la géographie de Ptolémée, à l'occasion de l'édition et de la traduction qu'en a données l'abbé Halma . . .* Paris: Imprimerie de Firmin-Didot frères.

Lettsom, John Coakley. 1774. *The Naturalist's and Traveller's Companion, Containing Instructions for Collecting and Preserving Objects of Natural History.* 2nd ed. London: E. and C. Dilly.

Levallois, Jean-Jacques. 1988. *Mesurer la terre. 300 ans de géodésie française. De la toise du Chatelet au satellite.* N.p.

Lévêque, Pierre. 1964. *L'Aventure grecque.* Paris: Armand Colin.

Liard, Louis. 1894. *L'Enseignement supérieur en France, 1789–1893.* 2 vols. Vol. 2. Paris: Armand Colin.

Linné, Carl von. 1738. *Classes plantarum; seu, Systema plantarum omnia a fructificatione desumta . . .* Lugduni Batavorum: apud C. Wishoff.

Livingstone, David N. 1984. Natural Theology and Neo-Lamarckism: The Changing Context of Nineteenth-Century Geography in the United States and Great Britain. *Annals of the Association of American Geographers* 74 (1): 9–28.

———. 1990. Geography, Tradition and the Scientific Revolution: An Interpretive Essay. *Transactions of the Institute of British Geographers,* n.s., 15 (3): 359–73.

———. 1992. *The Geographical Tradition: Episodes in the History of a Contested Enterprise.* Oxford, UK, and Cambridge, MA: Blackwell.

———. 1995. The Spaces of Knowledge: Contributions towards a Historical Geography of Science. *Environment and Planning D: Society and Space* 13:5–34.

———. 1995. Geographical Traditions. *Transactions of the Institute of British Geographers,* n.s., 20, 420–22.

———, and Charles Withers. 1999. *Geography and Enlightenment.* Chicago: University of Chicago Press.

Le Livre du centenaire du Journal des débats (1789–1889). 1889. Paris: Plon.

Lomet des Foucaux, Antoine-François. 1801–1802 (an X). Mémoire, sur l'emploi des machines aérostatiques aux reconnaissances militaires et à la construction des cartes géographiques. *Journal de l'École polytechnique* 4 (11th book): 252–59.

Loraux, Nicole, and Pierre Vidal-Naquet. 1979. La Formation de l'Athènes bourgeoise: Essai d'historiographie 1750–1870. In *Classical Influences on Western Thought A.D. 1650–1870,* edited by R. R. Bolgar. Cambridge: Cambridge University Press.

Lottin l'aîné, Augustin-Martin. 1789. *Catalogue chronologique des libraires et des libraires-imprimeurs de Paris, depuis l'an 1470, époque de l'établissement de l'imprimerie dans cette capitale, jusqu'à présent.* Paris: Chez Jean-Roch Lottin, de Saint-Germain.

Louandre, Charles-Léopold, and Felix Bourquelot. 1842–1857. *La Littérature française contemporaine, 1827–1844 (continuation de la France littéraire).* 6 vols. Paris: Félix Daguin (Delaroque aîné).

Lucas, Hippolyte. 1849. Histoire naturelle des animaux articulés, . . . In *Exploration scientifique de l'Algérie pendant les années 1840, 1841, 1842 . . . Sciences physiques. Zoologie,* edited by France. Commission scientifique de l'Algérie. Paris: Imprimerie nationale.

Lynn, John Albert. 1984. *The Bayonets of the Republic: Motivation and Tactics in the Army of Revolutionary France, 1791–1794.* Urbana: University of Illinois Press.

Mackay Quynn, Dorothy. 1945. The Art Confiscations of the Napoleonic Wars. *American Historical Review* 50 (3): 437–60.

MacKenzie, John M., ed. 1990. *Imperialism and the Natural World.* Manchester and New York: Manchester University Press.

Mackinder, Sir Halford. 1951 (1887). *On the Scope and Methods of Geography.* London: Royal Geographical Society.

Malina, Jaroslav, and Zdenk Vascaroníek 1990. *Archaeology Yesterday and Today. The Development of Archaeology in the Sciences and Humanities.* Cambridge: Cambridge University Press.

Malte-Brun, Conrad. *Letter from Malte-Brun to Napoleon dated 31 January 1813 requesting naturalisation.* Bibliothèque nationale, Département des manuscrits, N. A. Fr. 1306.

———. 1807. Review of *Géographie de Strabon, traduite de Grec en français.* In *Journal de l'empire* (1 novembre): 3–4.

———. 1807. Review of *Campagne des armées françaises en Saxe, en Prusse et en Pologne. Journal de l'empire* (11 mai): 3–4.

———. 1807. Review of Olivier (Membre de l'Institut) et Bruguière, *Voyage dans l'Empire Ottoman. Journal de l'empire* (28 mai): 1–2.

———. 1807. Sur la mnémonique et sur les rapports de cet art prétendu avec la mémoire et l'érudition. *Journal de l'empire* 18 (février): 2–4.

———. 1807. *Tableau de la Pologne ancienne et moderne . . .* 1st ed. Paris: Henri Tardieu.

———. 1808. Aperçu des agrandissements et des pertes de la monarchie prussienne. *Annales de géographie* 1: 204–52.

———. 1808. Review of "Coup d'Oeil sur la statistique de la Pologne," Varsovie 1807, brochure de 20 pages. *Annales des voyages* 1: 399–405.

———. 1809. *Correspondance de MM. Sevelinges et Malte-Brun.* N.p.

———. 1809. Nouvelle Description de la Kwarizmie . . . *Annales des voyages* 4: 373.

———. 1809. Review of *Voyage de découvertes aux Terres Australes. Annales des voyages* 4: 282–88.

———. 1810. "Analyse de quelques Mémoires hollandaises sur l'île de Formosa," with a map by Lapie, "Carte des îles Formose Madjicosemah et Lieu-Kieu avec une partie de la Chine, des Philippines et du Japon." *Annales des voyages* 8: 344–75.

———. 1810. Aperçu de la monarchie autrichienne. *Annales des voyages* 7: 281–355.

———. 1810. Description de l'île de Bornholm et des islots d'Ertholm situés dans la mer Baltique. *Annales des voyages* 8: 95–125.

———. 1810. Moeurs et usages des anciens habitans de l'Espagne, avant la réunion de ce pays a l'Europe Romain. *Annales des voyages* 5: 278, 316.

———. 1810. Périple de la Paphlagonie. Ou mémoire sur les lieux indiqués par les anciens et les modernes, sur la côte de la mer Noire . . . *Annales des voyages* 5: 210–31.

———. 1810. Review of *Campagnes des armées françaises en Espagne et en Portugal pendant les années 1808 et 1809, sous le commandement de S. M. l'Empereur* . . . *Annales des voyages* 7: 277–80.

———. 1810. Review of Lapie's *Carte réduite de la mer Méditerranée et de la mer Noire.* In *Annales des voyages* 5: 268–72.

———. 1810. Sur un voyage inédit fait aux États-Unis et aux Antilles, par M. Legris-Belle-Isle. *Annales des voyages* 6: 119–22.

———. 1810–1829. *Précis de la géographie universelle, ou Description de toutes les parties du monde, sur un plan nouveau, d'après les grandes divisions naturelles du globe, précédée de l'histoire de la géographie chez les peuples anciens et modernes, et d'une théorie générale de la géographie mathématique, physique et politique, et accompagnée de cartes, de tableaux analytiques, synoptiques et élémentaires, et d'une table alphabétique de noms de lieux.* 8 vols. Paris: Chez Fr. Buisson.

———. 1815. *Apologie de Louis XVIII.* Paris: Imprimerie de Poulet.

———. 1821. *Tableau politique de l'Europe au commencement de l'an 1821.* Paris: Gide.

———. 1828. *Mélanges scientifiques et littéraires de Malte-Brun, ou Choix de ses principaux articles sur la littérature, la géographie et l'histoire, recueilles et mis en ordre* . . . 3 vols. Paris: Aimé-André.

Malte-Brun, Conrad, ed. 1807–1814. *Annales des voyages de la géographie, et de l'histoire: ou, Collection des voyages nouveaux les plus estimés, traduit de toutes les langues européennes; des relations originales, inédites, communiquées par des voyageurs français et étrangers; et des mémoires historiques sur l'origine, la langue, les moeurs et les arts des peuples, ainsi que sur le climat, les productions et le commerce des pays jusqu'ici peu ou mal connus accompagnée d'un bulletin où l'on annonce toutes les découvertes, recherches et entreprises qui tendent à accélérer les progrès des sciences historiques, spécialement de la géographie, et où l'on donne des Nouvelles des Voyageurs et des extraits de leur correspondance.* 24 vols. Paris: Chez F. Buisson.

———. [1807–1808]. Explication géographique du traité de Tilsit (5 pages). Service historique de l'armée de terre, Vincennes. Manuscript, 1526.

———. [1823]. *Société de géographie. Discours sur les moyens de donner une direction méthodique aux travaux géographiques en général et à ceux de la Société de géographie en particulier, lu dans séance du 15 février* . . . Paris: Imprimerie de J. Smith.

———. N.d. *Analyse fidèle d'une diatribe de Jean-Gabriel Dentu, se disant éditeur de la Géographie de Pinkerton, contenant des lettres de désaveu contre J.-G. Dentu et des témoignages de plusieurs savans illustres* . . . Paris: Chez F. Buisson.

———. N.d. Manuscript materials concerning Conrad Malte-Brun in the Papers of the Société de géographie, at the Bibliothèque nationale. See Alfredo Fierro-Domenech *La Société de géographie 1821–1946.* Ouvrage publié avec le concours du Centre national de recherche scientifique.

———, and Jean-Baptiste-Gaspar Roux de Rochelle. 1824. Voyages de Marco Polo.

Première partie. Introduction, texte, glossaire et variantes. *Recueil de voyages et de mémoires* 1.

Manne, Louis-Charles-Joseph de. 1802 (an X). *Notice des ouvrages de M. d'Anville . . . Précédé de son éloge, . . . an X*. Paris: Chez Fuchs et Demanne.

Marcel, Gabriel. 1904. Lettres inédites du cardinal Passionei à d'Anville. *Bulletin de géographie historique et descriptive*, 418–38.

———. 1907. Correspondance de Michel Hennin et de d'Anville . . . *Bulletin de géographie historique et descriptive* (3): 441–83.

Marcel, Guy. 1981. L'Enseignement de l'histoire dans les écoles centrales (an IV–an XII). *Annales historiques de la Révolution française* (243): 89–122.

Martin, Jean-Pierre. 1987. *La Figure de la terre: Récit de l'expédition française en Laponie suédoise (1736–1737)*. Cherbourg: Isoète.

Martin, Pierre. 1818. Notice sur un grand monument souterrain à l'ouest de la ville d'Alexandrie. In *Description de l'Égypte*, edited by Edme-François Jomard, Antiquités, Descriptions, 2: 7–12. Paris: Imprimerie impériale.

Matless, David. 1995. Effects of History, *Transactions of the Institute of British Geographers*, n.s., 20, 405–9.

Maupertuis, Pierre-Louis Moreau de. 1740. *Élémens de géographie*. N.p.

———. 1750. *Essay de cosmologie . . .* N.p.

Maury, Alfred. 1834. Lapie, Pierre. In *Biographie universelle et portative des contemporains, ou Dictionnaire historique des hommes vivants et des hommes morts depuis 1788 jusqu'à nos jours . . .* , edited by A. Rabbe, C.-A. Vieilh de Boisjolin, and F.-G. Binet de Sainte-Preuve. Paris: Chez F. G. Levrault.

Mauskopf, Seymour H. 1976. Crystals and Compounds: Molecular Structure and Composition in Nineteenth-Century French Science. *Transactions of the American Philosophical Society*, n.s., 66 (3).

May, Joseph Austin. 1970. *Kant's Concept of Geography and Its Relation to Recent Geographical Thought*. Toronto: University of Toronto Press.

Mayhew, Robert. 1994. Contextualizing Practice in Intellectual History. *Journal of Historical Geography* 20 (3): 323–28.

McCloskey, Donald. 1985. *The Rhetoric of Economics*. Madison: University of Wisconsin Press.

McKay, Donald Vernon. 1943. Colonialism in the French Geographical Movement 1871–1881. *Geographical Review* 33: 214–32.

McKeon, Michael. 1994. The Origins of Interdisciplinary Studies. *Eighteenth-Century Studies* 28 (1): 17–28.

McNeill, William Hardy. 1982. *The Pursuit of Power: Technology, Armed Force and Society Since AD 1000*. Chicago and London: University of Chicago Press.

Méchain. 1938. Lettres inédites de Méchain adressées à J. Pons-Bernard (1785–1788). *Les Archives de Trans en Provence* 11:337.

Melling, Antoine Ignace. 1819. *Voyage pittoresque de Constantinople et des rives du Bosphore, d'après les dessins de M. Melling [rédigé par Lacretelle et Barbié du Bocage]*. Paris: Treuttel et Würtz.

Mémorial topographique et militaire. Rédigé au Dépôt général de la guerre. 1802/3–1825. Edited by France. Ministère de la guerre. Dépôt général de la guerre. Vol. 3. Paris: Imprimerie de Guiraudet.

Mendelsohn, Everett. 1966. The Context of Nineteenth-Century Science. In *The Golden Age of Science,* edited by B. Z. Jones. New York: Simon and Schuster.

Mentelle, Edme. A letter from Mentelle "Au conseiller d'Etat chargé de l'instruction publique, et membre de l'Institut." Bibliothèque nationale, Département des manuscrits, N. A. Fr. 1306.

———. 1778–1784. *Géographie comparée, ou Analyse de la géographie ancienne et moderne des peuples de tous les pays et de tous les âges; accompagnée de tableaux analytique et d'un grand nombre de cartes . . .* 7 vols. Paris: Chez l'auteur.

———. 1781. *Cosmographie élémentaire, divisée en parties astronomique et géographique. Ouvrage dans lequel on a tâché de mettre les vérités les plus intéressantes de la physique céleste à la partie de ceux même qui n'ont aucune notion de mathématiques, . . .* Paris: Chez l'auteur.

———. 1783. *Élémens de géographie, contenant: 1e les principales divisions des quatre parties du monde; 2e une description abrégée de la France, . . . à l'usage des commençans, avec des cartes, par . . .* Paris: Chez l'auteur.

———. 1795 (an IV). *La Géographie enseignée par une méthode nouvelle, ou Application de la synthèse à l'étude de la géographie. Ouvrage destiné aux Écoles primaires . . .* Paris: Chez l'auteur.

———. 1796 (an IV). A letter from Mentelle to the Commissaires des travaux publics, Paris le vendémiaire 18, an 4. Bibliothèque nationale, Département des manuscrits, N. A. Fr. 1306.

———. 1804 (an XII). *Géographie physique, historique, statistique et topographique de la France, en cent huit départements, et de ces colonies . . . Faut-titre "Abrégé élémentaire de géographie ancienne et moderne, 2e partie. Tome second."* Paris: Bernard, librairie de l' École polytechnique, et de celle des ponts-et-chaussées.

———, and Conrad Malte-Brun. 1805 (an XIII). *Géographie mathématique, physique et politique de toutes les parties du monde, rédigée d'après ce qui a été publié d'exact et de nouveau par les géographes, les naturalistes, les voyageurs et les auteurs de statistique des nations les plus éclairées . . .* Paris: Tardieu.

Merz, John Theodore. 1896–1914. *A History of European Thought in the Nineteenth Century.* 4 vols. Edinburgh and London: Wm. Blackwood and Sons.

Meschonnic, Henri. 1991. *Des mots et des mondes. Dictionnaires, encyclopédies, grammaires, nomenclatures.* Paris: Hatier.

Messedaglia, L. 1936. *Il catasto e la perequazione. Relazione parlamentare.* Bologna: Cappellii.

Meyendorff, Le colonel baron de, Pierre Lapie, and A Jaubert. N.d. Carte du Khanat de Boukhara et d'une partie des steppes des Kirghiz, edited by W. Blondeau. Paris: Imprimerie Sampier.

Minder, Thomas. 1990. A Dialectical Discussion on the Nature of Disciplines and Disciplinarity: The Thesis. *Social Epistemology* 4 (2): 201–13.

Minguet, Charles. 1969. *Alexandre de Humboldt, historien et géographe de l'Amérique espagnole, 1799–1804.* Paris: François Maspero.

Mitchell, Timothy. 1991. *Colonising Egypt.* Berkeley: University of California Press.

Mitchell, W. J. Thomas. 1986. *Iconology: Image, Text, Ideology.* Chicago: University of Chicago Press.

Moheau. 1912. *Recherches et considérations sur la population de la France 1778. Publié, avec introduction et table analytique,* . . . Paris: P. Geuthner.

Møller, Per Stig. 1971. *La Critique dramatique et littéraire de Malte-Brun.* Copenhagen: Munksgaard.

Monge, Gaspard. 1847. *Géométrie descriptive* . . . 7th ed. Paris: Bachelier.

Morachiello, Paolo, and Georges Teyssot. 1979. State Town: Colonization of the Territory during the First Empire. *Lotus International: Quarterly Architectural Review* 3.

Moravia, Sergio. 1967. Philosophie et géographie à la fin du XVIIIe siècle. *Studies on Voltaire and the Eighteenth Century* 57: 937–1011.

———. 1974. *Il pensiero degli idéologues. Scienza e filosofia in Francia (1780–1815).* Firenze: La nuova Italia.

———. 1989. La Méthode de Volney. *Corpus. Revue de philosophie* 11–12: 19–31.

Morgan, S. R. 1990. Schelling and the Origins of His *Naturphilosophie.* In *Romanticism and the Sciences,* edited by A. Cunningham and N. Jardine. Cambridge: Cambridge University Press.

Muchembled, Robert. 1985. *Popular Culture and Elite Culture in France 1400–1750.* Baton Rouge and London: Louisiana State University Press.

Muehrcke, Phillip. 1981. Maps in Geography. In *Cartographica Monograph, 27: Maps in Modern Geography: Geographical Perspectives on the New Cartography,* edited by L. Guelke. Toronto: University of Toronto Press.

Munier, Henri. 1943. *Tables de la Description de l'Égypte suivies d'une bibliographie sur l'expédition française de Bonaparte.* Cairo: Société royale de géographie d'Égypte.

Murray, George William. 1935. *Sons of Ishmael: A Study of the Egyptian Bedouin.* London: George Routledge and Sons, Ltd.

Musson, Albert Edward, ed. 1972. *Science, Technology, and Economic Growth in the Eighteenth Century.* London: Methuen.

Le Nain jaune, ou Journal des arts, des sciences et de la littérature.

Napoléon. 1858–1870. *Correspondance de Napoléon 1er, publiée par ordre de l'Empereur Napoléon III.* 32 vols. Paris: Henri Plon.

———. 1935. *Lettres inédites de Napoléon Ier à Marie-Louise, écrites de 1810 à 1814.* Edited by Bibliothèque nationale. Paris: Éditions des Bibliothèques nationales de France.

Nelson, John S., Allan Megill, and Donald N. McCloskey, eds. 1987. *The Rhetoric of the Human Sciences: Language and Argument in Scholarship and Public Affairs.* Madison: University of Wisconsin Press.

Newman, Fred D. 1968. *Explanation by Description. An Essay on Historical Methodology.* The Hague and Paris: Mouton.

Nicholson, Malcolm. 1990. Alexander von Humboldt and the Geography of Vegetation. In *Romanticism and the Sciences,* edited by A. Cunningham and N. Jardine. Cambridge: Cambridge University Press.

Nicolson, Marjorie Hope. 1959. *Mountain Gloom and Mountain Glory: The Development of the Aesthetics of the Infinite.* Ithaca, NY: Cornell University Press.

Nisbet, Robert A. 1943–1944. The French Revolution and the Rise of Sociology in France. *American Journal of Sociology* 49 (2): 156–64.

Nolin, Jean-Baptiste. 1700. Extrait d'une lettre de M. Nolin, Géographe ordinaire du Roi, et de son Altesse Royale, Monsieur. *Journal des savants*, 278.

———. 1700. Seconde lettre du S. Nolin Géographe ordinaire du Roi et de son Altesse Royale Monsieur. *Journal des sçavans*, 319–23.

Nordman, Daniel. 1984. Buache de La Neuville et la "frontière" des Pyrénées. In *Images de la montagne. De l'Artiste cartographe à l'ordinateur*, edited by Bibliothèque nationale. Paris: Bibliothèque nationale.

———. 1994. *Leçons d'histoire, de géographie, d'économie politique. Édition annoté des cours de Volney, Buache de la Neuville, Mentelle et Vandermonde avec introductions et notes par Alain Alcouffe, Giorgio Israel, Bethélémy Jobert, Gérard Jorland, François Labourie, Daniel Nordman, Jean-Claude Perrot, Denis Woronoff.* Paris: Dunod.

———, Marie-Vic Ozouf-Marignier, Roberto Gimeno, and Alexandra Laclau. 1989. Le Territoire (1). Réalités et représentations. In *Atlas de la Révolution française*, edited by S. Bonin and C. Langlois. Paris: Éditions de l'École des hautes études en sciences sociales.

———, Marie-Vic Ozouf-Marignier, and Alexandra Laclau. 1989. Le Territoire (2). Les Limites administratives. In *Atlas de la Révolution française*, edited by S. Bonin and C. Langlois. Paris: Éditions de l'École des hautes études en Sciences sociales.

Norry. 1818. Description de la colonne dite de Pompée, à Alexandrie. In *Description de l'Égypte*, edited by Edme-François Jomard, Antiquités, Descriptions, 2: 1–6. Paris: Imprimerie impériale.

O [*sic*]. 1804 (an XII). Review of Mentelle/Buache, *Géographie mathématique, physique et politique . . .* In the "Sciences et Arts" section—subheading "Géographie.—Statistique," in *La Décade philosophique littéraire et politique* 39 (vendémiaire—frimaire).

Omalius d'Halloy, Jean-Baptiste-Julien d'. 1823. *Observations sur un essai de carte géologique de la France, des Pays-Bas et des contrées voisines . . .* Paris: Imprimerie de Madame Huzard.

———. 1834. De la Classification des connaissances humaines . . . In *Nouveaux Mémoires de l'Académie royale des sciences et belles-lettres de Bruxelles.* Bruxelles: M. Hayez. Vol. 9, pp. 1–14, plus fold-out table.

———. 1838. Note additionnelle sur la classification des connaissances humaines . . . In *Nouveaux Mémoires de l'Académie royale des sciences et belles-lettres de Bruxelles.* Bruxelles: Imprimerie de M. Hayez, imprimeur de l'Académie royale. Vol. 2.

———. 1839. *Division de la terre en régions géographiques d'après les Éléments de géologie. Atlas.* Paris: Pitois-Levrault.

Orlove, Benjamin. 1993. The Ethnography of Maps: The Cultural and Social Contexts of Cartographic Representation in Peru. In *Cartographica Monograph, 44: Introducing Cultural and Social Cartography*, edited by R. A. Rundstrom. Toronto: University of Toronto Press.

Ospovat, Alexander M. 1982. Romanticism and German Geology: Five Students of Abraham Gottlob Werner. *Eighteenth Century Life* 7: 105–17.

Outram, Dorinda. 1983. The Ordeal of Vocation: The Paris Academy of Sciences and the Terror, 1793–95. *History of Science* 21 (3): 251–73.

———. 1984. *Georges Cuvier. Vocation, Science and Authority in Post-Revolutionay France.* Manchester: Manchester University Press.

Ozouf, Mona. 1966. Architecture et urbanisme: l'image de la ville chez Claude-Nicolas Ledoux. *Annales Economie Société Civilisations* 21 (novembre–décembre).

———. 1984. *L'École de la France. Essais sur la Révolution, l'utopie et l'enseignement.* Paris: Éditions Gallimard.

Palmer, Richard R. 1980. The Central Schools of the First French Republic: A Statistical Survey. *Historical Reflections-Réflexions Historiques* 7 (2 and 3): 223–47.

Palsky, Gilles. 1996. Aux origines de la cartographie thématique: les cartes spéciales avant 1800. In *La Cartografia Francesa,* Cicle de conferéncies sobre Història de la Cartografia, 5è curs, 21, 22, 23, 24 i 25 de febrer de 1994. Barcelona: Institut Cartogràfic de Catalunya, 129–45.

———. 1996. *Des Chiffres et des cartes. Naissance et développement de la cartographie quantitative au XIXe siècle.* Paris: Comité des travaux historiques et scientifiques.

Papayanis, Nicholas. 1996. *Horse-drawn Cabs and Omnibuses in Paris. The Idea of Circulation and the Business of Public Transit.* Baton Rouge and London: Louisiana State University Press.

Parigot, A. 1829. La Bataille de Leuthen gagnée par Le Roi de Prusse, le 5 décembre 1757, contre l'armée Impériale aux Ordres du Prince Charles de Lorraine. In *Mémorial du Dépôt de la guerre imprimé par ordre du Ministre 1802–03.* Paris: Ch. Picquet.

Paris, École polytechnique. 1894–1897. *Livre du centenaire 1794–1894 . . .* Edited by the École polytechnique. 3 vols. Vol. 1–3. Paris: Gauthier-Vilars et fils.

Pastoureau, Mireille. 1989. Histoire de la Bibliothèque nationale: Géographie et cartographie à la BN pendant la Révolution: un rendez-vous manqué. *Revue de la Bibliothèque nationale* 32 (été): 62–69.

Patterson, Elizabeth C. 1969. Mary Somerville. *British Journal for the History of Science* 4: 309–39.

———. 1970. Somerville, Mary Fairfax Greig. In *Dictionary of Scientific Biography,* edited by C. C. Gillespie. New York: Scribner.

Paulian, Le P. Aimé-Henri. 1761. *Dictionnaire de physique.* Avignon: Louis Chambreau.

Pauw, Cornelius de. 1768–1769. *Recherches philosophiques sur les Américains, ou Mémoires intéressants pour servir à l'histoire de l'espèce humaine, . . .* Vol. 2. Berlin: G. J. Decker.

Pedley, Mary Sponberg. 1992. *Bel et Utile. The Work of the Robert de Vaugondy Family of Mapmakers.* Tring: Map Collector Publications Ltd.

Pelet, Général Jean-Jacques-Germain. N.d. *Sur le projet de loi relatif à l'état-major général de l'armée, . . .* Paris: Imprimerie de Bourgogne et Martinet.

Pelletier, Monique. 1979. Jomard et le Département des cartes et plans. *Bulletin de la Bibliothèque nationale* 4, no. 1 (mars): 18–27.

———. 1990. *La Carte de Cassini. L'Extraordinaire aventure de la carte de France*. Paris: Presses de l'École nationale des ponts et chaussées.

Pennier, Ferdinan. 1886. *Les Noms topographiques devant la philologie . . .* Paris: F. Vieweg.

Pereire, Alfred. 1914 [Cover carries date of 1924]. *Le "Journal des Débats politiques et littéraires," 1814–1914 à-propos d'un document inédit, augmenté de la liste complète de ses collaborateurs, depuis sa fondation jusqu'à nos jours . . .* Paris: Librairie ancienne Edouard Champion.

Pérez-Gómez, Alberto. 1983. *Architecture and the Crisis of Modern Science*. Cambridge, MA, and London: MIT Press.

Perrot, Aristide Michel. [1845]. *Atlas géographique, statistique et progressif des départemens de la France et de ses colonies . . . Accompagné d'un texte historique sur la France par Bory de Saint-Vincent*. Paris: A. Boulland.

Perrot, Jean-Claude. 1977. *L'Âge d'or de la statistique régionale française (an IV–1804)*. Paris: Société des Études Robespierristes.

———. 1981. *La Statistique en France à l'époque napoléonienne*. Brussels: Centre Guillaume Jacquemyns.

———. 1984. The Golden Age of Regional Statistics (Year IV—1804). In *State and Statistics in France 1789–1815*, edited by J.-C. Perrot and S. J. Woolf. Chur, London, Paris, and New York: Harwood Academic Publishers.

Petot, Jean. 1958. *Histoire de l'administration des ponts et chaussées (1599–1815)*. Paris: Marcel Rivière.

Picard, O. 1990. Numismatique et épigraphie. In *Actes du colloque international du centenaire de l'année épigraphique. Paris: 19–21 octobre 1988. Un siècle d'épigraphie classique: aspects de l'oeuvre des savants français dans les pays du bassin méditerranéen de 1888 à nos jours*. Paris: Presses universitaires de France.

Picavet, François Joseph. 1888. *La philosophie de Kant en France de 1773 à 1814. Introduction à une nouvelle traduction de la critique de la raison pratique*. Paris: Félix Alcan.

———. 1891. *Les Idéologues. Essai sur l'histoire des idées et des théories scientifiques, philosophiques, religieuse, etc. en France depuis 1789*. Paris: Félix Alcan.

Pinkerton, John. 1807. *Modern Geography. A Description of the Empires, Kingdoms, States and Colonies with the Oceans, Seas and Isles; in all parts of the World . . . Digested on a new Plan* (London).

Pinkerton, John, Baron Charles Athanase Walckenaer, and Jean Baptiste Benoît Eyriés. 1828. *Abrégé de la géographie moderne ou description historique, politique, civile et naturelle des empires, royaumes, états et leurs colonies, avec celle des mers et des îles de toutes parties du monde*. Paris: Dentu, Imprimeur-Libraire.

Pinon, Pierre, and Annie Kriegel. 1981. L'Achèvement des canaux sous la Restauration et la Monarchie de Juillet. *Annales des ponts et chaussées*, n.s., 19:72–83.

Planhol, Xavier de. 1979. Saturation et sécurité: sur l'organisation des sociétés de pasteurs nomades. In *Pastoral Production and Society/Production Pastorale et*

Société, edited by L'Equipe écologie et anthropologie des sociétés pastorales. Cambridge, U.K.: Cambridge University Press.

Playfair, Lieut-Col. Sir Robert Lambert. 1889. A Bibliography of Algeria, from the Expedition of Charles V in 1541 to 1887. *Supplementary Papers of the Royal Geographical Society* 2, part 2: 129–430.

Playfair, William. 1801. *The Commercial and Political Atlas, Representing, by Means of Stained Copper-Plate Charts, the Progress of the Commerce, Revenues, Expenditure, and Debts of England, During the Whole of the Eighteenth Century.* 3rd ed. London: T. Burton for J. Wallis.

Plazaola, Juan. 1989. *Le Baron Taylor. Portrait d'un homme d'avenir.* Paris: Fondation Taylor.

Plongeron, Bernard. 1973. Nature, métaphysique et histoire chez les idéologues. *Dix-huitième siècle, revue annuelle publiée par la Société française d'Étude du XVIIIe siècle avec le concours du CNRS* 5: 375–412.

Pococke, Richard. 1746. *Aegypti ac nobilissimi ejus fluminis a cataractis usque ad ostia fidelis atque accurata descriptio . . .* Amsterdam: Covens and Mortier.

Poindron, Paul. 1943. Les cartes géographiques du Ministère des affaires étrangères (1780–1790): Jean-Denis Barbié du Bocage et la collection d'Anville. *Sources, études, recherches, informations des Bibliothèques nationales de France,* ser. 1 (1): 46–72.

Poirier, Lucien. 1985. *Les Voix de la stratégie. Généalogie de la stratégie militaire Guibert, Jomini.* Paris: Fayard.

Popper, Karl Raimund. 1957. *The Poverty of Historicism.* London: Routledge and Kegan Paul.

———. 1969. *Conjectures and Refutations. The Growth of Scientific Knowledge.* London: Routledge and Kegan Paul.

———. 1970. Normal Science and Its Dangers. In *Criticism and the Growth of Knowledge: Proceedings of the International Colloquium in the Philosophy of Science, London, 1965, Volume 4,* edited by I. Lakatos and A. Musgrave. Cambridge: Cambridge University Press.

Porter, Roy, and George Sebastian Rousseau, eds. 1980. *The Ferment of Knowledge: Studies in the Historiography of Eighteenth-Century Science.* New York: Cambridge University Press.

Pouilloux, J. 1990. L'Épigraphie grecque. In *Actes du colloque international du centenaire de l'année épigraphique. Paris: 19–21 octobre 1988. Un siècle d'épigraphie classique: aspects de l'oeuvre des savants français dans les pays du bassin méditerranéen de 1888 à nos jours.* Paris: Presses universitaires de France.

Pratt, Mary Louise. 1992. *Imperial Eyes. Travel Writing and Transculturation.* London and New York: Routledge.

Pringle, Sir John. 1793. *Observations sur les maladies des armées dans les camps et dans les garnisons, avec des mémoires sur les substances septiques et anti-septiques et la réponse à de Haen et à Gaber . . . 2e édition . . .* Paris: T. Barrois.

Proisy d'Eppe, César. 1815. *Dictionnaire des girouettes ou nos contemporains peint d'après eux-mêmes.* Paris: Alexis Eymery.

Prost, Antoine. 1968. *Histoire de l'enseignement en France, 1800–1967.* Paris: A. Colin.

Quarterly Review. 1853. Review of *Cosmos: Sketch of a Physical Description of the Universe*, vols. 2 and 3, by Alexander von Humboldt. In *Quarterly Review* 94 (187 [Dec.]): 49–79.

Quérard, Joseph-Marie. 1827–1839. *La France littéraire, ou Dictionnaire bibliographique des savants, historiens et gens de lettres de la France, ainsi que des littérateurs étrangers qui ont écrit en français, plus particulièrement pendant les XVIIIe et XIXe siècles. Ouvrage dans lequel on a inséré, afin d'en former une bibliographie nationale complète, l'indication: 1) des réimpressions des ouvrages français de tous les âges; 2) des diverses traductions en notre langue de tous les auteurs étrangers, anciens et modernes; 3) celle des réimpressions faites en France des ouvrages originaux de ces mêmes auteurs étrangers, pendant cette époque.* 10 vols. Paris: Firmin Didot père et fils.

Quesnay, François. 1762. *Questions intéressantes de la population, l'agriculture et le commerce proposées aux académies et autres sociétés sçavantes des provinces.* Hamburg: C. Hérold.

Quimby, Roland. 1957. *The Background to Napoleonic Warfare.* New York: Columbia University Press.

Rabbe, Alphonse, Claude-Augustin Vieilh de Boisjolin, and François-Georges Binet de Sainte-Preuve, eds. 1834. *Biographie universelle et portative des contemporains, ou Dictionnaire historique des hommes vivants et des hommes morts depuis 1788 jusqu'à nos jours . . .* 5 vols. Paris: Chez F. G. Levrault.

Raige, Remi. 1809. Mémoire sur le zodiaque nominal et primitif des anciens Égyptiens. In *Description de l'Égypte*, edited by Edme-François Jomard, Antiquités, Mémoires, 1: 169–80. Paris: Imprimerie impériale.

Ramel, M.-F.-B. 1802 (an X). *De l'Influence des marais et des étangs sur la santé de l'homme, ou Mémoire couronné par la ci-devant Société royale de médecine de Paris.* Marseille: Imprimerie de J. Massy.

Ramond de Carbonnières, Bon. Louis-François-Élisabeth. 1801 (an IX). *Voyages au Mont-Perdu et dans la partie adjacente des Hautes-Pyrénées . . .* Paris: Chez Belin.

Raoul-Rochette. 1820. Traduction française de Strabon; tome 5. *Journal des savants,* 234–42.

Rehbock, Philip F. 1990. Transcendental Anatomy. In *Romanticism and the Sciences,* edited by A. Cunningham and N. Jardine. Cambridge: Cambridge University Press.

Remusat, Jean Pierre Abel. 1827. Review of Balbi, Atlas Ethnographique du Globe . . . *Journal des savants* 139 (mai): 282–91.

Riccioli, Le Père Giovanni Battista. 1661. *Geographiae et hydrographiae reformatae libri duodecim . . .* Bononiae: Ex typ. Haeredis V. Benatii.

Ritterbush, Philip C. 1971. Environmental Studies: The Search for an Institutional Form. *Minerva* 9 (4): 493–509.

Rizzi-Zannoni, J. A. B. 1772. Carte de la Pologne divisée par provinces et palatinats et subdivisée par districts. Construite d'après quantité d'Arpentages d'Observations, et de mesures prises sur les lieux. Dédiée à son Altesse le Prince Prusse-Vindes Joseph Alexandre. 24 sheets.

Robert de Vaugondy, Didier. 1752. *Atlas universel complet en cent cartes géographiques.* Paris: A. Boudet.

———. 1755. *Essai sur l'histoire de la géographie, ou sur son origine, ses progrès et son état actuel,* . . . Paris: Chez Antoine Boudet.

———. 1760. *Mémoire sur les différens accroissemens de la ville de Paris, depuis César jusqu'à présent,* . . . Paris: Antoine Boudet.

———. 1762. *Nouvel Atlas portatif.*

———. 1766. *Institutions géographiques.*

———. 1775. *Mémoire sur une question de géographie pratique, si l'aplatissement de la terre peut être rendu sensible sur les cartes, et si les géographes peuvent la négliger, sans être taxés d'inexactitude? Lu à l'Académie royale des sciences, en juillet 1775* . . . Paris: Chez Antoine Boudet.

Robert, Louis. [1939]. *L'Épigraphie grecque au Collège de France. Leçon d'ouverture donnée le 25 avril 1939. Appendice Bibliographie de Paul Foucart.* Paris.

Roberts, R. H., and J. M. M. Good, eds. 1993. *The Recovery of Rhetoric. Persuasive Discourse and Disciplinarity in the Human Sciences.* Bristol: Bristol Classical Press.

Robinet, Jean-Baptiste-René. 1777. *Supplément à l'Encyclopédie ou dictionnaire raisonné des sciences, des arts, et des métiers par une société de gens de lettres.* Amsterdam: M.M. Roy.

Robinson, Arthur. 1982. *Early Thematic Mapping in the History of Cartography.* Chicago and London: University of Chicago Press.

Robinson, Arthur, and Hellen Wallis. 1967. Humboldt's Map of Isothermal Lines: A Milestone in Thematic Cartography. *Cartographic Journal* 4: 119–23.

Roger, Jean. 1963. *Les Sciences de la vie dans la pensée française du XVIIIe siècle.* Paris.

Role, André. 1973. *Un Destin hors série: La Vie aventureuse d'un savant, Bory de Saint Vincent 1778–1846.* Paris: La pensée universelle.

Roquette, De la. 1865. *Humboldt. Correspondance scientifique et littéraire recueillie, publié et précédée d'une notice et une introduction par M. De la Roquette* . . . Paris: E. Ducrocq.

Rose, Gillian. 1995. Tradition and Paternity: Same Difference? *Transactions of the Institute of British Geographers,* n.s., 20, 414–16.

Rossel, Élisabeth-Paul-Édouard de. 1808. *Voyage de Dentrecasteaux, envoyé à la recherche de La Pérouse,* . . . Paris: Imprimerie impériale.

———. 1825. *Rapport concernant l'exposition du système adopté par la Commission des phares pour éclairer les côtes de France. Avec une carte.* Paris: Imprimerie royale.

———. 1830. Rapport sur la navigation de l'Astrolabe, commandée par M. Dumont d'Urville, . . . In *Voyage de la corvette l'Astrolabe exécuté par ordre du roi pendant les années 1826–1829, sous le commandement de M. J. Dumont d'Urville. Histoire du voyage,* edited by J. S. C. Dumont d'Urville. Paris: J. Tastu.

———. 1830 (1826). Mémoire pour servir d'instruction à M. Dumont-d'Urville . . . pendant la campagne de découvertes dont le Roi lui a confié l'exécution. In *Voyage de la corvette l'Astrolabe exécuté par ordre du roi pendant les années 1826–1829, sous le commandement de M. J. Dumont d'Urville. Histoire du voyage,* edited by J. S. C. Dumont d'Urville. Paris: J. Tastu.

———. N.d. *Courants [extrait de Nouveau Dictionnaire d'histoire naturelle].* Paris.

Rousseau, Jean-Jacques. 1966. (1762). *Émile, ou de l'éducation.* Paris: Garnier-Flammarion.

Rouyer, P. C. 1809. Notice sur les embaumements des anciens Égyptiens. In *Description de l'Égypte,* edited by Edme-François Jomard, Antiquités, Mémoires, 1: 207–20. Paris: Imprimerie impériale.

Rozet, Commandant Claude-Antoine, and Antoine-Ernest-Hippolyte Carette. 1850. *L'Algérie.* Paris: Firmin-Didot frères.

Rozière, de. 1809. Description des carrières qui ont fourni les matériaux des monumens anciens, avec des observations sur la nature et l'emploi de ces materiaux. In *Description de l'Égypte,* edited by Edme-François Jomard, Antiquités, Descriptions, 1: 1–22. Paris: Imprimerie impériale.

———. 1809. De la Géographie comparée et de l'ancien état des côtes de la mer Rouge considérés par rapport au commerce des Égyptiens dans les differens âges. Seconde partie: Du commerce qui se fit par la voie de la Thébaïde, depuis Ptolémée Philadelphe jusqu'à la conquête des Arabes. Géographie comparée de la côte occidentale de la mer Rouge. In *Description de l'Égypte,* edited by Edme-François Jomard, Antiquités, Mémoires, 1: 221–50. Paris: Imprimerie impériale.

———. 1809. Mémoire sur les vases murrhins qu'on apportoit jadis en Égypte, et sur ceux qui s'y fabriquoient. In *Description de l'Égypte,* edited by Edme-François Jomard, Antiquités, Mémoires, 1: 115–26. Paris: Imprimerie impériale.

———. 1809. Notice sur les ruines d'un monument persépolitain découvert dans l'isthme de Suez. In *Description de l'Égypte,* edited by Edme-François Jomard, Antiquités, Mémoires, 1: 265–76. Paris: Imprimerie impériale.

Rudwick, Martin J. S. 1976. The Emergence of a Visual Language for Geological Science 1760–1840. *History of Science* 14:149–95.

———. 1985. *The Great Devonian Controversy. The Shaping of Scientific Knowledge among Gentlemanly Specialists.* Chicago and London: University of Chicago Press.

Rundstrom, Robert A. "The Role of Ethics, Mapping, and the Meaning of Place in Relations between Indians and Whites in the United States," in *Cartographica Monograph, 44: Introducing Cultural and Social Cartography,* edited by R. A. Rundstrom. Toronto: University of Toronto Press, 21–28

Rupke, Nicolaas A. 1990. Caves, Fossils and the History of the Earth. In *Romanticism and the Sciences,* edited by A. Cunningham and N. Jardine. Cambridge: Cambridge University Press.

———. 1997. Introduction to the 1997 Edition. In *Cosmos.* Baltimore and London: The Johns Hopkins Press.

Rupp-Eisenreich, Britta. 1984. Aux "origines" de la Völkerkunde allemande: de la *statistik* à l'*anthropologie* de Georg Forster. In *Histoires de l'anthropologie (XVI—XIX siècles): Colloque sur la pratique de l'anthropologie aujourd'hui, 19–21 novembre 1981, Sèvres,* edited by B. Rupp-Eisenreich. Paris: Klincksieck.

Russell, Edward Stuart. 1916. *Form and Function: A Contribution to the History of Animal Morphology*. London: John Murray.

Russo, François. 1964. L'Hydrographie en France aux XVIIe et XVIIIe siècles: Écoles et ouvrages d'enseignement. In *Enseignement et diffusion des sciences en France au dix-huitième siècle*, edited by R. Taton. Paris: Hermann.

Sainte-Croix. See Guilhem de Clermont-Lodève.

Saint-Genis. 1809. Description des ruines d'El-Kâb ou Elethyia. In *Description de l'Égypte*, edited by Edme-François Jomard, Antiquités, Descriptions, 1: 1–8. Paris: Imprimerie impériale.

———. 1818. Description des antiquités d'Alexandrie et de ses environs. In *Description de l'Égypte*, edited by Edme-François Jomard, Antiquités, Descriptions, 2: 1–95. Paris: Imprimerie impériale.

Sanderson, Marie. 1974. Mary Somerville: Her Work in Physical Geography. *Geographical Review* 64: 410–20.

Sandys, John Edwin. 1967. *The Eighteenth Century in Germany, and the Nineteenth Century in Europe and the United States of America*. Vol. 3, *A History of Classical Scholarship*. New York and London: Hafner Publishing Co.

———. 1967. *From the Revival of Learning to the End of the Eighteenth Century (In Italy, France, England, and the Netherlands)*. Vol. 2, *A History of Classical Scholarship*. New York and London: Hafner Publishing Co.

Sanson, le général. 1803. Lettre du général Sanson au général Mathieu Dumas, Conseiller d'Etat, le 25 prairial an 11 (14 juin 1803). Service historique de l'armée de terre, Vincennes. Manuscript, Mémoire, 1978.

Santarem, Manuel Francisco de Barrose Jonsa de Mesquita de Macedo Leitão e Carvalhosa, 2e Vde de. 1842. *Atlas composé de mappemondes et de cartes hydrographiques et historiques depuis le XIe jusqu'au XVIIe siècle pour la plupart inédites tirées de plusieurs bibliothèques de l'Europe devant servir de preuves à l'ouvrage sur la priorité de la découverte de la côte occidentale d'Afrique au-delà du cap Bajador par les Portugais et à l'histoire de la géographie du Moyen Age. Recueillies et gravées sous la direction du Vicomte de Santarem. Publié aux frais du gouvernement Portugais*. Paris: Imprimerie de Fain et Thunot.

———. 1847. *Examen des assertions contenues dans un opuscule intitulé: "Sur la publication des monuments de la géographie" publié au mois d'août 1847*. Paris: Imprimerie de Fain et Thunot.

———. 1848–1852. *Essai sur l'histoire de la cosmographie et de la cartographie pendant le moyen-âge, et sur les progrès de la géographie après les grandes découvertes du XVe siècle, pour servir d'introduction et d'explication à l'atlas composé de mappemondes et de portulans, et d'autres monuments géographiques, depuis le VIe siècle de notre ère jusqu'au XVIIe*. Paris: Imprimerie de Maulde et Renou.

Scherer, Le Père Henri. 1703, 1710, and 1730. Representatio totius Africae cujus partes quibus fides catholica illuxit, umbra carent reliquae omnes meris tenebris et umbra mortis involutae. In *Atlas novus exhibens orbem terraqueum per nature opera, historiae novaeac veteris monumenta, artisque geographicae leges et praecepta, hoc est geographia universa in septem partes contracta*: Auguste Vindel, Dilingae et Francofurti, apud J. C. Bencard.

Schneer, Cecil J., ed. 1969. *Toward a History of Geology: Proceedings*. Cambridge: MIT Press.

Séances des écoles normales, recueillies par des sténographes et revues par les professeurs. [1802]. Vol. 1. Paris: L. Reynier.

———. N.d. Débats. In *Séances des écoles normales, recueillies par des sténographes et revues par les professeurs.* Paris: L. Reynier.

Secord, James A. 1986. The Geological Survey of Great Britain as a Research School, 1839–1855. *History of Science* 24: 223–75.

Seine, Département de la. Administration. 1821. *Recherches statistiques sur la ville de Paris et le département de la Seine (publiées sous la direction du Cte. Chabrol de Volvic).* Vol. 1. Paris: Imprimerie de C. Ballard.

———. 1823. *Recherches statistiques sur la ville de Paris et le département de la Seine; recueil de tableaux dressés d'après les ordres de Monsieur le comte de Chabrol . . .* Vol. 2. Paris.

———. 1829. *Recherches statistiques sur la ville de Paris et le département de la Seine; recueil de tableaux dressés et réunis d'après les ordres de Monsieur le Comte de Chabrol . . .* Vol. 4. Paris: Imprimerie royale.

Serval, Pierre. 1965. *Alger fut a lui.* Paris: Calmann-Levy.

Seznec, Jean. 1979. Renan et la philologie classique. In *Classical Influences on Western Thought A.D. 1650–1870,* edited by R. R. Bolgar. Cambridge: Cambridge University Press.

Shaffer, Elinor S. 1990. Romantic Philosophy and the Organization of the Disciplines: The Founding of the Humboldt University of Berlin. In *Romanticism and the Sciences,* edited by A. Cunningham and N. Jardine. Cambridge: Cambridge University Press.

Shapiro, Fred R. 1981. On the Origin of the Term "Indo-Germanic." *Historiographia Linguistica* VIII (1): 165–70.

Sheridan, Alan. 1980. *Michel Foucault: The Will to Truth.* London and New York: Tavistock Publications.

Shinn, Terry. 1980. *L'École polytechnique, 1794–1914.* Paris: Presses de la fondation national des sciences politiques.

Shumway, David R., and Ellen Messer-Davidow. 1991. Disciplinarity: An Introduction. *Poetics Today* 12 (1): 201–25.

Sibenaler, Jean. 1992. *Il se faisait appeler Volney. Approche biographique de Constantin-François Chasseboeuf 1757–1820. Préface de Jean Gaulmier.* Paris: Herault-Éditions.

Simon, Jules. 1885. *Une Académie sous le Directoire . . .* Paris: Calmann Lévy.

Simon, Renée. 1955. *Les Débuts de la géographie en France. Les Sansons d'Abbeville, G. Delisle et Fréret [extrait du Bulletin de la Société d'antiquité de Picardie].*

———. 1961. *Nicolas Fréret, académicien.* Vol. 17, *Studies on Voltaire and the Eighteenth Century.* Geneva: Institut et musée Voltaire Les Délices.

Smith, Bernard. 1988. *European Vision and the South Pacific,* 2nd edition. New Haven and London: Yale University Press.

Smith, James R. 1986. *From Plane to Spheroid: Determining the Figure of the Earth from 3000 B.C. to the 18th Century Lapland and Peruvian Survey Expeditions.* Rancho Cordova, CA: Landmark Enterprises.

Snelders, H. M. A. 1970. Romanticism and Naturphilosophie and the Inorganic Natural Sciences, 1797–1840: An Introductory Survey. *Studies in Romanticism* 9: 193–215.

Société de Géographie, Paris. 1824–1866. *Recueil de voyages et de mémoires.* Paris: Bertrand.

————. 1913–20. *Exploration scientifique du Maroc, organisée par la Société de géographie de Paris.* 2 vols. Paris: Masson.

Société d'ethnographie, Paris. 1868. *Société d'ethnographie fondée en 1859. Exposé général. Actes constitutifs et statuts—liste des membres—catalogue des publications—récompenses décernées par la société—prix mis au concours—séances de la société—bibliothèque et collections—adresses des fonctionnaires—expositions et conférences—conditions à remplir pour faire membre de la société.* Paris.

Somerville, Mary. 1846. *On the Connection of the Physical Sciences.* From the Seventh London Edition New York: Harper and Brothers, Publishers.

————. 1867. *Physical Geography.* A new American edition from the third and revised London edition. New York: Sheldon and Co.

Soulavie (ingénieur-géographe). 1813. *Mémoire sur l'emploi des matériaux dans la construction de l'atlas napoléon.* Paris.

————. 1829. Notice sur la topographie considérée chez les diverses nations de l'Europe, avant et après la carte de la France par Cassini suivie d'un catalogue des meilleurs cartes. In *Mémorial du Dépôt de la guerre imprimé par ordre du Ministre 1802–03.* Paris: Ch. Picquet.

Spengler, Joseph John. 1942. *French Predecessors of Malthus: A Study in Eighteenth-Century Wage and Population Theory.* Durham, N.C.: Duke University Press.

Spillmann, General Georges. 1969. *Napoléon et l'Islam.* Paris: (Librairie académique) Perrin.

Stafford, Barbara Maria. 1984. *Voyage into Substance: Art, Science, Nature and the Illustrated Travel Account, 1760–1840.* Cambridge, MA: MIT Press.

Staum, Martin S. 1980. The Class of Moral and Political Sciences, 1795–1803. *French Historical Studies* 11, no. 3 (spring): 371–97.

————. 1985. Human, not Secular Sciences: Ideology in the Central Schools. *Historical Reflections / Réflexions Historiques* 12 (1): 49–76.

————. 1987. Human Geography in the French Institute: New Discipline or Missed Opportunity? *Journal of the History of the Behavioural Sciences* 23 (4): 332–40.

Stevens, Henry. 1863. *The Humboldt Library. A Catalogue of the Library of Alexander von Humboldt with a Bibliographical and Biographical Memoir.* London: Henry Stevens.

Stocking, William. 1955. French Anthropology in 1800. *Isis* 55: 134–50.

Stoddart, David Ross. 1986. *On Geography and Its History.* Oxford: Basil Blackwell.

Strabo [Strabon]. 1805–1819. *Géographie.* Paris: Imprimerie impériale.

————. 1969. *Géographie. Introduction par Germaine Aujac et François Lasserre. Texte établi et traduit par Germain Aujac.* Paris: Société d'édition "Les Belles Lettres."

Strahlenberg, Philipp Johann von. 1757. *Description historique de l'empire russien . . .* Translated by Barbeau de la Bruyère, Jean-Louis. 2 vols. Amsterdam and Paris.

Taaffe, Edward J. 1990. Some Thoughts on the Development of Urban Geography in the United States during the 1950s and 1960s. *Urban Geography* 11 (4): 422–31.

Taton, René. 1947. *Les Mathématiques dans le Bulletin de Férussac. [Extrait des Archives internationales d'histoire des sciences Numero 1.]*

———. 1951. *L'Oeuvre scientifique de Monge.* Paris: Presses universitaires de France.

———, ed. 1964. *Enseignement et diffusion des sciences en France au XVIIIe siècle, Histoire de la pensée, 11.* Paris: Hermann.

Taylor, Kenneth L. 1969. Nicolas Desmarest and Geology in the Eighteenth Century. Paper read at *Toward a History of Geology* (Proceedings of the New Hampshire Inter-Disciplinary Conference on the History of Geology), September 7–12, 1967, at Cambridge, MA.

Terracher, L. A. 1929. *Histoire des langues et la géographie linguistique.* Oxford: Clarendon Press.

Thomassy, Marie-Joseph-Raymond. N.d. *Des Recherches scientifiques sur l'Algérie et de sa colonisation.* N.p.

Thompson, R. D., A. M. Mannion, C. W. Mitchell, M. Parry, and J. R. G. Townshend. 1986. *Processes in Physical Geography.* London and New York: Longman.

Tinkler, Keith J. 1985. *A Short History of Geomorphology.* London: Croom Helm Ltd.

Todhunter, Isaac. 1865. *A History of the Mathematical Theory of Probability from the Time of Pascal to That of Laplace.* Cambridge and London: Macmillan and Co.

Tooley, Ronald Vere. 1967. *French Mapping of the Americas: The De l'Isle, Buache, Dezauche Succession (1700–1830).* Vol. 33. London: The Map Collectors' Circle.

Toulmin, Stephen. 1972. *Human Understanding.* Princeton: Princeton University Press.

Trenard, Louis. 1977. Histoire des sciences de l'éducation (période moderne). *Revue Historique* 257 (2): 429–72.

Trigger, Bruce. 1989. *A History of Archaeological Thought.* Cambridge: Cambridge University Press.

Troux, Albert. 1926. *L'École centrale du Doubs à Besançon (an IV–an XI).* Paris: F. Alcan.

Tufte, Edward R. 1983. *The Visual Display of Quantitative Information.* Cheshire, CT: Graphics Press.

———. 1990. *Envisioning Information.* Cheshire, CT: Graphics Press.

Tulard, Jean. 1976. *Paris et son administration, 1800–1830.* Paris.

Tulard, Jean, ed. 1987. *Dictionnaire Napoléon.* Paris: Fayard.

Vallaux, Camille. 1938. Deux précurseurs de la géographie humaine, Volney et Charles Darwin. *Revue de synthèse historique* 15: 83–93.

Vallongue, Pascal. 1829. Avertissement. In *Mémorial du Dépôt de la guerre imprimé par ordre du Ministre 1802–03.* Paris: Ch. Picquet.

———. 1831. Coup d'oeil sur les systèmes de géologie et sur le langage topographique. In *Mémorial du Dépôt de la guerre imprimé par ordre du Ministre 1803, 1805 et 1810.* Paris: Reprint. C. Picquet.

————. 1831. Essai sur les échelles graphiques. In *Mémorial du Dépôt de la guerre imprimé par ordre du Ministre 1803, 1805 et 1810*. Paris: Reprint. C. Picquet.

————. 1831. Explication des teintes et des signes conventionnels. In *Mémorial du Dépôt de la guerre imprimé par ordre du Ministre 1803, 1805 et 1810*. Paris: Reprint. C. Picquet.

————. 1831. Notice sur les caractères et les hauteurs des écritures pour les plans et cartes topographiques et géographiques. In *Mémorial du Dépôt de la guerre imprimé par ordre du Ministre 1803, 1805 et 1810*. Paris: Reprint. C. Picquet.

————. 1831. Procès-verbal des conférences de la commission chargée . . . à la perfection de la topographie . . . In *Mémorial du Dépôt de la guerre imprimé par ordre du Ministre 1803, 1805 et 1810*. Paris: Reprint. C. Picquet.

Van Duzer, Charles Hunter. 1935. *Contribution of the Ideologues to French Revolutionary Thought*. Baltimore: Johns Hopkins Press.

Varenius, Bernhard. 1736. *A Complete System of General Geography: Explaining the Nature and Properties of the Earth; viz. Its Figure, Magnitude, Motions, Situation, Contents, and Division into Land and Water, Mountains, Woods, Desarts, Lakes, Rivers, etc. Withe Particular Accounts of the Different Appearances of the Heavens in Different Countries; the Seasons of the Year over All the Globe; the Tides of the Sea; Bays, Capes, Islands, Rocks, Sand-Banks, and Shelves. The State of the Atmosphere; the Nature of Exhalations; Winds, Storms, Tornados, etc. The Origins of Springs, Mineral-Waters, Burning Mountains, Mines, etc. The Uses and Making of Maps, Globes, and Sea-Charts. The Foundations of Dialling; the Art of Measuring Heights and Distances; the Art of Ship-Building, Navigation, and the Ways of Finding the Longitude at Sea.* 3d ed., "with large additions." London.

Vauban, Sébastien Le Prestre, Chevalier, Mis de (Maréchal de France). 1881. (1696). Description géographique de l'élection de Vézelay. In *Mémoire des intendants sur l'état des généralités. Collection des documents inédits sur l'histoire de France*. Paris.

Vaumas, Étienne de. 1946. La Géographie. Essai sur sa nature et sa place parmi les sciences. *Revue géographique alpine* 34: 555–70.

Venture de Paradis, Jean Michel de. 1844. *Grammaire et dictionnaire abrégés de la langue berbère . . . ou, Recueil de voyages et de mémoires . . .* Vol. 7. Paris: Imprimerie royale.

Vergnaud, Colonel. 1936. Mémoires et documents: Souvenirs du Colonel Vergnaud. *Revue des études napoléoniennes* 15: 315–39.

Viel de Saint-Maux, Charles-François. 1813. *Inconvéniens de la communication des plans d'édifices avant leur exécution, suivis du détail de la construction de la vôute de la salle de vente de la succursale du Monte-de-Piété; plans, élévations et coupes du bâtiment . . .* Paris: Tilliard frères.

Vignon, Eugène Jean-Marie. 1862. *Études historiques sur l'administration des voies publiques en France au XVIIe et XVIIIe siècles, . . .* 3 vols. Paris: Dunod.

Villaret. 1749–1763. Carte géométrique du Haut-Dauphiné et du Comté de Nice.

Villoteau, Guillaume-André. 1809. Dissertation sur les diverses espèces d'instrumens de musique que l'on remarque parmi les sculptures qui décorent les antiques monumens de l'Égypte, et sur les noms que leur donnèrent, en leur langue propre, les premiers peuples de ce pays. In *Description de l'Égypte,* edited by Edme-François Jomard, Antiquités, Mémoires, 1: 181–206. Paris: Imprimerie impériale.

—. 1809. Mémoire sur la musique de l'antique Égypte. In *Description de l'Égypte,* edited by Edme-François Jomard, Antiquités, Mémoires, 1: 357–426. Paris: Imprimerie impériale.

Vincens Saint-Laurent, Jacques. N.d. Pascal-Vallongue, Joseph-Secret. In *Biographie universelle (Michaud) ancienne et moderne, ou histoire, par ordre alphabétique, de la vie publique et privée de tous les hommes qui se sont fait remarquer par leurs écrits, leurs actions, leurs talents, leurs vertus ou leurs crimes,* edited by C. Desplaces. Paris: Chez Madame C. Desplaces; and Leipzig: Librairie de F. A. Brockhaus.

Vincent de Touret. 1728. *Examen sur toutes les cartes générales des quatre parties de la terre, mises au jour par feu M. Delille, depuis l'année 1700 jusqu'en 1725, pour servir d'éclaircissement sur la géographie.* Paris: J.-B. Lamesle et Briasson.

Vivien de Saint-Martin, Louis. 1844. Notice sur les travaux de la Société de géographie, et sur le progrès des découvertes géographiques pendant l'année 1844. *Bulletin de la Société de géographie de Paris* 2: 345–99.

—. 1845. Notice sur le progrès des découvertes géographiques et les travaux de la Société de géographie pendant l'année 1845. *Bulletin de la Société de géographie de Paris* 4: 247–50.

—. 1847. Rapport sur les travaux de la Société de géographie de Paris et sur le progrès de la science pendant l'année 1847. *Bulletin de la Société de géographie de Paris* 8: 265–69.

—. 1873. *Histoire de la géographie et des découvertes géographiques depuis les temps les plus reculés jusqu'à nos jours . . .* Paris: Hachette.

Volney, Constantin-François de Chasseboeuf, Cte de. 1787. *Voyage en Syrie et en Égypte pendant les années 1783, 1784 et 1785 . . .* Paris: Volland; Desenne.

—. 1791. *Les Ruines, ou Méditation sur les révolutions des empires . . .* Paris: Desenne.

—. 1803 (an XII). *Tableau du climat et du sol des États-Unis d'Amérique. Suivi d'éclaircissements sur la Floride, sur la colonie française au Scioto, sur quelques colonies canadiennes et sur les sauvages . . .* Paris: Courcier.

—. 1813. *Questions de statistique à l'usage des voyageurs . . .* Paris: Vve Courcier.

—. 1819. *Histoire de Samuel, inventeur du Sacre des rois, fragment d'un voyageur américain, traduit sur le manuscrit anglais.* Paris: Brissot-Thivars.

—. 1821. Réponse de Volney au Docteur Priestly sur un pamphlet intitulé *Observations sur le progrès de l'infidélité, avec des remarques critiques sur les écrits de divers incrédules modernes, et particulièrement sur les Ruines de M. Volney . . .* In *Oeuvres complètes de C.-F. Volney,* edited by A. Bossange. Paris: Fayard.

—. [1821]. État physique de la Corse. In *Oeuvres complètes de C.-F. Volneyhbox . . . ,* edited by A. Bossange. Paris: Bossange frères.

—. [1821]. *Institut royale de France. Séance publique annuelle des quatre Académies du . . . 24 avril 1821, présidée par M. Walckenaer, . . . Prix fondé par M. le comte* de Volney. *Extrait du testament de M. Constantin Chasseboeuf de Volney, . . . reçu par Me Boulard, . . . le 22 avril 1820.* N.p.

Wagner, Moritz Friedrich. 1841. *Reisen in der Regentschaft Algier in den Jahren 1836, 1837 und 1838 . . .* Leipzig: L. Vass.

Wailly, Augustin-Jules de. *Discours prononcé aux funérailles de M. le Bon Walckenaer.* Paris.

Walch, Jean. 1974. Michel Chevalier. Economiste Saint-Simonien 1806–1879. PhD dissertation, Paris IV.

Walckenaer, Baron Charles Athanase. 1798. *Essai sur l'histoire de l'espèce humaine.* Paris: Chez du Pont.

———. 1802 (an XI). *Faune parisienne, insectes, ou Histoire abrégée des insectes des environs de Paris, classés d'après le système de Fabricius. Précédée d'un Discours sur les insectes en général pour servir d'introduction à l'étude de l'entomologie . . .* 2 vols. Paris: Dentu.

———. 1807. *Dicuili Liber de mensura orbis terrae ex duobus codd mss. bibliothecae imperialis nunc primum in lucem editus.* Parisiis: Ex typis Firmini Didot.

———. 1813. *L'Ile de Wight, ou Charles et Angelina.* Paris: Laurent-Beaupré.

———. 1815. *Cosmologie, ou Description générale de la terre, considérée sous ses rapports astronomiques, physiques, historiques, politiques et civils . . .* Paris: Déterville.

———. 1821. *Aperçu des recherches géographiques sur l'intérieur de l'Afrique septentrionale, . . . lu à la séance publique des quatres Académies de l'Institut, le 24 avril 1819.* Paris: Imprimerie de E.-N. Goetschy.

———. 1822. *Recherches sur la géographie ancienne et sur celle du Moyen-Âge . . .* Paris: Imprimerie royale.

———. 1825. Notice bibliographique, critique et géographique sur l'itinéraire de Bordeaux à Jérusalem. In *Histoire des croisades . . .*, edited by J.-F. Michaud. Paris: Imprimerie de L.-G. Michaud.

———. 1826–1831. *Histoire générale des voyages, ou Nouvelle Collection des relations de voyages par mer et par terre, mise en ordre et complétée jusqu'à nos jours . . .* 21 vols. Paris: Lefèvre.

———. 1838. *Rapport de M. Walckenaer fait dans la séance du 27 décembre 1833, au nom de la seconde commission nommée relativement à la lettre du ministre de la Guerre, en date du 18 novembre 1833. [Signé à la minute: Naudet, Raoul-Rochette, Ét. Quatremère, Dureau de la Malle, Jomard et Walckenaer, rapporteur.] . . . Faut titre: Académie royale des inscriptions et belles-lettres. Rapports sur les recherches géographiques, historiques, archéologiques à entreprendre dans l'Afrique septentrionale.* Paris: Imprimerie royale.

———. 1839. *Géographie ancienne historique et comparée des Gaules cisalpine et transalpine, suivie de l'Analyse géographique des itinéraires anciens, et accompagnée d'un atlas de 9 cartes . . .* 2 vols. Paris: P. Dufart.

———. 1839. *Introduction à l'analyse géographique des itinéraires anciens pour les Gaules. Cisalpine et transalpine, suivie de l'analyse géographique des itinéraires anciens, et accompagnée d'un atlas de neuf cartes.* Vol. 3. Paris: P. Dufart.

———. 1842. Dissertation sur les contes de fées . . . In *Mémoires, contes et autres oeuvres . . . Précédés d'une notice sur l'auteur par Paul L. Jacob, . . .* edited by C. Pernault. Paris: G. Gosselin.

————. 1847. *Paroles prononcées à la Société de géographie, . . . pour l'ouverture de la séance générale du 18 décembre 1846 à l'Hôtel-de-Ville.* Paris: Imprimerie de Martinet.

————. 1850. Notice historique sur la vie et les ouvrages de M. Letronne. Lu dans la séance publique annuelle de l'Académie royale des inscriptions et belles-lettres du 16 août 1850. In *Séance publique annuelle de l'Académie des inscriptions et belles-lettres,* edited by Institut national de la France. Paris: Firmin Didot.

————. 1850. *Rapport fait à l'Académie des inscriptions et belles-lettres au sujet des manuscrits inédits de Fréret . . .* Paris: Imprimerie nationale. Extract from the *Mémoires de l'Académie des inscriptions et belles-lettres* 16, 1 (1850), 70–216.

————. N.d. *Institut royal de France. Rapport fait à l'Académie des inscriptions et belles-lettres dans sa séance du 20 juillet 1821 [19 juillet 1822], par sa commission d'histoire et d'antiquités de la France, relativement aux trois médailles d'or accordées en prix . . . aux trois auteurs qui, . . . auraient composé les meilleurs mémoires sur nos antiquités. [Signé: Petit-Radel, comte de La Borde, Raoul-Rochette, Walckenaer (rapporteur).]* Paris: Imprimerie de F. Didot.

Weiss. N.d. Férussac, André-Étienne-Just-Pascal-Joseph-François d'Audebard, baron de. In *Biographie universelle (Michaud) ancienne et moderne, ou histoire, par ordre alphabétique, de la vie publique et privée de tous les hommes qui se sont fait remarquer par leurs écrits, leurs actions, leurs talents, leurs vertus ou leurs crimes,* edited by C. Desplaces. Paris: Chez Madame C. Desplaces; and Leipzig: Librairie de F. A. Brockhaus.

Withers, Charles W. J. 1993. Geography in Its Time: Geography and Historical Geography in Diderot and d'Alembert's *Encyclopédie. Journal of Historical Geography* 19 (3): 255–64.

Wood, Denis (with John Fels). 1992. *The Power of Maps.* New York and London: Guilford Press.

Woolf, Harry. 1959. *The Transits of Venus: A Study of Eighteenth-Century Science.* Princeton, NJ: Princeton University Press.

Woolf, Stuart Joseph. 1984. Towards the History of the Origins of Statistics: France, 1789–1815. In *State and Statistics in France 1789–1815,* edited by J.-C. Perrot and S. J. Woolf. Chur, London, Paris, and New York: Harwood Academic Publishers.

Wurtz, Charles-Adolphe. 1870. *Les Hautes études pratiques dans les universités allemandes.* Paris: Imprimerie impériale.

Yvon, Michel. 1987. Les "concours" de l'École des ponts et chaussées au XVIIIe siècle. In *Espace français. Vision et aménagement, XVIe–XIXe siècle,* edited by the Archives nationales. Paris: Archives nationales.

Ziman, John. 1981. What Are the Options? Social Determinants of Personal Research Plans. *Minerva* 19 (1): 1–42.

Zobel, M. 1961. Les Naturalistes voyageurs français et les grands voyages maritimes des XVIIIème et XIXème siècles. Thèse Médecine Paris, Paris.

Zuckerman, Harriet, and Robert K. Merton. 1971. Patterns of Evaluation in Science: Institutionalisation, Structure and Functions of the Referee System. *Minerva* 9 (1): 66–100.

Index

Abrégé de géographie . . . (Balbi), 222

Académie celtique, 193, 198

Académie des inscriptions et belles-lettres, 29, 133, 268, 271, 281, 284, 298, 306; Letronne's reform proposal, 295–96

Académie des sciences, 6, 27–29, 52, 60, 306; Férussac and, 170–71; geographical projects, 28; professionalism and, 318n.23; and reform of Paris Observatory, 73–74

Adanson, Michel, 12–13

advocacy, geographical, of François, 118–19

Aegyptiaca (Hamilton), 282

Allent, Pierre-Alexandre-Joseph, 162–63, 309, 339n.46; influence of, 171–72

Al-Rahim, A. Abd., 349n.71

ancient civilizations: and Barbié du Bocage's historical geography, 268–71; Gosselin's work on, 271–73; Letronne's work on, 282–85, 296–302; nineteenth-century passion for, 278–80, 302. *See also* Egypt

Andréossy, Antoine-François, 158, 208

Annales des voyages, 91, 110, 130, 167

Annales générales des sciences physiques (Bory de Saint-Vincent), 178

anthropology, of Volney, 197–98

Anville, Jean-Baptiste Bourguignon d', 28, 34–35, 47–48; and debate over shape of earth, 52–53; influence of, 291, 302; Malte-Brun and, 107

Artz, Frederick B., 32

astronomy, 68–71, 83–85, 98–99, 298–99

atlas, celestial, of Cassini IV, 72–73

atlas, ethnographic, of Balbi, 222–24, 227

atlas, facsimile: of Jomard, 142–46; of Santarem, 144–46

Atlas composé de mappemondes (Santarem), 144–46

Atlas ethnographique du globe (Balbi), 227–28, 230

Atlas géographique et physique du Royaume de la Nouvelle-Espagne (Humboldt), 252

authority: Strabo's claim to, 95; in works of Balbi, 229

Avezac-Macaya, Marie-Armand-Pascal d', 130, 277

Baker, Keith, 316n.17, 318n.23

Balbi, Adrien, 9, 110, 194–95, 221–31, 239, 273, 311, 343n.104, 343n.115, 350n.96; *Abrégé de géographie* . . . , 222; *Atlas ethnographique du globe*, 222–24, 227–28, 230; *Essai statistique sur le royaume de Portugal*, 221–23; *Essai statistique sur les bibliothèques de Vienne*, 221–23, 229; *Introduction à l'Atlas ethnographique du globe*, 229–30

Barbié du Bocage, Jean-Denis, 60, 130, 161, 267–71, 312, 350n.96; atlas for Barthélemy's *Voyage*, 269; Balbi and, 228–29; and Ste. Croix's *Examen*, 269–71

Barthélemy, Jean-Jacques, 279; *Voyage du jeune Anacharsis en Grèce*, 269

Beazley, C. Raymond, 130

Bedouins, 204

Bellin, Jacques Nicolas, 34–36, 48

Bernard, Pons Joseph, 133

Berthaut, Henri, 157
Berthelot, S., 350n.96
Berthier, Alexandre, 156
Best, Geoffrey, 153
bibliography, geographic, Férussac and, 173–75
Bibliothèque nationale, Paris, 40; Département des cartes et plans, 134; Jomard and, 143, 145; Letronne and, 281
Biot, Jean Baptiste, 133
Bonne, Rigobert, 53
borders and boundaries, Strabo's emphasis on, 93
Bory de Saint-Vincent, Jean-Baptiste-Geneviève-Marcellin, baron de, 61, 110, 165, 176–86, 271, 309; Bonapartist politics, 177–78; character of, 184–85; *Description du plateau de Saint-Pierre,* 178; *Dictionnaire classique d'histoire naturelle,* 177; *Guide du voyageur en Espagne,* 248; *L'Homme,* 182; military role and state service, 151, 177, 188–89, 242; research interests, 178–79; views on French colonialism, 182–84
Bret, Patrice, 157
Brian, Eric, 323n.2
Briet, Père Philippe, 23
Brittany, Chabrol de Volvic and, 212–13
Broc, Numa, 42
Buache, Jean-Nicolas (Buache de la Neuville), 35, 60–66
Buache, Philippe, 6, 28, 34–35, 82, 319n.39; "Cartes des nouvelles découvertes," 44; as royal tutor, 149
Bulletin de la Société de géographie, 130–31
Bulletin général et universel des annonces et des nouvelles scientifiques (Férussac), 173–75, 282, 294, 342n.104, 343n.112; Bory de Saint-Vincent and, 177

Bureau des longitudes, 60; Cassini IV and, 80
Butlin, Robin, 268

Cabanis, Georges, 197
cadastre, Egyptian, 210–11
Calon, Etienne-Nicolas de, 157–58
cameralism, 338n.27
Candolle, Augustin Pyramus de, 237; and plant geography, 245, 249–50
Carette, Ernest, 271, 300
"Carte des frontières Est de la France depuis Grenoble jusqu'à Marseilles" (Bourçet/d'Arçon), 44
"Cartes des nouvelles découvertes" (Buache), 44
cartography: cabinet, 57; Delisle's views on, 100; eighteenth-century procedures, 39–40; Expilly's views on, 101; François on, 115–16; Gosselin and, 359n.21; and hierarchy of geographic phenomena, 45–46; in historical geography, 267–71; of human activities, 255; Humboldt on history of, 277; Jomard and, 135–42, 146, 308–9; large-scale, 79–80, 105, 129–30; Mentelle's views on, 101; Société de géographie and, 130–32; thematic, 13–14, 264–65; topographic, 306; universal geographies and, 96. *See also* field mapping; geography, mathematical; language, geographical; maps
Cassini, César-François (Cassini de Thury) (Cassini III), 28, 34, 47, 52, 69–70, 82, 319n.44; *Description géométrique de la France,* 69
Cassini, Jacques (Cassini II), 28, 34, 51
Cassini, Jacques-Dominique (Cassini I), 34–35, 50–51, 82
Cassini, Jacques-Dominique, Comte de (Cassini IV), 66–85, 319n.40; banishment from science, 76–85, 307; and confiscation of Cassini map of France, 76–79; mapping

projects, 71–73; and reform of Paris Observatory, 70, 73–75; religious convictions, 80–82; and survey of France, 71; ties to astronomy, 68–71; views on post-Revolutionary science, 80–85

Cassini family: and French crown, 82–83; in Malte-Brun's universal geography, 106; ties to astronomy, 83–85

Cassini map of France, confiscation of, 76–79, 326n.66

Cassirer, Ernst, 11

causality, Humboldt on, 243

"censeur royal" for geographic works, 36, 320n.62

Chabrol de Volvic, Gilbert-Joseph-Gaspard, comte de, 133, 194, 209–20, 310–11; "Essai sur les moeurs des habitans modernes de l'Égypte," 211; as geographical engineer, 209–10; influence of, 232; as prefect of Montenotte, 213–15; as prefect of the Seine, 215–20; *Recherches statistiques sur la ville de Paris . . .* (with Fourier), 216–21; *Statistique des provinces . . .* , 213–14; as sub-prefect of Pontivy, 212

Champagny (minister of interior), 155

Champollion, Jean-François, 135, 287; influence of, 298

change, Humboldt's approach to, 244–46

Chateaubriand, François Auguste, vicomte de, 279

Châteauneuf, Louis-François Benoiston de, 136

Chevalier, Michel, 249

Chezy, Antoine-Léonard, 133

Choiseul-Gouffier, *Voyage pittoresque de la Grèce*, 279

citation: Balbi's use of footnotes, 229; eighteenth-century patterns, 35; Letronne's style of, 296

classification: Balbi and, 225–26;

Jomard and, 141–42; natural-history model for, 237

Claval, Paul, 249

climate, Volney on, 206–7

Collège de France, 155–56, 281, 283

colleges, military, 26, 284, 306; and geographical education, 32–33

colleges, religious, 26, 58, 306, 319n.29; and geographical education, 29–32

colonialism: Bory de Saint-Vincent on, 182–84; Humboldt and, 256–57

colonization, internal, 212, 349n.75

Columbus, Humboldt's study of, 275–78

Commission chargé . . . à la perfection de la topographie (Paris, 1802), 45

community, geographical: and Bory de Saint-Vincent, 179; eighteenth-century, 33–37, 306–7; and Malte-Brun's *Précis*, 109–11

Comparaison de plusieurs années d'observations (Jomard), 136

comparison, geography as, 226

Complete System of General Geography (Varenius), 329n.39

Comte, August, 133

Condillac, Étienne Bonnot de, 197

Condorcet, Marquis Marie Jean Antoine Nicolas de Caritat de, 197, 279

conscription, military, 153

Constant, Benjamin, 279

Cormack, Lesley, 149, 315n.1

Cosmographie élémentaire (Mentelle), 97–103

cosmography, 31, 69; Renaissance, 327n.2

Cosmos (Humboldt), 111–13, 119–27, 308

cost, of geographical research, 186

Coton, Père Pierre, 24

Crome, August, 253

Cuno, Kenneth M., 349n.71

Dainville, François de, 29, 319n.29; *Langage des géographes*, 42 data collection, Balbi and, 224, 228–29

Debs, Richard, 349n.71

Décade philosophique, 58, 64–65, 307

deforestation, Volney on, 207

Delamarre, Casimir, 130

Delambre, Jean-Baptiste-Joseph, *Histoire de l'astronomie au dix-huitième siècle,* 85

De La Renaudière, Philippe François, 110

De la Roncière, Charles, 130

Delisle, Guillaume, 28, 34–35, 47, 68, 97, 324n.37; *Introduction à la géographie,* 35–36, 97–103

Delisle, Joseph Nicolas, 28

De mensura orbis terrae (Dicuil), 282, 284–86, 292

Deneys, Anne, 348n.48

Dépôt de la guerre, 33, 45, 171, 208, 211; and geographical training, 157–59, 339n.42

description, Balbi's reliance on, 228–31

Description de l'Égypte, 133, 137, 211, 278, 282; Letronne on, 284–85

Description du plateau de Saint-Pierre (Bory de Saint-Vincent), 178

descriptive geography. *See* geography, descriptive

"Des lignes isothermes . . ." (Humboldt), 254

Desmarest, Nicolas, 24–25, 28, 158

Destutt de Tracy, Antoine Louis Claude, Comte, 197

Dettelbach, Michael, 241, 355n.25

Devic, Jean-François-Schlister, 72

dictionary, geographic, 38

Dictionnaire classique d'histoire naturelle (Bory de Saint-Vincent), 177

Dicuil, *De mensura orbis terrae,* 282, 284–86, 292

disease, Volney's approach to, 207

dissection, Humboldt and, 250

distribution, Humboldt's approach to, 244–46

Drapeyron, Ludovic, 130

Drouin, Jean-Marc, 179, 344n.134

Dupain-Triel, J. L., 253

Dutens, Joseph Michel, 133

earth, determination of shape of, 48–54

earth description: Humboldt on, 123; and universal geographies, 93

École d'application du Corps royal des ingénieurs-géographes, 158

École des chartes, 281, 299, 301

École des ponts et chaussées, 32, 150, 273

École du corps royal du génie, 32

École militaire, 32, 150

École normale, 58–66, 91, 208, 307

École polytechnique, 133, 150, 155, 209, 273, 283

education, geographical, 26, 29–32, 318n.16; Dépôt de la guerre and, 157–58, 339n.42; at École normale, 58–66, 307; Mentelle on, 102–3

Egypt, 349n.71; expedition to, 156, 209–12; growth of European interest in, 278–80; Letronne's work on, 286–91; survey of, 133, 326n.66; Volney and, 196, 207–8

Egyptology, 133; Jomard and, 134–35, 137–39

Egyptomania, 279

Élie de Beaumont, Léonce, 134–35

empiricism, Humboldt on, 123

Encyclopédie, 24–25, 27

Encyclopédie méthodique, 27

Encyclopédistes, Volney and, 196–98

Enfantin, Barthélemy, 133

engineers, geographical, 163–64, 187–88, 309; as government employees, 26, 149–50; Jacotin and, 77–78; mapping projects, 156, 159; reliance on field research, 162–63; and Société de géographie, 131–32, 334n.6; topographic memoirs, 159–62, 338n.40, 341n.66; training, 33, 157–58, 319n.44

engineers, scientists and, 73–74

engraving, cartographic, 48, 75

Enlightenment, influence of, 196–98, 240

enumeration, locational. *See* geography, descriptive

epigraphy, 286–95

episteme, the, 7–8

epistemic shift, nineteenth-century, 12–13

"esprit de système," Letronne's rejection of, 297–99

Essai statistique sur le royaume de Portugal (Balbi), 221–23

Essai statistique sur les bibliothèques de Vienne (Balbi), 221–23, 229

Essai sur la géographie des plantes (Humboldt), 261–63, 354n.6

"Essai sur les moeurs des habitans modernes de l'Égypte" (Chabrol de Volvic), 211

Essai sur l'histoire de la géographie (Robert de Vaugondy), 33–34

Essai sur l'histoire de l'espèce humaine (Walckenaer), 360n.32

"État physique de la Corse" (Volney), 198

Ethnographic Society of Paris, 134

ethnography: Balbi and, 223–24; Jomard and, 140–42

Examen des historiens d'Alexandre (Ste. Croix), 269–71

Expilly, Abbé Jean-Joseph, 97; *Polychrographie en six parties*, 97–103

exploration, 6, 129–30; and eighteenth-century cartography, 39; Jomard and, 133–34, 139–40; Volney and, 197

exploration accounts, 37–38. *See also* travel accounts

Fernandez de Navarette, D. Martin, 130, 277

Férussac, André de, 9, 151, 165–76, 241–42, 309, 354n.15; concern with contextualization of knowledge, 172–73; early life, 166–72; military role, 171–72, 177, 188; work on mollusks, 167–70

field mapping, 22, 46–47, 139

field research: Balbi and, 229; Letronne and, 292–93; in military geography, 162–63; Volney and, 200–201

Fierro-Domenech, Alfredo, 132, 334n.6

Fluctuations of Gold (Humboldt), 245

Foucault, Michel, 7–8; influence of, 1–2, 8–9, 12–13

Fourier, Jean-Baptiste-Joseph, 59, 65–66, 137; *Recherches statistiques sur la ville de Paris . . .* (with Chabrol de Volvic), 216–21

France, maps and surveys, 47, 68, 71; Cassini map, 76–79; 1:80,000 map, 132, 163, 353n.1

Francoeur, Louis Benjamin, 133

François, Père Jean, 24, 54, 100, 126; *Science de la géographie*, 111, 113–19, 308

Fréret, Nicolas, 274–75, 302, 360n.42

Gance, Abel, *Napoleon* (film), 338n.24

gatekeeper role, eighteenth-century, 35–36

geognosy, 239; Férussac and, 354n.15; Humboldt and, 246, 250–51, 259–60

"géographe de la ville de Paris," 36

"géographe du roi," 36

geographer, work of, Malte-Brun on, 331n.92

geographers: academic, 353n.148; amateur, 295–96; ancient, 39; cabinet, 39–40, 57–66, 263; Chabrol de Volvic and, 220; libraries of, 39–40, 274; military, 150–51, 263–64; participation in state aggression, 187; response to work of Volney, 208–9; and service to state, 149–50. *See also names of individuals*

geographical problems, as posed by Volney, 203–8

geographical research, Humboldt's innovation in, 242–63
geographical thought, history of, Humboldt on, 275–78
Géographie comparée (Mentelle), 97–103
Géographie des plantes (Humboldt), 238
Géographie enseignée par une methode nouvelle . . . (Mentelle), 62
Géographie royalle (Labbé), 24
geography, 41–54, 306–7; aims of, 117–18; author's approach to, 1–3, 7–8, 11–17; conceptual history, 8–11; as critical comparison, 106–7; criticisms of, 103–4; as earth description, 2–4, 21–26; epistemology of, 11; and history, in work of Letronne, 280–81; history of, 33–34, 142–46; loss of direction and status, 4–7, 57–86, 99–100, 307; "man" and nature in, 64; place of, among sciences, 118, 350n.96; as profession, 36–37; sociology of, 9–10; and state power, 149–52, 188–90, 309–10; Strabo's view of, 93–95; unity of, 102–3, 111–13, 116
geography, botanical, 108
geography, descriptive: Balbi and, 228–31; of Buache de Neuville and Mentelle, 61–66; and classification, 38; distinguished from cartography, 105–6; François and, 117; and Malte-Brun's rejection of theory, 107–9; military geography as, 157–64, 187; universal geographies as, 99
geography, historical, 267–78, 302, 312–13
geography, human: Bory de Saint-Vincent and, 181–82; Malte-Brun on, 108; of Volney, 196–201, 203–8
geography, linguistic, 352n.141; Balbi and, 223–24, 227–30, 352n.141
geography, mathematical, 46–48
geography, military, 156–66, 187. *See also* Bory de Saint-Vincent, Jean-Baptiste-Geneviève-Marcellin;

engineers, geographical; Férussac, André de
geography, natural, 235–42
geography, physical, 24–25, 236–37; Humboldt and, 111–13, 121–23; Malte-Brun on, 108–9; Varenius and, 96
geography, plant, 108; Candolle and, 245, 249–50; Humboldt and, 244–46, 250, 260–63
geography, practical: François on, 119; and universal geographies, 93, 101–2
geography, regional, and universal geographies, 93
geography, scientific: Delisle, Expilly, and Mentelle on, 98; and universal geographies, 100–101
geography, social, of Férussac, 176
geography, social-scientific: Balbi and, 222–24; Volney and, 198–203
geography, statistical, Chabrol de Volvic and, 211, 217–20
geography, terrain, 57; decline of, 66–85
geography, textual descriptive, 57; Société de géographie and, 131
geography, theoretical, Malte-Brun's rejection of, 107–9
Geography (Strabo), 92–96, 284
geology, Malte-Brun on, 109
geometry, 31, 47; ancient Egyptian, 137–39
Gérando, Joseph Marie, baron de, 197
Girault-Soulavie, Abbé Jean-Louis, 181, 249, 253
globe, terrestrial, 105–6, 115
Goethe, Johann Wolfgang von, 279
Gosselin, Pascal-François-Joseph, 158, 271–73, 312; maps of, 359n.21; *Rapports à l'Empereur*, 268; Walckenaer and, 359n.27
graphic representation, Humboldt's experimentation with, 253–63
graphic reproduction, technologies of, 264–65

Greece, 279
Guettard, Jean-Étienne, 253
Guide du voyageur en Espagne (Bory de Saint-Vincent), 248

Hahn, Roger, 28
Hamilton, William Richard, *Aegyptiaca*, 282
Hamy, Jules Théodore Ernest, 130
Harisse, Henry, 130
Harley, Brian, 149
Haüy, Abbé René, 59
Helvetius, Claude Adrien, 197
hierarchy: of geographic phenomena, 45–46; of sciences, 124
Histoire de l'astronomie au dix-huitième siècle (Delambre), 85
Holbach, Paul Henri Dietrich, baron d', 197
Homme, L' (Bory de Saint-Vincent), 182
human geography. *See* geography, human
Humboldt, Alexander von, 16, 178, 226, 236, 271, 311–12, 343n.112, 360n.53; *Atlas géographique et physique du Royaume de la Nouvelle-Espagne*, 252; Balbi and, 228–29; Bory de Saint-Vincent and, 344n.123; *Cosmos*, 111–13, 119–27, 308; "Des lignes isothermes . . . ," 254; embrace of geographic theory, 242–44; *Essai sur la géographie des plantes*, 261–63, 354n.6; *Examen critique*, 275–78; *Fluctuations of Gold*, 245; *Géographie des plantes*, 238; and historical geography, 268, 312; letter to Schiller (1794), 244–45; and thematic mapping, 251–63; *Voyage aux régions équinoxiales*, 356n.60
Humboldt, Wilhelm von, 353n.154
Huot, Jean-Jacques-Nicolas, 110

Ideologues, 58, 346n.13; Volney as, 196–98
ideology, Revolutionary, 152–53

ignorance, geographic: depiction of, 44; Strabo on, 94
Ile de Wight, L' (Walckenaer), 360n.32
illustration, scientific, 62; Humboldt's experimentation with, 253–63; relief depiction, 43–44. *See also* maps
Indians, North American: Letronne on, 364n.129; Volney on, 204–5
inscriptions, ancient, Letronne's work on, 286–95
Institut de France, 59–60, 198; Class of Moral and Political Sciences, 193, 209
Institut d'Égypte, 133
institutions, geographical, 36, 130–31. *See also names of institutions*
instrumentation, geographic, for field mapping, 46–47
intellectual property, Cassini IV's notion of, 77–79
interior structure and function, Humboldt's approach to, 249–51
Introduction à la géographie (Delisle), 35–36, 97–103
Introduction à l'Atlas ethnographique du globe (Balbi), 229–30
isolines, Humboldt and, 247, 253–54
Italy, Northern, Chabrol de Volvic and, 213–15

Jacob, Christian, 149
Jacotin, Pierre, 77–78, 132, 208
Jarry de Mancy, Adrien, 271
Jesuit order. *See* Society of Jesus
Jomard, Edme-François, 9, 130, 209, 216, 268, 271, 277, 308–9; *Comparaison de plusieurs années d'observations*, 136; as geographer, 132–35; "Mémoire sur la population comparée de l'Égypte ancienne et moderne," 211; "Mémoire sur le système métrique des anciens Égyptiens," 138–39; *Monuments de la géographie*, 142–46; views on cartography, 135–37; work on Egyptian measurement, 137–39

Journal de l'empire, 91
Journal des débats, 90, 327n.4
Journal des savants, 281, 294

knowledge, trees of, 237
Kretschmer, Konrad, 130

Labbé, Père Philippe, *Géographie royalle,* 24
Lakanal, Joseph, 197
Lalande, Joseph-Jérôme de, 69, 74, 82
landscape, Humboldt and, 248–49
Langage des géographes (Dainville), 42
language, geographical, 12–13; development of, 41–46, 160–61, 321n.75; Humboldt and, 239
language study, Balbi and, 223–24, 227–30
Lapie, Pierre, 267
Laplace, Pierre-Simon, 76, 80
La Renaudière, Philippe-François Lason, 131, 221
Latour, Bruno, 5
Lavallée, Théophile, 110
Legendre, Adrien Marie, 72
Lejeune, Dominique, 132
Lelewel, Joachim, 130
Lepetit, Bernard, 217–18
Lestringant, Frank, 327n.2
Letronne, Jean-Antoine, 9, 16, 61, 130, 268, 271, 278–301, 312–13, 359n.25, 364n.129; as engaged scholar, 281–83; geographic nature of his work, 300–301; geographic training of, 283–85; and Gosselin, 272–73; *Recueil des inscriptions,* 288, 291
Levesque, Pierre-Charles, 279
libraries, geographers', 39–40, 274
Linnaeus, Carolus, 12–13
lists, of Balbi, 225–28
Livingstone, David, 315n.5
location, Humboldt's approach to, 243–44
Loraux, Nicole, 279

Maistre, Joseph de, 279
Malte-Brun, Conrad, 9, 103–7, 130, 239, 271, 273, 307–8, 337n.11; Balbi and, 221, 228–29; Bory de Saint-Vincent and, 180; as geographer, 90–92; *Précis de la géographie universelle,* 91–92, 103–7, 109–11, 331n.92; rejection of theory, 107–9; *Société de géographie,* 331n.92
Malte-Brun, Victor, 110
Malus, Etienne Louis, 133
mapping. *See* cartography
maps, 39–41; Balbi's abandonment of, 226–28; collections of, 142–46; compiled, 47–48; field, 46–47; and globes, 115–16; Humboldt's knowledge of, 252–53; Humboldt's "map of error," 255–56; hydrographic charts, 48; increased access to, 129–30; legends, 46; quasi-explanatory, 322n.87; thematic, 106, 224, 251–52, 255, 313; topographic, 251–52
Marcel, Gabriel, 130
mathematics, and teaching of geography, 22, 30–31
mathematization, Humboldt and, 246–47
measurement, geographic, 22; and determination of shape of earth, 48–54
measurement system: ancient Egyptian, 137–39; ancient Greek, 271–73
Méchain, Pierre François André, 72
"Mémoire sur la population comparée de l'Égypte ancienne et moderne" (Jomard), 211
"Mémoire sur le système métrique des anciens Égyptiens" (Jomard), 138–39
"Mémoire sur une question de géographie pratique" (Robert de Vaugondy), 54

memoirs: military, 159; topographic, 159–62, 338n.40, 341n.66
Mémorial du Dépôt de la guerre, 158, 171
Mentelle, Edme, 60–66, 90, 97; *Cosmographie élémentaire*, 97–103; *Géographie comparée*, 97–103; *Géographie enseignée par une methode nouvelle* . . . , 62; and Letronne, 283
methodology: of Letronne, 291–300; of Volney, 199–203
Mexico, Humboldt's work on, 246, 252–53, 255, 259–61
militarization, of French society, 153, 309
military, the. *See* colleges, military; engineers, geographical; geography, military
military geography. *See* geography, military
military innovation, 152–54
miracles, François on, 113
Modern Geography (Pinkerton), 354n.6
Monge, Gaspard, 59
Morachiello, Paolo, 212
movement, Humboldt's approach to, 244–46
multidiscursivity, Humboldt and, 246–47

Nain jaune, 178
Napoleon: views on geography, 155–56, 338n.24; Volney's links to, 196
Napoleon (film by Gance), 338n.24
Napoleonic wars, and military innovation, 152–54
Napoléonville, Chabrol de Volvic and, 212–13
nationalism, French, 152–53
national surveys, eighteenth-century, 40–41
natural history: Bory de Saint-Vincent and, 176–77; Férussac and, 167–70, 176–77; Humboldt and, 238–40; and natural geography, 237–38; unity of, 102–3
natural region, Humboldt and, 248–49

nature: unity of, 123–24, 180–82, 239–40; Volney's view of, 206–8
Naturphilosophen, Humboldt and, 112, 120, 123, 125
Neptune François (1694), 28
Newton, Sir Isaac, 49; *Principia*, 50–51; and Varenius, 329n.39
Nicolson, Marjorie, 42
Nordenskiold, A. E., 130
Nouet, Nicolas-Antoine, 158
Nouvel atlas portatif (Robert de Vaugondy), 322n.87

observation: Humboldt and, 246–47; meteorological, of Cassini IV, 74–75
Observatory of Paris, 68, 70; reform of, 324n.36
"officiers de l'état-major," 163
Omalius d'Halloy, Jean-Baptiste-Julien d', 249
optics, 47

Papayanis, Nicholas, 215
Paris: Chabrol de Volvic and, 215–20
pasigraphy, 253; Humboldt and, 259–60
Pastoureau, Mireille, 60
Pauw, Cornelius de, 279
Pelet, Jean-Jacques, 164
Perny (astronomer), 158
philology, 362n.83; Letronne and, 293–94
philosopher, Chabrol de Volvic as, 349n.72
philosophy, and geography, 118, 319n.29; Strabo's view on, 95
physical geography. *See* geography, physical
physical region, 249
physical science, Humboldt and, 241–42, 355n.25
Pichot, Jean-Baptiste, 212–13
Pinkerton, John, 239; *Modern Geography*, 354n.6
Pius VII, Pope, Napoleon and, 214–15
Playfair, William, 253

Poinsot, Louis, 133
polychrographie, Expilly's use of term, 102
Polychrographie en six parties (Expilly), 97–103
Pompeii, 279
Portugal, Balbi on, 221–23, 228–29
Pratt, Mary Louise, 183
Précis de la géographie universelle (Malte-Brun), 103–7, 109–11, 331n.92
predation, institutionalized, warfare as, 153–54
"premier géographe du roi," 36, 149
Principia (Newton), 50–51
professionalism, Académie and, 318n.23
professionalization, Letronne's call for, 295–96
publishing, geographical, role of Férussac's *Bulletin,* 173–75

Questions de statistique à l'usage des voyageurs (Volney), 198, 206

race, monogenetic *vs.* polygenetic arguments, 230, 360n.32
Rapports à l'Empereur (Gosselin), 268
Recherches statistiques sur la ville de Paris . . . (Chabrol de Volvic and Fourier), 216–21, 310–11
Recueil des inscriptions (Letronne), 288, 291
Recueil de voyages et de mémoires, 131
relief depiction, 43–44
religion: of Cassini IV, 80–82; and Malte-Brun's rejection of theory, 107–9; underlying François's *Science de la géographie,* 113. *See also* Society of Jesus
Remusat, Jean Pierre Abel, 227–28
Renaissance, cosmographical tradition of, 327n.2
Renan, Ernest, 362n.83
Renouard, Jules, 109–10

representation, geographical: accuracy of, 46–48; language of, 41–46; limits of, 48–54
representation, scientific, 13
research guide, genre of, 346n.11
Revolution: Cassini IV and, 76–85; and military innovation, 152–54
rhetoric, and teaching of geography, 30–31
Robert, Louis, 301, 363n.124
Robert de Vaugondy, Didier, 22–23, 34, 36, 40, 319n.39, 322n.85; *Essai sur l'histoire de la géographie,* 33–34; on limits of geographic representation, 53–54; "Mémoire sur une question de géographie pratique," 54; *Nouvel atlas portatif,* 322n.87
Robinson, Arthur, 254, 317n.31
Roederer, Pierre Louis, Comte, 197
Roger, Jean, 42
Rose, Gillian, 315n.5
Rosetta stone, 287, 294
Rossel, Élisabeth-Paul-Édouard de, 130
Ruge, Sophus, 130
Ruines, Les (Volney), 202–3, 347n.36

Saint-Aulaire, Louis Claire de Beaupoil, comte de, 133
Sainte Croix, M. de, *Examen des historiens d'Alexandre,* 269–71
Saint-Martin, Louis Vivien de, 9
Santarem, Viscount Manuel Francisco de, 130, 271, 277; *Atlas composé de mappemondes,* 144–46
Say, Jean-Baptiste, 282
scale of analysis, Humboldt and, 248–49
Schiller, Friedrich, 279
science: Enlightenment, 240; post-Revolutionary, 80–85; and the state, 189–90, 193
Science de la géographie (François), 111, 113–19, 308
Scylax, periplus of, 296–97
Simon, Jules, 359n.18

skepticism, of Volney, 201–2
Smith, Adam, 282
Smith, Bernard, 240–41
social criticism, 193; of Volney, 196
social reform, 153
social sciences, emergence of, 193–95, 347n.37
Société anonyme du Bulletin universel . . . (Férussac), 175
Société de géographie, 91, 110, 274; founding of, 130–32; Jomard and, 134; membership trends, 131–32, 334n.6
Société de géographie (Malte-Brun), 331n.92
Société des arts, 28
Society of Jesus, 23–24, 29–32. *See also* colleges, religious
sociopolitical change, 154–55
Somerville, Mary, 312
Soulavie, 161, 339n.40, 340nn.49, 55
Spain, Bory de Saint-Vincent's work on, 179–82, 184
Stafford, Barbara, 5–6, 316n.13
state, science and, 149–52, 188–90, 193
state aggression, geographers' participation in, 187
statistical surveys, establishment of, 265
statistics: Balbi and, 224–27; Chabrol de Volvic and, 217–20; government reliance on, 154–55
Statistique des départements (Peuchet and Chanlaire), 283
Statistique des provinces . . . (Chabrol de Volvic), 213–14
Staum, Martin, 194, 271
Strabo, *Geography*, 92–96, 284
structures, geographical, eighteenth-century, 26–33
Suez Canal, 282
Supplément à l'Encyclopédie (Robinet), 27
symbology, cartographic, 45
synthesis, geographical, as aim of military geography, 158–59, 162

Syria, Volney and, 206
systems, Letronne's rejection of, 297–99

Tableau du climat et du sol des États-Unis (Volney), 198
Taton, René, 175
teaching, of geography, 23, 34, 318n.16; at École normale, 58–66
texts, geographical, eighteenth-century, 37–38. *See also* universal geographies
Teyssot, Georges, 212
theory, geographical: Humboldt's embrace of, 242–44; Malte-Brun's rejection of, 107–9; military geography and, 161–62, 187
topography, as field of geographical engineers, 164
tradition, concept of, 315n.5
travel accounts, 37–38; Humboldt and, 249, 356n.59; Letronne's use of, 292
"truth," scientific, Jomard on, 335n.22

United States, Volney's trip to, 205
unity of nature, 123–24, 180–82; Humboldt on, 239–40
universal geographies, 89–90, 225; of Balbi, 222, 351n.122; Barbié du Bocage and, 271; degeneration of, 96–103; of Delisle, 308; dismissal of mapping, 96; of Expilly, 308; of François, 113–19, 126; Humboldt's *Cosmos* as, 119–27; of Letronne, 284; of Malte-Brun, 103–7, 307–8; Malte-Brun as model for, 109–11; of Mentelle, 308; Strabo's approach, 92–96
universalities and particularities, François on, 116–17
Université de Paris, 29

Vallongue, Pascal, 158, 161–62, 309
Varenius, Bernhard, 54, 96, 322n.86; *Complete System of General Geography*, 329n.39
Vidal-Naquet, Pierre, 279
Vienna, Balbi on, 221–23

Villiers, André Jean François Marie Brochant de, 133

Vivien de Saint-Martin, Louis, 110

vocabulary. *See* language, geographical

Volney, Constantin-François Chasseboeuf de, 59, 194–209, 279, 310, 347n.37, 348n.38; "État physique de la Corse," 198; influence of, 232; interest in geography, 196–99; politics of, 209; *Questions de statistique à l'usage des voyageurs*, 198, 206; relation to state power, 195–96; *Ruines, Les*, 202–3, 347n.36; *Tableau du climat et du sol des États-Unis*, 198; *Voyage en Syrie et en Égypte*, 198–99, 208

Voltaire, 196–97

Voyage aux régions équinoxiales (Humboldt), 356n.60

Voyage du jeune Anacharsis en Grèce (Barthélemy), 269

Voyage en Syrie et en Égypte (Volney), 198–99, 208

Voyage pittoresque de la Grèce (Choiseul-Gouffier), 279

Wailly, Etienne Augustin de, 133

Walckenaer, Baron, 216, 337n.11

Walckenaer, Baron Charles Athanase, 9, 130, 133, 273–75, 277, 312, 349n.70, 360n.43; early life, 273; *Essai sur l'histoire de l'espèce humaine*, 360n.32; and Gosselin, 359n.27; *L'Ile de Wight*, 360n.32; obituary of Letronne, 296, 301, 359n.25; writings of, 274

Wallis, Helen, 254

wholeness, as aim of geography, 117–18

Wilkinson, Sir Gardner, 293

Wood, Denis, 251

zoology, Férussac and, 170